RANDOM SIGNALS:
DETECTION,
ESTIMATION
AND
DATA ANALYSIS

RANDOM SIGNALS:
DETECTION, ESTIMATION AND DATA ANALYSIS

K. Sam Shanmugan
University of Kansas

Arthur M. Breipohl
University of Oklahoma

WILEY

John Wiley & Sons,
New York · Chichester · Brisbane · Toronto · Singapore

Copyright © 1988, by John Wiley & Sons, Inc.

All rights reserved. Published simultaneously in Canada.

Reproduction or translation of any part of
this work beyond that permitted by Sections
107 and 108 of the 1976 United States Copyright
Act without the permission of the copyright
owner is unlawful. Requests for permission
or further information should be addressed to
the Permissions Department, John Wiley & Sons.

Library of Congress Cataloging in Publication Data:

Shanmugan, K. Sam, 1943–
   Random signals.

   Includes bibliographies and index.
   1. Signal detection. 2. Stochastic processes.
3. Estimation theory.   I. Breipohl, Arthur M.
II. Title
TK5102.5.S447   1988   621.38'043   87-37273
ISBN 0-471-81555-1

Printed and bound in the United States of America
by Braun-Brumfield, Inc.

10 9

CONTENTS

CHAPTER 1 **Introduction**

1.1	Historical Perspective	3
1.2	Outline of the Book	4
1.3	References	7

CHAPTER 2 **Review of Probability and Random Variables**

2.1	Introduction		8
2.2	Probability		9
	2.2.1	Set Definitions	9
	2.2.2	Sample Space	12
	2.2.3	Probabilities of Random Events	12
	2.2.4	Useful Laws of Probability	14
	2.2.5	Joint, Marginal, and Conditional Probabilities	15
2.3	Random Variables		21
	2.3.1	Distribution Functions	22
	2.3.2	Discrete Random Variables and Probability Mass Function	24
	2.3.3	Expected Values or Averages	26
	2.3.4	Examples of Probability Mass Functions	29
2.4	Continuous Random Variables		33
	2.4.1	Probability Density Functions	33
	2.4.2	Examples of Probability Density Functions	43
	2.4.3	Complex Random Variables	46
2.5	Random Vectors		47
	2.5.1	Multivariate Gaussian Distribution	50
	2.5.2	Properties of the Multivariate Gaussian Distribution	50
	2.5.3	Moments of Multivariate Gaussian pdf	53
2.6	Transformations (Functions) of Random Variables		55
	2.6.1	Scalar Valued Function of One Random Variable	57
	2.6.2	Functions of Several Random Variables	61
2.7	Bounds and Approximations		76
	2.7.1	Thebycheff Inequality	77
	2.7.2	Chernoff Bound	78
	2.7.3	Union Bound	79
	2.7.4	Approximating the Distribution of $Y = g(X_1, \ldots, X_n)$	81

vi CONTENTS

		2.7.5	Series Approximation of Probability Density Functions	83
		2.7.6	Approximations of Gaussian Probabilities	87
2.8	Sequences of Random Variables and Convergence			88
		2.8.1	Convergence Everywhere and Almost Everywhere	88
		2.8.2	Convergence in Distribution and Central Limit Theorem	89
		2.8.3	Convergence in Probability (in Measure) and the Law of Large Numbers	93
		2.8.4	Convergence in Mean Square	94
		2.8.5	Relationship Between Different Forms of Convergence	95
2.9	Summary			95
2.10	References			96
2.11	Problems			97

CHAPTER 3   Random Processes and Sequences

3.1	Introduction		111
3.2	Definition of Random Processes		113
	3.2.1	Concept of Random Processes	113
	3.2.2	Notation	114
	3.2.3	Probabilistic Structure	116
	3.2.4	Classification of Random Processes	117
	3.2.5	Formal Definition of Random Processes	119
3.3	Methods of Description		119
	3.3.1	Joint Distribution	119
	3.3.2	Analytical Description Using Random Variables	121
	3.3.3	Average Values	121
	3.3.4	Two or More Random Processes	124
3.4	Special Classes of Random Processes		125
	3.4.1	More Definitions	126
	3.4.2	Random Walk and Wiener Process	127
	3.4.3	Poisson Process	131
	3.4.4	Random Binary Waveform	132
3.5	Stationarity		135
	3.5.1	Strict-sense Stationarity	135
	3.5.2	Wide-sense Stationarity	136
	3.5.3	Examples	137
	3.5.4	Other Forms of Stationarity	141
	3.5.5	Tests for Stationarity	142

3.6	Autocorrelation and Power Spectral Density Functions of Real WSS Random Processes		142
	3.6.1	Autocorrelation Function of a Real WSS Random Process and Its Properties	143
	3.6.2	Cross correlation Function and its Properties	144
	3.6.3	Power Spectral Density Function of a WSS Random Process and Its Properties	145
	3.6.4	Cross-power Spectral Density Function and Its Properties	148
	3.6.5	Power Spectral Density Function of Random Sequences	149
3.7	Continuity, Differentiation, and Integration		160
	3.7.1	Continuity	161
	3.7.2	Differentiation	162
	3.7.3	Integration	165
3.8	Time Averaging and Ergodicity		166
	3.8.1	Time Averages	168
	3.8.2	Ergodicity	176
3.9	Spectral Decomposition and Series Expansion of Random Processes		185
	3.9.1	Ordinary Fourier Series Expansion	185
	3.9.2	Modified Fourier Series for Aperiodic Random Signals	187
	3.9.3	Karhunen–Loeve (K-L) Series Expansion	188
3.10	Sampling and Quantization of Random Signals		189
	3.10.1	Sampling of Lowpass Random Signals	190
	3.10.2	Quantization	196
	3.10.3	Uniform Quantizing	197
	3.10.4	Nonuniform Quantizing	200
3.11	Summary		202
3.12	References		203
3.13	Problems		204

CHAPTER 4 Response of Linear Systems to Random Inputs

4.1	Classification of Systems		216
	4.1.1	Lumped Linear Time-invariant Causal (LLTIVC) System	216
	4.1.2	Memoryless Nonlinear Systems	217
4.2	Response of LTIVC Discrete Time Systems		218
	4.2.1	Review of Deterministic System Analysis	218

viii   CONTENTS

	4.2.2	Mean and Autocorrelation of the Output	221
	4.2.3	Distribution Functions	222
	4.2.4	Stationarity of the Output	222
	4.2.5	Correlation and Power Spectral Density of the Output	223
4.3	Response of LTIVC Continuous Time Systems		227
	4.3.1	Mean and Autocorrelation Function of the Output	228
	4.3.2	Stationarity of the Output	229
	4.3.3	Power Spectral Density of the Output	230
	4.3.4	Mean-square Value of the Output	234
	4.3.5	Multiple Input–Output Systems	238
	4.3.6	Filters	239
4.4	Summary		242
4.5	References		243
4.6	Problems		244

CHAPTER 5   Special Classes of Random Processes

5.1	Introduction		249
5.2	Discrete Linear Models		250
	5.2.1	Autoregressive Processes	250
	5.2.2	Partial Autocorrelation Coefficient	262
	5.2.3	Moving Average Models	265
	5.2.4	Autoregressive Moving Average Models	271
	5.2.5	Summary of Discrete Linear Models	275
5.3	Markov Sequences and Processes		276
	5.3.1	Analysis of Discrete-time Markov Chains	278
	5.3.2	Continuous-time Markov Chains	289
	5.3.3	Summary of Markov Models	295
5.4	Point Processes		295
	5.4.1	Poisson Process	298
	5.4.2	Application of Poisson Process—Analysis of Queues	303
	5.4.3	Shot Noise	307
	5.4.4	Summary of Point Processes	312
5.5	Gaussian Processes		312
	5.5.1	Definition of Gaussian Process	313
	5.5.2	Models of White and Band-limited Noise	314
	5.5.3	Response of Linear Time-invariant Systems	317

CONTENTS ix

	5.5.4	Quadrature Representation of Narrowband (Gaussian) Processes	317
	5.5.5	Effects of Noise in Analog Communication Systems	322
	5.5.6	Noise in Digital Communication Systems	330
	5.5.7	Summary of Noise Models	331

5.6 Summary 331

5.7 References 332

5.8 Problems 333

CHAPTER 6 Signal Detection

6.1 Introduction 341

6.2 Binary Detection with a Single Observation 343

    6.2.1 Decision Theory and Hypothesis Testing 344
    6.2.2 MAP Decision Rule and Types of Errors 345
    6.2.3 Bayes' Decision Rule—Costs of Errors 348
    6.2.4 Other Decision Rules 351

6.3 Binary Detection with Multiple Observations 352

    6.3.1 Independent Noise Samples 353
    6.3.2 White Noise and Continuous Observations 355
    6.3.3 Colored Noise 361

6.4 Detection of Signals with Unknown Parameters 364

6.5 M-ary Detection 366

6.6 Summary 369

6.7 References 370

6.8 Problems 370

CHAPTER 7 Linear Minimum Mean-Square Error Filtering

7.1 Introduction 377

7.2 Linear Minimum Mean Squared Error Estimators 379

    7.2.1 Estimating a Random Variable with a Constant 379

x CONTENTS

		7.2.2	Estimating S with One Observation X	379
		7.2.3	Vector Space Representation	383
		7.2.4	Multivariable Linear Mean Squared Error Estimation	384
		7.2.5	Limitations of Linear Estimators	391
		7.2.6	Nonlinear Minimum Mean Squared Error Estimators	393
		7.2.7	Jointly Gaussian Random Variables	395
7.3	Innovations			397
		7.3.1	Multivariate Estimator Using Innovations	400
		7.3.2	Matrix Definition of Innovations	401
7.4	Review			406
7.5	Digital Wiener Filters			407
		7.5.1	Digital Wiener Filters with Stored Data	407
		7.5.2	Real-time Digital Wiener Filters	411
7.6	Kalman Filters			419
		7.6.1	Recursive Estimators	420
		7.6.2	Scalar Kalman Filter	421
		7.6.3	Vector Kalman Filter	432
7.7	Wiener Filters			442
		7.7.1	Stored Data (Unrealizable Filters)	444
		7.7.2	Real-time or Realizable Filters	448
7.8	Summary			465
7.9	References			466
7.10	Problems			467

CHAPTER 8 Statistics

8.1	Introduction			475
8.2	Measurements			476
		8.2.1	Definition of a Statistic	478
		8.2.2	Parametric and Nonparametric Estimators	478
8.3	Nonparametric Estimators of Probability Distribution and Density Functions			479
		8.3.1	Definition of the Empirical Distribution Function	479
		8.3.2	Joint Empirical Distribution Functions	480
		8.3.3	Histograms	481
		8.3.4	Parzen's Estimator for a pdf	484

8.4	Point Estimators of Parameters		485
	8.4.1	Estimators of the Mean	486
	8.4.2	Estimators of the Variance	486
	8.4.3	An Estimator of Probability	487
	8.4.4	Estimators of the Covariance	487
	8.4.5	Notation for Estimators	487
	8.4.6	Maximum Likelihood Estimators	488
	8.4.7	Bayesian Estimators	491
8.5	Measures of the Quality of Estimators		493
	8.5.1	Bias	493
	8.5.2	Minimum Variance, Mean Squared Error, RMS Error, and Normalized Errors	496
	8.5.3	The Bias, Variance, and Normalized RMS Errors of Histograms	497
	8.5.4	Bias and Variance of Parzen's Estimator	501
	8.5.5	Consistent Estimators	502
	8.5.6	Efficient Estimators	502
8.6	Brief Introduction to Interval Estimates		503
8.7	Distribution of Estimators		504
	8.7.1	Distribution of $\overline{X}$ with Known Variance	504
	8.7.2	Chi-square Distribution	505
	8.7.3	(Student's) t Distribution	508
	8.7.4	Distribution of S^2 and $\overline{X}$ with Unknown Variance	510
	8.7.5	F Distribution	511
8.8	Tests of Hypotheses		513
	8.8.1	Binary Detection	514
	8.8.2	Composite Alternative Hypothesis	517
	8.8.3	Tests of the Mean of a Normal Random Variable	518
	8.8.4	Tests of the Equality of Two Means	520
	8.8.5	Tests of Variances	522
	8.8.6	Chi-Square Tests	523
	8.8.7	Summary of Hypothesis Testing	528
8.9	Simple Linear Regression		529
	8.9.1	Analyzing the Estimated Regression	536
	8.9.2	Goodness of Fit Test	538
8.10	Multiple Linear Regression		540
	8.10.1	Two Controlled Variables	541
	8.10.2	Simple Linear Regression in Matrix Form	542
	8.10.3	General Linear Regression	543
	8.10.4	Goodness of Fit Test	545
	8.10.5	More General Linear Models	545
8.11	Summary		547

xii CONTENTS

8.12	References	548
8.13	Appendix 8-A	549
8.14	Problems	552

CHAPTER 9 Estimating the Parameters of Random Processes from Data

9.1	Introduction		560
9.2	Tests for Stationarity and Ergodicity		561
	9.1.1	Stationarity Tests	562
	9.2.2	Run Test for Stationarity	562
9.3	Model-free Estimation		565
	9.3.1	Mean Value Estimation	565
	9.3.2	Autocorrelation Function Estimation	566
	9.3.3	Estimation of the Power Spectral Density (psd) Functions	569
	9.3.4	Smoothing of Spectral Estimates	579
	9.3.5	Bias and Variance of Smoothed Estimators	584
9.4	Model-based Estimation of Autocorrelation Functions and Power Spectral Density Functions		584
	9.4.1	Preprocessing (Differencing)	587
	9.4.2	Order Identification	590
	9.4.3	Estimating the Parameters of Autoregressive Processes	594
	9.4.4	Estimating the Parameters of Moving Average Processes	600
	9.4.5	Estimating the Parameters of ARMA (p, q) Processes	605
	9.4.6	ARIMA Preliminary Parameter Estimation	606
	9.4.7	Diagnostic Checking	608
9.5	Summary		613
9.6	References		614
9.7	Problems		615

APPENDIXES

A.	Fourier Transforms	626
B.	Discrete Fourier Transforms	628
C.	Z Transforms	630
D.	Gaussian Probabilities	632
E.	Table of Chi-Square Distributions	633
F.	Table of Student's t Distribution	637
G.	Table of F Distributions	639
H.	Percentage Points of Run Distribution	649
I.	Critical Values of the Durbin–Watson Statistic	650
Index		651

About The Authors

Dr. Arthur M. Breipohl is currently the OG&E Professor of Electrical Engineering at the University of Oklahoma. He received his Sc. D. from the University of New Mexico in 1964. He has been on the electrical engineering faculties of Oklahoma State University and the University of Kansas, where he was also Chairman for nine years. He was a Visiting Professor in the Department of Engineering-Economic Systems at Stanford and has worked at Sandia Laboratory and Westinghouse. His research interests are in the area of applications of probabilistic models to engineering problems, and he is currently working on power system planning. He has published approximately 40 papers, and is the author of the textbook, *Probabilistic Systems Analysis* (Wiley, 1970), which is currently in its fifteenth printing.

Dr. K. Sam Shanmugan is currently the J. L. Constant Distinguished Professor of Telecommunications at the University of Kansas. He received the Ph. D. degree in Electrical Engineering from Oklahoma State University in 1970. Prior to joining the University of Kansas, Dr. Shanmugan was on the faculty of Wichita State University and served as a visiting scientist at AT&T Bell Laboratories. His research interests are in the areas of signal processing, satellite communications, and computer-aided analysis and design of communication systems. He has published more than 50 technical articles and is the author of a textbook on digital and analog communication systems (Wiley, 1979).

Dr. Shanmugan is a Fellow of the IEEE and has served as the editor of the IEEE Transactions on Communications.

PREFACE

Most electrical engineering curricula now require a course in probabilistic systems analysis and there are a number of excellent texts that are available for an introductory level course in applied probability. But these texts often ignore random processes or, at best, provide a brief coverage of them at the end.

Courses in signal analysis and communications require students to have a background in random processes. Texts for these courses usually review random processes only briefly.

In recent years most electrical engineering departments have started to offer a course in random processes that follows the probability course and precedes the signal analysis and communications courses. Although there are several advanced/graduate level textbooks on random processes that present a rigorous and theoretical view of random processes, we believe that there is a need for an intermediate level text that is written clearly in a manner which appeals to senior and beginning graduate students (as well as to their instructor).

This book is intended for use as a text for a senior/beginning graduate level course for electrical engineering students who have had some exposure to probability and to deterministic signals and systems analysis. Our intent was to select the material that would provide the foundation in random processes which would be needed in future courses in communication theory, signal processing, or control.

We have tried to present a logical development of the topics without emphasis on rigor. Proofs of theorems and statements are included only when we believed that they contribute sufficient insight into the problem being addressed. Proofs are omitted when they involve lengthy theoretical discourse of material that requires a level of mathematics beyond the scope of this text. In such cases, outlines of proofs with adequate reference are presented. We believe that it is often easier for engineering students to generalize specific results and examples than to specialize general results. Thus we devote considerable attention to examples and applications, and we have chosen the problems to illustrate further application of the theory.

The logical relation of the material in this text is shown in Figure i. The material in Chapters 2 to 4, 6, and 7 can be found in many other electrical engineering texts, which are referenced at the end of each chapter. This book differs from these other texts through its increased emphasis on random sequences (discrete time random processes), and of course by its selection of specific material, type of presentation, and examples and problems. Some of the material in Chapter 5, for example, has not usually been included in textbooks at this level, and (of course) we think that it is increasingly important material for electrical engineers. Chapter 8 is material that might be included in an engineering statistics course. We believe that such material is quite useful for practicing engineers and forms a basis for estimating the parameters of random processes. Such estimation is necessary to apply the theory of random processes to engineering design and analysis problems. Estimating random process parameters is the subject of Chapter 9. This material, though available in some textbooks, is often neglected in introductory texts on random processes for electrical engineers.

Some special features of the individual chapters follows. Chapter 2 is designed to be a very brief review of the material that is normally covered in an introductory probability course. This chapter also covers in more detail some aspects of probability theory that might not have been covered in an introductory level course. Chapters 3 and 4 are designed to balance presentation of discrete and continuous (in time) random processes, and the emphasis is on the second-order characteristics, that is, autocorrelation and power spectral density functions of random processes, because modern communication and control system design emphasizes these characteristics. Chapter 6 develops the idea of detecting a known signal in noise beginning with a simple example and progressing to more complex considerations in a way that our students have found easy to follow. Chapter 7 develops both Kalman and Wiener filters from the same two basic ideas: orthogonality and innovations. Chapter 9 introduces estimation of parameters of random sequences with approximately equal emphasis on estimating the parameters of an assumed model of the random sequence and on estimating

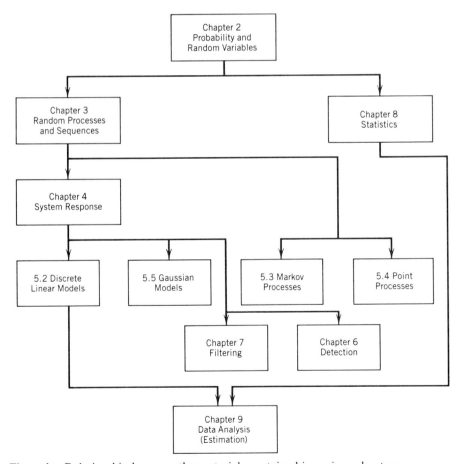

Figure i. Relationship between the materials contained in various chapters.

more general parameters such as the autocorrelation and power spectral density function directly from data without such a specific model.

There are several possible courses for which this book could be used as a text:

1. A two-semester class that uses the entire book.
2. A one-semester class for students with a good background in probability, which covers Chapters 3, 4, 6, 7, and selected sections of Chapter 5. This might be called a course in "Random Signals" and might be desired as a course to introduce senior students to the methods of analysis of random processes that are used in communication theory.
3. A one-semester class for students with limited background in probability using Chapters 2, 3, 4, and 5. This course might be called "Introduction to Random Variables and Random Processes." The instructor might supplement the material in Chapter 2.
4. A one-semester course that emphasized an introduction to random processes and estimation of the process parameters from data. This would use Chapter 2 as a review, and Chapters 3, 4, 5.2, 8, and 9. It might be called "Introduction to Random Processes and Their Estimation."

From the dependencies and independencies shown in Figure i, it is clear that other choices are possible.

We are indebted to many people who helped us in completing this book. We profited immensely from comments and reviews from our colleagues, J. R. Cruz, Victor Frost, and Bob Mulholland. We also made significant improvements as a result of additional reviews by Professors John Thomas, William Tranter, and Roger Ziemer. Our students at the University of Kansas and the University of Oklahoma suffered through earlier versions of the manuscript; their comments helped to improve the manuscript considerably.

The typing of the bulk of the manuscript was done by Ms. Karen Brunton. She was assisted by Ms. Jody Sadehipour and Ms. Cathy Ambler. We thank Karen, Jody, and Cathy for a job well done.

Finally we thank readers who find and report corrections and criticisms to either of us.

K. Sam Shanmugan
Arthur M. Breipohl

CHAPTER ONE

Introduction

Models in which there is uncertainty or randomness play a very important role in the analysis and design of engineering systems. These models are used in a variety of applications in which the signals, as well as the system parameters, may change randomly and the signals may be corrupted by noise. In this book we emphasize models of signals that vary with time and also are random (i.e., uncertain). As an example, consider the waveforms that occur in a typical data communication system such as the one shown in Figure 1.1, in which a number of terminals are sending information in binary format over noisy transmission links to a central computer. A transmitter in each link converts the binary data to an electrical waveform in which binary digits are converted to pulses of duration T and amplitudes ± 1. The received waveform in each link is a distorted and noisy version of the transmitted waveform where noise represents interfering electrical disturbances. From the received waveform, the receiver attempts to extract the transmitted binary digits. As shown in Figure 1.1, distortion and noise cause the receiver to make occasional errors in recovering the transmitted binary digit sequence.

As we examine the collection or "ensemble" of waveforms shown in Figure 1.1, randomness is evident in all of these waveforms. By observing one waveform, or one member of the ensemble, say $x_i(t)$, over the time interval $[t_1, t_2]$ we cannot, with certainty, predict the value of $x_i(t)$ for any other value of t outside the observation interval. Furthermore, knowledge of one member function, $x_i(t)$, will not enable us to know the value of another member function, $x_j(t)$. We will use a stochastic model called a *random process* to describe or

2 INTRODUCTION

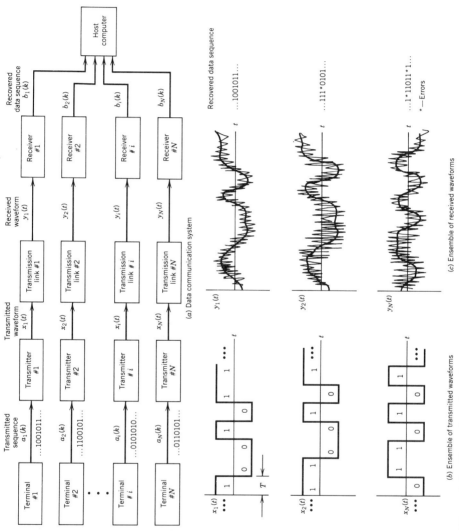

Figure 1.1 Examples of random processes and sequences.

characterize the ensemble of waveforms so that we can answer questions such as:

1. What are the spectral properties of the ensemble of waveforms shown in Figure 1.1?
2. How does the noise affect system performance as measured by the receiver's ability to recover the transmitted data correctly?
3. What is the optimum processing algorithm that the receiver should use?
4. How do we construct a model for the ensemble?

Another example of a random signal is the "noise" that one hears from an AM radio when it is tuned to a point on the dial where no stations are broadcasting. If the speaker is replaced by an oscilloscope so that it records the output voltage of the audio amplifier, then the trace on the oscilloscope will, in the course of time, trace an irregular curve that does not repeat itself precisely and cannot be predicted.

Signals or waveforms such as the two examples presented before are called random signals. Other examples of random signals are fluctuations in the instantaneous load in a power system, the fluctuations in the height of ocean waves at a given point, and the output of a microphone when someone is speaking into it. Waveforms that exhibit random fluctuations are called either signals or noise. Random *signals* are waveforms that contain some *information,* whereas *noise* that is also random is usually *unwanted* and interferes with our attempt to extract information.

Random signals and noise are described by random process models, and electrical engineers use such models to derive signal processing algorithms for recovering information from related physical observations. Typical examples include in addition to the recovery of data coming over a noisy communication channel, the estimation of the "trend" of a random signal such as the instantaneous load in a power system, the estimation of the location of an aircraft from radar data, the estimation of a state variable in a control system based on noisy measurements, and the decision as to whether a weak signal is a result of an incoming missile or is simply noise.

1.1 HISTORICAL PERSPECTIVE

The earliest stimulus for the application of probabilistic models to the physical world were provided by physicists who were discovering and describing our physical world by "laws." Most of the early studies involved experimentation, and physicists observed that when experiments were repeated under what were assumed to be identical conditions, the results were not always reproducible. Even simple experiments to determine the time required for an object to fall through a fixed distance produced different results on different tries due to slight changes in air resistance, gravitational anomalies, and other changes even though the conditions of the experiment were presumably unchanged. With a sufficiently fine scale of measurement almost any experiment becomes nonreproducible. Probabilistic models have proven successful in that they provide a useful description of the random nature of experimental results.

One of the earliest techniques for information extraction based on probabilistic models was developed by Gauss and Legendre around 1800 [2], [5]. This now familiar least-squares method was developed for studying the motion of planets and comets based upon measurements. The motion of these bodies is completely characterized by six parameters, and the least-squares method was developed for "estimating" the values of these parameters from telescopic measurements.

The study of time-varying and uncertain phenomena such as the motion of planets or the random motion of electrons and other charged particles led to the development of a stochastic model called a random process model. This model was developed in the later part of the nineteenth century. After the invention of radio at the beginning of the twentieth century, electrical engineers recognized that random process models can be used to analyze the effect of "noise" in radio communication links. Wiener [6] and Rice formulated the theory of random signals and applied them to devise signal processing (filtering) algorithms that can be used to extract weak radio signals that are masked by noise (1940–45). Shannon [4] used random process models to formulate a theory that has become the basis of digital communication theory (1948).

The invention of radar during World War II led to the development of many new algorithms for detecting weak signals (targets) and for navigation. The most significant algorithm for position locating and navigation was developed by Kalman [3] in the 1960s. The Kalman filtering algorithm made it possible to navigate precisely over long distances and time spans. Kalman's algorithm is used extensively in all navigation systems for deep-space exploration.

1.2 OUTLINE OF THE BOOK

This book introduces the theory of random processes and its application to the study of signals and noise and to the analysis of random data. After a review of probability and random variables, three important areas are discussed:

1. Fundamentals and examples of random process models.
2. Applications of random process models to signal detection and filtering.
3. Statistical estimation–analysis of measurements to estimate the structure and parameter values of probabilistic or stochastic models.

In the first part of the book, Chapters 2, 3, 4, and 5, we develop models for random signals and noise. These models are used in Chapters 6 and 7 to develop signal-processing algorithms that extract information from observations. Chapters 8 and 9 introduce methods of identifying the structure of probabilistic models, estimating the parameters of probabilistic models, and testing the resulting model with data.

It is assumed that the students have had some exposure to probabilities and, hence, Chapter 2, which deals with probability and random variables, is written as a review. Important introductory concepts in probabilities are covered thoroughly, but briefly. More advanced topics that are covered in more detail include random vectors, sequences of random variables, convergence and limiting distributions, and bounds and approximations.

In Chapters 3, 4, and 5 we present the basic theory of random processes, properties of random processes, and special classes of random processes and their applications. The basic theory of random processes is developed in Chapter 3. Fundamental properties of random processes are discussed, and second-order time domain and frequency domain models are emphasized because of their importance in design and analysis. Both discrete-time and continuous-time models are emphasized in Chapter 3.

The response of systems to random input signals is covered in Chapter 4. Time domain and frequency domain methods of computing the response of systems are presented with emphasis on linear time invariant systems. The concept of filtering is introduced and some examples of filter design for signal extraction are presented.

Several useful random process models are presented in Chapter 5. The first part of this chapter introduces discrete time models called autoregressive moving average (ARMA) models which are becoming more important because of their use in data analysis. Other types of models for signals and noise are presented next, and their use is illustrated through a number of examples. The models represent Markov processes, point processes, and Gaussian processes; once again, these types of models are chosen because of their importance to electrical engineering.

Chapters 6 and 7 make use of the models developed in Chapter 5 for developing optimum algorithms for signal detection and estimation. Consider the problem of detecting the presence and estimating the location of an object in space using a radar that sends out a packet of electromagnetic energy in the direction of the target and observes the reflected waveform. We have two problems to consider. First we have to decide whether an object is present and then we have to determine its location. If there is no noise or distortion, then by observing the peak in the received waveform we can determine the presence of the object, and by observing the time delay between the transmitted waveform and the received waveform, we can determine the relative distance between the radar and the object.

In the presence of noise (or interference), the peaks in the received waveform may be masked by the noise, making it difficult to detect the presence and estimate the location of the peaks. Noise might also introduce erroneous peaks, which might lead us to incorrect conclusions. Similar problems arise when we attempt to determine the sequence of binary digits transmitted over a communication link. In these kinds of problems we are interested in two things. First of all, we might be interested in analyzing how well a particular algorithm for signal extraction is performing. Second, we might want to design an "optimum" signal-extraction algorithm.

Analysis and design of signal-extraction algorithms are covered in Chapters 6 and 7. The models for signals and noise developed in Chapters 3 and 5 and the analysis of the response of systems to random signals developed in Chapter 4 are used to develop signal-extraction algorithms. Signal-detection algorithms are covered in Chapter 6 from a decision theory point of view. Maximum A Posterori (MAP), Maximum Likelihood (ML), Neyman Person (NP), and Min-max decision rules are covered first, followed by the matched filter approach for detecting known signals corrupted by additive white noise. The emphasis here is on detecting discrete signals.

In Chapter 7 we discuss the problem of estimating the value of a random signal from observations of a related random process [for example, estimating (i.e., filtering) an audio signal that is corrupted with noise]. Estimating the value of one random variable on the basis of observing other random variables is introduced first. This is followed by the discrete Weiner and the discrete Kalman filter (scalar and vector versions), and finally the classical continuous Wiener filter is discussed. All developments are based on the concepts of orthogonality and innovations. A number of examples are presented to illustrate their applications.

In order to apply signal extraction algorithms, we need models of the underlying random processes, and in Chapters 6 and 7, we assume that these models are known. However, in many practical applications, we might have only a partial knowledge of the models. Some aspects of the model structure and some parameter values might not be known.

Techniques for estimating the structure and parameter values of random process models from data are presented in Chapters 8 and 9. Parameter estimation is the focus of Chapter 8, where we develop procedures for estimating unknown parameter(s) of a model using data. Procedures for testing assumptions about models using data (i.e., hypothesis testing) are also presented in Chapter 8.

Chapter 9 deals with estimating the time domain and frequency domain structure of random process models. A treatment of techniques that are relatively model-free, for example, computing a sample autocorrelation function from a sample signal, is followed by a technique for identifying a model of a certain type and estimating the parameters of the model. Here, we rely very heavily on the ARMA models developed in Chapter 5 for identifying the structure and estimating the parameters of random process models. Digital processing techniques for data analysis are emphasized throughout this chapter.

Throughout the book we present a large number of examples and exercises for the student. Proofs of theorems and statements are included only when it is felt that they contribute sufficient insight into the problem being addressed. Proofs are omitted when they involve lengthy theoretical discourse of material at a level beyond the scope of this text. In such cases, outlines of proofs with adequate references to outside materials are presented. Supplementary material including tables of mathematical relationships and other numerical data are included in the appendices.

1.3 REFERENCES

[1] Davenport, W. B., and Root, W. L., *Introduction to Random Signals and Noise,* McGraw-Hill, New York, 1958.

[2] Gauss, K. G., *Theory of Motion of the Heavenly Bodies* (translated), Dover, New York, 1963.

[3] Kalman, R. E., "A New Approach to Linear Filtering and Prediction Problems," *J. Basic Eng.,* Vol. 82D, March 1960, pp. 35–45.

[4] Shannon, C. E., "A Mathematical Theory of Communication," *Bell Systems Tech. J.,* Vol. 27, 1948, pp. 379–423, 623–656.

[5] Sorenson, H. W., "Least-Squares Estimation: From Gauss to Kalman," *Spectrum,* July, 1970, pp. 63–68.

[6] Wiener, N., *Cybernetics,* MIT Press, Cambridge, Mass., 1948.

CHAPTER TWO

Review of Probability and Random Variables

2.1 INTRODUCTION

The purpose of this chapter is to provide a review of probability for those electrical engineering students who have already completed a course in probability. We assume that course covered at least the material that is presented here in Sections 2.2 through 2.4. Thus, the material in these sections is particularly brief and includes very few examples. Sections 2.5 through 2.8 may or may not have been covered in the prerequisite course; thus, we elaborate more in these sections. Those aspects of probability theory and random variables used in later chapters and in applications are emphasized. The presentation in this chapter relies heavily on intuitive reasoning rather than on mathematical rigor. A bulk of the proofs of statements and theorems are left as exercises for the reader to complete. Those wishing a detailed treatment of this subject are referred to several well-written texts listed in Section 2.10.

We begin our review of probability and random variables with an introduction to basic sets and set operations. We then define probability measure and review the two most commonly used probability measures. Next we state the rules governing the calculation of probabilities and present the notion of multiple or joint experiments and develop the rules governing the calculation of probabilities associated with joint experiments.

The concept of *random variable* is introduced next. A random variable is characterized by a probabilistic model that consists of (1) the probability space, (2) the set of values that the random variable can have, and (3) a rule for computing the probability that the random variable has a value that belongs to a subset of the set of all permissible values. The use of probability distribution functions and density functions are developed. We then discuss summary measures (averages or expected values) that frequently prove useful in characterizing random variables.

Vector-valued random variables (or random vectors, as they are often referred to) and methods of characterizing them are introduced in Section 2.5. Various multivariate distribution and density functions that form the basis of probability models for random vectors are presented.

As electrical engineers, we are often interested in calculating the response of a system for a given input. Procedures for calculating the details of the probability model for the output of a system driven by a random input are developed in Section 2.6.

In Section 2.7, we introduce inequalities for computing probabilities, which are often very useful in many applications because they require less knowledge about the random variables. A series approximation to a density function based on some of its moments is introduced, and an approximation to the distribution of a random variable that is a nonlinear function of other (known) random variables is presented.

Convergence of sequences of random variable is the final topic introduced in this chapter. Examples of convergence are the law of large numbers and the central limit theorem.

2.2 PROBABILITY

In this section we outline mathematical techniques for describing the results of an experiment whose outcome is not known in advance. Such an experiment is called a *random experiment*. The mathematical approach used for studying the results of random experiments and random phenomena is called probability theory. We begin our review of probability with some basic definitions and axioms.

2.2.1 Set Definitions

A set is defined to be a *collection of elements*. Notationally, capital letters A, B, ..., will designate sets; and the small letters a, b, ..., will designate elements or members of a set. The symbol, $\in$, is read as "is an element of," and the symbol, $\notin$, is read "is not an element of." Thus $x \in A$ is read "x is an element of A."

Two special sets are of some interest. A set that has no elements is called the *empty* set or *null* set and will be denoted by $\emptyset$. A set having at least one element is called *nonempty*. The *whole* or *entire space* S is a set that contains all other sets under consideration in the problem.

A set is *countable* if its elements can be put into one-to-one correspondence with the integers. A countable set that has a finite number of elements and the null set are called *finite* sets. A set that is not countable is called *uncountable*. A set that is not finite is called an *infinite* set.

Subset. Given two sets A and B, the notation

$$A \subset B$$

or equivalently

$$B \supset A$$

is read A is *contained* in B, or A is a subset of B, or B contains A. Thus A is contained in B or $A \subset B$ if and only if every element of A is an element of B.

There are three results that follow from the foregoing definitions. For an arbitrary set, A

$$A \subset S$$
$$\emptyset \subset A$$
$$A \subset A$$

Set Equality. Two arbitrary sets, A and B, are called equal if and only if they contain exactly the same elements, or equivalently,

$$A = B \quad \text{if and only if} \quad A \subset B \quad \text{and} \quad B \subset A$$

Union. The Union of two arbitrary sets, A and B, is written as

$$A \cup B$$

and is the set of all elements that belong to A *or* belong to B (or to both). The union of N sets is obtained by repeated application of the foregoing definition and is denoted by

$$A_1 \cup A_2 \cup \cdots \cup A_N = \bigcup_{i=1}^{N} A_i$$

Intersection. The intersection of two arbitrary sets, A and B, is written as

$$A \cap B$$

and is the set of all elements that belong to both *A and B*. $A \cap B$ is also written AB. The intersection of N sets is written as

$$A_1 \cap A_2 \cap \cdots \cap A_N = \bigcap_{i=1}^{N} A_i$$

Mutually Exclusive. Two sets are called mutually exclusive (or disjoint) if they have no common elements; that is, two arbitrary sets A and B are mutually exclusive if

$$A \cap B = AB = \emptyset$$

where $\emptyset$ is the null set.

The n sets $A_1, A_2, \ldots, A_n$ are called mutually exclusive if

$$A_i \cap A_j = \emptyset \quad \text{for all } i, j, \quad i \neq j$$

Complement. The complement, $\overline{A}$, of a set A relative to S is defined as the set of all elements of S that are not in A.

Let S be the whole space and let A, B, C be arbitrary subsets of S. The following results can be verified by applying the definitions and verifying that each is a subset of the other. Note that the operator precedence is (1) parentheses, (2) complement, (3) intersection, and (4) union.

Commutative Laws.

$$A \cup B = B \cup A$$
$$A \cap B = B \cap A$$

Associative Laws.

$$(A \cup B) \cup C = A \cup (B \cup C) = A \cup B \cup C$$
$$(A \cap B) \cap C = A \cap (B \cap C) = A \cap B \cap C$$

Distributive Laws.

$$A \cap (B \cup C) = (A \cap B) \cup (A \cap C)$$
$$A \cup (B \cap C) = (A \cup B) \cap (A \cup C)$$

DeMorgan's Laws.

$$\overline{(A \cup B)} = \overline{A} \cap \overline{B}$$
$$\overline{(A \cap B)} = \overline{A} \cup \overline{B}$$

2.2.2 Sample Space

When applying the concept of sets in the theory of probability, the whole space will consist of elements that are outcomes of an experiment. In this text an experiment is a sequence of actions that produces outcomes (that are not known in advance). This definition of experiment is broad enough to encompass the usual scientific experiment and other actions that are sometimes regarded as observations.

The totality of all possible outcomes is the sample space. Thus, in applications of probability, outcomes correspond to elements and the sample space corresponds to S, the whole space. With these definitions an event may be defined as a collection of outcomes. Thus, an event is a set, or subset, of the sample space. An event A is said to have occurred if the experiment results in an outcome that is an element of A.

For mathematical reasons, one defines a completely additive family of subsets of S to be events where the class, **S**, of sets defined on S is called completely additive if

1. $S \subset \mathbf{S}$

2. If $A_k \subset \mathbf{S}$ for $k = 1, 2, 3, \ldots$, then $\bigcup_{k=1}^{n} A_k \subset \mathbf{S}$ for $n = 1, 2, 3, \ldots$

3. If $A \subset \mathbf{S}$, then $\overline{A} \subset \mathbf{S}$, where $\overline{A}$ is the complement of A

2.2.3 Probabilities of Random Events

Using the simple definitions given before, we now proceed to define the probabilities (of occurrence) of random events. The probability of an event A, denoted by $P(A)$, is a number assigned to this event. There are several ways in which probabilities can be assigned to outcomes and events that are subsets of the sample space. In order to arrive at a satisfactory theory of probability (a theory that does not depend on the method used for assigning probabilities to events), the probability measure is required to obey a set of axioms.

Definition. A *probability measure* is a set function whose domain is a completely additive class **S** of events defined on the sample space S such that the measure satisfies the following conditions:

1. $P(S) = 1$ (2.1)

2. $P(A) \geq 0$ for all $A \subset S$ (2.2)

3. $P\left(\bigcup_{k=1}^{N} A_k\right) = \sum_{k=1}^{N} P(A_k)$ (2.3)

if $A_i \cap A_j = \emptyset$ for $i \neq j$,
and N may be infinite
($\emptyset$ is the empty or null set)

A random experiment is completely described by a sample space, a probability measure (i.e., a rule for assigning probabilities), and the class of sets forming the domain set of the probability measure. The combination of these three items is called a *probabilistic model*.

By assigning numbers to events, a probability measure distributes numbers over the sample space. This intuitive notion has led to the use of *probability distribution* as another name for a probability measure. We now present two widely used definitions of the probability measure.

Relative Frequency Definition. Suppose that a random experiment is repeated n times. If the event A occurs n_A times, then its probability $P(A)$ is defined as the limit of the relative frequency n_A/n of the occurrence of A. That is

$$P(A) \triangleq \lim_{n \to \infty} \frac{n_A}{n}$$ (2.4)

For example, if a coin (fair or not) is tossed n times and heads show up n_H times, then the probability of heads equals the limiting value of n_H/n.

Classical Definition. In this definition, the probability $P(A)$ of an event A is found without experimentation. This is done by counting the total number, N, of the possible outcomes of the experiment, that is, the number of outcomes in S (S is finite). If N_A of these outcomes belong to event A, then $P(A)$ is defined to be

$$P(A) \triangleq \frac{N_A}{N}$$ (2.5)

If we use this definition to find the probability of a tail when a coin is tossed, we will obtain an answer of $\frac{1}{2}$. This answer is correct when we have a fair coin. If the coin is not fair, then the classical definition will lead to incorrect values for probabilities. We can take this possibility into account and modify the definition as: the probability of an event A consisting of N_A outcomes equals the ratio N_A/N provided the outcomes are equally likely to occur.

The reader can verify that the two definitions of probabilities given in the preceding paragraphs indeed satisfy the axioms stated in Equations 2.1–2.3. The difference between these two definitions is illustrated by Example 2.1.

EXAMPLE 2.1. (Adapted from Shafer [9]).

DIME-STORE DICE: Willard H. Longcor of Waukegan, Illinois, reported in the late 1960s that he had thrown a certain type of plastic die with drilled pips over one million times, using a new die every 20,000 throws because the die wore down. In order to avoid recording errors, Longcor recorded only whether the outcome of each throw was odd or even, but a group of Harvard scholars who analyzed Longcor's data and studied the effects of the drilled pips in the die guessed that the chances of the six different outcomes might be approximated by the relative frequencies in the following table:

Upface	1	2	3	4	5	6	Total
Relative Frequency	.155	.159	.164	.169	.174	.179	1.000
Classical	$\frac{1}{6}$	$\frac{1}{6}$	$\frac{1}{6}$	$\frac{1}{6}$	$\frac{1}{6}$	$\frac{1}{6}$	1.000

They obtained these frequencies by calculating the excess of even over odd in Longcor's data and supposing that each side of the die is favored in proportion to the extent that is has more drilled pips than the opposite side. The 6, since it is opposite the 1, is the most favored.

2.2.4 Useful Laws of Probability

Using any of the many definitions of probability that satisfies the axioms given in Equations 2.1, 2.2, and 2.3, we can establish the following relationships:

1. If $\emptyset$ is the null event, then

$$P(\emptyset) = 0 \qquad (2.6)$$

2. For an arbitrary event, A

$$P(A) \leq 1 \tag{2.7}$$

3. If $A \cup \overline{A} = S$ and $A \cap \overline{A} = \emptyset$, then $\overline{A}$ is called the *complement* of A and

$$P(\overline{A}) = 1 - P(A) \tag{2.8}$$

4. If A is a subset of B, that is, $A \subset B$, then

$$P(A) \leq P(B) \tag{2.9}$$

5. $P(A \cup B) = P(A) + P(B) - P(A \cap B)$ (2.10.a)
6. $P(A \cup B) \leq P(A) + P(B)$ (2.10.b)
7. If $A_1, A_2, \ldots, A_n$ are random events such that

$$A_i \cap A_j = \emptyset \quad \text{for} \quad i \neq j \tag{2.10.c}$$

and

$$A_1 \cup A_2 \cup \cdots \cup A_n = S \tag{2.10.d}$$

then

$$\begin{aligned} P(A) = P(A \cap S) &= P[A \cap (A_1 \cup A_2 \cup \cdots \cup A_n)] \\ &= P[(A \cap A_1) \cup (A \cap A_2) \cup \cdots \cup (A \cap A_n)] \\ &= P(A \cap A_1) + P(A \cap A_2) + \cdots + P(A \cap A_n) \end{aligned} \tag{2.10.e}$$

The sets $A_1, A_2, \ldots, A_n$ are said to be *mutually exclusive and exhaustive* if Equations 2.10.c and 2.10.d are satisfied.

8. $P\left(\bigcup_{i=1}^{n} A_i\right) = P(A_1) + P(\overline{A}_1 A_2) + P(\overline{A}_1 \overline{A}_2 A_3) + \cdots$

$$+ P\left(A_n \cap \bigcap_{i=1}^{n-1} \overline{A}_i\right) \tag{2.11}$$

Proofs of these relationships are left as an exercise for the reader.

2.2.5 Joint, Marginal, and Conditional Probabilities

In many engineering applications we often perform an experiment that consists of many subexperiments. Two examples are the simultaneous observation of the input and output digits of a binary communication system, and simultaneous observation of the trajectories of several objects in space. Suppose we have a random experiment E that consists of two subexperiments E_1 and E_2 (for example, E: toss a die and a coin; E_1: toss a die; and E_2: toss a coin). Now if the sample space S_1 of E_1 consists of outcomes $a_1, a_2, \ldots, a_{n_1}$ and the sample space

S_2 of E_2 consists of outcomes $b_1, b_2, \ldots, b_{n_2}$, then the sample space S of the combined experiment is the *Cartesian product* of S_1 and S_2. That is

$$\begin{aligned} S &= S_1 \times S_2 \\ &= \{(a_i, b_j): \quad i = 1, 2, \ldots, n_1, \quad j = 1, 2, \ldots, n_2\} \end{aligned}$$

We can define probability measures on S_1, S_2 and $S = S_1 \times S_2$. If events A_1, $A_2, \ldots, A_n$ are defined for the first subexperiment E_1, and the events B_1, $B_2, \ldots, B_m$ are defined for the second subexperiment E_2, then event A_iB_j is an event of the total experiment.

Joint Probability. The probability of an event such as $A_i \cap B_j$ that is the intersection of events from subexperiments is called the joint probability of the event and is denoted by $P(A_i \cap B_j)$. The abbreviation A_iB_j is often used to denote $A_i \cap B_j$.

Marginal Probability. If the events $A_1, A_2, \ldots, A_n$ associated with subexperiment E_1 are mutually exclusive and exhaustive, then

$$\begin{aligned} P(B_j) = P(B_j \cap S) &= P[B_j \cap (A_1 \cup A_2 \cup \cdots \cup A_n)] \\ &= \sum_{i=1}^{n} P(A_iB_j) \end{aligned} \quad (2.12)$$

Since B_j is an event associated with subexperiment E_2, $P(B_j)$ is called a marginal probability.

Conditional Probability. Quite often, the probability of occurrence of event B_j may depend on the occurrence of a related event A_i. For example, imagine a box containing six resistors and one capacitor. Suppose we draw a component from the box. Then, without replacing the first component, we draw a second component. Now, the probability of getting a capacitor on the second draw depends on the outcome of the first draw. For if we had drawn a capacitor on the first draw, then the probability of getting a capacitor on the second draw is zero since there is no capacitor left in the box! Thus, we have a situation where the occurrence of event B_j (a capacitor on the second draw) on the second subexperiment is conditional on the occurrence of event A_i (the component drawn first) on the first subexperiment. We denote the probability of event B_j given that event A_i is known to have occurred by the conditional probability $P(B_j|A_i)$.

An expression for the conditional probability $P(B|A)$ in terms of the joint probability $P(AB)$ and the marginal probabilities $P(A)$ and $P(B)$ can be obtained as follows using the classical definition of probability. Let N_A, N_B, and

N_{AB} be the number of outcomes belonging to events A, B, and AB, respectively, and let N be the total number of outcomes in the sample space. Then,

$$P(AB) = \frac{N_{AB}}{N}$$

$$P(A) = \frac{N_A}{N} \qquad (2.13)$$

Given that the event A has occurred, we know that the outcome is in A. There are N_A outcomes in A. Now, for B to occur given that A has occurred, the outcome should belong to A and B. There are N_{AB} outcomes in AB. Thus, the probability of occurrence of B given A has occurred is

$$P(B|A) = \frac{N_{AB}}{N_A} = \frac{N_{AB}/N}{N_A/N}$$

The implicit assumption here is that $N_A \neq 0$. Based on this motivation we define conditional probability by

$$P(B|A) \triangleq \frac{P(AB)}{P(A)}, \ P(A) \neq 0 \qquad (2.14)$$

One can show that $P(B|A)$ as defined by Equation 2.14 is a probability measure, that is, it satisfies Equations 2.1, 2.2, and 2.3.

Relationships Involving Joint, Marginal, and Conditional Probabilities. The reader can use the results given in Equations 2.12 and 2.14 to establish the following useful relationships.

1. $P(AB) = P(A|B)P(B) = P(B|A)P(A)$ (2.15)
2. If $AB = \emptyset$, then $P(A \cup B|C) = P(A|C) + P(B|C)$ (2.16)
3. $P(ABC) = P(A)P(B|A)P(C|AB)$ (Chain Rule) (2.17)
4. If $B_1, B_2, \ldots, B_m$ are a set of mutually exclusive and exhaustive events, then

$$P(A) = \sum_{j=1}^{m} P(A|B_j)P(B_j) \qquad (2.18)$$

18 REVIEW OF PROBABILITY AND RANDOM VARIABLES

EXAMPLE 2.2.

An examination of records on certain components showed the following results when classified by manufacturer and class of defect:

Manufacturer	Class of Defect					Totals
	$B_1 =$ none	$B_2 =$ critical	$B_3 =$ serious	$B_4 =$ minor	$B_5 =$ incidental	
M_1	124	6	3	1	6	140
M_2	145	2	4	0	9	160
M_3	115	1	2	1	1	120
M_4	101	2	0	5	2	110
Totals	485	11	9	7	18	530

What is the probability of a component selected at random from the 530 components (a) being from manufacturer M_2 and having no defects, (b) having a critical defect, (c) being from manufacturer M_1, (d) having a critical defect given the component is from manufacturer M_2, (e) being from manufacturer M_1, given it has a critical defect?

SOLUTION:

(a) This is a joint probability and is found by assuming that each component is equally likely to be selected. There are 145 components from M_2 having no defects out of a total of 530 components. Thus

$$P(M_2 B_1) = \frac{145}{530}$$

(b) This calls for a marginal probability.

$$P(B_2) = P(M_1 B_2) + P(M_2 B_2) + P(M_3 B_2) + P(M_4 B_2)$$

$$= \frac{6}{530} + \frac{2}{530} + \frac{1}{530} + \frac{2}{530} = \frac{11}{530}$$

Note that $P(B_2)$ can also be found in the bottom *margin* of the table, that is

$$P(B_2) = \frac{11}{530}$$

(c) Directly from the right margin

$$P(M_1) = \frac{140}{530}$$

(d) This conditional probability is found by the interpretation that given the component is from manufacturer M_2, there are 160 outcomes in the space, two of which have critical defects. Thus

$$P(B_2|M_2) = \frac{2}{160}$$

or by the formal definition, Equation 2.14

$$P(B_2|M_2) = \frac{P(B_2 M_2)}{P(M_2)} = \frac{\frac{2}{530}}{\frac{160}{530}} = \frac{2}{160}$$

(e)
$$P(M_1|B_2) = \frac{6}{11}$$

Bayes' Rule. Sir Thomas Bayes applied Equations 2.15 and 2.18 to arrive at the form

$$P(B_j|A) = \frac{P(A|B_j)P(B_j)}{\sum_{j=1}^{m} P(A|B_j)P(B_j)} \qquad (2.19)$$

which is used in many applications and particularly in interpreting the impact of additional information A on the probability of some event $P(B_j)$. An example illustrates another application of Equation 2.19, which is called Bayes' rule.

EXAMPLE 2.3.

A binary communication channel is a system that carries data in the form of one of two types of signals, say, either zeros or ones. Because of noise, a transmitted zero is sometimes received as a one and a transmitted one is sometimes received as a zero.

We assume that for a certain binary communication channel, the probability a transmitted zero is received as a zero is .95 and the probability that a transmitted

20 REVIEW OF PROBABILITY AND RANDOM VARIABLES

one is received as a one is .90. We also assume the probability a zero is transmitted is .4. Find

(a) Probability a one is received.
(b) Probability a one was transmitted given a one was received.

SOLUTION: Defining

$$A = \text{one transmitted}$$
$$\overline{A} = \text{zero transmitted}$$
$$B = \text{one received}$$
$$\overline{B} = \text{zero received}$$

From the problem statement

$$P(A) = .6, \quad P(B|A) = .90, \quad P(B|\overline{A}) = .05$$

(a) With the use of Equation 2.18

$$P(B) = P(B|A)P(A) + P(B|\overline{A})P(\overline{A})$$
$$= .90(.6) + .05(.4)$$
$$= .56.$$

(b) Using Bayes' rule, Equation 2.19

$$P(A|B) = \frac{P(B|A)P(A)}{P(B)} = \frac{(.90)(.6)}{.56} = \frac{27}{28}$$

Statistical Independence. Suppose that A_i and B_j are events associated with the outcomes of two experiments. Suppose that the occurrence of A_i does not influence the probability of occurrence of B_j and vice versa. Then we say that the events are statistically independent (sometimes, we say probabilistically independent or simply independent). More precisely, we say that two events A_i and B_j are statistically independent if

$$P(A_i B_j) = P(A_i)P(B_j) \qquad (2.20.\text{a})$$

or when

$$P(A_i|B_j) = P(A_i) \qquad (2.20.\text{b})$$

Equation 2.20.a implies Equation 2.20.b and conversely. Observe that statistical independence is quite different from mutual exclusiveness. Indeed, if A_i and B_j are mutually exclusive, then $P(A_iB_j) = 0$ by definition.

2.3 RANDOM VARIABLES

It is often useful to describe the outcome of a random experiment by a number, for example, the number of telephone calls arriving at a central switching station in an hour, or the lifetime of a component in a system. The numerical quantity associated with the outcomes of a random experiment is called loosely a random variable. Different repetitions of the experiment may give rise to different observed values for the random variable. Consider tossing a coin ten times and observing the number of heads. If we denote the number of heads by X, then X takes integer values from 0 through 10, and X is called a random variable.

Formally, a random variable is a *function* whose domain is the set of outcomes $\lambda \in S$, and whose range is R_1, the real line. For every outcome $\lambda \in S$, the random variable assigns a number, $X(\lambda)$ such that

1. The set $\{\lambda : X(\lambda) \leq x\}$ is an event for every $x \in R_1$.
2. The probabilities of the events $\{\lambda : X(\lambda) = \infty\}$, and $\{\lambda : X(\lambda) = -\infty\}$ equal zero, that is,

$$P(X = \infty) = P(X = -\infty) = 0$$

Thus, a random variable maps S onto a set of real numbers $S_X \subset R_1$, where S_X is the range set that contains all permissible values of the random variable. Often S_X is also called the *ensemble* of the random variable. This definition guarantees that to every set $A \subset S$ there corresponds a set $T \subset R_1$ called the image (under X) of A. Also for every (Borel) set $T \subset R_1$ there exists in S the inverse image $X^{-1}(T)$ where

$$X^{-1}(T) = \{\lambda \in S : X(\lambda) \in T\}$$

and this set is an event which has a probability, $P[X^{-1}(T)]$.

We will use uppercase letters to denote random variables and lowercase letters to denote fixed values of the random variable (i.e., numbers).

Thus, the random variable X induces a probability measure on the real line as follows

$$P(X = x) = P\{\lambda : X(\lambda) = x\}$$
$$P(X \leq x) = P\{\lambda : X(\lambda) \leq x\}$$
$$P(x_1 < X \leq x_2) = P\{\lambda : x_1 < X(\lambda) \leq x_2\}$$

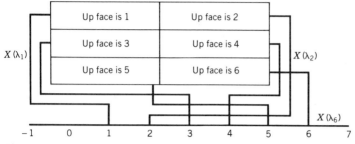

Figure 2.1 Mapping of the sample space by a random variable.

EXAMPLE 2.4.

Consider the toss of one die. Let the random variable X represent the value of the up face. The mapping performed by X is shown in Figure 2.1. The values of the random variable are 1, 2, 3, 4, 5, 6.

2.3.1 Distribution Functions

The probability $P(X \leq x)$ is also denoted by the function $F_X(x)$, which is called the *distribution function* of the random variable X. Given $F_X(x)$, we can compute such quantities as $P(X > x_1)$, $P(x_1 \leq X \leq x_2)$, and so on, easily.

A distribution function has the following properties

1. $F_X(-\infty) = 0$
2. $F_X(\infty) = 1$
3. $\lim_{\substack{\epsilon \to 0 \\ \epsilon > 0}} F_X(x + \epsilon) = F_X(x)$
4. $F_X(x_1) \leq F_X(x_2)$ if $x_1 < x_2$
5. $P[x_1 < X \leq x_2] = F_X(x_2) - F_X(x_1)$

EXAMPLE 2.5.

Consider the toss of a fair die. Plot the distribution function of X where X is a random variable that equals the number of dots on the up face.

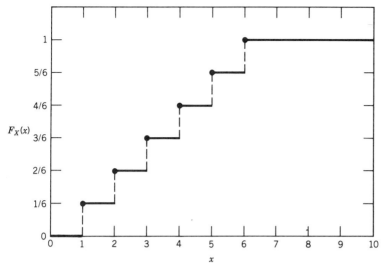

Figure 2.2 Distribution function of the random variable X shown in Figure 2.1.

SOLUTION: The solution is given in Figure 2.2.

Joint Distribution Function. We now consider the case where two random variables are defined on a sample space. For example, both the voltage and current might be of interest in a certain experiment.

The probability of the joint occurrence of two events such as A and B was called the joint probability $P(A \cap B)$. If the event A is the event $(X \leq x)$ and the event B is the event $(Y \leq y)$, then the joint probability is called the joint distribution function of the random variables X and Y; that is

$$F_{X,Y}(x, y) = P[(X \leq x) \cap (Y \leq y)]$$

From this definition it can be noted that

$$F_{X,Y}(-\infty, -\infty) = 0, \quad F_{X,Y}(-\infty, y) = 0, \quad F_{X,Y}(\infty, y) = F_Y(y),$$
$$F_{X,Y}(x, -\infty) = 0, \quad F_{X,Y}(\infty, \infty) = 1, \quad F_{X,Y}(x, \infty) = F_X(x) \qquad (2.21)$$

A random variable may be discrete or continuous. A discrete random variable can take on only a countable number of distinct values. A continuous random variable can assume any value within one or more intervals on the real line. Examples of discrete random variables are the number of telephone calls arriving at an office in a finite interval of time, or a student's numerical score on an examination. The exact time of arrival of a telephone call is an example of a continuous random variable.

2.3.2 Discrete Random Variables and Probability Mass Functions

A discrete random variable X is characterized by a set of allowable values x_1, $x_2, \ldots, x_n$ and the probabilities of the random variable taking on one of these values based on the outcome of the underlying random experiment. The probability that $X = x_i$ is denoted by $P(X = x_i)$ for $i = 1, 2, \ldots, n$, and is called the *probability mass function*.

The probability mass function of a random variable has the following important properties:

1. $P(X = x_i) > 0, \quad i = 1, 2, \ldots, n$ \hfill (2.22.a)

2. $\sum_{i=1}^{n} P(X = x_i) = 1$ \hfill (2.22.b)

3. $P(X \leq x) = F_X(x) = \sum_{\text{all } x_i \leq x} P(X = x_i)$ \hfill (2.22.c)

4. $P(X = x_i) = \lim_{\substack{\epsilon \to 0 \\ \epsilon > 0}} [F_X(x_i) - F_X(x_i - \epsilon)]$ \hfill (2.22.d)

Note that there is a one-to-one correspondence between the probability distribution function and the probability mass function as given in Equations 2.22c and 2.22d.

EXAMPLE 2.6.

Consider the toss of a fair die. Plot the probability mass function.

SOLUTION: See Figure 2.3.

Two Random Variables—Joint, Marginal, and Conditional Distributions and Independence. It is of course possible to define two or more random variables on the sample space of a single random experiment or on the combined sample spaces of many random experiments. If these variables are all discrete, then they are characterized by a joint probability mass function. Consider the example of two random variables X and Y that take on the values $x_1, x_2, \ldots, x_n$ and $y_1, y_2, \ldots, y_m$. These two variables can be characterized by a joint probability

RANDOM VARIABLES 25

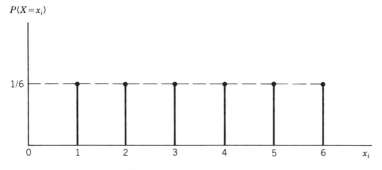

Figure 2.3 Probability mass function for Example 2.6.

mass function $P(X = x_i, Y = y_j)$, which gives the probability that $X = x_i$ and $Y = y_j$.

Using the probability rules stated in the preceding sections, we can prove the following relationships involving joint, marginal and conditional probability mass functions:

1. $$P(X \le x, Y \le y) = \sum_{x_i \le x} \sum_{y_j \le y} P(X = x_i, Y = y_j) \quad (2.23)$$

2. $$P(X = x_i) = \sum_{j=1}^{m} P(X = x_i, Y = y_j)$$

   $$= \sum_{j=1}^{m} P(X = x_i | Y = y_j) P(Y = y_j) \quad (2.24)$$

3. $$P(X = x_i | Y = y_j) = \frac{P(X = x_i, Y = y_j)}{P(Y = y_j)}, \quad P(Y = y_j) \ne 0 \quad (2.25)$$

   $$= \frac{P(Y = y_j | X = x_i) P(X = x_i)}{\sum_{i=1}^{n} P(Y = y_j | X = x_i) P(X = x_i)} \quad \text{(Bayes' rule)} \quad (2.26)$$

4. Random variables X and Y are *statistically independent* if
   $$P(X = x_i, Y = y_j) = P(X = x_i) P(Y = y_j)$$
   $$i = 1, 2, \ldots, n; \quad j = 1, 2, \ldots, m \quad (2.27)$$

EXAMPLE 2.7.

Find the joint probability mass function and joint distribution function of X, Y associated with the experiment of tossing two fair dice where X represents the number appearing on the up face of one die and Y represents the number appearing on the up face of the other die.

SOLUTION:

$$P(X = i, Y = j) = \frac{1}{36}, \qquad i = 1, 2, \ldots, 6; \quad j = 1, 2, \ldots, 6$$

$$F_{X,Y}(x, y) = \sum_{i=1}^{x} \sum_{j=1}^{y} \frac{1}{36}, \qquad x = 1, 2, \ldots, 6; \quad y = 1, 2, \ldots, 6$$

$$= \frac{xy}{36}$$

If x and y are not integers and are between 0 and 6, $F_{X,Y}(x, y) = F_{X,Y}([x], [y])$ where $[x]$ is the greatest integer less than or equal to x. $F_{X,Y}(x, y) = 0$ for $x < 1$ or $y < 1$. $F_{X,Y}(x, y) = 1$ for $x \geq 6$ and $y \geq 6$. $F_{X,Y}(x, y) = F_X(x)$ for $y \geq 6$. $F_{X,Y}(x, y) = F_Y(y)$ for $x \geq 6$.

2.3.3 Expected Values or Averages

The probability mass function (or the distribution function) provides as complete a description as possible for a discrete random variable. For many purposes this description is often too detailed. It is sometimes simpler and more convenient to describe a random variable by a few characteristic numbers or summary measures that are representative of its probability mass function. These numbers are the various *expected values* (sometimes called statistical averages). The expected value or the average of a function $g(X)$ of a discrete random variable X is defined as

$$E\{g(X)\} \triangleq \sum_{i=1}^{n} g(x_i) P(X = x_i) \tag{2.28}$$

It will be seen in the next section that the expected value of a random variable is valid for all random variables, not just for discrete random variables. The form of the average simply appears different for continuous random variables. Two expected values or *moments* that are most commonly used for characterizing a random variable X are its *mean* μ_X and its *variance* σ_X^2. The mean and variance are defined as

$$E\{X\} = \mu_X = \sum_{i=1}^{n} x_i P(X = x_i) \tag{2.29}$$

$$E\{(X - \mu_X)^2\} = \sigma_X^2 = \sum_{i=1}^{n} (x_i - \mu_X)^2 P(X = x_i) \qquad (2.30)$$

The square-root of variance is called the *standard deviation*. The mean of a random variable is its average value and the variance of a random variable is a measure of the "spread" of the values of the random variable.

We will see in a later section that when the probability mass function is not known, then the mean and variance can be used to arrive at bounds on probabilities via the *Tchebycheff's inequality*, which has the form

$$P[|X - \mu_X| > k] \leq \frac{\sigma_X^2}{k^2} \qquad (2.31)$$

The Tchebycheff's inequality can be used to obtain bounds on the probability of finding X outside of an interval $\mu_X \pm k\sigma_X$.

The expected value of a function of two random variables is defined as

$$E\{g(X, Y)\} = \sum_{i=1}^{n} \sum_{j=1}^{m} g(x_i, y_j) P(X = x_i, Y = y_j) \qquad (2.32)$$

A useful expected value that gives a measure of dependence between two random variables X and Y is the *correlation coefficient* defined as

$$\rho_{XY} = \frac{E\{(X - \mu_X)(Y - \mu_Y)\}}{\sigma_X \sigma_Y} = \frac{\sigma_{XY}}{\sigma_X \sigma_Y} \qquad (2.33)$$

The numerator of the right-hand side of Equation 2.33 is called the *covariance* (σ_{XY}) of X and Y. The reader can verify that if X and Y are statistically independent, then $\rho_{XY} = 0$ and that in the case when X and Y are linearly dependent (i.e., when $Y = (b + kX)$, then $|\rho_{XY}| = 1$. Observe that $\rho_{XY} = 0$ does not imply statistical independence.

Two random variables X and Y are said to be *orthogonal* if

$$E\{XY\} = 0$$

The relationship between two random variables is sometimes described in terms of conditional expected values, which are defined as

$$E\{g(X, Y)|Y = y_j\} = \sum_i g(x_i, y_j) P(X = x_i | Y = y_j) \qquad (2.34.a)$$

$$E\{g(X, Y)|X = x_i\} = \sum_j g(x_i, y_j) P(Y = y_j | X = x_i) \qquad (2.34.b)$$

28 REVIEW OF PROBABILITY AND RANDOM VARIABLES

The reader can verify that

$$E\{g(X, Y)\} \triangleq E_{X,Y}\{g(X, Y)\}$$
$$= E_X\{E_{Y|X}[g(X, Y)|X]\} \quad (2.34.c)$$

where the subscripts denote the distributions with respect to which the expected values are computed.

One of the important conditional expected values is the conditional mean:

$$E\{X|Y = y_j\} = \mu_{X|Y=y_j} = \sum_i x_i P(X = x_i | Y = y_j) \quad (2.34.d)$$

The conditional mean plays an important role in estimating the value of one random variable given the value of a related random variable, for example, the estimation of the weight of an individual given the height.

Probability Generating Functions. When a random variable takes on values that are uniformly spaced, it is said to be a *lattice* type random variable. The most common example is one whose values are the nonnegative integers, as in many applications that involve counting. A convenient tool for analyzing probability distributions of non-negative integer-valued random variables is the probability generating function defined by

$$G_X(z) = \sum_{k=0}^{\infty} z^k P(X = k) \quad (2.35.a)$$

The reader may recognize this as the z transform of a sequence of probabilities $\{p_k\}$, $p_k = P(X = k)$, except that z^{-1} has been replaced by z. The probability generating function has the following useful properties:

1. $G_X(1) = \sum_{k=0}^{\infty} P(X = k) = 1$ \quad (2.35.b)

2. If $G_X(z)$ is given, p_k can be obtained from it either by expanding it in a power series or from

$$P(X = k) = \frac{1}{k!} \frac{d^k}{dz^k}[G_X(z)]|_{z=0} \quad (2.35.c)$$

3. The derivatives of the probability generating function evaluated at $z = 1$ yield the *factorial moments* C_n, where

$$C_n = E\{X(X - 1)(X - 2) \cdots (X - n + 1)\}$$
$$= \frac{d^n}{dz^n}[G_X(z)]|_{z=1} \quad (2.35.d)$$

RANDOM VARIABLES

From the factorial moments, we can obtain ordinary moments, for example, as

$$\mu_X = C_1$$

and

$$\sigma_X^2 = C_2 + C_1 - C_1^2$$

2.3.4 Examples of Probability Mass Functions

The probability mass functions of some random variables have convenient analytical forms. Several examples are presented. We will encounter these probability mass functions very often in analysis of communication systems.

The Uniform Probability Mass Function. A random variable X is said to have a uniform probability mass function (or distribution) when

$$P(X = x_i) = 1/n, \quad i = 1, 2, 3, \ldots, n \tag{2.36}$$

The Binomial Probability Mass Function. Let p be the probability of an event A, of a random experiment E. If the experiment is repeated n times and the n outcomes are independent, let X be a random variable that represents the number of times A occurs in the n repetitions. The probability that event A occurs k times is given by the binomial probability mass function

$$P(X = k) = \binom{n}{k} p^k (1-p)^{n-k}, \quad k = 0, 1, 2, \ldots, n \tag{2.37}$$

where

$$\binom{n}{k} \triangleq \frac{n!}{k!(n-k)!} \quad \text{and} \quad m! \triangleq m(m-1)(m-2)\cdots(3)(2)(1); \quad 0! \triangleq 1.$$

The reader can verify that the mean and variance of the binomial random variable are given by (see Problem 2.13)

$$\mu_X = np \tag{2.38.a}$$

$$\sigma_X^2 = np(1-p) \tag{2.38.b}$$

Poisson Probability Mass Function. The Poisson random variable is used to model such things as the number of telephone calls received by an office and the number of electrons emitted by a hot cathode. In situations like these if we make the following assumptions:

1. The number of events occurring in a small time interval $\Delta t \to \lambda' \Delta t$ as $\Delta t \to 0$.
2. The number of events occurring in nonoverlapping time intervals are independent.

then the number of events in a time interval of length T can be shown (see Chapter 5) to have a Poisson probability mass function of the form

$$P(X = k) = \frac{\lambda^k}{k!} e^{-\lambda}, \quad k = 0, 1, 2, \ldots \quad (2.39.a)$$

where $\lambda = \lambda'T$. The mean and variance of the Poisson random variable are given by

$$\mu_X = \lambda \quad (2.39.b)$$

$$\sigma_X^2 = \lambda \quad (2.39.c)$$

Multinomial Probability Mass Function. Another useful probability mass function is the multinomial probability mass function that is a generalization of the binomial distribution to two or more variables. Suppose a random experiment is repeated n times. On each repetition, the experiment terminates in but one of k mutually exclusive and exhaustive events $A_1, A_2, \ldots, A_k$. Let p_i be the probability that the experiment terminates in A_i and let p_i remain constant throughout n independent repetitions of the experiment. Let X_i, $i = 1, 2, \ldots, k$ denote the number of times the experiment terminates in event A_i. Then

$$P(X_1 = x_1, X_2 = x_2, \ldots, X_k = x_k)$$
$$= \frac{n!}{x_1!x_2!\cdots x_{k-1}!x_k!} p_1^{x_1} p_2^{x_2} \cdots p_k^{x_k} \quad (2.40)$$

where $x_1 + x_2 + \cdots + x_k = n$, $p_1 + p_2 + \cdots + p_k = 1$, and $x_i = 0, 1, 2, \ldots, n$. The probability mass function given by Equation 2.40 is called a multinomial probability mass function.

Note that with $A_1 = A$, and $A_2 = \bar{A}$, $p_1 = p$, and $p_2 = 1 - p$, the multinomial probability mass function reduces to the binomial case.

Before we proceed to review continuous random variables, let us look at three examples that illustrate the concepts described in the preceding sections.

EXAMPLE 2.8.

The input to a binary communication system, denoted by a random variable X, takes on one of two values 0 or 1 with probabilities $\frac{3}{4}$ and $\frac{1}{4}$, respectively. Due to errors caused by noise in the system, the output Y differs from the input X occasionally. The behavior of the communication system is modeled by the conditional probabilities

$$P(Y = 1|X = 1) = \frac{3}{4} \quad \text{and} \quad P(Y = 0|X = 0) = \frac{7}{8}$$

(a) Find $P(Y = 1)$ and $P(Y = 0)$.
(b) Find $P(X = 1|Y = 1)$.

(Note that this is similar to Example 2.3. The primary difference is the use of random variables.)

SOLUTION:

(a) Using Equation 2.24, we have

$$P(Y = 1) = P(Y = 1|X = 0)P(X = 0)$$
$$+ P(Y = 1|X = 1)P(X = 1)$$
$$= \left(1 - \frac{7}{8}\right)\left(\frac{3}{4}\right) + \left(\frac{3}{4}\right)\left(\frac{1}{4}\right) = \frac{9}{32}$$

$$P(Y = 0) = 1 - P(Y = 1) = \frac{23}{32}$$

(b) Using Bayes' rule, we obtain

$$P(X = 1|Y = 1) = \frac{P(Y = 1|X = 1)P(X = 1)}{P(Y = 1)}$$

$$= \frac{\left(\frac{3}{4}\right)\frac{1}{4}}{\frac{9}{32}} = \frac{2}{3}$$

$P(X = 1|Y = 1)$ is the probability that the input to the system is 1 when the output is 1.

32 REVIEW OF PROBABILITY AND RANDOM VARIABLES

EXAMPLE 2.9.

Binary data are transmitted over a noisy communication channel in blocks of 16 binary digits. The probability that a received binary digit is in error due to channel noise is 0.1. Assume that the occurrence of an error in a particular digit does not influence the probability of occurrence of an error in any other digit within the block (i.e., errors occur in various digit positions within a block in a statistically independent fashion).

(a) Find the average (or expected) number of errors per block.
(b) Find the variance of the number of errors per block.
(c) Find the probability that the number of errors per block is greater than or equal to 5.

SOLUTION:

(a) Let X be the random variable representing the number of errors per block. Then, X has a binomial distribution

$P\{M\} = \binom{n}{k} p^k (1-p)^{n-k}, \quad k = 0, 1, 2, \ldots, n$

$$P(X = k) = \binom{16}{k}(.1)^k(.9)^{16-k}, \quad k = 0, 1, \ldots, 16$$

and using Equation 2.38.a $\binom{16}{k}(\cdot 1)^k (1 - 0 \cdot 1)^{16-k}$

$$E\{X\} = np = (16)(.1) = 1.6$$

(b) The variance of X is found from Equation 2.38.b:

$$\sigma_X^2 = np(1 - p) = (16)(.1)(.9) = 1.44$$

(c)
$$P(X \geq 5) = 1 - P(X \leq 4)$$

$$= 1 - \sum_{k=0}^{4} \binom{16}{k}(0.1)^k(0.9)^{16-k}$$

$$\approx 0.017$$

EXAMPLE 2.10.

The number N of defects per plate of sheet metal is Poisson with $\lambda = 10$. The inspection process has a constant probability of .9 of finding each defect and the successes are independent, that is, if M represents the number of found defects

$$P(M = i | N = n) = \binom{n}{i}(.9)^i(.1)^{n-i}, \quad i \leq n$$

Find

(a) The joint probability mass function of M and N.
(b) The marginal probability mass function of M.
(c) The condition probability mass function of N given M.
(d) $E\{M|N\}$.
(e) $E\{M\}$ from part (d).

SOLUTION:

(a) $P(M = i, N = n) = \dfrac{e^{-10}}{n!}(10)^n \binom{n}{i}(.9)^i(.1)^{n-i}$, $\quad \begin{matrix} n = 0, 1, \ldots \\ i = 0, 1, \ldots, n \end{matrix}$

(b) $P(M = i) = \sum\limits_{n=i}^{\infty} \dfrac{e^{-10}(10)^n(.1)^n}{n!} \dfrac{n!}{i!(n-i)!}(.9)^i(.1)^{-i}$

$= \dfrac{e^{-10}(9)^i}{i!} \sum\limits_{n=i}^{\infty} \dfrac{1}{(n-i)!}$

$= \dfrac{e^{-9}(9)^i}{i!}, \quad i = 0, 1, \ldots$

(c) $P(N = n | M = i) = \dfrac{e^{-10}10^n}{n!}\binom{n}{i}(.9)^i(.1)^{n-i} \dfrac{i!}{e^{-9}(9)^i}$

$= e^{-1}/(n-i)!, \quad \begin{matrix} n = i, i+1, \ldots \\ i = 0, 1, \ldots \end{matrix}$

(d) Using Equation 2.38.a

$$E\{M|N = n\} = .9n$$

Thus

$$E\{M|N\} = .9N$$

(e) $E\{M\} = E_N\{E\{M|N\}\} = E_N(.9N) = (.9)E_N\{N\} = 9$

This may also be found directly using the results of part (b) if these results are available.

2.4 CONTINUOUS RANDOM VARIABLES

2.4.1 Probability Density Functions

A continuous random variable can take on more than a countable number of values in one or more intervals on the real line. The probability law for a

34 REVIEW OF PROBABILITY AND RANDOM VARIABLES

continuous random variable X is defined by a *probability density function* (pdf) $f_X(x)$ where

$$f_X(x) = \frac{dF_X(x)}{dx} \tag{2.41}$$

With this definition the probability that the observed value of X falls in a small interval of length Δx containing the point x is approximated by $f_X(x)\Delta x$. With such a function, we can evaluate probabilities of events by integration. As with a probability mass function, there are properties that $f_X(x)$ must have before it can be used as a density function for a random variable. These properties follow from Equation 2.41 and the properties of a distribution function.

1. $f_X(x) \geq 0$ \hfill (2.42.a)

2. $\int_{-\infty}^{\infty} f_X(x)\, dx = 1$ \hfill (2.42.b)

3. $P(X \leq a) = F_X(a) = \int_{-\infty}^{a} f_X(x)\, dx$ \hfill (2.42.c)

4. $P(a \leq X \leq b) = \int_{a}^{b} f_X(x)\, dx$ \hfill (2.42.d)

Furthermore, from the definition of integration, we have

$$P(X = a) = \int_{a}^{a} f_X(x)\, dx = \lim_{\Delta x \to 0} f_X(a)\, \Delta x = 0 \tag{2.42.e}$$

for a continuous random variable.

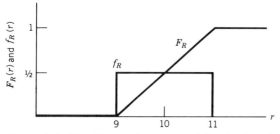

Figure 2.4 Distribution function and density function for Example 2.11.

EXAMPLE 2.11.

Resistors are produced that have a nominal value of 10 ohms and are ±10% resistors. Assume that any possible value of resistance is equally likely. Find the density and distribution function of the random variable R, which represents resistance. Find the probability that a resistor selected at random is between 9.5 and 10.5 ohms.

SOLUTION: The density and distribution functions are shown in Figure 2.4. Using the distribution function,

$$P(9.5 < R \leq 10.5) = F_R(10.5) - F_R(9.5) = \frac{3}{4} - \frac{1}{4} = \frac{1}{2}$$

or using the density function,

$$P(9.5 < R \leq 10.5) = \int_{9.5}^{10.5} \frac{1}{2} \, dr = \frac{10.5 - 9.5}{2} = \frac{1}{2}$$

Mixed Random Variable. It is possible for a random variable to have a distribution function as shown in Figure 2.5. In this case, the random variable and the distribution function are called mixed, because the distribution function consists of a part that has a density function and a part that has a probability mass function.

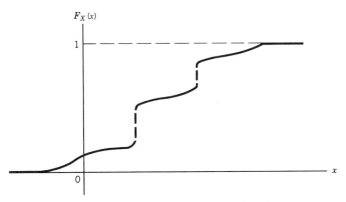

Figure 2.5 Example of a mixed distribution function.

Two Random Variables—Joint, Marginal, and Conditional Density Functions and Independence.

If we have a multitude of random variables defined on one or more random experiments, then the probability model is specified in terms of a *joint probability density function*. For example, if there are two random variables X and Y, they may be characterized by a joint probability density function $f_{X,Y}(x, y)$. If the joint distribution function, $F_{X,Y}$, is continuous and has partial derivatives, then a joint density function is defined by

$$f_{X,Y}(x, y) = \frac{\partial^2 F_{X,Y}(x, y)}{\partial x \, \partial y}$$

It can be shown that

$$f_{X,Y}(x, y) \geq 0$$

From the fundamental theorem of integral calculus

$$F_{X,Y}(x, y) = \int_{-\infty}^{y} \int_{-\infty}^{x} f_{X,Y}(\mu, \nu) \, d\mu \, d\nu$$

Since $F_{X,Y}(\infty, \infty) = 1$

$$\int\int_{-\infty}^{\infty} f_{X,Y}(\mu, \nu) \, d\mu \, d\nu = 1$$

A joint density function may be interpreted as

$$\lim_{\substack{dx \to 0 \\ dy \to 0}} P[(x < X \leq x + dx) \cap (y < Y \leq y + dy)] = f_{X,Y}(x, y) \, dx \, dy$$

From the joint probability density function one can obtain marginal probability density functions $f_X(x)$, $f_Y(y)$, and conditional probability density functions $f_{X|Y}(x|y)$ and $f_{Y|X}(y|x)$ as follows:

$$f_X(x) = \int_{-\infty}^{\infty} f_{X,Y}(x, y) \, dy \qquad (2.43.\text{a})$$

$$f_Y(y) = \int_{-\infty}^{\infty} f_{X,Y}(x, y) \, dx \qquad (2.43.\text{b})$$

CONTINUOUS RANDOM VARIABLES

$$f_{X|Y}(x|y) \triangleq \frac{f_{X,Y}(x, y)}{f_Y(y)}, \quad f_Y(y) > 0 \qquad (2.44.\text{a})$$

$$f_{Y|X}(y|x) \triangleq \frac{f_{X,Y}(x, y)}{f_X(x)}, \quad f_X(x) > 0 \qquad (2.44.\text{b})$$

$$f_{Y|X}(y|x) = \frac{f_{X|Y}(x|y)f_Y(y)}{\int_{-\infty}^{\infty} f_{X|Y}(x|\lambda)f_Y(\lambda)\,d\lambda} \quad \text{Bayes' rule} \qquad (2.44.\text{c})$$

Finally, random variables X and Y are said to be *statistically independent* if

$$f_{X,Y}(x, y) = f_X(x)f_Y(y) \qquad (2.45)$$

EXAMPLE 2.12.

The joint density function of X and Y is

$$f_{X,Y}(x, y) = axy, \quad 1 \le x \le 3,\ 2 \le y \le 4$$
$$= 0 \quad \text{elsewhere}$$

Find a, $f_X(x)$, and $F_Y(y)$

SOLUTION: Since the area under the joint pdf is 1, we have

$$1 = \int_2^4 \int_1^3 axy\, dx\, dy = a \int_2^4 y \left[\frac{x^2}{2}\right]\Big|_1^3 dy$$
$$= a \int_2^4 4y\, dy = 4a \left[\frac{y^2}{2}\right]\Big|_2^4 = 24a$$

or

$$a = \frac{1}{24}$$

The marginal pdf of X is obtained from Equation 2.43.a as

$$f_X(x) = \frac{1}{24} \int_2^4 xy\, dy = \frac{x}{24}[8 - 2] = \frac{x}{4}, \quad 1 \le x \le 3$$
$$= 0 \qquad \qquad \text{elsewhere}$$

And the distribution function of Y is

$$F_Y(y) = 0, \quad y \leq 2$$
$$= 1, \quad y > 4$$
$$= \frac{1}{24} \int_2^y \int_1^3 xv \, dx \, dv = \frac{1}{6} \int_2^y v \, dv$$
$$= \frac{1}{12} [y^2 - 4], \quad 2 \leq y \leq 4$$

Expected Values. As in the case of discrete random variables, continuous random variables can also be described by statistical averages or expected values. The expected values of functions of continuous random variables are defined by

$$E\{g(X, Y)\} = \int_{-\infty}^{\infty} \int_{-\infty}^{\infty} g(x, y) f_{X,Y}(x, y) \, dx \, dy \tag{2.46}$$

$$\mu_X = E\{X\} = \int_{-\infty}^{\infty} x f_X(x) \, dx \tag{2.47.a}$$

$$\sigma_X^2 = E\{(X - \mu_X)^2\} = \int_{-\infty}^{\infty} (x - \mu_X)^2 f_X(x) \, dx \tag{2.47.b}$$

$$\sigma_{XY} = E\{(X - \mu_X)(Y - \mu_Y)\} \tag{2.47.c}$$

$$= \int_{-\infty}^{\infty} \int_{-\infty}^{\infty} (x - \mu_X)(y - \mu_Y) f_{X,Y}(x, y) \, dx \, dy$$

and

$$\rho_{XY} = \frac{E\{(X - \mu_X)(Y - \mu_Y)\}}{\sigma_X \sigma_Y} \tag{2.47.d}$$

It can be shown that $-1 \leq \rho_{XY} \leq 1$. The Tchebycheff's inequality for a continuous random variable has the same form as given in Equation 2.31.

Conditional expected values involving continuous random variables are defined as

$$E\{g(X, Y) | Y = y\} = \int_{-\infty}^{\infty} g(x, y) f_{X|Y}(x|y) \, dx \tag{2.48}$$

Finally, if X and Y are independent, then

$$E\{g(X)h(Y)\} = E\{g(X)\}E\{h(Y)\} \tag{2.49}$$

It should be noted that the concept of the expected value of a random variable is equally applicable to discrete and continuous random variables. Also, if generalized derivatives of the distribution function are defined using the Dirac delta function $\delta(x)$, then discrete random variables have generalized density functions. For example, the generalized density function of die tossing as given in Example 2.6, is

$$f_X(x) = \frac{1}{6}[\delta(x-1) + \delta(x-2) + \delta(x-3)$$
$$+ \delta(x-4) + \delta(x-5) + \delta(x-6)]$$

If this approach is used then, for example, Equations 2.29 and 2.30 are special cases of Equations 2.47.a and 2.47.b, respectively.

Characteristic Functions and Moment Generating Functions. In calculus we use a variety of transform techniques to help solve various analysis problems. For example, Laplace and Fourier transforms are used extensively for solving linear differential equations. In probability theory we use two similar "transforms" to aid in the analysis. These transforms lead to the concepts of characteristic and moment generating functions.

The characteristic function $\Psi_X(\omega)$ of a random variable X is defined as the expected value of $\exp(j\omega X)$

$$\Psi_X(\omega) = E\{\exp(j\omega X)\}, \quad j = \sqrt{-1}$$

For a continuous random variable (and using δ functions also for a discrete random variable) this definition leads to

$$\Psi_X(\omega) = \int_{-\infty}^{\infty} f_X(x)\exp(j\omega x)\, dx \tag{2.50.a}$$

which is the complex conjugate of the Fourier transform of the pdf of X. Since $|\exp(j\omega x)| \leq 1$,

$$\int_{-\infty}^{\infty} |f_X(x)\exp(j\omega x)|\, dx \leq \int_{-\infty}^{\infty} f_X(x)\, dx = 1$$

and hence the characteristic function always exists.

Using the inverse Fourier transform, we can obtain $f_X(x)$ from $\Psi_X(\omega)$ as

$$f_X(x) = \frac{1}{2\pi} \int_{-\infty}^{\infty} \Psi_X(\omega) \exp(-j\omega x) \, d\omega \qquad (2.50.\text{b})$$

Thus, $f_X(x)$ and $\Psi_X(\omega)$ form a Fourier transform pair. The characteristic function of a random variable has the following properties.

1. The characteristic function is unique and determines the pdf of a random variable (except for points of discontinuity of the pdf). Thus, if two continuous random variables have the same characteristic function, they have the same pdf.
2. $\Psi_X(0) = 1$, and

$$E\{X^k\} = \frac{1}{j^k} \left[\frac{d^k \Psi_X(\omega)}{d\omega^k} \right] \quad \text{at } \omega = 0 \qquad (2.51.\text{a})$$

Equation (2.51.a) can be established by differentiating both sides of Equation (2.50.a) k times with respect to ω and setting $\omega = 0$.

The concept of characteristic functions can be extended to the case of two or more random variables. For example, the characteristic function of two random variables X_1 and X_2 is given by

$$\Psi_{X_1, X_2}(\omega_1, \omega_2) = E\{\exp(j\omega_1 X_1 + j\omega_2 X_2)\} \qquad (2.51.\text{b})$$

The reader can verify that

$$\Psi_{X_1, X_2}(0, 0) = 1$$

and

$$E\{X_1^m X_2^n\} = j^{-(m+n)} \frac{\partial^m \partial^n}{\partial \omega_1^m \partial \omega_2^n} [\Psi_{X_1, X_2}(\omega_1, \omega_2)] \quad \text{at } (\omega_1, \omega_2) = (0, 0) \quad (2.51.\text{c})$$

The real-valued function $M_X(t) = E\{\exp(tX)\}$ is called the *moment generating function*. Unlike the characteristic function, the moment generating function need not always exist, and even when it exists, it may be defined for only some values of t within a region of convergence (similar to the existence of the Laplace transform). If $M_X(t)$ exists, then $M_X(t) = \Psi_X(t/j)$.

We illustrate two uses of characteristic functions.

EXAMPLE 2.13. $N(\mu_x, \sigma_x^2)$

X_1 and X_2 are two independent (Gaussian) random variables with means μ_1 and μ_2 and variances σ_1^2 and σ_2^2. The pdfs of X_1 and X_2 have the form

$$f_{X_i}(x_i) = \frac{1}{\sqrt{2\pi}\,\sigma_i} \exp\left[-\frac{(x_i - \mu_i)^2}{2\sigma_i^2}\right], \quad i = 1, 2$$

(a) Find $\Psi_{X_1}(\omega)$ and $\Psi_{X_2}(\omega)$
(b) Using $\Psi_X(\omega)$ find $E\{X^4\}$ where X is a Gaussian random variable with mean zero and variance σ^2.
(c) Find the pdf of $Z = a_1 X_1 + a_2 X_2$

SOLUTION:

(a) $\Psi_{X_1}(\omega) = \int_{-\infty}^{\infty} \frac{1}{\sqrt{2\pi}\,\sigma_1} \exp[-(x_1 - \mu_1)^2/2\sigma_1^2]\exp(j\omega x_1)\,dx_1$

We can combine the exponents in the previous equation and write it as

$$\exp[j\mu_1\omega + (\sigma_1 j\omega)^2/2]\exp\{-[x_1 - (\mu_1 + \sigma_1^2 j\omega)]^2/2\sigma_1^2\}$$

and hence

$$\Psi_{X_1}(\omega) = \exp[j\mu_1\omega + (\sigma_1 j\omega)^2/2] \cdot \int_{-\infty}^{\infty} \frac{1}{\sqrt{2\pi}\,\sigma_1} \\ \times \exp[-(x_1 - \mu_1')^2/2\sigma_1^2]\,dx_1$$

where $\mu_1' = \mu_1 + \sigma_1^2 j\omega$.
The value of the integral in the preceding equation is 1 and hence

$$\Psi_{X_1}(\omega) = \exp[j\mu_1\omega + (\sigma_1 j\omega)^2/2]$$

Similarly

$$\Psi_{X_2}(\omega) = \exp[j\mu_2\omega + (\sigma_2 j\omega)^2/2]$$

(b) From part (a) we have

$$\Psi_X(\omega) = \exp(-\sigma^2\omega^2/2)$$

and from Equation 2.51.a

$$E\{X^4\} = \frac{1}{j^4} \{\text{Fourth derivative of } \Psi_X(\omega) \text{ at } \omega = 0\}$$
$$= 3\sigma^4$$

Following the same procedure it can be shown for X a normal random variable with mean zero and variance σ^2 that

$$E[X^n] = \begin{cases} 0 & n = 2k + 1 \\ 1\cdot 3 \cdots (n-1)\sigma^n & n = 2k, \ k \text{ an integer.} \end{cases}$$

(c) $\Psi_Z(\omega) = E\{\exp(j\omega Z)\} = E\{\exp(j\omega[a_1 X_1 + a_2 X_2])\}$
$= E\{\exp(j\omega a_1 X_1)\exp(j\omega a_2 X_2)\}$
$= E\{\exp(j\omega a_1 X_1)\}E\{\exp(j\omega a_2 X_2)\}$

since X_1 and X_2 are independent. Hence,

$$\Psi_Z(\omega) = \Psi_{X_1}(\omega a_1)\Psi_{X_2}(\omega a_2)$$
$$= \exp[j(a_1\mu_1 + a_2\mu_2)\omega + (\sigma_1^2 a_1^2 + \sigma_2^2 a_2^2)(j\omega)^2/2]$$

which shows that Z is Gaussian with

$$\mu_Z = a_1\mu_1 + a_2\mu_2$$

and

$$\sigma_Z^2 = a_1^2\sigma_1^2 + a_2^2\sigma_2^2$$

Cumulant Generating Function. The cumulant generating function C_X of X is defined by

$$C_X(\omega) = \ln \Psi_X(\omega) \qquad (2.52.a)$$

Thus

$$\exp\{C_X(\omega)\} = \Psi_X(\omega)$$

Using series expansions on both sides of this equation results in

$$\exp\left\{K_1(j\omega) + K_2\frac{(j\omega)^2}{2!} + \cdots K_n\frac{(j\omega)^n}{n!} + \cdots\right\}$$
$$= 1 + E[X](j\omega) + E[X^2]\frac{(j\omega)^2}{2!} + \cdots + E[X^n]\frac{(j\omega)^n}{n!} + \cdots \quad (2.52.b)$$

The *cumulants* K_i are defined by the identity in ω given in Equation 2.52.b. Expanding the left-hand side of Equation 2.52.b as the product of the Taylor series expansions of

$$\exp\{K_1 j\omega\}\exp\left\{K_2\frac{(j\omega)^2}{2}\right\}\cdots\exp\left\{K_n\frac{(j\omega)^n}{n!}\right\}\cdots$$

and equating like powers of ω results in

$$E[X] = K_1 \quad (2.52.\text{c})$$

$$E[X^2] = K_2 + K_1^2 \quad (2.52.\text{d})$$

$$E[X^3] = K_3 + 3K_2 K_1 + K_1^3 \quad (2.52.\text{e})$$

$$E[X^4] = K_4 + 4K_3 K_1 + 3K_2^2 + 6K_2 K_1^2 + K_1^4 \quad (2.52.\text{f})$$

Reference [5] contains more information on cumulants. The cumulants are particularly useful when independent random variables are summed because the individual cumulants are directly added.

2.4.2 Examples of Probability Density Functions

We now present three useful models for continuous random variables that will be used later. Several additional models are given in the problems included at the end of the chapter.

Uniform Probability Density Functions. A random variable X is said to have a uniform pdf if

$$f_X(x) = \begin{cases} 1/(b-a), & a \leq x \leq b \\ 0 & \text{elsewhere} \end{cases} \quad (2.53.\text{a})$$

The mean and variance of a uniform random variable can be shown to be

$$\mu_X = \frac{b+a}{2} \quad (2.53.\text{b})$$

$$\sigma_X^2 = \frac{(b-a)^2}{12} \quad (2.53.\text{c})$$

Gaussian Probability Density Function. One of the most widely used pdfs is the Gaussian or normal probability density function. This pdf occurs in so many applications partly because of a remarkable phenomenon called the central limit theorem and partly because of a relatively simple analytical form. The central limit theorem, to be proved in a later section, implies that a random variable that is determined by the sum of a large number of independent causes tends to have a Gaussian probability distribution. Several versions of this theorem have been proven by statisticians and verified experimentally from data by engineers and physicists.

One primary interest in studying the Gaussian pdf is from the viewpoint of using it to model random electrical noise. Electrical noise in communication

44 REVIEW OF PROBABILITY AND RANDOM VARIABLES

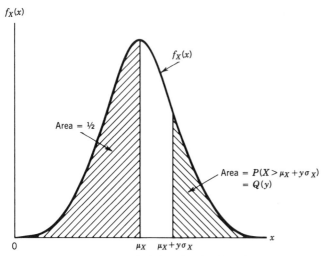

Figure 2.6 Gaussian probability density function.

systems is often due to the cumulative effects of a large number of randomly moving charged particles and hence the instantaneous value of the noise will tend to have a Gaussian distribution—a fact that can be tested experimentally. (The reader is cautioned that there are examples of noise that cannot be modeled by Gaussian pdfs. Such examples include pulse type disturbances on a telephone line and the electrical noise from nearby lightning discharges.)

The Gaussian pdf shown in Figure 2.6 has the form

$$f_X(x) = \frac{1}{\sqrt{2\pi\sigma_X^2}} \exp\left[-\frac{(x-\mu_X)^2}{2\sigma_X^2}\right] \tag{2.54}$$

The family of Gaussian pdfs is characterized by only two parameters, μ_X and σ_X^2, which are the mean and variance of the random variable X. In many applications we will often be interested in probabilities such as

$$P(X > a) = \int_a^\infty \frac{1}{\sqrt{2\pi\sigma_X^2}} \exp\left[-\frac{(x-\mu_X)^2}{2\sigma_X^2}\right] dx$$

By making a change of variable $z = (x - \mu_X)/\sigma_X$, the preceding integral can be reduced to

$$P(X > a) = \int_{(a-\mu_X)/\sigma_X}^\infty \frac{1}{\sqrt{2\pi}} \exp(-z^2/2)\, dz$$

CONTINUOUS RANDOM VARIABLES   45

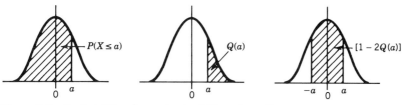

Figure 2.7 Probabilities for a standard Gaussian pdf.

Unfortunately, this integral cannot be evaluated in closed form and requires numerical evaluation. Several versions of the integral are tabulated, and we will use tabulated values (Appendix D) of the Q function, which is defined as

$$Q(y) = \frac{1}{\sqrt{2\pi}} \int_y^\infty \exp(-z^2/2)\, dz, \qquad y > 0 \qquad (2.55)$$

In terms of the values of the Q functions we can write $P(X > a)$ as

$$P(X > a) = Q[(a - \mu_X)/\sigma_X] \qquad (2.56)$$

Various tables give any of the areas shown in Figure 2.7, so one must observe which is being tabulated. However, any of the results can be obtained from the others by using the following relations for the standard ($\mu = 0$, $\sigma = 1$) normal random variable X:

$$P(X \leq x) = 1 - Q(x)$$
$$P(-a \leq X \leq a) = 2P(-a \leq X \leq 0) = 2P(0 \leq X \leq a)$$
$$P(X \leq 0) = \frac{1}{2} = Q(0)$$

EXAMPLE 2.14.

The voltage X at the output of a noise generator is a standard normal random variable. Find $P(X > 2.3)$ and $P(1 \leq X \leq 2.3)$.

SOLUTION: Using one of the tables of standard normal distributions

$$P(X > 2.3) = Q(2.3) \approx .011$$
$$P(1 \leq X \leq 2.3) = 1 - Q(2.3) - [1 - Q(1)] = Q(1) - Q(2.3) \approx .148$$

EXAMPLE 2.15.

The velocity V of the wind at a certain location is normal random variable with $\mu = 2$ and $\sigma = 5$. Determine $P(-3 \leq V \leq 8)$.

SOLUTION:

$$P(-3 \leq V \leq 8) = \int_{-3}^{8} \frac{1}{\sqrt{2\pi(25)}} \exp\left[-\frac{(v-2)^2}{2(25)}\right] dv$$

$$= \int_{(-3-2)/5}^{(8-2)/5} \frac{1}{\sqrt{2\pi}} \exp\left[-\frac{x^2}{2}\right] dx$$

$$= 1 - Q(1.2) - [1 - Q(-1)] \approx .726$$

Bivariate Gaussian pdf. We often encounter the situation when the instantaneous amplitude of the input signal to a linear system has a Gaussian pdf and we might be interested in the joint pdf of the amplitude of the input and the output signals. The bivariate Gaussian pdf is a valid model for describing such situations. The bivariate Gaussian pdf has the form

$$f_{X,Y}(x, y) = \frac{1}{2\pi\sigma_X\sigma_Y\sqrt{1-\rho^2}} \exp\left\{\frac{-1}{2(1-\rho^2)}\left[\left(\frac{x-\mu_X}{\sigma_X}\right)^2 + \left(\frac{y-\mu_Y}{\sigma_Y}\right)^2 - \frac{2\rho(x-\mu_X)(y-\mu_Y)}{\sigma_X\sigma_Y}\right]\right\} \quad (2.57)$$

The reader can verify that the marginal pdfs of X and Y are Gaussian with means μ_X, μ_Y, and variances σ_X^2, σ_Y^2, respectively, and

$$\rho = \rho_{XY} = \frac{E\{(X-\mu_X)(Y-\mu_Y)\}}{\sigma_X\sigma_Y} = \frac{\sigma_{XY}}{\sigma_X\sigma_Y}$$

2.4.3 Complex Random Variables

A complex random variable Z is defined in terms of the real random variables X and Y by

$$Z = X + jY$$

The expected value of $g(Z)$ is defined as

$$E\{g(Z)\} \triangleq \int_{-\infty}^{\infty} \int_{-\infty}^{\infty} g(z) f_{X,Y}(x, y)\, dx\, dy$$

Thus the mean, μ_Z, of Z is

$$\mu_Z = E\{Z\} = E\{X\} + jE\{Y\} = \mu_X + j\mu_Y$$

The variance, σ_Z^2, is defined as

$$\sigma_Z^2 \triangleq E\{|Z - \mu_Z|^2\}$$

The covariance of two complex random variables Z_m and Z_n is defined by

$$C_{Z_m Z_n} \triangleq E\{(Z_m - \mu_{Z_m})^*(Z_n - \mu_{Z_n})\}$$

where * denotes complex conjugate.

2.5 RANDOM VECTORS

In the preceding sections we concentrated on discussing the specification of probability laws for one or two random variables. In this section we shall discuss the specification of probability laws for many random variables (i.e., random vectors). Whereas scalar-valued random variables take on values on the real line, the values of "vector-valued" random variables are points in a real-valued higher (say m) dimensional space (R_m). An example of a three-dimensional random vector is the location of a space vehicle in a Cartesian coordinate system.

The probability law for vector-valued random variables is specified in terms of a joint distribution function

$$F_{X_1,\ldots,X_m}(x_1, \ldots, x_m) = P[(X_1 \leq x_1) \ldots (X_m \leq x_m)]$$

or by a joint probability mass function (discrete case) or a joint probability density function (continuous case). We treat the continuous case in this section leaving details of the discrete case for the reader.

The joint probability density function of an m-dimensional random vector is the partial derivative of the distribution function and is denoted by

$$f_{X_1,X_2,\ldots,X_m}(x_1, x_2, \ldots, x_m)$$

From the joint pdf, we can obtain the marginal pdfs as

$$f_{X_1}(x_1) = \underbrace{\int_{-\infty}^{\infty}\int_{-\infty}^{\infty}\cdots\int_{-\infty}^{\infty}}_{m-1 \text{ integrals}} f_{X_1,X_2,\ldots,X_m}(x_1, x_2, \ldots, x_m)\, dx_2 \cdots dx_m$$

and

$$f_{X_1,X_2}(x_1, x_2) = \underbrace{\int_{-\infty}^{\infty}\int_{-\infty}^{\infty}\cdots\int_{-\infty}^{\infty}}_{m-2 \text{ integrals}} f_{X_1,X_2,\ldots,X_m}(x_1, x_2, x_3, \ldots, x_m)\, dx_3\, dx_4 \cdots dx_m \quad (2.58)$$

Note that the marginal pdf of any subset of the m variables is obtained by "integrating out" the variables not in the subset.

The conditional density functions are defined as (using $m = 4$ as an example),

$$f_{X_1,X_2,X_3|X_4}(x_1, x_2, x_3|x_4) = \frac{f_{X_1,X_2,X_3,X_4}(x_1, x_2, x_3, x_4)}{f_{X_4}(x_4)} \quad (2.59)$$

and

$$f_{X_1,X_2|X_3,X_4}(x_1, x_2|x_3, x_4) = \frac{f_{X_1,X_2,X_3,X_4}(x_1, x_2, x_3, x_4)}{f_{X_3,X_4}(x_3, x_4)} \quad (2.60)$$

Expected values are evaluated using multiple integrals. For example,

$$E\{g(X_1, X_2, X_3, X_4)\}$$
$$= \int_{-\infty}^{\infty}\int_{-\infty}^{\infty}\int_{-\infty}^{\infty}\int_{-\infty}^{\infty} g(x_1, x_2, x_3, x_4) f_{X_1,X_2,X_3,X_4}(x_1, x_2, x_3, x_4)\, dx_1\, dx_2\, dx_3\, dx_4$$
$$(2.61)$$

where g is a scalar-valued function. Conditional expected values are defined for example, as

$$E\{g(X_1, X_2, X_3, X_4)|X_3 = x_3, X_4 = x_4\}$$
$$= \int_{-\infty}^{\infty}\int_{-\infty}^{\infty} g(x_1, x_2, x_3, x_4) f_{X_1,X_2|X_3,X_4}(x_1, x_2|x_3, x_4)\, dx_1\, dx_2 \quad (2.62)$$

Important parameters of the joint distribution are the means and the *co-variances*

$$\mu_{X_i} = E\{X_i\}$$

and

$$\sigma_{X_i X_j} = E\{X_i X_j\} - \mu_{X_i}\mu_{X_j}$$

Note that $\sigma_{X_i X_i}$ is the variance of X_i. We will use both $\sigma_{X_i X_i}$, and $\sigma_{X_i}^2$ to denote the variance of X_i. Sometimes the notations E_{X_i}, $E_{X_i X_j}$, $E_{X_i|X_j}$ are used to denote expected values with respect to the marginal distribution of X_i, the joint distribution of X_i and X_j, and the conditional distribution of X_i given X_j, respectively. We will use subscripted notation for the expectation operator only when there is ambiguity with the use of unsubscripted notation.

The probability law for random vectors can be specified in a concise form using the vector notation. Suppose we are dealing with the joint probability law for m random variables $X_1, X_2, \ldots, X_m$. These m variables can be represented as components of an $m \times 1$ column vector $\mathbf{X}$,

$$\mathbf{X} = \begin{bmatrix} X_1 \\ X_2 \\ \vdots \\ X_m \end{bmatrix} \quad \text{or } \mathbf{X}^T = (X_1, X_2, \ldots, X_m)$$

where T indicates the transpose of a vector (or matrix). The values of $\mathbf{X}$ are points in the m-dimensional space R_m. A specific value of $\mathbf{X}$ is denoted by

$$\mathbf{x}^T = (x_1, x_2, \ldots, x_m)$$

Then, the joint pdf is denoted by

$$f_\mathbf{X}(\mathbf{x}) = f_{X_1, X_2, \ldots, X_m}(x_1, x_2, \ldots, x_m)$$

The mean vector is defined as

$$\mu_\mathbf{X} = E(\mathbf{X}) = \begin{bmatrix} E(X_1) \\ E(X_2) \\ \vdots \\ E(X_m) \end{bmatrix}$$

and the "covariance-matrix", Σ_X, an $m \times m$ matrix is defined as

$$\Sigma_X = E\{XX^T\} - \mu_X \mu_X^T$$

$$= \begin{bmatrix} \sigma_{X_1 X_1} & \sigma_{X_1 X_2} & \cdots & \sigma_{X_1 X_m} \\ \sigma_{X_2 X_1} & \sigma_{X_2 X_2} & \cdots & \sigma_{X_2 X_m} \\ \vdots & & & \\ \sigma_{X_m X_1} & \sigma_{X_m X_2} & \cdots & \sigma_{X_m X_m} \end{bmatrix} \quad m \times m$$

The covariance matrix describes the second-order relationship between the components of the random vector X. The components are said to be "uncorrelated" when

$$\sigma_{X_i X_j} = \sigma_{ij} = 0, \quad i \neq j$$

and independent if

$$f_{X_1, X_2, \ldots, X_m}(x_1, x_2, \ldots, x_m) = \prod_{i=1}^{m} f_{X_i}(x_i) \tag{2.63}$$

2.5.1 Multivariate Gaussian Distribution

An important extension of the bivariate Gaussian distribution is the multivariate Gaussian distribution, which has many applications. A random vector X is multivariate Gaussian if it has a pdf of the form

$$f_X(x) = [(2\pi)^{m/2} |\Sigma_X|^{1/2}]^{-1} \exp\left[-\frac{1}{2}(x - \mu_X)^T \Sigma_X^{-1} (x - \mu_X)\right] \tag{2.64}$$

where μ_X is the mean vector, Σ_X is the covariance matrix, Σ_X^{-1} is its inverse, $|\Sigma_X|$ is the determinant of Σ_X, and X is of dimension m.

2.5.2 Properties of the Multivariate Gaussian Distribution

We state next some of the important properties of the multivariate Gaussian distribution. Proofs of these properties are given in Reference [6].

1. Suppose **X** has an m-dimensional multivariate Gaussian distribution. If we partition **X** as

$$\mathbf{X} = \begin{bmatrix} \mathbf{X}_1 \\ \mathbf{X}_2 \end{bmatrix} \qquad \mathbf{X}_1 = \begin{bmatrix} X_1 \\ X_2 \\ \vdots \\ X_k \end{bmatrix} \qquad \mathbf{X}_2 = \begin{bmatrix} X_{k+1} \\ X_{k+2} \\ \vdots \\ X_m \end{bmatrix}$$

and

$$\boldsymbol{\mu}_\mathbf{X} = \begin{bmatrix} \boldsymbol{\mu}_{\mathbf{X}_1} \\ \hline \boldsymbol{\mu}_{\mathbf{X}_2} \end{bmatrix} \qquad \boldsymbol{\Sigma}_\mathbf{X} = \begin{bmatrix} \boldsymbol{\Sigma}_{11} & \boldsymbol{\Sigma}_{12} \\ \hline \boldsymbol{\Sigma}_{21} & \boldsymbol{\Sigma}_{22} \end{bmatrix}$$

where $\boldsymbol{\mu}_{\mathbf{X}_1}$ is $k \times 1$ and $\boldsymbol{\Sigma}_{11}$ is $k \times k$, then $\mathbf{X}_1$ has a k-dimensional multivariate Gaussian distribution with a mean $\boldsymbol{\mu}_{\mathbf{X}_1}$ and covariance $\boldsymbol{\Sigma}_{11}$.

2. If $\boldsymbol{\Sigma}_\mathbf{X}$ is a diagonal matrix, that is,

$$\boldsymbol{\Sigma}_\mathbf{X} = \begin{bmatrix} \sigma_1^2 & 0 & 0 & \cdots & 0 \\ 0 & \sigma_2^2 & 0 & \cdots & 0 \\ \vdots & & \vdots & & \vdots \\ 0 & 0 & 0 & \cdots & \sigma_m^2 \end{bmatrix}$$

then the components of **X** are independent (i.e., uncorrelatedness implies independence. However, this property does not hold for other distributions).

3. If **A** is a $k \times m$ matrix of rank k, then $\mathbf{Y} = \mathbf{AX}$ has a k-variate Gaussian distribution with

$$\boldsymbol{\mu}_\mathbf{Y} = \mathbf{A}\boldsymbol{\mu}_\mathbf{X} \qquad (2.65.\text{a})$$

$$\boldsymbol{\Sigma}_\mathbf{Y} = \mathbf{A}\boldsymbol{\Sigma}_\mathbf{X}\mathbf{A}^T \qquad (2.65.\text{b})$$

4. With a partition of **X** as in (1), the conditional density of $\mathbf{X}_1$ given $\mathbf{X}_2 = \mathbf{x}_2$ is a k-dimensional multivariate Gaussian with

$$\boldsymbol{\mu}_{\mathbf{X}_1|\mathbf{X}_2} = E[\mathbf{X}_1|\mathbf{X}_2 = \mathbf{x}_2] = \boldsymbol{\mu}_{\mathbf{X}_1} + \boldsymbol{\Sigma}_{12}\boldsymbol{\Sigma}_{22}^{-1}(\mathbf{x}_2 - \boldsymbol{\mu}_{\mathbf{X}_2}) \qquad (2.66.\text{a})$$

and

$$\boldsymbol{\Sigma}_{\mathbf{X}_1|\mathbf{X}_2} = \boldsymbol{\Sigma}_{11} - \boldsymbol{\Sigma}_{12}\boldsymbol{\Sigma}_{22}^{-1}\boldsymbol{\Sigma}_{21} \qquad (2.66.\text{b})$$

Properties (1), (3), and (4) state that marginals, conditionals, as well as linear transformations derived from a multivariate Gaussian distribution all have multivariate Gaussian distributions.

EXAMPLE 2.15.

Suppose $\mathbf{X}$ is four-variate Gaussian with

$$\mu_{\mathbf{X}} = \begin{bmatrix} 2 \\ 1 \\ 1 \\ 0 \end{bmatrix}$$

and

$$\Sigma_{\mathbf{X}} = \begin{bmatrix} 6 & 3 & 2 & 1 \\ 3 & 4 & 3 & 2 \\ 2 & 3 & 4 & 3 \\ 1 & 2 & 3 & 3 \end{bmatrix}$$

Let

$$\mathbf{X}_1 = \begin{bmatrix} X_1 \\ X_2 \end{bmatrix} \qquad \mathbf{X}_2 = \begin{bmatrix} X_3 \\ X_4 \end{bmatrix}$$

(a) Find the distribution of $\mathbf{X}_1$.
(b) Find the distribution of

$$\mathbf{Y} = \begin{bmatrix} 2X_1 \\ X_1 + 2X_2 \\ X_3 + X_4 \end{bmatrix}$$

(c) Find the distribution of $\mathbf{X}_1$ given $\mathbf{X}_2 = (x_3, x_4)^T$.

SOLUTION:

(a) $\mathbf{X}_1$ has a bivariate Gaussian distribution with

$$\mu_{\mathbf{X}_1} = \begin{bmatrix} 2 \\ 1 \end{bmatrix} \quad \text{and} \quad \Sigma_{\mathbf{X}_1} = \begin{bmatrix} 6 & 3 \\ 3 & 4 \end{bmatrix}$$

(b) We can express $\mathbf{Y}$ as

$$\mathbf{Y} = \begin{bmatrix} 2 & 0 & 0 & 0 \\ 1 & 2 & 0 & 0 \\ 0 & 0 & 1 & 1 \end{bmatrix} \begin{bmatrix} X_1 \\ X_2 \\ X_3 \\ X_4 \end{bmatrix} = \mathbf{AX}$$

Hence **Y** has a trivariate Gaussian distribution with

$$\mu_Y = A\mu_X = \begin{bmatrix} 2 & 0 & 0 & 0 \\ 1 & 2 & 0 & 0 \\ 0 & 0 & 1 & 1 \end{bmatrix} \begin{bmatrix} 2 \\ 1 \\ 1 \\ 0 \end{bmatrix} = \begin{bmatrix} 4 \\ 4 \\ 1 \end{bmatrix}$$

and

$$\Sigma_Y = A\Sigma_X A^T$$

$$= \begin{bmatrix} 2 & 0 & 0 & 0 \\ 1 & 2 & 0 & 0 \\ 0 & 0 & 1 & 1 \end{bmatrix} \begin{bmatrix} 6 & 3 & 2 & 1 \\ 3 & 4 & 3 & 2 \\ 2 & 3 & 4 & 3 \\ 1 & 2 & 3 & 3 \end{bmatrix} \begin{bmatrix} 2 & 1 & 0 \\ 0 & 2 & 0 \\ 0 & 0 & 1 \\ 0 & 0 & 1 \end{bmatrix}$$

$$= \begin{bmatrix} 24 & 24 & 6 \\ 24 & 34 & 13 \\ 6 & 13 & 13 \end{bmatrix}$$

(c) $\mathbf{X}_1$ given $\mathbf{X}_2 = (x_3, x_4)^T$ has a bivariate Gaussian distribution with

$$\mu_{\mathbf{X}_1|\mathbf{X}_2} = \begin{pmatrix} 2 \\ 1 \end{pmatrix} + \begin{pmatrix} 2 & 1 \\ 3 & 2 \end{pmatrix} \begin{pmatrix} 4 & 3 \\ 3 & 3 \end{pmatrix}^{-1} \begin{pmatrix} x_3 - 1 \\ x_4 - 0 \end{pmatrix}$$

$$= \begin{bmatrix} x_3 - \dfrac{2}{3} x_4 + 1 \\ x_3 - \dfrac{1}{3} x_4 \end{bmatrix}$$

and

$$\Sigma_{\mathbf{X}_1|\mathbf{X}_2} = \begin{bmatrix} 6 & 3 \\ 3 & 4 \end{bmatrix} - \begin{bmatrix} 2 & 1 \\ 3 & 2 \end{bmatrix} \begin{bmatrix} 4 & 3 \\ 3 & 3 \end{bmatrix}^{-1} \begin{bmatrix} 2 & 3 \\ 1 & 2 \end{bmatrix}$$

$$= \begin{bmatrix} 14/3 & 4/3 \\ 4/3 & 5/3 \end{bmatrix}$$

2.5.3 Moments of Multivariate Gaussian pdf

Although Equation 2.65 gives the moments of a linear combination of multivariate Gaussian variables, there are many applications where we need to compute moments such as $E\{X_1^2 X_2^2\}$, $E\{X_1 X_2 X_3 X_4\}$, and so on. These moments can be calculated using the joint characteristic function of the multivariate Gaussian density function, which is defined by

$$\Psi_X(\omega_1, \omega_2, \ldots, \omega_n) = E\{\exp[j(\omega_1 X_1 + \omega_2 X_2 + \cdots \omega_n X_n)]\}$$
$$= \exp\left[j\mu_X^T\omega - \frac{1}{2}\omega^T \Sigma_X \omega\right] \qquad (2.67)$$

where $\omega^T = (\omega_1, \omega_2, \ldots, \omega_n)$. From the joint characteristic function, the moments can be obtained by partial differentiation. For example,

$$E\{X_1 X_2 X_3 X_4\} = \frac{\partial^4 \Psi_X(\omega_1, \omega_2, \omega_3, \omega_4)}{\partial \omega_1 \partial \omega_2 \partial \omega_3 \partial \omega_4} \quad \text{at } \omega = (0) \qquad (2.68)$$

To simplify the illustrative calculations, let us assume that all random variables have zero means. Then,

$$\Psi_X(\omega_1, \omega_2, \omega_3, \omega_4) = \exp\left(-\frac{1}{2}\omega^T \Sigma_X \omega\right)$$

Expanding the characteristic function as a power series prior to differentiation, we have

$$\Psi_X(\omega_1, \omega_2, \omega_3, \omega_4) = 1 - \frac{1}{2}\omega^T \Sigma_X \omega$$
$$+ \frac{1}{8}(\omega^T \Sigma_X \omega)^2 + R$$

where R contains terms of ω raised to the sixth and higher power. When we take the partial derivatives and set $\omega_1 = \omega_2 = \omega_3 = \omega_4 = 0$, the only nonzero terms come from terms proportional to $\omega_1 \omega_2 \omega_3 \omega_4$ in

$$\frac{1}{8}(\omega^T \Sigma_X \omega)^2 = \frac{1}{8}\{\sigma_{11}\omega_1^2 + \sigma_{22}\omega_2^2 + \sigma_{33}\omega_3^2 + \sigma_{44}\omega_4^2$$
$$+ 2\sigma_{12}\omega_1\omega_2 + 2\sigma_{13}\omega_1\omega_3 + 2\sigma_{14}\omega_1\omega_4$$
$$+ 2\sigma_{23}\omega_2\omega_3 + 2\sigma_{24}\omega_2\omega_4 + 2\sigma_{34}\omega_3\omega_4\}^2$$

When we square the quadradic term, the only terms proportional to $\omega_1\omega_2\omega_3\omega_4$ will be

$$\frac{1}{8}\{8\sigma_{12}\sigma_{34}\omega_1\omega_2\omega_3\omega_4 + 8\sigma_{23}\sigma_{14}\omega_2\omega_3\omega_1\omega_4 + 8\sigma_{24}\sigma_{13}\omega_2\omega_4\omega_1\omega_3\}$$

Taking the partial derivative of the preceding expression and setting $\omega = (0)$, we have

$$\begin{aligned} E\{X_1X_2X_3X_4\} &= \sigma_{12}\sigma_{34} + \sigma_{23}\sigma_{14} + \sigma_{24}\sigma_{13} \\ &= E\{X_1X_2\}E\{X_3X_4\} + E\{X_2X_3\}E\{X_1X_4\} \\ &\quad + E\{X_2X_4\}E\{X_1X_3\} \end{aligned} \quad (2.69)$$

The reader can verify that for the zero mean case

$$E\{X_1^2X_2^2\} = E\{X_1^2\}E\{X_2^2\} + 2[E\{X_1X_2\}]^2 \quad (2.70)$$

2.6 TRANSFORMATIONS (FUNCTIONS) OF RANDOM VARIABLES

In the analysis of electrical systems we are often interested in finding the properties of a signal after it has been "processed" by the system. Typical processing operations include integration, weighted averaging, and limiting. These signal processing operations may be viewed as transformations of a set of input variables to a set of output variables. If the input is a set of random variables, then the output will also be a set of random variables. In this section, we develop techniques for obtaining the probability law (distribution) for the set of output random variables given the transformation and the probability law for the set of input random variables.

The general type of problem we address is the following. Assume that X is a random variable with ensemble S_X and a known probability distribution. Let g be a scalar function that maps each $x \in S_X$ to $y = g(x)$. The expression

$$Y = g(X)$$

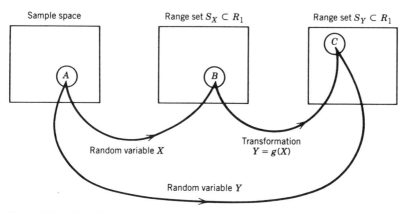

Figure 2.8 Transformation of a random variable.

defines a new random variable* as follows (see Figure 2.8). For a given outcome λ, $X(\lambda)$ is a number x, and $g[X(\lambda)]$ is another number specified by $g(x)$. This number is the value of the random variable Y, that is, $Y(\lambda) = y = g(x)$. The ensemble S_Y of Y is the set

$$S_Y = \{y = g(x) : x \in S_X\}$$

We are interested in finding the probability law for Y.

The method used for identifying the probability law for Y is to equate the probabilities of equivalent events. Suppose $C \subset S_Y$. Because the function $g(x)$ maps $S_X \to S_Y$, there is an equivalent subset B, $B \subset S_X$, defined by

$$B = \{x : g(x) \in C\}$$

Now, B corresponds to event A, which is a subset of the sample space S (see Figure 2.8). It is obvious that A maps to C and hence

$$P(C) = P(A) = P(B)$$

---

*For Y to be a random variable, the function $g : X \to Y$ must have the following properties:

1. Its domain must include the range of the random variable X.
2. It must be a Baire function, that is, for every y, the set I_y such that $g(x) \leq y$ must consist of the union and intersection of a countable number of intervals in S_X. Only then $\{Y \leq y\}$ is an event.
3. The events $\{\lambda : g(X(\lambda)) = \pm\infty\}$ must have zero probability.

Now, suppose that g is a continuous function and $C = (-\infty, y]$. If $B = \{x : g(x) \le y\}$, then

$$P(C) = P(Y \le y) = F_Y(y)$$
$$= \int_B f_X(x)\,dx$$

which gives the distribution function of Y in terms of the density function of X. The density function of Y (if Y is a continuous random variable) can be obtained by differentiating $F_Y(y)$.

As an alternate approach, suppose I_y is a small interval of length Δy containing the point y. Let $I_x = \{x : g(x) \in I_y\}$. Then, we have

$$P(Y \in I_y) \approx f_Y(y)\,\Delta y$$
$$\approx \int_{I_x} f_X(x)\,dx$$

which shows that we can derive the density of Y from the density of X.

We will use the principles outlined in the preceding paragraphs to find the distribution of scalar-valued as well as vector-valued functions of random variables.

2.6.1 Scalar-valued Function of One Random Variable

Discrete Case. Suppose X is a discrete random variable that can have one of n values $x_1, x_2, \ldots, x_n$. Let $g(x)$ be a scalar-valued function. Then $Y = g(X)$ is a discrete random variable that can have one of m, $m \le n$, values $y_1, y_2, \ldots, y_m$. If $g(X)$ is a one-to-one mapping, then m will be equal to n. However, if $g(x)$ is a many-to-one mapping, then m will be smaller than n. The probability mass function of Y can be obtained easily from the probability mass function of X as

$$P(Y = y_j) = \sum P(X = x_i)$$

where the sum is over all values of x_i that map to y_j.

Continuous Random Variables. If X is a continuous random variable, then the pdf of $Y = g(X)$ can be obtained from the pdf of X as follows. Let y be a particular value of Y and let $x^{(1)}, x^{(2)}, \ldots, x^{(k)}$ be the roots of the equation $y = g(x)$. That is $y = g(x^{(1)}) = \ldots = g(x^{(k)})$. (For example, if $y = x^2$, then the

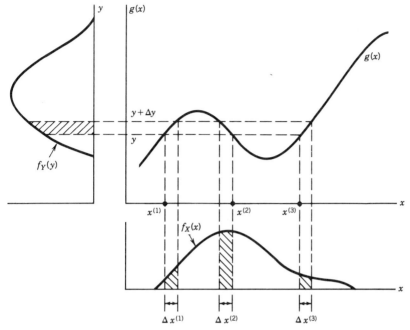

Figure 2.9 Transformation of a continuous random variable.

two roots are $x^{(1)} = +\sqrt{y}$ and $x^{(2)} = -\sqrt{y}$; also see Figure 2.9 for another example.) We know that

$$P(y < Y \le y + \Delta y) = f_Y(y)\, \Delta y \quad \text{as} \quad \Delta y \to 0$$

Now if we can find the set of values of x such that $y < g(x) \le y + \Delta y$, then we can obtain $f_Y(y)\, \Delta y$ from the probability that X belongs to this set. That is

$$P(y < Y \le y + \Delta y) = P[\{x : y < g(x) \le y + \Delta y\}]$$

For the example shown in Figure 2.9, this set consists of the following three intervals:

$$x^{(1)} < x \le x^{(1)} + \Delta x^{(1)}$$
$$x^{(2)} + \Delta x^{(2)} < x \le x^{(2)}$$
$$x^{(3)} < x \le x^{(3)} + \Delta x^{(3)}$$

TRANSFORMATIONS (FUNCTIONS) OF RANDOM VARIABLES

where $\Delta x^{(1)} > 0$, $\Delta x^{(3)} > 0$ but $\Delta x^{(2)} < 0$. From the foregoing it follows that

$$P(y < Y < y + \Delta y) = P(x^{(1)} < X < x^{(1)} + \Delta x^{(1)})$$
$$+ P(x^{(2)} + \Delta x^{(2)} < X < x^{(2)})$$
$$+ P(x^{(3)} < X < x^{(3)} + \Delta x^{(3)})$$

We can see from Figure 2.9 that the terms in the right-hand side are given by

$$P(x^{(1)} < X < x^{(1)} + \Delta x^{(1)}) = f_X(x^{(1)}) \, \Delta x^{(1)}$$
$$P(x^{(2)} + \Delta x^{(2)} < X < x^{(2)}) = f_X(x^{(2)}) |\Delta x^{(2)}|$$
$$P(x^{(3)} < X < x^{(3)} + \Delta x^{(3)}) = f_X(x^{(3)}) \, \Delta x^{(3)}$$

Since the slope $g'(x)$ of $g(x)$ is $\Delta y / \Delta x$, we have

$$\Delta x^{(1)} = \Delta y / g'(x^{(1)})$$
$$\Delta x^{(2)} = \Delta y / g'(x^{(2)})$$
$$\Delta x^{(3)} = \Delta y / g'(x^{(3)})$$

Hence we conclude that, when we have three roots for the equation $y = g(x)$,

$$f_Y(y) \Delta y = \frac{f_X(x^{(1)})}{g'(x^{(1)})} \Delta y + \frac{f_X(x^{(2)})}{|g'(x^{(2)})|} \Delta y + \frac{f_X(x^{(3)})}{g'(x^{(3)})} \Delta y$$

Canceling the Δy and generalizing the result, we have

$$f_Y(y) = \sum_{i=1}^{k} \frac{f_X(x^{(i)})}{|g'(x^{(i)})|} \tag{2.71}$$

$g'(x)$ is also called the Jacobian of the transformation and is often denoted by $J(x)$. Equation 2.71 gives the pdf of the transformed variable Y in terms of the pdf of X, which is given. The use of Equation 2.71 is limited by our ability to find the roots of the equation $y = g(x)$. If $g(x)$ is highly nonlinear, then the solutions of $y = g(x)$ can be difficult to find.

EXAMPLE 2.16.

Suppose X has a Gaussian distribution with a mean of 0 and variance of 1 and $Y = X^2 + 4$. Find the pdf of Y.

SOLUTION: $y = g(x) = x^2 + 4$ has two roots:

$$x^{(1)} = +\sqrt{y - 4}$$
$$x^{(2)} = -\sqrt{y - 4}$$

and hence

$$g'(x^{(1)}) = 2\sqrt{y - 4}$$
$$g'(x^{(2)}) = -2\sqrt{y - 4}$$

The density function of Y is given by

$$f_Y(y) = \frac{f_X(x^{(1)})}{|g'(x^{(1)})|} + \frac{f_X(x^{(2)})}{|g'(x^{(2)})|}$$

With $f_X(x)$ given as

$$f_X(x) = \frac{1}{\sqrt{2\pi}} \exp(-x^2/2),$$

we obtain

$$f_Y(y) = \begin{cases} \frac{1}{\sqrt{2\pi(y-4)}} \exp(-(y-4)/2), & y \geq 4 \\ 0 & y < 4 \end{cases}$$

Note that since $y = x^2 + 4$, and the domain of X is $(-\infty, \infty)$, the domain of Y is $[4, \infty)$.

EXAMPLE 2.17

Using the pdf of X and the transformation shown in Figure 2.10.a and 2.10.b, find the distribution of Y.

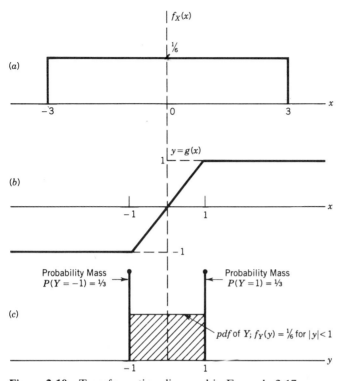

Figure 2.10 Transformation discussed in Example 2.17.

SOLUTION: For $-1 < x < 1$, $y = x$ and hence

$$f_Y(y) = f_X(y) = \frac{1}{6}, \quad -1 < y < 1$$

All the values of $x > 1$ map to $y = 1$. Since $x > 1$ has a probability of $\frac{1}{3}$, the probability that $Y = 1$ is equal to $P(X > 1) = \frac{1}{3}$. Similarly $P(Y = -1) = \frac{1}{3}$. Thus, Y has a *mixed distribution* with a continuum of values in the interval $(-1, 1)$ and a discrete set of values from the set $\{-1, 1\}$. The continuous part is characterized by a pdf and the discrete part is characterized by a probability mass function as shown in Figure 2.10.c.

2.6.2 Functions of Several Random Variables

We now attempt to find the joint distribution of n random variables $Y_1, Y_2, \ldots, Y_n$ given the distribution of n related random variables $X_1, X_2, \ldots, X_n$

62   REVIEW OF PROBABILITY AND RANDOM VARIABLES

and the relationship between the two sets of random variables,

$$Y_i = g_i(X_1, X_2, \ldots, X_n), \quad i = 1, 2, \ldots, n$$

Let us start with a mapping of two random variables onto two other random variables:

$$Y_1 = g_1(X_1, X_2)$$
$$Y_2 = g_2(X_1, X_2)$$

Suppose $(x_1^{(i)}, x_2^{(i)})$, $i = 1, 2, \ldots, k$ are the k roots of $y_1 = g_1(x_1, x_2)$ and $y_2 = g_2(x_1, x_2)$. Proceeding along the lines of the previous section, we need to find the region in the x_1, x_2 plane such that

$$y_1 < g_1(x_1, x_2) < y_1 + \Delta y_1$$

and

$$y_2 < g_2(x_1, x_2) < y_2 + \Delta y_2$$

There are k such regions as shown in Figure 2.11 ($k = 3$). Each region consists of a parallelogram and the area of each parallelogram is equal to $\Delta y_1 \Delta y_2 /$

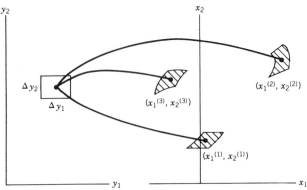

Figure 2.11   Transformation of two random variables.

$|J(x_1^{(i)}, x_2^{(i)})|$ where $J(x_1, x_2)$ is the *Jacobian* of the transformation defined as

$$J(x_1, x_2) = \begin{vmatrix} \dfrac{\partial g_1}{\partial x_1} & \dfrac{\partial g_1}{\partial x_2} \\ \dfrac{\partial g_2}{\partial x_1} & \dfrac{\partial g_2}{\partial x_2} \end{vmatrix} \qquad (2.72)$$

By summing the contribution from all regions, we obtain the joint pdf of Y_1 and Y_2 as

$$f_{Y_1,Y_2}(y_1, y_2) = \sum_{i=1}^{k} \dfrac{f_{X_1,X_2}(x_1^{(i)}, x_2^{(i)})}{|J(x_1^{(i)}, x_2^{(i)})|} \qquad (2.73)$$

Using the vector notation, we can generalize this result to the n-variate case as

$$f_\mathbf{Y}(\mathbf{y}) = \sum_{i=1}^{k} \dfrac{f_\mathbf{X}(\mathbf{x}^{(i)})}{|J(\mathbf{x}^{(i)})|} \qquad (2.74.\text{a})$$

where $\mathbf{x}^{(i)} = [x_1^{(i)}, x_2^{(i)}, \ldots, x_n^{(i)}]^T$ is the ith solution to $\mathbf{y} = \mathbf{g}(\mathbf{x}) = [g_1(\mathbf{x}), g_2(\mathbf{x}), \ldots, g_n(\mathbf{x})]^T$, and the Jacobian J is defined by

$$J[\mathbf{x}^{(i)}] = \begin{vmatrix} \dfrac{\partial g_1}{\partial x_1} & \dfrac{\partial g_1}{\partial x_2} & \cdots & \dfrac{\partial g_1}{\partial x_n} \\ \vdots & & & \vdots \\ \dfrac{\partial g_n}{\partial x_1} & \dfrac{\partial g_n}{\partial x_2} & \cdots & \dfrac{\partial g_n}{\partial x_n} \end{vmatrix} \text{ at } \mathbf{x}^{(i)} \qquad (2.74.\text{b})$$

Suppose we have n random variables with known joint pdf, and we are interested in the joint pdf of $m < n$ functions of them, say

$$y_i = g_i(x_1, x_2, \ldots, x_n), \qquad i = 1, 2, \ldots, m$$

Now, we can define $n - m$ additional functions

$$y_j = g_j(x_1, x_2, \ldots, x_n), \qquad j = m+1, \ldots, n$$

in any convenient way so that the Jacobian is nonzero, compute the joint pdf of $Y_1, Y_2, \ldots, Y_n$, and then obtain the marginal pdf of $Y_1, Y_2, \ldots, Y_m$ by

64 REVIEW OF PROBABILITY AND RANDOM VARIABLES

integrating out $Y_{m+1}, \ldots, Y_n$. If the additional functions are carefully chosen, then the inverse can be easily found and the resulting integration can be handled, but often with great difficulty.

EXAMPLE 2.18.

Let two resistors, having independent resistances, X_1 and X_2, uniformly distributed between 9 and 11 ohms, be placed in parallel. Find the probability density function of resistance Y_1 of the parallel combination.

SOLUTION: The resistance of the parallel combination is

$$y_1 = \frac{x_1 x_2}{(x_1 + x_2)}$$

Introducing the variable

$$y_2 = x_2$$

and solving for x_1 and x_2 results in the unique solution

$$x_1 = \frac{y_1 y_2}{y_2 - y_1}; \quad x_2 = y_2$$

Thus, Equation 2.73 reduces to

$$f_{Y_1,Y_2}(y_1, y_2) = f_{X_1,X_2}\left(\frac{y_1 y_2}{y_2 - y_1}, y_2\right) \bigg/ |J(x_1, x_2)|$$

where

$$J(x_1, x_2) = \begin{vmatrix} \frac{x_2^2}{(x_1 + x_2)^2} & \frac{x_1^2}{(x_1 + x_2)^2} \\ 0 & 1 \end{vmatrix} = \frac{x_2^2}{(x_1 + x_2)^2}$$

$$= \frac{(y_2 - y_1)^2}{y_2^2}$$

TRANSFORMATIONS (FUNCTIONS) OF RANDOM VARIABLES

We are given

$$f_{X_1,X_2}(x_1, x_2) = f_{X_1}(x_1)f_{X_2}(x_2) = \frac{1}{4}, \quad 9 \leq x_1 \leq 11, \quad 9 \leq x_2 \leq 11$$
$$= 0 \quad \text{elsewhere}$$

Thus

$$f_{Y_1,Y_2}(y_1, y_2) = \frac{1}{4}\frac{y_2^2}{(y_2 - y_1)^2}, \quad y_1, y_2 = ?$$
$$= 0 \quad \text{elsewhere}$$

We must now find the region in the y_1, y_2 plane that corresponds to the region $9 \leq x_1 \leq 11$, $9 \leq x_2 \leq 11$. Figure 2.12 shows the mapping and the resulting region in the y_1, y_2 plane.

Now to find the marginal density of Y_1, we "integrate out" y_2.

$$f_{Y_1}(y_1) = \int_9^{9y_1/(9-y_1)} \frac{y_2^2}{4(y_2 - y_1)^2} dy_2, \quad 4\frac{1}{2} \leq y_1 \leq 4\frac{19}{20}$$

$$= \int_{11y_1/(11-y_1)}^{11} \frac{y_2^2}{4(y_2 - y_1)^2} dy_2, \quad 4\frac{19}{20} \leq y_1 \leq 5\frac{1}{2}$$

$$= 0 \quad \text{elsewhere}$$

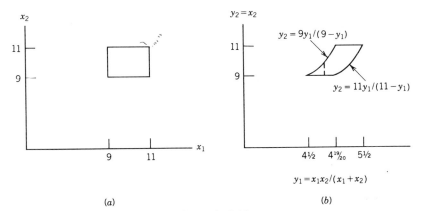

(a)   (b)

Figure 2.12 Transformation of Example 2.18.

Carrying out the integration results in

$$f_{Y_1}(y_1) = \frac{y_1 - 9}{2} + \frac{y_1^2}{2(9 - y_1)} + y_1 \ln \frac{y_1}{9 - y_1}, \qquad 4\frac{1}{2} \le y_1 \le 4\frac{19}{20}$$

$$= \frac{11 - y_1}{2} - \frac{y_1^2}{2(11 - y_1)} + y_1 \ln \frac{11 - y_1}{y_1}, \qquad 4\frac{19}{20} \le y_1 \le 5\frac{1}{2}$$

$$= 0 \qquad \text{elsewhere}$$

Special-case: Linear Transformations. One of the most frequently used type of transformation is the affine transformation, where each of the new variables is a linear combination of the old variables plus a constant. That is

$$Y_1 = a_{1,1}X_1 + a_{1,2}X_2 + \cdots + a_{1,n}X_n + b_1$$
$$Y_2 = a_{2,1}X_1 + a_{2,2}X_2 + \cdots + a_{2,n}X_n + b_2$$
$$\vdots \qquad \vdots \qquad \vdots$$
$$Y_n = a_{n,1}X_1 + a_{n,2}X_2 + \cdots + a_{n,n}X_n + b_n$$

where the $a_{i,j}$'s and b_i's are all constants. In matrix notation we can write this transformation as

$$\begin{bmatrix} Y_1 \\ Y_2 \\ \vdots \\ Y_n \end{bmatrix} = \begin{bmatrix} a_{1,1} & a_{1,2} & \cdots & a_{1,n} \\ a_{2,1} & a_{2,2} & \cdots & a_{2,n} \\ \vdots & \vdots & & \vdots \\ a_{n,1} & a_{n,2} & \cdots & a_{n,n} \end{bmatrix} \begin{bmatrix} X_1 \\ X_2 \\ \vdots \\ X_n \end{bmatrix} + \begin{bmatrix} b_1 \\ b_2 \\ \vdots \\ b_n \end{bmatrix}$$

or

$$\mathbf{Y} = \mathbf{AX} + \mathbf{B} \qquad (2.75)$$

where $\mathbf{A}$ is $n \times n$, $\mathbf{Y}$, $\mathbf{X}$, and $\mathbf{B}$ are $n \times 1$ matrices. If $\mathbf{A}$ is nonsingular, then the inverse transformation exists and is given by

$$\mathbf{X} = \mathbf{A}^{-1}\mathbf{Y} - \mathbf{A}^{-1}\mathbf{B}$$

The Jacobian of the transformation is

$$J = \begin{vmatrix} a_{1,1} & a_{1,2} & \cdots & a_{1,n} \\ a_{2,1} & a_{2,2} & \cdots & a_{2,n} \\ \vdots & & & \\ a_{n,1} & a_{n,2} & \cdots & a_{n,n} \end{vmatrix} = |\mathbf{A}|$$

Substituting the preceding two equations into Equation 2.71, we obtain the pdf of **Y** as

$$f_\mathbf{Y}(\mathbf{y}) = f_\mathbf{X}(\mathbf{A}^{-1}\mathbf{y} - \mathbf{A}^{-1}\mathbf{B}) \|\mathbf{A}\|^{-1} \tag{2.76}$$

Sum of Random Variables. We consider $Y_1 = X_1 + X_2$ where X_1 and X_2 are independent random variables. As suggested before, let us introduce an additional function $Y_2 = X_2$ so that the transformation is given by

$$\begin{bmatrix} Y_1 \\ Y_2 \end{bmatrix} = \begin{bmatrix} 1 & 1 \\ 0 & 1 \end{bmatrix} \begin{bmatrix} X_1 \\ X_2 \end{bmatrix}$$

From Equation 2.76 it follows that

$$\begin{aligned} f_{Y_1,Y_2}(y_1, y_2) &= f_{X_1,X_2}(y_1 - y_2, y_2) \\ &= f_{X_1}(y_1 - y_2) f_{X_2}(y_2) \end{aligned}$$

since X_1 and X_2 are independent.
The pdf of Y_1 is obtained by integration as

$$f_{Y_1}(y_1) = \int_{-\infty}^{\infty} f_{X_1}(y_1 - y_2) f_{X_2}(y_2) \, dy_2 \tag{2.77.a}$$

The relationship given in Equation 2.77.a is said to be the *convolution* of f_{X_1} and f_{X_2}, which is written symbolically as

$$f_{Y_1} = f_{X_1} * f_{X_2} \tag{2.77.b}$$

Thus, the density function of the sum of two independent random variables is given by the convolution of their densities. This also implies that the charac-

68 REVIEW OF PROBABILITY AND RANDOM VARIABLES

teristic functions are multiplied, and the cumulant generating functions as well as individual cumulants are summed.

EXAMPLE 2.19.

X_1 and X_2 are independent random variables with identical uniform distributions in the interval $[-1, 1]$. Find the pdf of $Y_1 = X_1 + X_2$.

SOLUTION: See Figure 2.13

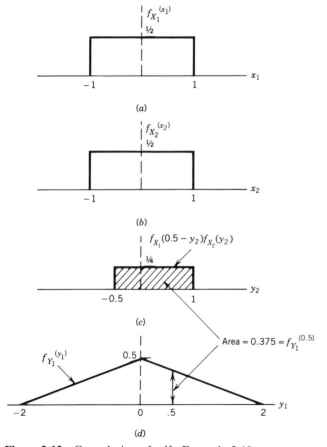

Figure 2.13 Convolution of pdfs–Example 2.19.

EXAMPLE 2.20.

Let $Y = X_1 + X_2$ where X_1 and X_2 are independent, and

$$f_{X_1}(x_1) = \exp(-x_1), \quad x_1 \geq 0; \qquad f_{X_2}(x_2) = 2\exp(-2x_2), \quad x_2 \geq 0,$$
$$= 0 \qquad\qquad x_1 < 0 \qquad\qquad\qquad = 0 \qquad\qquad\qquad x_2 < 0.$$

Find the pdf of Y.

SOLUTION: (See Figure 2.14)

$$f_Y(y) = \int_0^y \exp(-x_1) 2\exp[-2(y - x_1)]\, dx_1$$
$$= 2\exp(-2y) \int_0^y \exp(x_1)\, dx_1 = 2\exp(-2y)[\exp(y) - 1]$$
$$f_Y(y) = 2[\exp(-y) - \exp(-2y)], \quad y \geq 0$$
$$= 0 \qquad\qquad\qquad\qquad\qquad y < 0$$

EXAMPLE 2.21.

X has an n-variate Gaussian density function with $E\{X_i\} = 0$, and a covariance matrix of Σ_X. Find the pdf of $\mathbf{Y} = \mathbf{AX}$ where $\mathbf{A}$ is an $n \times n$ nonsingular matrix.

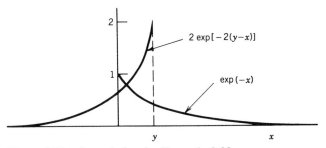

Figure 2.14 Convolution for Example 2.20.

70   REVIEW OF PROBABILITY AND RANDOM VARIABLES

SOLUTION: We are given

$$f_\mathbf{X}(\mathbf{x}) = [(2\pi)^{n/2}|\Sigma_\mathbf{X}|^{1/2}]^{-1} \exp\left[-\frac{1}{2}\mathbf{x}^T \Sigma_\mathbf{X}^{-1}\mathbf{x}\right]$$

With $\mathbf{x} = \mathbf{A}^{-1}\mathbf{y}$, and $J = |A|$, we obtain

$$f_\mathbf{Y}(\mathbf{y}) = [(2\pi)^{n/2}|\Sigma_\mathbf{X}|^{1/2}]^{-1} \exp\left[-\frac{1}{2}\mathbf{y}^T \mathbf{A}^{-1T} \Sigma_\mathbf{X}^{-1} \mathbf{A}^{-1}\mathbf{y}\right] \|\mathbf{A}\|^{-1}$$

Now if we define $\Sigma_\mathbf{Y} = \mathbf{A}\Sigma_\mathbf{X}\mathbf{A}^T$, then the exponent in the pdf of $\mathbf{Y}$ has the form

$$\exp\left(-\frac{1}{2}\mathbf{y}^T \Sigma_\mathbf{Y}^{-1}\mathbf{y}\right)$$

which corresponds to a multivariate Gaussian pdf with zero means and a covariance matrix of $\Sigma_\mathbf{Y}$. Hence, we conclude that $\mathbf{Y}$, which is a linear transformation of a multivariate Gaussian vector $\mathbf{X}$, also has a Gaussian distribution. (*Note:* This cannot be generalized for any arbitrary distribution.)

Order Statistics.   Ordering, comparing, and finding the minimum and maximum are typical statistical or data processing operations. We can use the techniques outlined in the preceding sections for finding the distribution of minimum and maximum values within a group of independent random variables.

Let $X_1, X_2, X_3, \ldots, X_n$ be a group of independent random variables having a common pdf, $f_X(x)$, defined over the interval (a, b). To find the distribution of the smallest and largest of these X_is, let us define the following transformation:

$$\text{Let } Y_1 = \text{smallest of } (X_1, X_2, \ldots, X_n)$$
$$Y_2 = \text{next } X_i \text{ in order of magnitude}$$
$$\vdots$$
$$Y_n = \text{largest of } (X_1, X_2, \ldots, X_n)$$

That is $Y_1 < Y_2 < \cdots < Y_n$ represent $X_1, X_2, \ldots, X_n$ when the latter are arranged in ascending order of magnitude. Then Y_i is called the *ith order statistic* of the group. We will now show that the joint pdf of $Y_1, Y_2, \ldots, Y_n$ is given by

$$f_{Y_1,Y_2,\ldots,Y_n}(y_1, y_2, \ldots, y_n) = n! f_X(y_1) f_X(y_2) \cdots f_X(y_n)$$
$$a < y_1 < y_2 < \cdots < y_n < b$$

We shall prove this for $n = 3$, but the argument can be entirely general. With $n = 3$

$$f_{X_1,X_2,X_3}(x_1, x_2, x_3) = f_X(x_1) f_X(x_2) f_X(x_3)$$

and the transformation is

$$Y_1 = \text{smallest of } (X_1, X_2, X_3)$$
$$Y_2 = \text{middle value of } (X_1, X_2, X_3)$$
$$Y_3 = \text{largest of } (X_1, X_2, X_3)$$

A given set of values x_1, x_2, x_3 may fall into one of the following six possibilities:

$x_1 < x_2 < x_3$ or $y_1 = x_1$, $y_2 = x_2$, $y_3 = x_3$
$x_1 < x_3 < x_2$ or $y_1 = x_1$, $y_2 = x_3$, $y_3 = x_2$
$x_2 < x_1 < x_3$ or $y_1 = x_2$, $y_2 = x_1$, $y_3 = x_3$
$x_2 < x_3 < x_1$ or $y_1 = x_2$, $y_2 = x_3$, $y_3 = x_1$
$x_3 < x_1 < x_2$ or $y_1 = x_3$, $y_2 = x_1$, $y_3 = x_2$
$x_3 < x_2 < x_1$ or $y_1 = x_3$, $y_2 = x_2$, $y_3 = x_1$

(Note that $x_1 = x_2$, etc., occur with a probability of 0 since X_1, X_2, X_3 are continuous random variables.)

Thus, we have six or 3! inverses. If we take a particular inverse, say, $y_1 = x_3$, $y_2 = x_1$, and $y_3 = x_2$, the Jacobian is given by

$$J = \begin{vmatrix} 0 & 0 & 1 \\ 1 & 0 & 0 \\ 0 & 1 & 0 \end{vmatrix} = 1$$

The reader can verify that, for all six inverses, the Jacobian has a magnitude of 1, and using Equation 2.71, we obtain the joint pdf of Y_1, Y_2, Y_3 as

$$f_{Y_1,Y_2,Y_3}(y_1, y_2, y_3) = 3! f_X(y_1) f_X(y_2) f_X(y_3), \qquad a < y_1 < y_2 < y_3 < b$$

Generalizing this to the case of n variables we obtain

$$f_{Y_1,Y_2,\ldots,Y_n}(y_1, y_2, \ldots, y_n) = n! f_X(y_1) f_X(y_2) \cdots f_X(y_n)$$
$$a < y_1 < y_2 < \cdots < y_n < b \quad (2.78.\text{a})$$

The marginal pdf of Y_n is obtained by integrating out $y_1, y_2, \ldots, y_{n-1}$,

$$f_{Y_n}(y_n) = \int_a^{y_n} \int_a^{y_{n-1}} \cdots \int_a^{y_3} \int_a^{y_2} n! f_X(y_1) f_X(y_2) \cdots f_X(y_n) \, dy_1 \, dy_2 \cdots dy_{n-1}$$

The innermost integral on y_1 yields $F_X(y_2)$, and the next integral is

$$\int_a^{y_3} F_X(y_2) f_X(y_2) \, dy_2 = \int_a^{y_3} F_X(y_2) d[F_X(y_2)]$$
$$= \frac{[F_X(y_3)]^2}{2}$$

Repeating this process $(n - 1)$ times, we obtain

$$f_{Y_n}(y_n) = n[F_X(y_n)]^{n-1} f_X(y_n), \quad a < y_n < b \quad (2.78.\text{b})$$

Proceeding along similar lines, we can show that

$$f_{Y_1}(y_1) = n[1 - F_X(y_1)]^{n-1} f_X(y_1), \quad a < y_1 < b \quad (2.78.\text{c})$$

Equations 2.78.b and 2.78.c can be used to obtain and analyze the distribution of the largest and smallest among a group of random variables.

EXAMPLE 2.22.

A peak detection circuit processes 10 identically distributed random samples and selects as its output the sample with the largest value. Find the pdf of the peak detector output assuming that the individual samples have the pdf

$$f_X(x) = ae^{-ax}, \quad x \geq 0$$
$$= 0 \quad x < 0$$

SOLUTION: From Equation 2.78.b, we obtain

$$f_{Y_{10}}(y) = 10[1 - e^{-ay}]^9 a e^{-ay}, \quad y \geq 0$$
$$= 0 \quad y < 0$$

Nonlinear Transformations. While it is relatively easy to find the distribution of $Y = g(\mathbf{X})$ when g is linear or affine, it is usually very difficult to find the distribution of Y when g is nonlinear. However, if X is a scalar random variable, then Equation 2.71 provides a general solution. The difficulties when $\mathbf{X}$ is two-dimensional are illustrated by Example 2.18, and this example suggests the difficulties when $\mathbf{X}$ is more than two-dimensional and g is nonlinear.

For general nonlinear transformations, two approaches are common in practice. One is the Monte Carlo approach, which is outlined in the next subsection. The other approach is based upon an approximation involving moments and is presented in Section 2.7. We mention here that the mean, the variance, and higher moments of Y can be obtained easily (at least conceptually) as follows. We start with

$$E\{\mathbf{h}(\mathbf{Y})]\} = \int_{\mathbf{y}} \mathbf{h}(\mathbf{y}) f_{\mathbf{Y}}(\mathbf{y}) d\mathbf{y}$$

However, $\mathbf{Y} = \mathbf{g}(\mathbf{X})$, and hence we can compute $E\{\mathbf{h}(\mathbf{Y})\}$ as

$$E_{\mathbf{Y}}\{\mathbf{h}(\mathbf{Y})\} = E_{\mathbf{X}}\{\mathbf{h}(\mathbf{g}(\mathbf{X}))\}$$

Since the right-hand side is a function of $\mathbf{X}$ alone, its expected value is

$$E_{\mathbf{X}}\{h(g(\mathbf{X}))\} = \int_{\mathbf{x}} h(g(\mathbf{x})) f_{\mathbf{X}}(\mathbf{x}) \, d\mathbf{x} \tag{2.79}$$

Using the means and covariances, we may be able to approximate the distribution of Y as discussed in the next section.

Monte Carlo (Synthetic Sampling) Technique. We seek an approximation to the distribution or pdf of Y when

$$Y = g(X_1, \ldots, X_n)$$

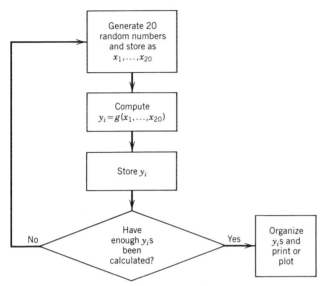

Figure 2.15 Simple Monte Carlo simulation.

It is assumed that $Y = g(X_1, \ldots, X_n)$ is known and that the joint density $f_{X_1, X_2, \ldots, X_n}$ is known. Now if a sample value of each random variable were known (say $X_1 = x_{1,1}$, $X_2 = x_{1,2}, \ldots, X_n = x_{1,n}$), then a sample value of Y could be computed [say $y_1 = g(x_{1,1}, x_{1,2}, \ldots, x_{1,n})$]. If another set of sample values were chosen for the random variables (say $X_1 = x_{2,1}, \ldots, X_n = x_{2,n}$), then $y_2 = g(x_{2,1}, x_{2,2}, \ldots, x_{2,n})$ could be computed.

Monte Carlo techniques simply consist of computer algorithms for selecting the samples $x_{i,1}, \ldots, x_{i,n}$, a method for calculating $y_i = g(x_{i,1}, \ldots, x_{i,n})$, which often is just one or a few lines of code, and a method of organizing and displaying the results of a large number of repetitions of the procedure.

Consider the case where the components of **X** are independent and uniformly distributed between zero and one. This is a particularly simple example because computer routines that generate pseudorandom numbers uniformly distributed between zero and one are widely available. A Monte Carlo program that approximates the distribution of Y when **X** is of dimension 20 is shown in Figure 2.15. The required number of samples is beyond the scope of this introduction. However, the usual result of a Monte Carlo routine is a histogram, and the errors of histograms, which are a function of the number of samples, are discussed in Chapter 8.

If the random variable X_i is not uniformly distributed between zero and one, then random sampling is somewhat more difficult. In such cases the following procedure is used. Select a random sample of U that is uniformly distributed between 0 and 1. Call this random sample u_1. Then $F_{X_i}^{-1}(u_1)$ is the random sample of X_i.

TRANSFORMATIONS (FUNCTIONS) OF RANDOM VARIABLES 75

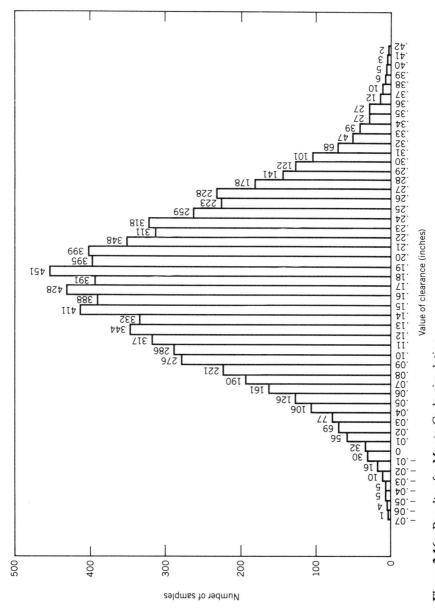

Figure 2.16 Results of a Monte Carlo simulation.

For example, suppose that X is uniformly distributed between 10 and 20. Then

$$\begin{aligned} F_{X_i}(x) &= 0 & x < 10 \\ &= (x - 10)/10, & 10 \leq x < 20 \\ &= 1 & x \geq 20 \end{aligned}$$

Notice $F_{X_i}^{-1}(u) = 10u + 10$. Thus, if the value .250 were the random sample of U, then the corresponding random sample of X would be 12.5.

The reader is asked to show using Equation 2.71 that if X_i has a density function and if $X_i = F_i^{-1}(U) = g(U)$ where U is uniformly distributed between zero and one then F_i^{-1} is unique and

$$f_{X_i}(x) = \frac{dF_i(x)}{dx} \quad \text{where} \quad F_i = (F_i^{-1})^{-1}$$

If the random variables X_i are dependent, then the samples of $X_2, \ldots, X_n$ are based upon the conditional density function $f_{X_2|X_1}, \ldots, f_{X_n|X_{n-1}}, \ldots, X_1$.

The results of an example Monte Carlo simulation of a mechanical tolerance application where Y represents clearance are shown in Figure 2.16. In this case Y was a somewhat complex trigonometric function of 41 dimensions on a production drawing. The results required an assumed distribution for each of the 41 individual dimensions involved in the clearance, and all were assumed to be uniformly distributed between their tolerance limits. This quite nonlinear transformation resulted in results that appear normal, and interference, that is, negative clearance, occurred 71 times in 8000 simulations. This estimate of the probability of interference was verified by results of the assembly operation.

2.7 BOUNDS AND APPROXIMATIONS

In many applications requiring the calculations of probabilities we often face the following situations:

1. The underlying distributions are not completely specified—only the means, variances, and some of the higher order moments $E\{(X - \mu_X)^k\}$, $k > 2$ are known.
2. The underlying density function is known but integration in closed form is not possible (example: the Gaussian pdf).

In these cases we use several approximation techniques that yield upper and/or lower bounds on probabilities.

2.7.1 Tchebycheff Inequality

If only the mean and variance of a random variable X are known, we can obtain upper bounds on $P(|X| \geq \epsilon)$ using the Tchebycheff inequality, which we prove now. Suppose X is a random variable, and we define

$$Y_\epsilon = \begin{cases} 1 & \text{if } |X| \geq \epsilon \\ 0 & \text{if } |X| < \epsilon \end{cases}$$

where ϵ is a positive constant. From the definition of Y_ϵ it follows that

$$X^2 \geq X^2 Y_\epsilon \geq \epsilon^2 Y_\epsilon$$

and thus

$$E\{X^2\} \geq E\{X^2 Y_\epsilon\} \geq \epsilon^2 E\{Y_\epsilon\} \tag{2.80}$$

However,

$$E\{Y_\epsilon\} = 1 \cdot P(|X| \geq \epsilon) + 0 \cdot P(|X| < \epsilon) = P(|X| \geq \epsilon) \tag{2.81}$$

Combining Equations 2.80 and 2.81, we obtain the Tchebycheff inequality as

$$P(|X| \geq \epsilon) \leq \frac{1}{\epsilon^2} E[X^2] \tag{2.82.a}$$

(Note that the foregoing inequality does not require the complete distribution of X, that is, it is *distribution free*.)

Now, if we let $X = (Y - \mu_Y)$, and $\epsilon = k$, Equation 2.82.a takes the form

$$P(|(Y - \mu_Y)| \geq k\sigma_Y) \leq \frac{1}{k^2} \tag{2.82.b}$$

or

$$P(|Y - \mu_Y| \geq k) \leq \frac{\sigma_Y^2}{k^2} \tag{2.82.c}$$

Equation 2.82.b gives an upper bound on the probability that a random variable has a value that deviates from its mean by more than k times its standard deviation. Equation 2.82.b thus justifies the use of the standard deviation as a measure of variability for any random variable.

2.7.2 Chernoff Bound

The Tchebycheff inequality often provides a very "loose" upper bound on probabilities. The Chernoff bound provides a "tighter" bound. To derive the Chernoff bound, define

$$Y_\epsilon = \begin{cases} 1 & X \geq \epsilon \\ 0 & X < \epsilon \end{cases}$$

Then, for all $t \geq 0$, it must be true that

$$e^{tX} \geq e^{t\epsilon} Y_\epsilon$$

and, hence,

$$E\{e^{tX}\} \geq e^{t\epsilon} E\{Y_\epsilon\} = e^{t\epsilon} P(X \geq \epsilon)$$

or

$$P(X \geq \epsilon) \leq e^{-t\epsilon} E\{e^{tX}\}, \qquad t \geq 0$$

Furthermore,

$$\begin{aligned} P(X \geq \epsilon) &\leq \min_{t \geq 0} e^{-t\epsilon} E\{e^{tX}\} \\ &\leq \min_{t \geq 0} \exp[-t\epsilon + \ln E\{e^{tX}\}] \end{aligned} \qquad (2.83)$$

Equation 2.83 is the Chernoff bound. While the advantage of the Chernoff bound is that it is tighter than the Tchebycheff bound, the disadvantage of the Chernoff bound is that it requires the evaluation of $E\{e^{tX}\}$ and thus requires more extensive knowledge of the distribution. The Tchebycheff bound does not require such knowledge of the distribution.

2.7.3 Union Bound

This bound is very useful in approximating the probability of union of events, and it follows directly from

$$P(A \cup B) = P(A) + P(B) - P(AB) \le P(A) + P(B)$$

since $P(AB) \ge 0$. This result can be generalized as

$$P\left(\bigcup_i A_i\right) \le \sum_i P(A_i) \tag{2.84}$$

We now present an example to illustrate the use of these bounds.

EXAMPLE 2.23.

X_1 and X_2 are two independent Gaussian random variables with $\mu_{X_1} = \mu_{X_2} = 0$ and $\sigma^2_{X_1} = 1$ and $\sigma^2_{X_2} = 4$.

(a) Find the Tchebycheff and Chernoff bounds on $P(X_1 \ge 3)$ and compare it with the exact value of $P(X_1 \ge 3)$.
(b) Find the union bound on $P(X_1 \ge 3 \text{ or } X_2 \ge 4)$ and compare it with the actual value.

SOLUTION:

(a) The Tchebycheff bound on $P(X_1 \ge 3)$ is obtained using Equation 2.82.c as

$$P(X_1 \ge 3) \le P(|X_1| \ge 3) \le \frac{1}{9} \approx 0.111$$

To obtain the Chernoff bound we start with

$$\begin{aligned} E\{e^{tX_1}\} &= \int_{-\infty}^{\infty} e^{tx_1} \frac{1}{\sqrt{2\pi}} e^{-x_1^2/2} \, dx_1 \\ &= e^{t^2/2} \int_{-\infty}^{\infty} \frac{1}{\sqrt{2\pi}} \exp[-(x_1 - t)^2/2] \, dx_1 \\ &= e^{t^2/2} \end{aligned}$$

Hence,

$$P(X_1 \geq \epsilon) \leq \min_{t>0} \exp\left(-t\epsilon + \frac{t^2}{2}\right)$$

The minimum value of the right-hand side occurs with $t = \epsilon$ and

$$P(X_1 \geq \epsilon) \leq e^{-\epsilon^2/2}$$

Thus, the Chernoff bound on $P(X_1 \geq 3)$ is given by

$$P(X_1 \geq 3) \leq e^{-9/2} \approx 0.0111$$

From the tabulated values of the $Q(\ )$ function (Appendix D), we obtain the value of $P(X_1 \geq 3)$ as

$$P(X_1 \geq 3) = Q(3) = .0013$$

Comparison of the exact value with the Chernoff and Tchebycheff bounds indicates that the Tchebycheff bound is much looser than the Chernoff bound. This is to be expected since the Tchebycheff bound does not take into account the functional form of the pdf.

(b) $P(X_1 \geq 3 \text{ or } X_2 \geq 4)$
$= P(X_1 \geq 3) + P(X_2 \geq 4) - P(X_1 \geq 3 \text{ and } X_2 \geq 4)$
$= P(X_1 \geq 3) + P(X_2 \geq 4) - P(X_1 \geq 3)P(X_2 \geq 4)$

since X_1 and X_2 are independent. The union bound consists of the sum of the first two terms of the right-hand side of the preceding equation, and the union bound is "off" by the value of the third term. Substituting the value of these probabilities, we have

$$P(X_1 \geq 3 \text{ or } X_2 \geq 4) \approx (.0013) + (.0228) - (.0013)(.0228)$$
$$\approx .02407$$

The union bound is given by

$$P(X_1 \geq 3 \text{ or } X_2 \geq 4) \leq P(X_1 \geq 3) + P(X_2 \geq 4) = .0241$$

The union bound is usually very tight when the probabilities involved are small and the random variables are independent.

2.7.4 Approximating the Distribution of $Y = g(X_1, X_2, \ldots, X_n)$

A practical approximation based on the first-order Taylor series expansion is discussed. Consider

$$Y = g(X_1, X_2, \ldots, X_n)$$

If Y is represented by its first-order Taylor series expansion about the point $\mu_1, \mu_2, \ldots, \mu_n$

$$Y \approx g(\mu_1, \mu_2, \ldots, \mu_n) + \sum_{i=1}^{n} \left[\frac{\partial g}{\partial X_i} (\mu_1, \mu_2, \ldots, \mu_n) \right] [X_i - \mu_i]$$

then

$$E[Y] \approx g(\mu_1, \mu_2, \ldots, \mu_n)$$

$$\sigma_Y^2 = E[(Y - \mu_Y)^2]$$

$$\approx \sum_{i=1}^{n} \left[\frac{\partial Y}{\partial X_i} (\mu_1, \ldots, \mu_n) \right]^2 \sigma_{X_i}^2$$

$$+ \sum_{i=1}^{n} \sum_{\substack{j=1 \\ j \neq i}}^{n} \frac{\partial Y}{\partial X_i} (\mu_1, \ldots, \mu_n) \frac{\partial Y}{\partial X_j} (\mu_1, \ldots, \mu_n) \rho_{X_i X_j} \sigma_{X_i} \sigma_{X_j}$$

where

$$\mu_i = E[X_i]$$
$$\sigma_{X_i}^2 = E[X_i - \mu_i)^2]$$
$$\rho_{X_i X_j} = \frac{E[(X_i - \mu)(X_j - \mu_j)]}{\sigma_{X_i} \sigma_{X_j}}$$

If the random variables, $X_1, \ldots, X_n$, are uncorrelated ($\rho_{X_i X_j} = 0$), then the double sum is zero.

Furthermore, as will be explained in Section 2.8.2, the central limit theorem

82 REVIEW OF PROBABILITY AND RANDOM VARIABLES

suggests that if n is reasonably large, then it may not be too unreasonable to assume that Y is normal if the X_is meet certain conditions.

EXAMPLE 2.24.

$$Y = \frac{X_1}{X_2} + X_3 X_4 - X_5^2$$

The X_is are independent.

$$\mu_{X_1} = 10 \qquad \sigma_{X_1}^2 = 1$$

$$\mu_{X_2} = 2 \qquad \sigma_{X_2}^2 = \frac{1}{2}$$

$$\mu_{X_3} = 3 \qquad \sigma_{X_3}^2 = \frac{1}{4}$$

$$\mu_{X_4} = 4 \qquad \sigma_{X_4}^2 = \frac{1}{3}$$

$$\mu_{X_5} = 1 \qquad \sigma_{X_5}^2 = \frac{1}{5}$$

Find approximately (a) μ_Y, (b) σ_Y^2, and (c) $P(Y \le 20)$.

SOLUTION:

(a) $\mu_Y \approx \dfrac{10}{2} + (3)(4) - 1 = 16$

(b) $\sigma_Y^2 \approx \left(\dfrac{1}{2}\right)^2 (1) + \left(-\dfrac{10}{4}\right)^2 \left(\dfrac{1}{2}\right) + 4^2 \left(\dfrac{1}{4}\right) + 3^2 \left(\dfrac{1}{3}\right) + 2^2 \left(\dfrac{1}{5}\right)$
≈ 11.2

(c) With only five terms in the approximate linear equation, we assume, for an approximation, that Y is normal. Thus

$$P(Y \le 20) \approx \int_{-\infty}^{1.2} \frac{1}{\sqrt{2\pi}} \exp(-z^2/2)\, dz = 1 - Q(1.2) \approx .885$$

2.7.5 Series Approximation of Probability Density Functions

In some applications, such as those that involve nonlinear transformations, it will not be possible to calculate the probability density functions in closed form. However, it might be easy to calculate the expected values. As an example, consider $Y = X^3$. Even if the pdf of Y cannot be specified in analytical form, it might be possible to calculate $E\{Y^k\} = E\{X^{3k}\}$ for $k \leq m$. In the following paragraphs we present a method for approximating the unknown pdf $f_Y(y)$ of a random variable Y whose moments $E\{Y^k\}$ are known. To simplify the algebra, we will assume that $E\{Y\} = 0$ and $\sigma_Y^2 = 1$.

The readers have seen the Fourier series expansion for periodic functions. A similar series approach can be used to expand probability density functions. A commonly used and mathematically tractable series approximation is the Gram–Charlier series, which has the form:

$$f_Y(y) = h(y) \sum_{j=0}^{\infty} C_j H_j(y) \qquad (2.85)$$

where

$$h(y) = \frac{1}{\sqrt{2\pi}} \exp(-y^2/2) \qquad (2.86)$$

and the basis functions of the expansion, $H_j(y)$, are the Tchebycheff–Hermite (T-H) polynomials. The first eight T-H polynomials are

$$\begin{aligned}
H_0(y) &= 1 \\
H_1(y) &= y \\
H_2(y) &= y^2 - 1 \\
H_3(y) &= y^3 - 3y \\
H_4(y) &= y^4 - 6y^2 + 3 \\
H_5(y) &= y^5 - 10y^3 + 15y \\
H_6(y) &= y^6 - 15y^4 + 45y^2 - 15 \\
H_7(y) &= y^7 - 21y^5 + 105y^3 - 105y \\
H_8(y) &= y^8 - 28y^6 + 210y^4 - 420y^2 + 105
\end{aligned} \qquad (2.87)$$

and they have the following properties:

1. $H_k(y)h(y) = -\dfrac{d(H_{k-1}(y)h(y))}{dy}, \qquad k \geq 1$

84 REVIEW OF PROBABILITY AND RANDOM VARIABLES

2. $H_k(y) - yH_{k-1}(y) + (k-1)H_{k-2}(y) = 0, \quad k \geq 2$

3. $\int_{-\infty}^{\infty} H_m(y)H_n(y)h(y)\,dy = 0, \quad m \neq n$

$\qquad\qquad\qquad\qquad\qquad\qquad = n!, \quad m = n$ (2.88)

The coefficients of the series expansion are evaluated by multiplying both sides of Equation 2.85 by $H_k(y)$ and integrating from $-\infty$ to ∞. By virtue of the orthogonality property given in Equation 2.88, we obtain

$$C_k = \frac{1}{k!}\int_{-\infty}^{\infty} H_k(y)f_Y(y)\,dy$$

$$= \frac{1}{k!}\left[\mu_k - \frac{k^{[2]}}{(2)1!}\mu_{k-2} + \frac{k^{[4]}}{2^2 2!}\mu_{k-4} - \cdots\right] \quad (2.89.\text{a})$$

where

$$\mu_m = E\{Y^m\}$$

and

$$k^{[m]} = \frac{k!}{(k-m)!} = k(k-1)\cdots[k-(m-1)], \quad k \geq m$$

The first eight coefficients follow directly from Equations 2.87 and 2.89.a and are given by

$$C_0 = 1$$
$$C_1 = \mu_1$$
$$C_2 = \frac{1}{2}(\mu_2 - 1)$$
$$C_3 = \frac{1}{6}(\mu_3 - 3\mu_1)$$
$$C_4 = \frac{1}{24}(\mu_4 - 6\mu_2 + 3)$$
$$C_5 = \frac{1}{120}(\mu_5 - 10\mu_3 + 15\mu_1)$$
$$C_6 = \frac{1}{720}(\mu_6 - 15\mu_4 + 45\mu_2 - 15)$$
$$C_7 = \frac{1}{5040}(\mu_7 - 21\mu_5 + 105\mu_3 - 105\mu_1)$$
$$C_8 = \frac{1}{40320}(\mu_8 - 28\mu_6 + 210\mu_4 - 420\mu_2 + 105) \quad (2.89.\text{b})$$

BOUNDS AND APPROXIMATIONS

Substituting Equation 2.89 into Equation 2.85 we obtain the series expansion for the pdf of a random variable in terms of the moments of the random variable and the *T-H* polynomials.

The Gram–Charlier series expansion for the pdf of a random variable X with mean μ_X and variance σ_X^2 has the form:

$$f_X(x) = \frac{1}{\sqrt{2\pi}\sigma_X} \exp\left[-\frac{(x-\mu_X)^2}{2\sigma_X^2}\right] \sum_{j=0}^{\infty} C_j H_j\left(\frac{x-\mu_X}{\sigma_X}\right) \quad (2.90)$$

where the coefficients C_j are given by Equation 2.89 with μ_k' used for μ_k where

$$\mu_k' = E\left\{\left[\frac{X-\mu_X}{\sigma_X}\right]^k\right\}$$

EXAMPLE 2.25.

For a random variable X

$$\mu_1 = 3, \quad \mu_2 = 13, \quad \mu_3 = 59, \quad \mu_4 = 309$$

Find $P(X \leq 5)$ using four terms of a Gram–Charlier series.

SOLUTION:

$$\sigma_X^2 = E(X^2) - [E(X)]^2 = \mu_2 - \mu_1^2 = 4$$

Converting to the standard normal form

$$Z = \frac{X-3}{2}$$

Then the moments of Z are

$$\mu_1' = 0 \qquad \mu_2' = 1$$

$$\mu_3' = \frac{\mu_3 - 9\mu_2 + 27\mu_1 - 27}{8} = -.5$$

$$\mu_4' = \frac{\mu_4 - 12\mu_3 + 54\mu_2 - 108\mu_1 + 81}{16} = 3.75$$

86 REVIEW OF PROBABILITY AND RANDOM VARIABLES

Then for the random variable Z, using Equation 2.89,

$$C_0 = 1$$
$$C_1 = 0$$
$$C_2 = 0$$
$$C_3 = \frac{1}{6}(-.5) = -.08333$$
$$C_4 = \frac{1}{24}(3.75 - 6 + 3) = .03125$$

Now $P(X \leq 5) = P(Z \leq 1)$

$$= \int_{-\infty}^{1} \frac{1}{\sqrt{2\pi}} \exp(-z^2/2) \left[\sum_{j=0}^{4} C_j H_j(z)\right] dz$$

$$= \int_{-\infty}^{1} \frac{1}{\sqrt{2\pi}} \exp(-z^2/2) \, dz + \int_{-\infty}^{1} (-.0833) h(z) H_3(z) \, dz$$

$$+ \int_{-\infty}^{1} .03125 h(z) H_4(z) \, dz$$

Using the property (1) of the T-H polynomials yields

$P(Z \leq 1)$
$= .8413 + .0833 h(1) H_2(1) - .03125 h(1) H_3(1)$
$= .8413 + .0833 \dfrac{1}{\sqrt{2\pi}} \exp\left(-\dfrac{1}{2}\right)(0) - .03125 \dfrac{1}{\sqrt{2\pi}} \exp\left(-\dfrac{1}{2}\right)(-2)$
$= .8413 + .0151 = .8564$

Equation 2.90 is a series approximation to the pdf of a random variable X whose moments are known. If we know only the first two moments, then the series approximation reduces to

$$\tilde{f}_X(x) = \frac{1}{\sqrt{2\pi}\sigma_X} \exp[-(x - \mu_X)^2/2\sigma_X^2]$$

which says that (if only the first and second moments of a random variable are known) the Gaussian pdf is used as an approximation to the underlying pdf. As we add more terms, the higher order terms will force the pdf to take a more proper shape.

A series of the form given in Equation 2.90 is useful only if it converges rapidly and the terms can be calculated easily. This is true for the Gram–Charlier series when the underlying pdf is nearly Gaussian or when the random variable X is the sum of many independent components. Unfortunately, the Gram–Charlier series is not uniformly convergent, thus adding more terms does not guarantee increased accuracy. A rule of thumb suggests four to six terms for many practical applications.

2.7.6 Approximations of Gaussian Probabilities

The Gaussian pdf plays an important role in probability theory. Unfortunately, this pdf cannot be integrated in closed form. Several approximations have been developed for evaluating

$$Q(y) = \int_y^\infty \frac{1}{\sqrt{2\pi}} \exp(-x^2/2)\, dx$$

and are given in the *Handbook of Mathematical* functions edited by Abramowitz and Stegun (pages 931–934). For large values of y, ($y > 4$), an approximation for $Q(y)$ is

$$Q(y) \approx \frac{1}{\sqrt{2\pi} y} \exp\left(\frac{-y^2}{2}\right) \qquad (2.91.\text{a})$$

For $0 \leq y$, the following approximation is excellent as measured by $|\epsilon(y)|$, the magnitude of the error.

$$Q(y) = h(y)(b_1 t + b_2 t^2 + b_3 t^3 + b_4 t^4 + b_5 t^5) + \epsilon(y) \qquad (2.91.\text{b})$$

where

$$h(y) = \frac{1}{\sqrt{2\pi}} \exp(-y^2/2)$$

$$t = \frac{1}{1 + py} \qquad\qquad b_2 = -.356563782$$

$$|\epsilon(y)| < 7.5 \times 10^{-8} \qquad\qquad b_3 = 1.781477937$$

$$p = .2316419 \qquad\qquad b_4 = -1.821255978$$

$$b_1 = .319381530 \qquad\qquad b_5 = 1.330274429$$

2.8 SEQUENCES OF RANDOM VARIABLES AND CONVERGENCE

One of the most important concepts in mathematical analysis is the concept of convergence and the existence of a limit. Fundamental operations of calculus such as differentiation, integration, and summation of infinite series are defined by means of a limiting process. The same is true in many engineering applications, for example, the steady state of a dynamic system or the asymptotic trajectory of a moving object. It is similarly useful to study the convergence of random sequences.

With real continuous functions, we use the notation

$$x(t) \to a \quad \text{as} \quad t \to t_0 \quad \text{or} \quad \lim_{t \to t_0} x(t) = a$$

to denote that $x(t)$ converges to a as t approaches t_0 where t is continuous. The corresponding statement for t a discrete variable is

$$x(t_n) \to a \quad \text{as} \quad t_n \to t_0 \quad \text{or} \quad \lim_{n \to 0} x(t_n) = a$$

for any discrete sequence such that

$$t_n \to t_0 \quad \text{as} \quad n \to \infty$$

With this remark in mind, let us proceed to investigate the convergence of sequences of random variables, or random sequences. A random sequence is denoted by $X_1, X_2, \ldots, X_n, \ldots$. For a specific outcome, λ, $X_n(\lambda) = x_n$ is a sequence of numbers that might or might not converge. The concept of convergence of a random sequence may be concerned with the convergence of individual sequences, $X_n(\lambda) = x_n$, or the convergence of the probabilities of some sequence of events determined by the entire ensemble of sequences or both. Several definitions and criteria are used for determining the convergence of random sequences, and we present four of these criteria.

2.8.1 Convergence Everywhere and Almost Everywhere

For every outcome λ, we have a sequence of numbers

$$X_1(\lambda), X_2(\lambda), \ldots, X_n(\lambda), \ldots$$

and hence the random sequence $X_1, X_2, \ldots, X_n$ represents a family of sequences. If each member of the family converges to a limit, that is, $X_1(\lambda), X_2(\lambda),$

..., converges for every $\lambda \in S$, then we say that the random sequence converges *everywhere*. The limit of each sequence can depend upon λ, and if we denote the limit by X, then X is a random variable.

Now, there may be cases where the sequence does not converge for every outcome. In such cases if the set of outcomes for which the limit exists has a probability of 1, that is, if

$$P\{\lambda : \lim_{n \to \infty} X_n(\lambda) = X(\lambda)\} = 1$$

then we say that the sequence converges *almost everywhere* or almost surely. This is written as

$$P\{X_n \to X\} = 1 \quad \text{as} \quad n \to \infty \tag{2.92}$$

2.8.2 Convergence in Distribution and Central Limit Theorem

Let $F_n(x)$ and $F(x)$ denote the distribution functions of X_n and X, respectively. If

$$F_n(x) \to F(x) \quad \text{as} \quad n \to \infty \tag{2.93}$$

for all x at which $F(x)$ is continuous, then we say that the sequence X_n converges in distribution to X.

Central Limit Theorem. Let $X_1, X_2, \ldots, X_n$ be a sequence of independent, identically distributed random variables, each with mean μ and variance σ^2. Let

$$Z_n = \sum_{i=1}^{n} (X_i - \mu)/\sqrt{n\sigma^2}$$

Then Z_n has a limiting (as $n \to \infty$) distribution that is Gaussian with mean 0 and variance 1.

The central limit theorem can be proved as follows. Suppose we assume that the moment-generating function $M(t)$ of X_k exists for $|t| < h$. Then the function $m(t)$

$$m(t) \triangleq E\{\exp[t(X_k - \mu)]\} = \exp(-\mu t) M(t)$$

exists for $-h < t < h$. Furthermore, since X_k has a finite mean and variance, the first two derivatives of $M(t)$ and hence the derivatives of $m(t)$ exist at $t =$

0. We can use Taylor's formula and expand $m(t)$ as

$$m(t) = m(0) + m'(0)t + m''(\zeta)t^2/2, \qquad 0 \leq \zeta < t$$
$$= 1 + \frac{\sigma^2 t^2}{2} + \frac{[m''(\zeta) - \sigma^2]t^2}{2}$$

Next consider

$$M_n(\tau) = E\{\exp(\tau Z_n)\}$$
$$= E\left\{\exp\left(\tau \frac{X_1 - \mu}{\sigma\sqrt{n}}\right) \exp\left(\tau \frac{X_2 - \mu}{\sigma\sqrt{n}}\right) \cdots \exp\left(\tau \frac{X_n - \mu}{\sigma\sqrt{n}}\right)\right\}$$
$$= E\left\{\exp\left(\tau \frac{X_1 - \mu}{\sigma\sqrt{n}}\right)\right\} \cdots E\left\{\exp\left(\tau \frac{X_n - \mu}{\sigma\sqrt{n}}\right)\right\}$$
$$= \left[E\left\{\exp\left(\tau \frac{X - \mu}{\sigma\sqrt{n}}\right)\right\}\right]^n$$
$$= \left[m\left(\frac{\tau}{\sigma\sqrt{n}}\right)\right]^n, \qquad -h < \frac{\tau}{\sigma\sqrt{n}} < h$$

In $m(t)$, replace t by $\tau/(\sigma\sqrt{n})$ to obtain

$$m\left(\frac{\tau}{\sigma\sqrt{n}}\right) = 1 + \frac{\tau^2}{2n} + \frac{[m''(\zeta) - \sigma^2]\tau^2}{2n\sigma^2}$$

where now ζ is between 0 and $\tau/(\sigma\sqrt{n})$. Accordingly,

$$M_n(\tau) = \left\{1 + \frac{\tau^2}{2n} + \frac{[m''(\zeta) - \sigma^2]\tau^2}{2n\sigma^2}\right\}^n, \qquad 0 \leq \zeta < \frac{\tau}{\sigma\sqrt{n}}$$

Since $m''(t)$ is continuous at $t = 0$ and since $\zeta \to 0$ as $n \to \infty$, we have

$$\lim_{n\to\infty}[m''(\zeta) - \sigma^2] = 0$$

and

$$\lim_{n\to\infty} M_n(\tau) = \lim_{n\to\infty}\left\{1 + \frac{\tau^2}{2n}\right\}^n$$
$$= \exp(\tau^2/2) \qquad (2.94)$$

(The last step follows from the familiar formula of calculus $\lim_{n\to\infty}[1 + a/n]^n = e^a$). Since $\exp(\tau^2/2)$ is the moment-generating function of a Gaussian random variable with 0 mean and variance 1, and since the moment-generating function uniquely determines the underlying pdf at all points of continuity, Equation 2.94 shows that Z_n converges to a Gaussian distribution with 0 mean and variance 1.

In many engineering applications, the central limit theorem and hence the Gaussian pdf play an important role. For example, the output of a linear system is a weighted sum of the input values, and if the input is a sequence of random variables, then the output can be approximated by a Gaussian distribution. Another example is the total noise in a radio link that can be modeled as the sum of the contributions from a large number of independent sources. The central limit theorem permits us to model the total noise by a Gaussian distribution.

We had assumed that X_i's are independent and identically distributed and that the moment-generating function exists in order to prove the central limit theorem. The theorem, however, holds under a variety of weaker conditions (Reference [6]):

1. The random variables $X_1, X_2, \ldots$, in the original sequence are independent with the same mean and variance but not identically distributed.
2. $X_1, X_2, \ldots$, are independent with different means, same variance, and not identically distributed.
3. Assume $X_1, X_2, X_3, \ldots$ are independent and have variances $\sigma_1^2, \sigma_2^2, \sigma_3^2, \ldots$. If there exist positive constants ϵ and ζ such that $\epsilon < \sigma_i^2 < \zeta$ for all i, then the distribution of the standardized sum converges to the standard Gaussian; this says in particular that the variances must exist and be neither too large nor too small.

The assumption of finite variances, however, is essential for the central limit theorem to hold.

Finite Sums. The central limit theorem states that an infinite sum, Y, has a normal distribution. For a finite sum of independent random variables, that is,

$$Y = \sum_{i=1}^{n} X_i$$

then

$$f_Y = f_{X_1} * f_{X_2} * \ldots * f_{X_n}$$

$$\psi_Y(\omega) = \prod_{i=1}^{n} \psi_{X_i}(\omega)$$

and

$$C_Y(\omega) = \sum C_{X_i}(\omega)$$

where Ψ is the characteristic function and C is the cumulant-generating function. Also, if K_i is the ith cumulant where K_i is the coefficient of $(j\omega)^i/i!$ in a power series expansion of C, then it follows that

$$K_{i,Y} = \sum_{j=1}^{n} K_{i,X_j}$$

and in particular the first cumulant is the mean, thus

$$\mu_Y = \sum_{i=1}^{n} \mu_{X_i}$$

and the second cumulant is the variance

$$\sigma_Y^2 = \sum_{i=1}^{n} \sigma_{X_i}^2$$

and the third cumulant, $K_{3,X}$ is $E\{(X - \mu_X)^3\}$, thus

$$E\{(Y - \mu_Y)^3\} = \sum_{i=1}^{n} E\{(X_i - \mu_{X_i})^3\}$$

and $K_{4,X}$ is $E\{(X - \mu_X)^4\} - 3K_{2,X}$, thus

$$K_{4,Y} = \sum_{i=1}^{n} K_{4,X_i} = \sum_{i=1}^{n} (E\{(X - \mu_X)^4\} - 3K_{2,X})$$

For finite sums the normal distribution is often rapidly approached; thus a Gaussian approximation or a Gram–Charlier approximation is often appropriate. The following example illustrates the rapid approach to a normal distribution.

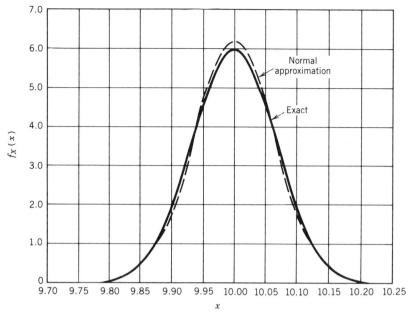

Figure 2.17 Density and approximation for Example 2.26.

EXAMPLE 2.26.

Find the resistance of a circuit consisting of five independent resistances in series. All resistances are assumed to have a uniform density function between 1.95 and 2.05 ohms (2 ohms ± 2.5%). Find the resistance of the series combination and compare it with the normal approximation.

SOLUTION: The exact density is found by four convolutions of uniform density functions. The mean value of each resistance is 2 and the standard deviation is $(20\sqrt{3})^{-1}$. The exact density function of the resistance of the series circuit is plotted in Figure 2.17 along with the normal density function, which has the same mean (10) and the same variance (1/240). Note the close correspondence.

2.8.3 Convergence in Probability (in Measure) and the Law of Large Numbers

The probability $P\{|X - X_n| > \epsilon\}$ of the event $\{|X - X_n| > \epsilon\}$ is a sequence of numbers depending on n and ϵ. If this sequence tends to zero as $n \to \infty$, that is, if

$$P\{|X - X_n| > \epsilon\} \to 0 \quad \text{as} \quad n \to \infty$$

for any $\epsilon > 0$, then we say that X_n converges to the random variable X in probability. This is also called stochastic convergence. An important application of convergence in probability is the law of large numbers.

Law of Large Numbers. Assume that $X_1, X_2, \ldots, X_n$ is a sequence of independent random variables each with mean μ and variance σ^2. Then, if we define

$$\overline{X}_n = \frac{1}{n} \sum_{i=1}^{n} X_i \qquad (2.95.\text{a})$$

$$\lim_{n \to \infty} P\{|\overline{X}_n - \mu| \geq \epsilon\} = 0 \quad \text{for each} \quad \epsilon > 0 \qquad (2.95.\text{b})$$

The law of large numbers can be proved directly by using Tchebycheff's inequality.

2.8.4 Convergence in Mean Square

A sequence X_n is said to converge in mean square if there exists a random variable X (possibly a constant) such that

$$E[(X_n - X)^2] \to 0 \quad \text{as} \quad n \to \infty \qquad (2.96)$$

If Equation 2.96 holds, then the random variable X is called the mean square limit of the sequence X_n and we use the notation

$$\text{l.i.m.} \ X_n = X$$

where l.i.m. is meant to suggest the phrase *limit in mean* (square) to distinguish it from the symbol lim for the ordinary limit of a sequence of numbers.

Although the verification of some modes of convergences is difficult to establish, the Cauchy criterion can be used to establish conditions for mean-square convergence. For deterministic sequences the Cauchy criterion establishes convergence of x_n to x without actually requiring the value of the limit, that is, x. In the deterministic case, $x_n \to x$ if

$$|x_{n+m} - x_n| \to 0 \quad \text{as} \quad n \to \infty \quad \text{for any} \quad m > 0$$

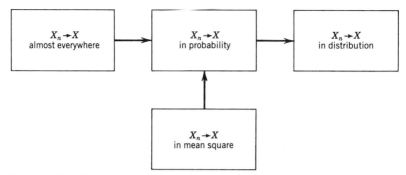

Figure 2.18 Relationship between various modes of convergence.

For random sequences the following version of the Cauchy criterion applies.

$$E\{(X_n - X)^2\} \to 0 \quad \text{as} \quad n \to \infty$$

if and only if

$$E\{|X_{n+m} - X_n|^2\} \to 0 \quad \text{as} \quad n \to \infty \quad \text{for any} \quad m > 0 \qquad (2.97)$$

2.8.5 Relationship between Different Forms of Convergence

The relationship between various modes of convergence is shown in Figure 2.18. If a sequence converges in MS sense, then it follows from the application of Tchebycheff's inequality that the sequence also converges in probability. It can also be shown that almost everywhere convergence implies convergence in probability, which in turn implies convergence in distribution.

2.9 SUMMARY

The reviews of probability, random variables, distribution function, probability mass function (for discrete random variables), and probability density functions (for continuous random variables) were brief, as was the review of expected value. Four particularly useful expected values were briefly discussed: the characteristic function $E\{\exp(j\omega X)\}$; the moment generating function $E\{\exp(tX)\}$; the cumulative generating function $\ln E\{\exp(tX)\}$; and the probability generating function $E\{z^X\}$ (non-negative integer-valued random variables).

The review of random vectors, that is, vector random variables, extended the ideas of marginal, joint, and conditional density function to n dimensions, and vector notation was introduced. Multivariate normal random variables were emphasized.

Transformations of random variables were reviewed. The special cases of a function of one random variable and a sum (or more generally an affine transformation) of random variables were considered. Order statistics were considered as a special transformation. The difficulty of a general nonlinear transformations was illustrated by an example, and the Monte Carlo technique was introduced.

We reviewed the following bounds: the Tchebycheff inequality, the Chernoff bound, and the union bound. We also discussed the Gram–Charlier series approximation to a density function using moments. Approximating the distribution of $Y = g(X_1, \ldots, X_n)$ using a linear approximation with the first two moments was also reviewed. Numerical approximations to the Gaussian distribution function were suggested.

Limit concepts for sequences of random variables were introduced. Convergence almost everywhere, in distribution, in probability and in mean square were defined. The central limit theorem and the law of large numbers were introduced. Finite sum convergence was also discussed.

These concepts will prove to be essential in our study of random signals.

2.10 REFERENCES

The material presented in this chapter was intended as a review of probability and random variables. For additional details, the reader may refer to one of the following books. Reference [2], particularly Vol. 1, has become a classic text for courses in probability theory. References [8] and the first edition of [7] are widely used for courses in applied probability taught by electrical engineering departments. References [1], [3], and [10] also provide an introduction to probability from an electrical engineering perspective. Reference [4] is a widely used text for statistics and the first five chapters are an excellent introduction to probability. Reference [5] contains an excellent treatment of series approximations and cumulants. Reference [6] is written at a slightly higher level and presents the theory of many useful applications. Reference [9] describes a theory of probable reasoning that is based on a set of axioms that differs from those used in probability.

[1] A. M. Breipohl, *Probabilistic Systems Analysis,* John Wiley & Sons, New York, 1970.

[2] W. Feller, *An Introduction to Probability Theory and Applications,* Vols. I, II, John Wiley & Sons, New York, 1957, 1967.

[3] C. H. Helstrom, *Probability and Stochastic Processes for Engineers,* Macmillan, New York, 1977.

[4] R. V. Hogg and A. T. Craig, *Introduction to Mathematical Statistics,* Macmillan, New York, 1978.

[5] M. Kendall and A. Stuart, *The Advanced Theory of Statistics*, Vol. 1, 4th ed., Macmillan, New York, 1977.

[6] H. L. Larson and B. O. Shubert, *Probabilistic Models in Engineering Sciences*, Vol. I, John Wiley & Sons, New York, 1979.

[7] A. Papoulis, *Probability, Random Variables and Stochastic Processes*, McGraw-Hill, New York, 1984.

[8] P. Z. Peebles, Jr., *Probability, Random Variables, and Random Signal Principles*, 2nd ed., McGraw-Hill, New York, 1987.

[9] G. Shafer, *A Mathematical Theory of Evidence*, Princeton University Press, Princeton, N.J., 1976.

[10] J. B. Thomas, *An Introduction to Applied Probability and Random Processes*, John Wiley & Sons, New York, 1971.

2.11 PROBLEMS

2.1 Suppose we draw four cards from an ordinary deck of cards. Let

A_1: an ace on the first draw

A_2: an ace on the second draw

A_3: an ace on the third draw

A_4: an ace on the fourth draw.

    a. Find $P(A_1 \cap A_2 \cap A_3 \cap A_4)$ assuming that the cards are drawn with replacement (i.e., each card is replaced and the deck is reshuffled after a card is drawn and observed).

    b. Find $P(A_1 \cap A_2 \cap A_3 \cap A_4)$ assuming that the cards are drawn without replacement.

2.2 A random experiment consists of tossing a die and observing the number of dots showing up. Let

A_1: number of dots showing up = 3

A_2: even number of dots showing up

A_3: odd number of dots showing up

    a. Find $P(A_1)$ and $P(A_1 \cap A_3)$.

    b. Find $P(A_2 \cup A_3)$, $P(A_2 \cap A_3)$, $P(A_1|A_3)$.

    c. Are A_2 and A_3 disjoint?

    d. Are A_2 and A_3 independent?

2.3 A box contains three 100-ohm resistors labeled R_1, R_2, and R_3 and two 1000-ohm resistors labeled R_4 and R_5. Two resistors are drawn from this box without replacement.

98 REVIEW OF PROBABILITY AND RANDOM VARIABLES

a. List all the outcomes of this random experiment. [A typical outcome may be listed as (R_1, R_5) to represent that R_1 was drawn first followed by R_5.]

b. Find the probability that both resistors are 100-ohm resistors.

c. Find the probability of drawing one 100-ohm resistor and one 1000-ohm resistor.

d. Find the probability of drawing a 100-ohm resistor on the first draw and a 1000-ohm resistor on the second draw.

Work parts (b), (c), and (d) by counting the outcomes that belong to the appropriate events.

2.4 With reference to the random experiment described in Problem 2.3, define the following events.

A_1: 100-ohm resistor on the first draw

A_2: 1000-ohm resistor on the first draw

B_1: 100-ohm resistor on the second draw

B_2: 1000-ohm resistor on the second draw

a. Find $P(A_1 B_1)$, $P(A_2 B_1)$, and $P(A_2 B_2)$.

b. Find $P(A_1)$, $P(A_2)$, $P(B_1|A_1)$, and $P(B_1|A_2)$. Verify that $P(B_1) = P(B_1|A_1)P(A_1) + P(B_1|A_2)P(A_2)$.

2.5 Show that:

a. $P(A \cup B \cup C) = P(A) + P(B) + P(C) - P(AB) - P(BC) - P(CA) + P(ABC)$.

b. $P(A|B) = P(A)$ implies $P(B|A) = P(B)$.

c. $P(ABC) = P(A)P(B|A)P(C|AB)$.

2.6 A_1, A_2, A_3 are three mutually exclusive and exhaustive sets of events associated with a random experiment E_1. Events B_1, B_2, and B_3 are mutually

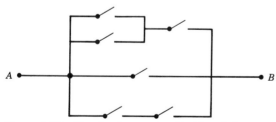

Figure 2.19 Circuit diagram for Problem 2.8.

exclusive and exhaustive sets of events associated with a random experiment E_2. The joint probabilities of occurrence of these events and some marginal probabilities are listed in the table:

A_i \ B_i	B_1	B_2	B_3
A_1	3/36	*4/36	5/36
A_2	5/36	4/36	5/36
A_3	*4/36	6/36	*
$P(B_i)$	12/36	14/36	*

a. Find the missing probabilities (*) in the table.

b. Find $P(B_3|A_1)$ and $P(A_1|B_3)$.

c. Are events A_1 and B_1 statistically independent?

2.7 There are two bags containing mixtures of blue and red marbles. The first bag contains 7 red marbles and 3 blue marbles. The second bag contains 4 red marbles and 5 blue marbles. One marble is drawn from bag one and transferred to bag two. Then a marble is taken out of bag two. Given that the marble drawn from the second bag is red, find the probability that the color of the marble transferred from the first bag to the second bag was blue.

2.8 In the diagram shown in Figure 2.19, each switch is in a closed state with probability p, and in the open state with probability $1 - p$. Assuming that the state of one switch is independent of the state of another switch, find the probability that a closed path can be maintained between A and B (*Note:* There are many closed paths between A and B.)

2.9 The probability that a student passes a certain exam is .9, given that he studied. The probability that he passes the exam without studying is .2. Assume that the probability that the student studies for an exam is .75 (a somewhat lazy student). Given that the student passed the exam, what is the probability that he studied?

2.10 A fair coin is tossed four times and the faces showing up are observed.

a. List all the outcomes of this random experiment.

b. If X is the number of heads in each of the outcomes of this experiment, find the probability mass function of X.

REVIEW OF PROBABILITY AND RANDOM VARIABLES

2.11 Two dice are tossed. Let X be the sum of the numbers showing up. Find the probability mass function of X.

2.12 A random experiment can terminate in one of three events A, B, or C with probabilities 1/2, 1/4, and 1/4, respectively. The experiment is repeated three times. Find the probability that events A, B, and C each occur exactly one time.

2.13 Show that the mean and variance of a binomial random variable X are $\mu_X = np$ and $\sigma_X^2 = npq$, where $q = 1 - p$.

2.14 Show that the mean and variance of a Poisson random variable are $\mu_X = \lambda$ and $\sigma_X^2 = \lambda$.

2.15 The probability mass function of a *geometric* random variable has the form

$$P(X = k) = pq^{k-1}, \quad k = 1, 2, 3, \ldots\,; p, q > 0, p + q = 1.$$

  a. Find the mean and variance of X.
  b. Find the probability-generating function of X.

2.16 Suppose that you are trying to market a digital transmission system (modem) that has a bit error probability of 10^{-4} and the bit errors are independent. The buyer will test your modem by sending a known message of 10^4 digits and checking the received message. If more than two errors occur, your modem will be rejected. Find the probability that the customer will buy your modem.

2.17 The input to a communication channel is a random variable X and the output is another random variable Y. The joint probability mass functions of X and Y are listed:

Y \ X	−1	0	1
−1	$\frac{1}{4}$	$\frac{1}{8}$	0
0	0	$\frac{1}{4}$	0
1	0	$\frac{1}{8}$	$\frac{1}{4}$

  a. Find $P(Y = 1 | X = 1)$.
  b. Find $P(X = 1 | Y = 1)$.
  c. Find ρ_{XY}.

2.18 Show that the expected value operator has the following properties.

a. $E\{a + bX\} = a + bE\{X\}$

b. $E\{aX + bY\} = aE\{X\} + bE\{Y\}$

c. Variance of $aX + bY = a^2 \text{Var}[X] + b^2 \text{Var}[Y] + 2ab \text{Covar}[X, Y]$

2.19 Show that $E_{X,Y}\{g(X, Y)\} = E_X\{E_{Y|X}[g(X, Y)]\}$ where the subscripts denote the distributions with respect to which the expected values are computed.

2.20 A thief has been placed in a prison that has three doors. One of the doors leads him on a one-day trip, after which he is dumped on his head (which destroys his memory as to which door he chose). Another door is similar except he takes a three-day trip before being dumped on his head. The third door leads to freedom. Assume he chooses a door immediately and with probability 1/3 when he has a chance. Find his expected number of days to freedom. (*Hint:* Use conditional expectation.)

2.21 Consider the circuit shown in Figure 2.20. Let the time at which the ith switch closes be denoted by X_i. Suppose X_1, X_2, X_3, X_4 are independent, identically distributed random variables each with distribution function F. As time increases, switches will close until there is an electrical path from A to C. Let

U = time when circuit is first completed from A to B

V = time when circuit is first completed from B to C

W = time when circuit is first completed from A to C

Find the following:

a. The distribution function of U.

b. The distribution function of W.

c. If $F(x) = x$, $0 \leq x \leq 1$ (i.e., uniform), what are the mean and variance of X_i, U, and W?

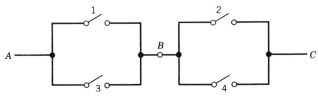

Figure 2.20 Circuit diagram for Problem 2.21.

2.22 Prove the following inequalities

  a. $(E\{XY\})^2 \leq E\{X^2\}E\{Y^2\}$ (Schwartz or cosine inequality)
  b. $\sqrt{E\{(X+Y)^2\}} \leq \sqrt{E\{X^2\}} + \sqrt{E\{Y^2\}}$ (triangle inequality)

2.23 Show that the mean and variance of a random variable X having a uniform distribution in the interval $[a, b]$ are $\mu_X = (a + b)/2$ and $\sigma_X^2 = (b - a)^2/12$.

2.24 X is a Gaussian random variable with $\mu_X = 2$ and $\sigma_X^2 = 9$. Find $P(-4 < X \leq 5)$ using tabulated values of $Q(\ )$.

2.25 X is a zero mean Gaussian random variable with a variance of σ_X^2. Show that

$$E\{X^n\} = \begin{cases} (\sigma_X)^n \, 1 \cdot 3 \cdot 5 \cdots (n-1), & n \text{ even} \\ 0 & n \text{ odd} \end{cases}$$

2.26 Show that the characteristic function of a random variable can be expanded as

$$\Psi_X(\omega) = \sum_{k=0}^{\infty} \frac{(j\omega)^k}{k!} E\{X^k\}$$

(*Note:* The series must be terminated by a remainder term just before the first infinite moment, if any exist).

2.27 a. Show that the characteristic function of the sum of two independent random variables is equal to the product of the characteristic functions of the two variables.

  b. Show that the cumulant generating function of the sum of two independent random variables is equal to the sum of the cumulant generating function of the two variables.

  c. Show that Equations 2.52.c through 2.52.f are correct by equating coefficients of like powers of $j\omega$ in Equation 2.52.b.

2.28 The probability density function of Cauchy random variable is given by

$$f_X(x) = \frac{\alpha}{\pi(x^2 + \alpha^2)}, \qquad \alpha > 0,$$

  a. Find the characteristic function of X.
  b. Comment about the first two moments of X.

2.29 The joint pdf of random variables X and Y is

$$f_{X,Y}(x, y) = \tfrac{1}{2}, \qquad 0 \leq x \leq y, \quad 0 \leq y \leq 2$$

a. Find the marginal pdfs, $f_X(x)$ and $f_Y(y)$.

b. Find the conditional pdfs $f_{X|Y}(x|y)$ and $f_{Y|X}(y|x)$.

c. Find $E\{X|Y = 1\}$ and $E\{X|Y = 0.5\}$.

d. Are X and Y statistically independent?

e. Find ρ_{XY}.

2.30 The joint pdf of two random variables is

$$f_{X_1, X_2}(x_1\ x_2) = 1, \qquad 0 \le x_1 \le 1, \quad 0 \le x_2 \le 1$$

Let $Y_1 = X_1 X_2$ and $Y_2 = X_1$

a. Find the joint pdf of $f_{Y_1, Y_2}(y_1, y_2)$; clearly indicate the domain of y_1, y_2.

b. Find $f_{Y_1}(y_1)$ and $f_{Y_2}(y_2)$.

c. Are Y_1 and Y_2 independent?

2.31 X and Y have a bivariate Gaussian pdf given in Equation 2.57.

a. Show that the marginals are Gaussian pdfs.

b. Find the conditional pdf $f_{X|Y}(x|y)$. Show that this conditional pdf has a mean

$$E\{X|Y = y\} = \mu_X + \rho \frac{\sigma_X}{\sigma_Y}(y - \mu_Y)$$

and a variance

$$\sigma_X^2(1 - \rho^2)$$

2.32 Let $Z = X + Y - c$, where X and Y are independent random variables with variances σ_X^2 and σ_Y^2 and c is constant. Find the variance of Z in terms of σ_X^2, σ_Y^2, and c.

2.33 X and Y are independent zero mean Gaussian random variables with variances σ_X^2, and σ_Y^2. Let

$$Z = \tfrac{1}{2}(X + Y) \quad \text{and} \quad W = \tfrac{1}{2}(X - Y)$$

a. Find the joint pdf $f_{Z,W}(z, w)$.

b. Find the marginal pdf $f_Z(z)$.

c. Are Z and W independent?

2.34 $X_1, X_2, \ldots, X_n$ are n independent zero mean Gaussian random variables with equal variances, $\sigma_{X_i}^2 = \sigma^2$. Show that

$$Z = \frac{1}{n}[X_1 + X_2 + \cdots + X_n]$$

is a Gaussian random variable with $\mu_Z = 0$ and $\sigma_Z^2 = \sigma^2/n$. (Use the result derived in Problem 2.32.)

2.35 X is a Gaussian random variable with mean 0 and variance σ_X^2. Find the pdf of Y if:

    a. $Y = X^2$

    b. $Y = |X|$

    c. $Y = \frac{1}{2}[X + |X|]$

    d. $Y = \begin{cases} 1 & \text{if } X > \sigma_X \\ X & \text{if } |X| \le \sigma_X \\ -1 & \text{if } X < -\sigma_X \end{cases}$

2.36 X is a zero-mean Gaussian random variable with a variance σ_X^2. Let $Y = aX^2$.

    a. Find the characteristic function of Y, that is, find
$$\Psi_Y(\omega) = E\{\exp(j\omega Y)\} = E\{\exp(j\omega aX^2)\}$$

    b. Find $f_Y(y)$ by inverting $\Psi_Y(\omega)$.

2.37 X_1 and X_2 are two identically distributed independent Gaussian random variables with zero mean and variance σ_X^2. Let
$$R = \sqrt{X_1^2 + X_2^2}$$
and
$$\Theta = \tan^{-1}[X_2/X_1]$$

    a. Find $f_{R,\Theta}(r, \theta)$.

    b. Find $f_R(r)$, and $f_\Theta(\theta)$.

    c. Are R and Θ statistically independent?

2.38 X_1 and X_2 are two independent random variables with uniform pdfs in the interval $[0, 1]$. Let
$$Y_1 = X_1 + X_2 \quad \text{and} \quad Y_2 = X_1 - X_2$$

    a. Find the joint pdf $f_{Y_1,Y_2}(y_1, y_2)$ and clearly identify the domain where this joint pdf is nonzero.

    b. Find $\rho_{Y_1 Y_2}$ and $E\{Y_1|Y_2 = 0.5\}$.

2.39 X_1 and X_2 are two independent random variables each with the following density function:
$$f_{X_i}(x) = e^{-x}, \quad x > 0$$
$$= 0 \quad\quad\quad x \le 0$$

Let $Y_1 = X_1 + X_2$ and $Y_2 = X_1/(X_1 + X_2)$

    a. Find $f_{Y_1,Y_2}(y_1, y_2)$.

    b. Find $f_{Y_1}(y_1)$, $f_{Y_2}(y_2)$ and show that Y_1 and Y_2 are independent.

2.40 $X_1, X_2, X_3, \ldots, X_n$ are n independent Gaussian random variables with zero means and unit variances. Let

$$Y = \sum_{i=1}^{n} X_i^2$$

Find the pdf of Y.

2.41 X is uniformly distributed in the interval $[-\pi, \pi]$. Find the pdf of $Y = a \sin(X)$.

2.42 X is multivariate Gaussian with

$$\mu_X = \begin{bmatrix} 6 \\ 0 \\ 8 \end{bmatrix} \quad \Sigma_X = \begin{bmatrix} \frac{1}{2} & \frac{1}{4} & \frac{1}{3} \\ \frac{1}{4} & 2 & \frac{2}{3} \\ \frac{1}{3} & \frac{2}{3} & 1 \end{bmatrix}$$

Find the mean vector and the covariance matrix of $Y = [Y_1, Y_2, Y_3]^T$, where

$Y_1 = X_1 - X_2$
$Y_2 = X_1 + X_2 - 2X_3$
$Y_3 = X_1 + X_3$

2.43 X is a four-variate Gaussian with

$$\mu_X = \begin{bmatrix} 0 \\ 0 \\ 0 \\ 0 \end{bmatrix} \quad \text{and} \quad \Sigma_X = \begin{bmatrix} 4 & 3 & 2 & 1 \\ 3 & 4 & 3 & 2 \\ 2 & 3 & 4 & 3 \\ 1 & 2 & 3 & 4 \end{bmatrix}$$

Find $E\{X_1|X_2 = 0.5, X_3 = 1.0, X_4 = 2.0\}$ and the variance of X_1 given $X_2 = X_3 = X_4 = 0$.

2.44 Show that a necessary condition for Σ_X to be a covariance matrix is that for all

$$V = \begin{bmatrix} v_1 \\ v_2 \\ \vdots \\ v_n \end{bmatrix}$$

$$V^T \Sigma_X V \geq 0$$

(This is the condition for positive semidefiniteness of a matrix.)

2.45 Consider the following 3 × 3 matrices

$$A = \begin{bmatrix} 10 & 3 & 1 \\ 2 & 5 & 0 \\ 1 & 0 & 2 \end{bmatrix}, \quad B = \begin{bmatrix} 10 & 5 & 2 \\ 5 & 3 & 1 \\ 2 & 1 & 2 \end{bmatrix}, \quad C = \begin{bmatrix} 10 & 5 & 2 \\ 5 & 3 & 3 \\ 2 & 3 & 2 \end{bmatrix}$$

Which of the three matrices can be covariance matrices?

2.46 Suppose $\mathbf{X}$ is an n-variate Gaussian with zero means and a covariance matrix $\Sigma_\mathbf{X}$. Let $\lambda_1, \lambda_2, \ldots, \lambda_n$ be n distinct eigenvalues of $\Sigma_\mathbf{X}$ and let $\mathbf{V}_1, \mathbf{V}_2, \ldots, \mathbf{V}_n$ be the corresponding normalized eigenvectors. Show that

$$\mathbf{Y} = \mathbf{A}\mathbf{X}$$

where

$$\mathbf{A} = [\mathbf{V}_1, \mathbf{V}_2, \mathbf{V}_3, \ldots, \mathbf{V}_n]_{n \times n}^T$$

has an n variate Gaussian density with zero means and

$$\Sigma_\mathbf{Y} = \begin{bmatrix} \lambda_1 & & & 0 \\ & \lambda_2 & & \\ & & \ddots & \\ 0 & & & \lambda_n \end{bmatrix}$$

2.47 $\mathbf{X}$ is bivariate Gaussian with

$$\mu_\mathbf{X} = \begin{bmatrix} 0 \\ 0 \end{bmatrix} \quad \text{and} \quad \Sigma_\mathbf{X} = \begin{bmatrix} 3 & 1 \\ 1 & 3 \end{bmatrix}$$

a. Find the eigenvalues and eigenvectors of $\Sigma_\mathbf{X}$.

b. Find the transformation $\mathbf{Y} = [Y_1, Y_2]^T = \mathbf{A}\mathbf{X}$ such that the components of $\mathbf{Y}$ are uncorrelated.

2.48 If $U(x) \geq 0$ for all x and $U(x) > a > 0$ for all $x \in \zeta$ where ζ is some interval, show that

$$P[U(X) \geq a] \leq \frac{1}{a} E\{U(X)\}$$

2.49 Plot the Tchebycheff and Chernoff bounds as well as the exact values for $P(X \geq a)$, $a > 0$, if X is

a. Uniform in the interval $[0, 1]$.

b. Exponential, $f_X(x) = \exp(-x)$, $x > 0$.

c. Gaussian with zero mean and unit variance.

2.50 Compare the Tchebycheff and Chernoff bounds on $P(Y \geq a)$ with exact values for the Laplacian pdf

$$f_Y(y) = \frac{1}{2} \exp(-|y|)$$

2.51 In a communication system, the received signal **Y** has the form

$$\mathbf{Y} = \mathbf{X} + \mathbf{N}$$

where **X** is the "signal" component and **N** is the noise. **X** can have one of eight values shown in Figure 2.21, and **N** has an uncorrelated bivariate Gaussian distribution with zero means and variances of $\frac{1}{9}$. The signal **X** and noise **N** can be assumed to be independent.

The receiver observes **Y** and determines an estimated value $\hat{\mathbf{X}}$ of **X** according to the algorithm

$$\text{if } \mathbf{y} \in A_i \text{ then } \hat{\mathbf{X}} = \mathbf{x}_i$$

The decision regions A_i for $i = 1, 2, 3, \ldots, 8$ are illustrated by A_1 in Figure 2.21. Obtain an upper bound on $P(\hat{\mathbf{X}} \neq \mathbf{X})$ assuming that $P(\mathbf{X} = \mathbf{x}_i) = \frac{1}{8}$ for $i = 1, 2, \ldots, 8$.

Hint:

1. $P(\hat{\mathbf{X}} \neq \mathbf{X}) = \sum_{i=1}^{8} P(\hat{\mathbf{X}} \neq \mathbf{X} | \mathbf{X} = \mathbf{x}_i) P(\mathbf{X} = \mathbf{x}_i)$

2. Use the union bound.

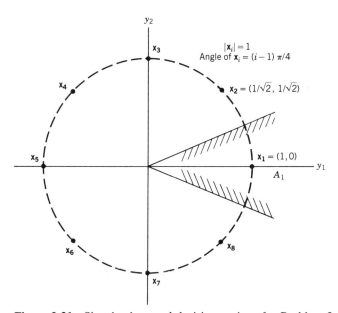

Figure 2.21 Signal values and decision regions for Problem 2.51.

2.52 Show that the Tchebycheff–Hermite polynomials satisfy

$$(-1)^k \frac{d^k h(y)}{dy^k} = H_k(y)h(y), \quad k = 1, 2, \ldots$$

2.53 X has a triangular pdf centered in the interval $[-1, 1]$. Obtain a Gram–Charlier approximation to the pdf of X that includes the first six moments of X and sketch the approximation for values of X ranging from -2 to 2.

2.54 Let p be the probability of obtaining heads when a coin is tossed. Suppose we toss the coin N times and form an estimate of p as

$$\hat{p} = \frac{N_H}{N}$$

where $N_H =$ number of heads showing up in N tosses. Find the smallest value of N such that

$$P[|\hat{p} - p| \geq 0.01 p) \leq 0.1$$

(Assume that the unknown value of p is in the range 0.4 to 0.6.)

2.55 $X_1, X_2, \ldots, X_n$ are n independent samples of a continuous random variable X, that is

$$f_{X_1, X_2, \ldots, X_n}(x_1, x_2, \ldots, x_n) = \prod_{i=1}^{n} f_X(x_i)$$

Assume that $\mu_X = 0$ and σ_X^2 is finite.

a. Find the mean and variance of

$$\overline{X} = \frac{1}{n} \sum_{i=1}^{n} X_i$$

b. Show that $\overline{X}$ converges to 0 in MS, that is, $\text{l.i.m. } \overline{X} = 0$.

2.56 Show that if X_is are of continuous type and independent, then for sufficiently large n the density of $\sin(X_1 + X_2 + \cdots + X_n)$ is nearly equal to the density of $\sin(X)$ where X is a random variable with uniform distribution in the interval $(-\pi, \pi)$.

2.57 Using the Cauchy criterion, show that a sequence X_n tends to a limit in the MS sense if and only if $E\{X_m X_n\}$ exists as $m, n \to \infty$.

2.58 A box has a large number of 1000-ohm resistors with a tolerance of ± 100 ohms (assume a uniform distribution in the interval 900 to 1100 ohms). Suppose we draw 10 resistors from this box and connect them in series and let R be the resistive value of the series combination. Using the Gaussian approximation for R find

$$P[9000 \leq R \leq 11000]$$

2.59 Let

$$Y_n = \frac{1}{n} \sum_{i=1}^{n} X_i$$

where X_i, $i = 1, 2, \ldots, n$ are statistically independent and identically distributed random variables each with a Cauchy pdf

$$f_X(x) = \frac{a/\pi}{x^2 + a^2}$$

a. Determine the characteristic function Y_n.

b. Determine the pdf of Y_n.

c. Consider the pdf of Y_n in the limit as $n \to \infty$. Does the central limit theorem hold? Explain.

2.60 Y is a Guassian random variable with zero mean and unit variance and

$$X_n = \begin{cases} \sin(Y/n) & \text{if } y > 0 \\ \cos(Y/n) & \text{if } y \leq 0 \end{cases}$$

Discuss the convergence of the sequence X_n. (Does the series converge, if so, in what sense?)

2.61 Let Y be the number of dots that show up when a die is tossed, and let

$$X_n = \exp[-n(Y - 3)]$$

Discuss the convergence of the sequence X_n.

2.62 Y is a Gaussian random variable with zero mean and unit variance and

$$X_n = \exp(-Y/n)$$

Discuss the convergence of the sequence X_n.

CHAPTER THREE

Random Processes and Sequences

In electrical systems we use voltage or current waveforms as signals for collecting, transmitting and processing information, as well as for controlling and providing power to a variety of devices. Signals, whether they are voltage or current waveforms, are functions of time and belong to one of two important classes: deterministic and random. Deterministic signals can be described by functions in the usual mathematical sense with time t as the independent variable. In contrast with a deterministic signal, a random signal always has some element of uncertainty associated with it and hence it is not possible to determine exactly its value at any given point in time.

Examples of random signals include the audio waveform that is transmitted over a telephone channel, the data waveform transmitted from a space probe, the navigational information received from a submarine, and the instantaneous load in a power system. In all of these cases, we cannot precisely specify the value of the random signal in advance. However, we may be able to describe the random signal in terms of its average properties such as the average power in the random signal, its spectral distribution on the average, and the probability that the signal amplitude exceeds a given value. The probabilistic model used for characterizing a random signal is called a *random process* (also referred to as a *stochastic process* or *time series*). In this and the following four chapters, we will study random process models and their applications.

Basic properties of random processes and analysis of linear systems driven by random signals are dealt with in this chapter and in Chapter 4. Several classes of random process models that are commonly used in various applications are presented in Chapter 5. The use of random process models in the design of communication and control systems is introduced in Chapters 6 and 7. Finally,

INTRODUCTION   111

techniques for deriving or building random process models by collecting and analyzing data are discussed in Chapters 8 and 9. We assume that the reader has a background in deterministic systems and signal analysis, including analysis in the frequency domain.

3.1 INTRODUCTION

In many engineering problems, we deal with time-varying waveforms that have some element of chance or randomness associated with them. As an example, consider the waveforms that occur in a typical data communication system such as the one shown in Figure 3.1 in which a number of terminals are sending information in binary format over noisy transmission links to a central computer. A transmitter in each link converts the binary data to an electrical waveform in which binary digits are converted to pulses of duration T and amplitudes ± 1. The received waveform in each link is a distorted and noisy version of the transmitted waveform where noise represents interfering electrical disturbances. From the received waveform, the receiver attempts to extract the transmitted binary digits. As shown in Figure 3.1, distortion and noise cause the receiver to make occasional errors in recovering the transmitted binary digit sequence.

As we examine the collection or "ensemble" of waveforms shown in Figure 3.1, randomness is evident in all of these waveforms. By observing one waveform, say $x_i(t)$, over the time interval $[t_1, t_2]$ we cannot, with certainty, predict the value of $x_i(t)$ for any other value of $t \notin [t_1, t_2]$. Furthermore, knowledge of one member function $x_i(t)$ will not enable us to know the value of another member function $x_j(t)$. We will use a probabilistic model to describe or characterize the ensemble of waveforms so that we can answer questions such as:

1. What are the spectral properties of the ensemble of waveforms shown in Figure 3.1?
2. How does the noise affect system performance as measured by the receiver's ability to recover the transmitted data correctly?
3. What is the optimum processing algorithm that the receiver should use?

By extending the concept of a *random variable* to include time, we can build a *random process* model for characterizing an ensemble of time functions. For the waveforms shown in Figure 3.1, consider a random experiment that consists of tossing N coins simultaneously and repeating the N tossings once every T seconds. If we label the outcomes of the experiment by "1" when a coin flip results in a head and "0" when the toss results in a tail, then we have a probabilistic model for the bit sequences transmitted by the terminals. Now, by representing 1s and 0s by pulses of amplitude ± 1 and duration T, we can model the transmitted waveform $x_i(t)$. If the channel is linear, its impulse response $h(t)$ is known, and the noise is additive, then we can express $y_i(t)$ as $x_i(t) * h_i(t) + n_i(t)$, where $n_i(t)$ is the additive channel "noise," and * indicates convolution.

112 RANDOM PROCESSES AND SEQUENCES

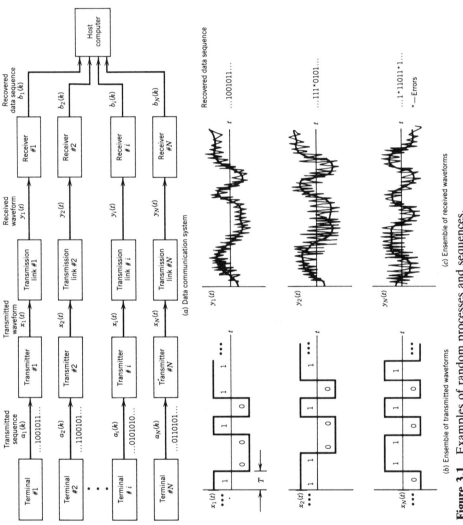

Figure 3.1 Examples of random processes and sequences.

By processing $y_i(t)$ the receiver can generate the output sequence $b_i(k)$. Thus, by extending the concept of random variables to include time and using the results from deterministic systems analysis, we can model random signals and analyze the response of systems to random inputs.

The validity of the random-process model suggested in the previous paragraph for the signals shown in Figure 3.1 can be decided only by collecting and analyzing sample waveforms. Model building and validation fall into the realm of statistics and will be the subject of coverage in Chapters 8 and 9. For the time being, we will assume that appropriate probabilistic models are given and proceed with the analysis.

We start our study of random process models with an introduction to the notation, terminology, and definitions. Then, we present a number of examples and develop the idea of using certain averages to characterize random processes. Basic signal-processing operations such as differentiation, integration, and limiting will be discussed next. Both time-domain and frequency-domain techniques will be used in the analysis, and the concepts of power spectral distribution and bandwidth will be discussed in detail. Finally, we develop series approximations to random processes that are analogous to Fourier and other series representations for deterministic signals.

3.2 DEFINITION OF RANDOM PROCESSES

3.2.1 Concept of Random Processes

A random variable maps the outcomes of a random experiment to a set of real numbers. In a similar vein, a random process can be viewed as a mapping of the outcomes of a random experiment to a set of waveforms or functions of time. While in some applications it may not be possible to explicitly define the underlying random experiment and the associated mapping to waveforms, we can still use the random process as a model for characterizing a collection of waveforms. For example, the waveforms in the data communication system shown in Figure 3.1 were the result of programmers pounding away on terminals. Although the underlying random experiment (what goes through the minds of programmers) that generates the waveforms is not defined, we can use a hypothetical experiment such as tossing N coins and define the waveforms based on the outcomes of the experiment.

By way of another example of a random process, consider a random experiment that consists of tossing a die at $t = 0$ and observing the number of dots showing on the top face. The sample space of the experiment consists of the outcomes 1, 2, 3, 4, 5, and 6. For each outcome of the experiment, let us arbitrarily assign the following functions of time, t, $0 \le t < \infty$.

Outcome	Waveform
1	$x_1(t) = -4$
2	$x_2(t) = -2$

114 RANDOM PROCESSES AND SEQUENCES

Outcome	Waveform
3	$x_3(t) = +2$
4	$x_4(t) = +4$
5	$x_5(t) = -t/2$
6	$x_6(t) = t/2$

The set of waveforms $\{x_1(t), x_2(t), \ldots, x_6(t)\}$, which are shown in Figure 3.2, represents this random process and are called the *ensemble*.

3.2.2 Notation

A random process, which is a collection or ensemble of waveforms, can be denoted by $X(t, \Lambda)$, where t represents time and Λ is a variable that represents an outcome in the sample space S of some underlying random experiment E. Associated with each specific outcome*, say λ_i, we have a specific *member function* $x_i(t)$ of the ensemble. Each member function, also referred to as a *sample function* or a *realization* of the process, is a deterministic function of time even though we may not always be able to express it in closed form.

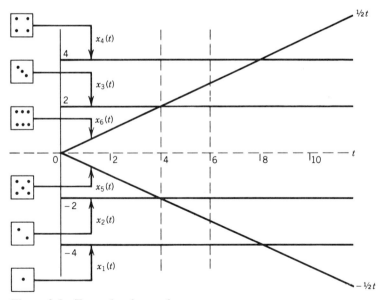

Figure 3.2 Example of a random process.

---

*If the number of outcomes is countable, then we will use the subscripted notation λ_i and $x_i(t)$ to denote a particular outcome and the corresponding member function. Otherwise, we will use λ and $x(t)$ to denote a specific outcome and the corresponding member function.

DEFINITION OF RANDOM PROCESSES

For a specific value of time $t = t_0$, $X(t_0, \Lambda)$ represents a collection of numerical values of the various member functions at $t = t_0$. The actual value depends on the outcome of the random experiment and the member function associated with that outcome. Hence, $X(t_0, \Lambda)$ is a random variable and the probability distribution of the random variable, $X(t_0, \Lambda)$, is derived from the probabilities of the various outcomes of the random experiment E.

When t and Λ are fixed at say $t = t_0$, and $\Lambda = \lambda_i$, then $X(t_0, \lambda_i)$ represents a single numerical value of the ith member function of the process at $t = t_0$. That is $X(t_0, \lambda_i) = x_i(t_0)$. Thus, $X(t, \Lambda)$ can denote the following quantities:

1. $X(t, \Lambda) = \{X(t, \lambda_i) | \lambda_i \in S\} = \{x_1(t), x_2(t), \cdots\}$, a collection of functions of time.
2. $X(t, \lambda_i) = x_i(t)$, a specific member function or deterministic function of time.
3. $X(t_0, \Lambda) = \{X(t_0, \lambda_i) | \lambda_i \in S\} = \{x_1(t_0), x_2(t_0), \ldots\}$, a collection of the numerical values of the member functions at $t = t_0$, that is, a random variable.
4. $X(t_0, \lambda_i) = x_i(t_0)$, numerical value of the ith member function at $t = t_0$.

While the notation given in the preceding paragraphs is well defined, convention adds an element of confusion for the sake of conformity with the notation for deterministic signals by using $X(t)$ rather than $X(t, \Lambda)$ to denote a random process. $X(t)$ may represent a family of time functions, a single time function, a random variable, or a single number. Fortunately, the specific interpretation of $X(t)$ usually can be understood from the context.

EXAMPLE 3.1 (NOTATION)

For the random process shown in Figure 3.2, the random experiment E consists of tossing a die and observing the number of dots on the up face.

$$\Lambda = \{1, 2, 3, 4, 5, 6\} = \{\lambda_1, \lambda_2, \lambda_3, \lambda_4, \lambda_5, \lambda_6\}$$

$$X(t, \lambda_1) = X(t, \Lambda = 1) = x_1(t) = -4, \quad 0 \leq t$$

$$X(t, \lambda_5) = X(t, \Lambda = 5) = x_5(t) = -\frac{1}{2}t, \quad 0 \leq t$$

$X(6, \Lambda) = X(6)$ is a random variable that has values from the set $\{-4, -3, -2, 2, 3, 4\}$

$X(t = 6, \Lambda = 5) = -3$, a constant

3.2.3 Probabilistic Structure

The probabilistic structure of a random process comes from the underlying random experiment E. Knowing the probability of each outcome of E and the time function it maps to, we can derive probability distribution functions for $P[X(t_1) \leq a_1]$, $P[X(t_1) \leq a_1$ and $X(t_2) \leq a_2]$, and so on. If A_1 is a subset of the sample space S of E and it contains all the outcomes λ for which $X(t_1, \lambda) \leq a_1$, then

$$P[X(t_1) \leq a_1] = P(A_1)$$

Note that A_1 is an event associated with E and its probability is derived from the probability structure of the random experiment E. In a similar fashion, we can define joint and conditional probabilities also by first identifying the event that is the inverse image of a given set of values of $X(t)$ and then calculating the probability of this event.

EXAMPLE 3.2 (PROBABILISTIC STRUCTURE)

For the random process shown in Figure 3.2, find (a) $P[X(4) = -2]$; (b) $P[X(4) \leq 0]$; (c) $P[X(0) = 0, X(4) = -2]$; and (d) $P[X(4) = -2 | X(0) = 0]$.

SOLUTION:

(a) Let A be the set of outcomes such that for every $\lambda_i \in A$, $X(4, \lambda_i) = -2$. It is clear from Figure 3.2 that $A = \{2, 5\}$. Hence, $P[X(4) = -2] = P(A) = \frac{2}{6} = \frac{1}{3}$.

(b) $P[X(4) \leq 0] = P[\text{set of outcomes such that } X(4) \leq 0] = \frac{3}{6} = \frac{1}{2}$.

(c) Let B be the set of outcomes that maps to $X(0) = 0$ and $X(4) = -2$. Then $B = \{5\}$, and hence $P[X(0) = 0, X(4) = -2] = P(B) = \frac{1}{6}$.

(d) $P[X(4) = -2 | X(0) = 0] = \dfrac{P[X(4) = -2, X(0) = 0]}{P[X(0) = 0]}$

$= \dfrac{(1/6)}{(2/6)} = \dfrac{1}{2}$

We can attach a relative frequency interpretation to the probabilities as follows. In the case of the previous example, we toss the die n times and observe a time function at each trial. We note the values of these functions at, say, time $t = 4$. Let k be the total number of trials such that at time $t = 4$ the values of the functions are equal to -2. Then,

$$P[X(4) = -2] = \lim_{n\to\infty} \frac{k}{n}$$

We can use a similar interpretation for joint and conditional probabilities.

3.2.4 Classification of Random Processes

Random processes are classified according to the characteristics of t and the random variable $X(t)$ at time t. If t has a continuum of values in one or more intervals on the real line R_1, then $X(t)$ is called a continuous-time random process, examples of which are shown in Figures 3.1 and 3.2. If t can take on a finite, or countably infinite, number of values, say $\{\cdots, t_{-2}, t_{-1}, t_0, t_1, t_2, \cdots\}$ then $X(t)$ is called a discrete-time random process or a *random sequence*, an example of which is the ensemble of random binary digits shown in Figure 3.1. We often denote a random sequence by $X(n)$ where n represents t_n.

$X(t)$ [or $X(n)$] is a discrete-state or discrete-valued process (or sequence) if its values are countable. Otherwise, it is a continuous-state or continuous-valued random process (or sequence). The ensemble of binary waveforms $X(t)$ shown in Figure 3.1 is a discrete-state, continuous-time, random process. From here on, we will use a somewhat abbreviated terminology shown in Table 3.1 to refer to these four classes of random processes. Note that "continuous" or "discrete" will be used to refer to the nature of the amplitude distribution of $X(t)$, and "process" or "sequence" is used to distinguish between continuous time or discrete time, respectively. Additional classification of random processes given in the following sections apply to both random processes and random sequences.

Another attribute that is used to classify random processes is the dependence of the probabilistic structure of $X(t)$ on t. If certain probability distributions or averages do not depend on t, then the process is called *stationary*. Otherwise it is called *nonstationary*. The random process shown in Figure 3.1 is stationary if

TABLE 3.1 CLASSIFICATION OF RANDOM PROCESSES

$X(t)$ \ t	Continuous	Discrete
Continuous	Continuous random process	Continuous random sequence
Discrete	Discrete random process	Discrete random sequence

the noise is stationary, whereas the process shown in Figure 3.2 is nonstationary, that is, $X(0)$ has a different distribution than $X(4)$. More concrete definitions of stationarity and several examples will be presented in Section 3.5 of this chapter.

A random process may be either real-valued or complex-valued. In many applications in communication systems, we deal with real-valued bandpass random processes of the form

$$Z(t) = A(t)\cos[2\pi f_c t + \Theta(t)]$$

where f_c is the carrier or center frequency, and $A(t)$ and $\Theta(t)$ are real-valued random processes. $Z(t)$ can also be written as

$$\begin{aligned} Z(t) &= \text{Real part of } \{A(t)\exp[j\Theta(t)] \exp(j2\pi f_c t)\} \\ &= \text{Real part of } \{W(t)\exp(j2\pi f_c t)\} \end{aligned}$$

where the *complex envelope* $W(t)$ is given by

$$\begin{aligned} W(t) &= A(t)\cos \Theta(t) + jA(t)\sin \Theta(t) \\ &= X(t) + jY(t) \end{aligned}$$

$W(t)$ is a complex-valued random process whereas $X(t)$, $Y(t)$, and $Z(t)$ are real-valued random processes.

Finally, a random process can be either predictable or unpredictable based on observations of its past values. In the case of the ensemble of binary waveforms $X(t)$ shown in Figure 3.1, randomness is evident in each member function, and future values of a member function cannot be determined in terms of past values taken during the preceding T seconds, or earlier. Hence, the process is unpredictable. On the other hand, all member functions of the random process $X(t)$ shown in Figure 3.2 are completely predictable if past values are known. For example, future values of a member function can be determined completely for $t > t_0 > 0$ if past values are known for $0 \le t \le t_0$. We know the six member functions, and the uncertainty results from not knowing which outcome (and hence the corresponding member function) is being observed. The member function as well as the outcome can be determined from two past values. Note that we cannot uniquely determine the member function from one observed value, say at $t = 4$, since $X(4) = 2$ could result from either $x_3(t)$ or $x_6(t)$. If we observe $X(t)$ at two values of t, then we can determine the member function uniquely.

3.2.5 Formal Definition of Random Processes

Let S be the sample space of a random experiment and let t be a variable that can have values in the set $\Gamma \subset R_1$, the real line. A real-valued random process $X(t)$, $t \in \Gamma$, is then a measurable function* on $\Gamma \times S$ that maps $\Gamma \times S$ onto R_1. If the set Γ is a union of one or more intervals on the real line, then $X(t)$ is a random process, and if Γ is a subset of integers, then $X(t)$ is a random sequence.

A real-valued random process $X(t)$ is described by its nth order distribution functions.

$$F_{X(t_1), X(t_2), \ldots, X(t_n)} (x_1, x_2, \ldots, x_n)$$
$$= P[X(t_1) \leq x_1, \ldots, X(t_n) \leq x_n]$$
$$\text{for all } n \text{ and } t_1, \ldots, t_n \in \Gamma \qquad (3.1)$$

These functions satisfy all the requirements of joint probability distribution functions.

Note that if Γ consists of a finite number of points, say $t_1, t_2, \ldots, t_n$, then the random sequence is completely described by the joint distribution function of the n-dimensional random vector, $[X(t_1), X(t_2), \ldots, X(t_n)]^T$, where T denotes the transpose of a vector.

3.3 METHODS OF DESCRIPTION

A random process can be described in terms of a random experiment and the associated mapping. While such a description is a natural extension of the concept of random variables, there are alternate methods of characterizing random processes that will be of use in analyzing random signals and in the design of systems that process random signals for various applications.

3.3.1 Joint Distribution

Since we defined a random process as an indexed set of random variables, we can obviously use joint probability distribution functions to describe a random process. For a random process $X(t)$, we have many joint distribution functions

---

*It is necessary only to assume that $X(t)$ is measurable on S for every $t \in \Gamma$. A random process is sometimes also defined as a family of indexed random variables, denoted by $[X(t, \cdot); t \in \Gamma]$, where the index set Γ represents the set of observation times.

of the form given in Equation 3.1. This leads to a formidable description of the process because at least one n-variate distribution function is required for each value of n. However, the first-order distribution function(s) $P[X(t_1) \leq a_1]$ and the second-order distribution function(s) $P[X(t_1) \leq a_1, X(t_2) \leq a_2]$ are primarily used. The first-order distribution function describes the instantaneous amplitude distribution of the process and the second-order distribution function tells us something about the structure of the signal in the time-domain and thus the spectral content of the signal. The higher-order distribution functions describe the process in much finer detail.

While the joint distribution functions of a process can be derived from a description of the random experiment and the mapping, there is no technique for constructing member functions from joint distribution functions. Two different processes may have the same nth order distribution but the member functions need not have a one-to-one correspondence.

EXAMPLE 3.3.

For the random process shown in Figure 3.2, obtain the joint probabilities $P[X(0) \text{ and } X(6)]$ and the marginal probabilities $P[X(0)]$ and $P[X(6)]$.

SOLUTION: We know that $X(0)$ and $X(6)$ are discrete random variables and hence we can obtain the distribution functions from probability mass functions, which can be obtained by inspection from Table 3.2.

TABLE 3.2 JOINT AND MARGINAL PROBABILITIES OF $X(t)$ AT $t = 0$ AND $t = 6$

Values of $X(0)$	Values of $X(6)$							
	−4	−3	−2	2	3	4		
−4	1/6	0	0	0	0	0	1/6	
−2	0	0	1/6	0	0	0	1/6	Marginal probabilities of $X(0)$
0	0	1/6	0	0	1/6	0	2/6	
2	0	0	0	1/6	0	0	1/6	
4	0	0	0	0	0	1/6	1/6	
	1/6	1/6	1/6	1/6	1/6	1/6		

Marginal probabilities of $X(6)$

Joint probabilities of $X(0)$ and $X(6)$

METHODS OF DESCRIPTION 121

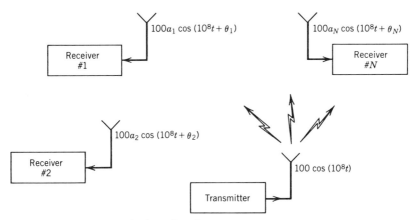

Figure 3.3 Example of a broadcasting system.

3.3.2 Analytical Description Using Random Variables

We are used to expressing deterministic signals in simple analytical forms such as $x(t) = 20 \sin(10t)$ or $y(t) = \exp(-t^2)$. It is sometimes possible to express a random process in an analytical form using one or more random variables. Consider for example an FM station that is broadcasting a "tone," $x(t) = 100 \cos(10^8 t)$, to a large number of receivers distributed randomly in a metropolitan area (see Figure 3.3). The amplitude and phase of the waveform received by the ith receiver will depend on the distance between the transmitter and the receiver. Since we have a large number of receivers distributed randomly over an area, we can model the distance as a continuous random variable. Since the attenuation and the phase are functions of distance, they are also random variables, and we can represent the ensemble of received waveforms by a random process $Y(t)$ of the form

$$Y(t) = A \cos(10^8 t + \Theta)$$

where A and Θ are random variables representing the amplitude and phase of the received waveforms. It might be reasonable to assume uniform distributions for A and Θ.

Representation of a random process in terms of one or more random variables whose probability law is known is used in a variety of applications in communication systems.

3.3.3 Average Values

As in the case of random variables, random processes can be described in terms of averages or expected values. In many applications, only certain averages derived from the first- and second-order distributions of $X(t)$ are of interest. For real- or complex-valued random processes, these averages are defined as follows:

Mean. The mean of $X(t)$ is the expected value of the random variable $X(t)$

$$\mu_X(t) \triangleq E\{X(t)\} \quad (3.2)$$

Autocorrelation. The autocorrelation of $X(t)$, denoted by $R_{XX}(t_1, t_2)$, is the expected value of the product $X^*(t_1) X(t_2)$

$$R_{XX}(t_1, t_2) \triangleq E\{X^*(t_1) X(t_2)\} \quad (3.3)$$

where * denotes conjugate.

Autocovariance. The autocovariance of $X(t)$ is defined as

$$C_{XX}(t_1, t_2) \triangleq R_{XX}(t_1, t_2) - \mu_X^*(t_1)\mu_X(t_2) \quad (3.4)$$

Correlation Coefficient. The autocorrelation coefficient of $X(t)$ is defined as

$$r_{XX}(t_1, t_2) \triangleq \frac{C_{XX}(t_1, t_2)}{\sqrt{C_{XX}(t_1, t_1) C_{XX}(t_2, t_2)}} \quad (3.5)$$

The mean of the random process is the "ensemble" average of the values of all the member functions at time t, and the autocovariance function $C_{XX}(t_1, t_1)$ is the variance of the random variable $X(t_1)$. For $t_1 \neq t_2$, the second moments $R_{XX}(t_1, t_2)$, $C_{XX}(t_1, t_2)$, and $r_{XX}(t_1, t_2)$ partially describe the time domain structure of the random process. We will see later that we can use these functions to derive the spectral properties of $X(t)$.

For random sequences the argument n is substituted for t, and n_1 and n_2 are substituted for t_1 and t_2, respectively. In this case the four functions defined above are also discrete time functions.

EXAMPLE 3.4.

Find $\mu_X(t)$, $R_{XX}(t_1, t_2)$, $C_{XX}(t_1, t_2)$, and $r_{XX}(t_1, t_2)$ for the random process shown in Figure 3.2.

SOLUTION: We compute these expected values by averaging the appropriate ensemble values.

$$\mu_X(t) = E\{X(t)\} = \frac{1}{6} \sum_{i=1}^{6} x_i(t) = 0$$

$$R_{XX}(t_1, t_2) = E\{X(t_1) X(t_2)\} = \frac{1}{6} \sum_{i=1}^{6} x_i(t_1) x_i(t_2)$$

$$= \frac{1}{6}\left\{16 + 4 + 4 + 16 + \frac{1}{4}t_1 t_2 + \frac{1}{4}t_1 t_2\right\}$$

$$= \frac{1}{6}\left\{40 + \frac{1}{2}t_1 t_2\right\}$$

METHODS OF DESCRIPTION   123

Note that because X is real, complex conjugates are omitted.

$$C_{XX}(t_1, t_2) = R_{XX}(t_1, t_2)$$

and

$$r_{XX}(t_1, t_2) = \frac{40 + \frac{1}{2}t_1 t_2}{\sqrt{\left(40 + \frac{1}{2}t_1^2\right)\left(40 + \frac{1}{2}t_2^2\right)}}$$

EXAMPLE 3.5.

A random process $X(t)$ has the functional form

$$X(t) = A \cos(100t + \Theta)$$

where A is a normal random variable with a mean of 0 and variance of 1, and Θ is uniformly distributed in the interval $[-\pi, \pi]$. Assuming A and Θ are independent random variables, find $\mu_X(t)$ and $R_{XX}(t, t + \tau)$.

SOLUTION:

$$\mu_X(t) = E\{A\} E\{\cos(100t + \Theta)\} = 0$$

$$\begin{aligned}
R_{XX}(t, t + \tau) &= E\{X(t_1) X(t_2)\} \quad \text{with} \quad t_1 = t \quad \text{and} \quad t_2 = t + \tau \\
&= E\{A \cos(100t + \Theta) A \cos(100t + 100\tau + \Theta)\} \\
&= E\left\{\frac{A^2}{2}[\cos(100\tau) + \cos(200t + 100\tau + 2\Theta)]\right\} \\
&= \frac{A^2}{2}\cos(100\tau) + \frac{A^2}{2}E\{\cos(200t + 100\tau + 2\Theta)\} \\
&= \frac{A^2}{2}\cos(100\tau),
\end{aligned}$$

since $E\{\cos(200t + 100\tau + 2\Theta)\} = 0$

Note that $R_{XX}(t, t + \tau)$ is a function only of τ and is periodic in τ. In general, if a process has a periodic component, its autocorrelation function will also have a periodic component with the same period.

3.3.4 Two or More Random Processes

When we deal with two or more random processes, we can use joint distribution functions, analytical descriptions, or averages to describe the relationship between the random processes. Consider two random processes $X(t)$ and $Y(t)$ whose joint distribution function is denoted by

$$P[X(t_1) \le x_1, \ldots, X(t_n) \le x_n, Y(t_1') \le y_1, \ldots, Y(t_m') \le y_m]$$

Three averages or expected values that are used to describe the relationship between $X(t)$ and $Y(t)$ are

Cross-correlation Function.

$$R_{XY}(t_1, t_2) \triangleq E\{X^*(t_1)Y(t_2)\} \tag{3.6}$$

Cross-covariance Function.

$$C_{XY}(t_1, t_2) \triangleq R_{XY}(t_1, t_2) - \mu_X^*(t_1)\mu_Y(t_2) \tag{3.7}$$

Correlation Coefficient (Also called cross-correlation coefficient).

$$r_{XY}(t_1, t_2) \triangleq \frac{C_{XY}(t_1, t_2)}{\sqrt{C_{XX}(t_1, t_1) C_{YY}(t_2, t_2)}} \tag{3.8}$$

Using the joint and marginal distribution functions as well as the expected values, we can determine the degree of dependence between two random processes. As above, the same definitions are used for random sequences with n_1 and n_2 replacing the arguments t_1 and t_2.

Equality. Equality of two random processes will mean that their respective member functions are identical for each outcome $\lambda \in S$. Note that equality also implies that the processes are defined on the same random experiment.

Uncorrelated. Two processes $X(t)$ and $Y(t)$ are uncorrelated when

$$C_{XY}(t_1, t_2) = 0, \quad t_1, t_2 \in \Gamma \tag{3.9}$$

Orthogonal. $X(t)$ and $Y(t)$ are said to be orthogonal if

$$R_{XY}(t_1, t_2) = 0, \quad t_1, t_2 \in \Gamma \tag{3.10}$$

Independent. Random processes $X(t)$ and $Y(t)$ are independent if

$$P[X(t_1) \le x_1, \ldots, X(t_n) \le x_n, Y(t_1') \le y_1, \ldots, Y(t_m') \le y_m]$$
$$= P[X(t_1) \le x_1, \ldots, X(t_n) \le x_n] P[Y(t_1') \le y_1, \ldots, Y(t_m') \le y_m]$$
$$\text{for all } n, m \text{ and } t_1, t_2, \ldots, t_n, t_1', t_2', \ldots, t_m' \in \Gamma. \tag{3.11}$$

As in the case of random variables, "independent" implies uncorrelated but not conversely.

EXAMPLE 3.6.

Let E_1 be a random experiment that consists of tossing a die at $t = 0$ and observing the number of dots on the up face, and let E_2 be a random experiment that consists of tossing a coin and observing the up face. Define random processes $X(t)$, $Y(t)$, and $Z(t)$ as follows:

Outcome of Experiment	$X(t)$	$Y(t)$	Outcome of Experiment	$Z(t)$
E_1	$0 < t < \infty$	$0 < t < \infty$	E_2	$0 < t < \infty$
λ_i	$x_i(t)$	$y_i(t)$	q_j	$z_j(t)$
1	-4	2	1 (head)	$\cos t$
2	-2	-4	2 (tail)	$\sin t$
3	2	4		
4	4	-2		
5	$-t/2$	0		
6	$t/2$	0		

Random processes $X(t)$ and $Y(t)$ are defined on the same random experiment E_1. However, $X(t) \neq Y(t)$ since $x_i(t) \neq y_i(t)$ for every outcome, λ_i. These two processes are orthogonal to each other since

$$E\{X(t_1)Y(t_2)\} = \sum_{i=1}^{6} x_i(t_1) \, y_i(t_2) \, P[\lambda_i]$$
$$= 0$$

They are also uncorrelated because $C_{XY}(t_1, t_2) = 0$. However, $X(t)$ and $Y(t)$ are clearly not independent. On the other hand, $X(t)$ and $Z(t)$ are independent processes since these processes are defined on two unrelated random experiments E_1 and E_2, and hence for any pair of outcomes $\lambda_i \in S_1$ and $q_j \in S_2$,

$$P(\lambda_i \text{ and } q_j) = P(\lambda_i) \, P(q_j)$$

3.4 SPECIAL CLASSES OF RANDOM PROCESSES

In deterministic signal analysis, we use elementary signals such as sinusoidal, exponential, and step signals as building blocks from which other more complicated signals can be constructed. A number of random processes with special properties are also used in a similar fashion in random signal analysis. In this section, we introduce examples of a few specific processes. These processes and their applications will be studied in detail in Chapter 5, and they are presented here only as examples to illustrate some of the important and general properties of random processes.

3.4.1 More Definitions

Markov. A random process $X(t)$, $t \in \Gamma$, is called a first-order Markov (or Markoff) process if for all sequences of times $t_1 < t_2 < \cdots < t_k \in \Gamma$ and $k = 1, 2, \ldots$ we have

$$P[X(t_k) \leq x_k | X(t_{k-1}), \ldots, X(t_1)]$$
$$= P[X(t_k) \leq x_k | X(t_{k-1})] \qquad (3.12)$$

Equation 3.12 says that the conditional probability distribution of $X(t_k)$ given all past values $X(t_1) = x_1, \ldots, X(t_{k-1}) = x_{k-1}$ depends only upon the most recent value $X(t_{k-1}) = x_{k-1}$.

Independent Increments. A random process $X(t)$, $t \in \Gamma$ is said to have independent increments if for all times $t_1 < t_2 \cdots < t_k \in \Gamma$, and $k = 3, 4, \ldots$, the random variables $X(t_2) - X(t_1)$, $X(t_3) - X(t_2), \ldots$, and $X(t_k) - X(t_{k-1})$ are mutually independent.

The probability distribution of a process with independent increments is completely specified by the distribution of an increment, $X(t) - X(t')$, for all $t' < t$ and by the first-order distribution $P[X(t_0) \leq x_0]$ at some single time instant, $t_0 \in \Gamma$, since there is a simple linear relationship between $X(t_1), \ldots, X(t_k)$ and the increments $X(t_2) - X(t_1), \ldots, X(t_k) - X(t_{k-1})$, and since the joint distribution of the increments is equal to the product of the marginal distributions.

Two processes with independent increments play a central role in the theory of random processes. One is the Poisson process that has a Poisson distribution for the increments, and the second one is the Wiener process with a Gaussian distribution for the increments. We will study these two processes in detail later.

Martingale. A random process $X(t)$, $t \in \Gamma$, is called a Martingale if $E\{|X(t)|\} < \infty$ for all $t \in \Gamma$, and

$$E\{X(t_2) | X(t_1), t_1 \leq t_2\} = X(t_1) \qquad \text{for all} \quad t_1 \leq t_2 \qquad (3.13)$$

Martingales have several interesting properties such as having a constant mean, and they play an important role in the theory of prediction of future values of random processes based on past observations.

Gaussian. A random process $X(t)$, $t \in \Gamma$ is called a Gaussian process if all its nth order distributions $F_{X_1, X_2, \ldots, X_n}(x_1, x_2, \ldots, x_n)$ are n-variate Gaussian distributions $[t_1, t_2, \ldots, t_n \in \Gamma,$ and $X_i = X(t_i)]$.

Gaussian random processes are widely used to model signals that result from the sum of a large number of independent sources, for example, the noise in a low-frequency communication channel caused by a large number of independent sources such as automobiles, power lines, lightning, and other atmospheric phenomena. Since a k-variate Gaussian density is specified by a set of means and a covariance matrix, knowledge of the mean $\mu_X(t)$, $t \in \Gamma$, and the correlation function $R_{XX}(t_1, t_2)$, $t_1, t_2 \in \Gamma$, are sufficient to completely specify the probability distribution of a Gaussian process.

If a Gaussian process is also a Markov process, then it is called a *Gauss–Markov process.*

3.4.2 Random Walk and Wiener Process

In the theory and applications of random processes, the Wiener process, which provides a model for Brownian motion and thermal noise in electrical circuits, plays a fundamental role. In 1905, Einstein showed that a small particle (of say diameter 10^{-4} cm) immersed in a medium moves randomly due to the continual bombardment of the molecules of the medium, and in 1923, Wiener derived a random process model for this random Brownian motion. The Wiener process can be derived easily as a limiting operation on a related random process called a random walk.

Random Walk. A discrete version of the Wiener process used to model the random motion of a particle can be constructed as follows: Assume that a particle is moving along a horizontal line until it collides with another molecule, and that each collision causes the particle to move "up" or "down" from its previous path by a distance "d." Furthermore, assume that the collision takes place once every T seconds and that the movement after a collision is independent of all previous jumps and hence independent of its position. This model, which is analogous to tossing a coin once every T seconds and taking a step "up" if heads show and "down" if tails show, is called a random walk. The position of the particle at $t = nT$ is a random sequence $X(n)$ where in this notation for a sequence, $X(n)$ corresponds with the process $X(nT)$, and one member function of the sequence is shown in Figure 3.4. We will assume that we start observing the particle at $t = 0$, its initial location $X(0) = 0$ and that the jump of $\pm d$ appears instantly after each toss.

If k heads show up in the first n tosses, then the position of the particle at $t = nT$ is given by

$$X(n) = kd + (n - k)(-d)$$
$$= (2k - n)d \qquad (3.14)$$

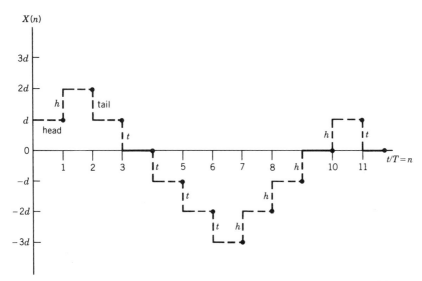

Figure 3.4 Sample function of the random walk process. Values of $X(n)$ are shown as "●".

and $X(n)$ is a discrete random variable having values md, where m equals $-n$, $-n+2, \ldots, n-2, n$. If we denote the sequence of jumps by a sequence of random variables $\{J_i\}$, then we can express $X(n)$ as

$$X(n) = J_1 + J_2 + \cdots + J_n$$

The random variables J_i, $i = 1, 2, \ldots, n$, are independent and have identical distributions with

$$P(J_i = d) = \frac{1}{2}, \quad P(J_i = -d) = \frac{1}{2}$$

$$E\{J_i\} = 0 \quad\quad E\{J_i^2\} = d^2$$

From Equation 3.14 it follows that

$$P[X(n) = md] = P[k \text{ heads in } n \text{ tosses}], \quad k = \frac{m+n}{2}$$

SPECIAL CLASSES OF RANDOM PROCESSES   129

Since the number of heads in n tosses has a binomial distribution, we have

$$P[X(n) = md] = \binom{n}{k}\left(\frac{1}{2}\right)^n, \quad k = 0, 1, 2, \ldots, n; \quad m = 2k - n$$

and

$$E\{X(n)\} = 0$$
$$E\{X(n)^2\} = E\{[J_1 + J_2 + \cdots + J_n]^2\}$$
$$= nd^2$$

We can obtain the autocorrelation function of the random walk sequence as

$$R_{XX}(n_1, n_2) = E\{X(n_1) X(n_2)\}$$
$$= E\{X(n_1) [X(n_1) + X(n_2) - X(n_1)]\}$$
$$= E\{X(n_1)^2\} + E\{X(n_1) [X(n_2) - X(n_1)]\}$$

Now, if we assume $n_2 > n_1$, then $X(n_1)$ and $[X(n_2) - X(n_1)]$ are independent random variables since the number of heads from the first to n_1th tossing is independent of the number of heads from $(n_1 + 1)$th tossing to the n_2th tossing. Hence,

$$R_{XX}(n_1, n_2) = E\{X(n_1)^2\} + E\{X(n_1)\} E\{[X(n_2) - X(n_1)]\}$$
$$= E\{X(n_1)^2\} = n_1 d^2$$

If $n_1 > n_2$, then $R_{XX}(n_1, n_2) = n_2 d^2$ and in general we can express $R_{XX}(n_1, n_2)$ as

$$R_{XX}(n_1, n_2) = \min(n_1, n_2) d^2 \qquad (3.15)$$

It is left as an exercise for the reader to show that $X(n)$ is a Markov sequence and a Martingale.

Wiener Process. Suppose we define a continuous-time random process $Y(t)$, $t \in \Gamma = [0, \infty)$ from the random sequence $X(n)$ as

$$Y(t) = \begin{cases} 0, & t = 0 \\ X(n), & (n-1)T < t \leq nT, \quad n = 1, 2, \ldots \end{cases}$$

130 RANDOM PROCESSES AND SEQUENCES

A sample function of $Y(t)$ is shown as a broken line in Figure 3.4. The mean and variance of $Y(t)$ at $t = nT$ are given by

$$E\{Y(t)\} = 0 \quad \text{and} \quad E\{Y^2(t)\} = \frac{td^2}{T} = nd^2 \tag{3.16}$$

The Wiener process is obtained from $Y(t)$ by letting both the time (T) between jumps and the step size (d) approach zero with the constraint $d^2 = \alpha T$ to assure that the variance will remain finite and nonzero for finite values of t. As a result of the limiting, we have the Wiener process $W(t)$ with the following properties:

1. $W(t)$ is a continuous-amplitude, continuous-time, independent-increment process.
2. $E\{W(t)\} = 0$, and $E\{W^2(t)\} = \alpha t$.
3. $W(t)$ will have a Gaussian distribution since the total displacement or position can be regarded as the sum of a large number of small independent displacements and hence the central limit theorem applies. The probability density function of W is given by

$$f_W(w) = \frac{1}{\sqrt{2\pi\alpha t}} \exp\left(\frac{-w^2}{2\alpha t}\right)$$

4. For any value of t', $0 \le t' < t$, the increment $W(t) - W(t')$ has a Gaussian pdf with zero mean and a variance of $\alpha(t - t')$.
5. The autocorrelation of $W(t)$ is

$$R_{WW}(t_1, t_2) = \alpha \min(t_1, t_2) \tag{3.17}$$

A sample function of the Wiener process, which is also referred to as the Wiener–Levy process, is shown in Figure 3.5. The reader can verify that the Wiener process is a (nonstationary) Markov process and a Martingale.

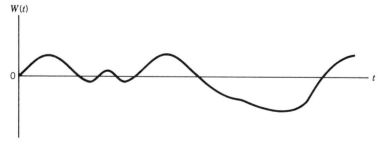

Figure 3.5 Sample function of the Wiener–Levy process.

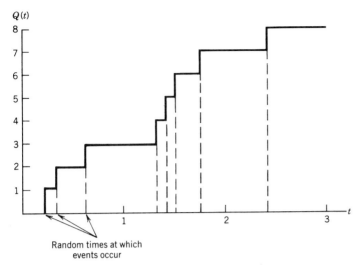

Figure 3.6 Sample function of the Poisson random process.

3.4.3 Poisson Process

The Poisson process is a continuous time, discrete-amplitude random process that is used to model phenomena such as the emission of photons from a light-emitting diode, the arrival of telephone calls at a central exchange, the occurrence of component failures, and other events. We can describe these events by a counting function $Q(t)$, defined for $t \in \Gamma = [0, \infty)$, which represents the number of "events" that have occurred during the time period 0 to t. A typical realization $Q(t)$ is shown in Figure 3.6. The initial value $Q(0)$ of the process is assumed to be equal to zero.

$Q(t)$ is an integer-valued random process and is said to be a Poisson process if the following assumptions hold:

1. For any times $t_1, t_2 \in \Gamma$ and $t_2 > t_1$, the number of events $Q(t_2) - Q(t_1)$ that occur in the interval t_1 to t_2 is Poisson distributed according to the probability law

$$P[Q(t_2) - Q(t_1) = k] = \frac{[\lambda(t_2 - t_1)]^k}{k!} \exp[-\lambda(t_2 - t_1)]$$

$$k = 0, 1, 2, \ldots \quad (3.18)$$

2. The number of events that occur in any interval of time is independent of the number of events that occur in other nonoverlapping time intervals.

From Equation 3.18 we obtain

$$P[Q(t) = k] = \frac{(\lambda t)^k}{k!} \exp(-\lambda t), \quad k = 0, 1, 2, \ldots$$

and hence the mean and variance of $Q(t)$ are

$$E\{Q(t)\} = \lambda t; \quad \text{var}\{Q(t)\} = \lambda t \qquad (3.19)$$

Using the property of independent increments, we find the autocorrelation of $Q(t)$ as

$$\begin{aligned}
R_{QQ}(t_1, t_2) &= E\{Q(t_1) Q(t_2)\} \\
&= E\{Q(t_1)[Q(t_1) + Q(t_2) - Q(t_1)]\} \quad \text{for } t_2 \geq t_1 \\
&= E\{Q^2(t_1)\} + E\{Q(t_1)\} E\{Q(t_2) - Q(t_1)\} \\
&= [\lambda t_1 + \lambda^2 t_1^2] + \lambda t_1[\lambda(t_2 - t_1)] \\
&= \lambda t_1[1 + \lambda t_2] \quad \text{for } t_2 \geq t_1 \\
&= \lambda^2 t_1 t_2 + \lambda \cdot \min(t_1, t_2) \quad \text{for all } t_1, t_2 \in \Gamma \qquad (3.20)
\end{aligned}$$

The reader can verify that the Poisson process is a Markov process and is nonstationary. Unlike the Wiener–Levy process, the Poisson process is not a Martingale since its mean is time varying. Additional properties of the Poisson process and its applications are discussed in Chapter 5.

3.4.4 Random Binary Waveform

Waveforms used in data communication systems are modeled by a random sequence of pulses with the following properties:

1. Each pulse has a rectangular shape with a fixed duration of T and a random amplitude of ± 1.
2. Pulse amplitudes are equally likely to be ± 1.
3. All pulse amplitudes are statistically independent.
4. The start times of the pulse sequences are arbitrary; that is, the starting time of the first pulse following $t = 0$ is equally likely to be any value between 0 and T.

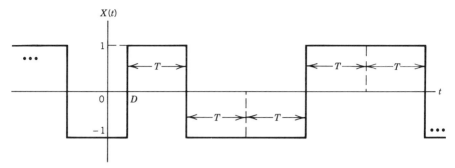

Figure 3.7 Random binary waveform.

The random sequence of pulses shown in Figure 3.7 is called a random binary waveform, and it can be expressed as

$$X(t) = \sum_{k=-\infty}^{\infty} A_k p(t - kT - D)$$

where $p(t)$ is a unit amplitude pulse of duration T, A_k is a binary random variable that represents the amplitude of the kth pulse, and D is the random start time with a uniform distribution in the interval $[0, T]$. The sample function of $X(t)$ shown in Figure 3.7 is defined by a specific amplitude sequence $\{\cdots 1, -1, 1, -1, -1, 1, 1, -1, \cdots\}$ and a specific value of delay $D = T/4$.

For any value of t, $X(t)$ has one of two values, ± 1, with equal probability, and hence the mean and variance of $X(t)$ are

$$E\{X(t)\} = 0 \quad \text{and} \quad E\{X^2(t)\} = 1 \tag{3.21}$$

To calculate the autocorrelation function of $X(t)$, let us choose two values of time t_1 and t_2 such that $0 < t_1 < t_2 < T$. After finding $R_{XX}(t_1, t_2)$ with $0 < t_1 < t_2 < T$, we will generalize the result for arbitrary values of t_1 and t_2. From Figure 3.8 we see that when $0 < D < t_1$ or $t_2 < D < T$, t_1 and t_2 lie in the same pulse interval and hence $X(t_1) = X(t_2)$ and the product $X(t_1) X(t_2) = 1$. On the other hand, when $t_1 < D < t_2$, t_1 and t_2 lie in different pulse intervals and the product of pulse amplitudes $X(t_1) X(t_2)$ has the value $+1$ or -1 with equal probability. Hence we have

$$X(t_1) X(t_2) = \begin{cases} 1 & \text{if } 0 < D < t_1 \quad \text{or} \quad t_2 < D < T \\ \pm 1 & \text{if } t_1 < D < t_2 \end{cases}$$

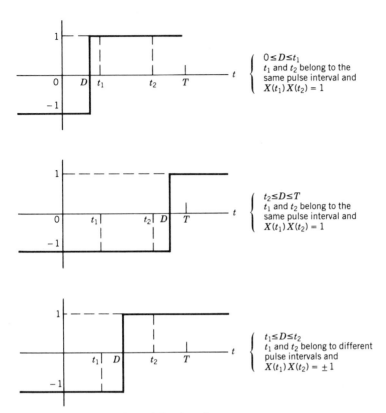

Figure 3.8 Calculation of $R_{XX}(t_1, t_2)$.

The random variable D has a uniform distribution in the interval $[0, T]$ and hence $P[0 < D < t_1 \text{ or } t_2 < D < T] = 1 - (t_2 - t_1)/T$, and $P(t_1 < D < t_2) = (t_2 - t_1)/T$. Using these probabilities and conditional expectations, we obtain

$$R_{XX}(t_1, t_2) = E\{X(t_1) X(t_2)\}$$
$$= E\{X(t_1) X(t_2) | 0 < D < t_1 \text{ or } t_2 < D < T\}$$
$$\cdot P[0 < D < t_1 \text{ or } t_2 < D < T]$$
$$+ E\{X(t_1) X(t_2) | t_1 \leq D \leq t_2\} \cdot P(t_1 \leq D \leq t_2)$$
$$= 1 \cdot \left[1 - \frac{(t_2 - t_1)}{T}\right] + 1 \cdot \frac{1}{2} \cdot \frac{(t_2 - t_1)}{T} - 1 \cdot \frac{1}{2} \cdot \frac{(t_2 - t_1)}{T}$$
$$= 1 - \frac{(t_2 - t_1)}{T}$$

To generalize this result to arbitrary values of t_1 and t_2, we note that $R_{XX}(t_1, t_2) = R_{XX}(t_2, t_1)$, and that $R_{XX}(t_1, t_2) = 0$ when $|t_2 - t_1| > T$. Furthermore, $R_{XX}(t_1 + kT, t_2 + kT) = R_{XX}(t_1, t_2)$, and hence

$$R_{XX}(t_1, t_2) = \begin{cases} 1 - \dfrac{|t_2 - t_1|}{T}, & |t_2 - t_1| < T \\ 0 & \text{elsewhere} \end{cases} \quad (3.22)$$

The reader can verify that the random binary waveform is not an independent increment process and is not a Martingale.

A general version of the random binary waveform with multiple and correlated amplitude levels is widely used as a model for digitized speech and other signals. We will discuss this generalized model and its application in Chapters 5 and 6.

3.5 STATIONARITY

Time-invariant systems and steady-state analysis are familiar terms to electrical engineers. These terms portray certain time-invariant properties of systems and signals. Stationarity plays a similar role in the description of random processes, and it describes the time invariance of certain properties of a random process. Whereas individual member functions of a random function may fluctuate rapidly as a function of time, the ensemble averaged values such as the mean of the process might remain constant with respect to time. Loosely speaking, a process is called stationary if its distribution functions or certain expected values are invariant with respect to a translation of the time axis.

There are several degrees of stationarity ranging from stationarity in a strict sense to a less restrictive form of stationarity called wide-sense stationarity. We define different forms of stationarity and present a number of examples in this section.

3.5.1 Strict-sense Stationarity

A random process $X(t)$ is called time stationary or *stationary in the strict sense* (abbreviated as SSS) if all of the distribution functions describing the process are invariant under a translation of time. That is, for all $t_1, t_2, \ldots, t_k, t_1 + \tau, t_2 + \tau, \ldots, t_k + \tau \in \Gamma$ and all $k = 1, 2, \ldots,$

$$P[X(t_1) \leq x_1, X(t_2) \leq x_2, \ldots, X(t_k) \leq x_k] \\ = P[X(t_1 + \tau) \leq x_1, X(t_2 + \tau) \leq x_2, \ldots, X(t_k + \tau) \leq x_k] \quad (3.23)$$

If the foregoing definition holds for all kth order distribution functions $k =$

1, ..., N but not necessarily for $k > N$, then the process is said to be *Nth order stationary*.

From Equation 3.23 it follows that for a SSS process

$$P[X(t) \le x] = P[X(t + \tau) \le x] \quad (3.24)$$

for any τ. Hence, the first-order distribution is independent of t. Similarly

$$P[X(t_1) \le x_1, X(t_2) \le x_2] = P[X(t_1 + \tau) \le x_1, X(t_2 + \tau) \le x_2] \quad (3.25)$$

for any τ implies that the second-order distribution is strictly a function of the time difference $t_2 - t_1$. As a consequence of Equations 3.24 and 3.25, we conclude that for a SSS process

$$E\{X(t)\} = \mu_X = \text{constant} \quad (3.26)$$

and the autocorrelation function will be a function of the time difference $t_2 - t_1$. We denote the autocorrelation of a SSS process by $R_{XX}(t_2 - t_1)$, defined as

$$E\{X^*(t_1)X(t_2)\} = R_{XX}(t_2 - t_1) \quad (3.27)$$

It should be noted here that a random process with a constant mean and an autocorrelation function that depends only on the time difference $t_2 - t_1$ need not even be first-order stationary.

Two real-valued processes $X(t)$ and $Y(t)$ are *jointly stationary in the strict sense* if the joint distributions of $X(t)$ and $Y(t)$ are invariant under a translation of time, and a complex process $Z(t) = X(t) + jY(t)$ is SSS if the processes $X(t)$ and $Y(t)$ are jointly stationary in the strict sense.

3.5.2 Wide-sense Stationarity

A less restrictive form of stationarity is based on the mean and the autocorrelation function. A process $X(t)$ is said to be *stationary in the wide sense* (WSS or weakly stationary) if its mean is a constant and the autocorrelation function depends only on the time difference:

$$E\{X(t)\} = \mu_X \quad (3.28.\text{a})$$

$$E\{X^*(t)X(t + \tau)\} = R_{XX}(\tau) \quad (3.28.\text{b})$$

STATIONARITY   137

Two processes $X(t)$ and $Y(t)$ are jointly WSS if each process satisfies Equation 3.28 and for all $t \in \Gamma$.

$$E[X^*(t)Y(t + \tau)] = R_{XY}(\tau) \qquad (3.29)$$

For random sequences, the conditions for WSS are

$$E\{X(k)\} = \mu_X \qquad (3.30.\text{a})$$

and

$$E\{X^*(n)X(n + k)\} = R_{XX}(k) \qquad (3.30.\text{b})$$

It is easy to show that SSS implies WSS; however, the converse is not true in general.

3.5.3 Examples

EXAMPLE 3.7.

Two random processes $X(t)$ and $Y(t)$ are shown in Figures 3.9 and 3.10. Find the mean and autocorrelation functions of $X(t)$ and $Y(t)$ and discuss their stationarity properties.

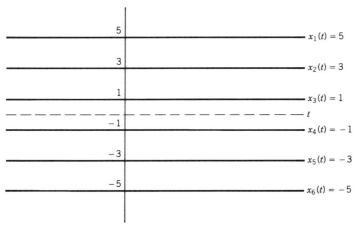

Figure 3.9 Example of a stationary random process. (Assume equal probabilities of occurrence for the six outcomes in sample space.)

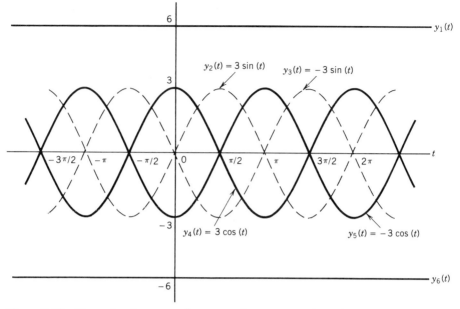

Figure 3.10 Example of a nonstationary random process. (Member function are assumed to have equal probabilities.)

SOLUTION:

$$E\{X(t)\} = 0 \quad \text{for all} \quad t \in \Gamma = (-\infty, \infty)$$

$$R_{XX}(t_1, t_2) = \frac{1}{6}(25 + 9 + 1 + 1 + 9 + 25) = \frac{70}{6}$$

Furthermore, a translation of the time axis does not result in any change in any member function, and hence, Equation 3.23 is satisfied and $X(t)$ is stationary in the strict sense.

For the random process $Y(t)$, $E\{Y(t)\} = 0$, and

$$\begin{aligned} R_{YY}(t_1, t_2) &= \frac{1}{6}\{36 + 9 \sin t_1 \sin t_2 + 9 \sin t_1 \sin t_2 \\ &\quad + 9 \cos t_1 \cos t_2 + 9 \cos t_1 \cos t_2 + 36\} \\ &= \frac{1}{6}\{72 + 18 \cos(t_2 - t_1)\} \\ &= R_{YY}(t_2 - t_1) \end{aligned}$$

Since the mean of the random process $Y(t)$ is constant and the autocorrelation function depends only on the time difference $t_2 - t_1$, $Y(t)$ is stationary in the wide sense. However, $Y(t)$ is not strict-sense stationary since the values that $Y(t)$ can have at $t = 0$ and $t = \pi/4$ are different and hence even the first-order distribution is not time invariant.

EXAMPLE 3.8.

A binary-valued Markov sequence $X(n)$, $n \in I = \{\ldots, -2, -1, 0, 1, 2, \ldots\}$ has the following joint probabilities:

$$P[X(n) = 0, X(n + 1) = 0] = 0.2,$$
$$P[X(n) = 0, X(n + 1) = 1] = 0.2$$
$$P[X(n) = 1, X(n + 1) = 0] = 0.2,$$
$$P[X(n) = 1, X(n + 1) = 1] = 0.4$$

Find $\mu_x(n)$, $R_{XX}(n, n+1)$, $R_{XX}(n, n+2)$, $\ldots$, and show that the sequence is wide-sense stationary.

SOLUTION:

$$\begin{aligned} P[X(n) = 0] &= P[X(n) = 0, X(n+1) = 0] \\ &\quad + P[X(n) = 0, X(n+1) = 1] \\ &= 0.4 \\ P[X(n) = 1] &= 0.6 \end{aligned}$$

Hence, $E\{X(n)\} = 0.6$, and $E\{[X(n)]^2\} = 0.6$.

$$\begin{aligned} E\{X(n)X(n+1)\} &= 1 \cdot P[X(n) = 1, X(n+1) = 1] = 0.4 \\ E\{X(n)X(n+2)\} &= 1 \cdot P[X(n) = 1, X(n+2) = 1] \\ &= 1 \cdot P[X(n) = 1, X(n+1) = 1, X(n+2) = 1] \\ &\quad + 1 \cdot P[X(n) = 1, X(n+1) = 0, X(n+2) = 1] \\ &= 1 \cdot P[X(n) = 1]P[X(n+1) = 1|X(n) = 1] \\ &\quad \times P[X(n+2) = 1|X(n) = 1, X(n+1) = 1] \\ &\quad + 1 \cdot P[X(n) = 1]P[X(n+1) = 0|X(n) = 1] \\ &\quad \times P[X(n+2) = 1|X(n) = 1, X(n+1) = 0] \end{aligned}$$

The Markov property implies that

$$P[X(n+2) = 1 | X(n) = 1, X(n+1) = 1]$$
$$= P[X(n+2) = 1 | X(n+1) = 1]$$
$$P[X(n+2) = 1 | X(n) = 1, X(n+1) = 0]$$
$$= P[X(n+2) = 1 | X(n+1) = 0]$$

and hence

$$E\{X(n)X(n+2)\} = (0.6)\left(\frac{0.4}{0.6}\right)\left(\frac{0.4}{0.6}\right) + (0.6)\left(\frac{0.2}{0.6}\right)\left(\frac{0.2}{0.4}\right)$$
$$\approx 0.367$$

Thus we have

$$\left.\begin{aligned} E\{X(n)\} &= 0.6 \\ R_{XX}(n, n) &= 0.6 \\ R_{XX}(n, n+1) &= 0.4 \\ R_{XX}(n, n+2) &\approx 0.367 \end{aligned}\right\} \text{ all independent of } n$$

Proceeding in a similar fashion, we can show that $R_{XX}(n, n+k)$ will be independent of n, and hence this Markov sequence is wide-sense stationary.

EXAMPLE 3.9.

A_i and B_i, $i = 1, 2, 3, \ldots, n$, is a set of $2n$ random variables that are uncorrelated and have a joint Gaussian distribution with $E\{A_i\} = E\{B_i\} = 0$, and $E\{A_i^2\} = E\{B_i^2\} = \sigma^2$. Let

$$X(t) = \sum_{i=1}^{n} (A_i \cos \omega_i t + B_i \sin \omega_i t)$$

Show that $X(t)$ is a SSS Gaussian random process.

SOLUTION:

$$E\{X(t)\} = \sum_{i=1}^{n} [E\{A_i\} \cos \omega_i t + E\{B_i\} \sin \omega_i t] = 0$$

$$E\{X(t)X(t+\tau)\} = E\left\{\sum_{i=1}^{n}\sum_{j=1}^{n} [A_i \cos \omega_i t + B_i \sin \omega_i t]\right.$$
$$\left. \times [A_j \cos \omega_j(t+\tau) + B_j \sin \omega_j(t+\tau)]\right\}$$

Since $E\{A_i A_j\}$, $E\{A_i B_i\}$, $E\{A_i B_j\}$, and $E(B_i B_j)$, $i \neq j$ are all zero, we have

$$E\{X(t)X(t+\tau)\} = \sum_{i=1}^{n} [E\{A_i^2\} \cos \omega_i t \cos \omega_i(t+\tau)$$
$$+ E\{B_i^2\} \sin \omega_i t \sin \omega_i(t+\tau)]$$
$$= \sigma^2 \sum_{i=1}^{n} \cos \omega_i \tau = R_{XX}(\tau)$$

Since $E\{X(t)\}$ and $E\{X(t)X(t+\tau)\}$ do not depend on t, the process $X(t)$ is WSS.

This process $X(t)$ for any values of $t_1, t_2, \ldots, t_k$ is a weighted sum of $2n$ Gaussian random variables, A_i and B_i, $i = 1, 2, \ldots, n$. Since A_i's and B_i's have a joint Gaussian distribution, any linear combinations of these variables will also have a Gaussian distribution. That is, the joint distribution of $X(t_1)$, $X(t_2), \ldots, X(t_k)$ will be Gaussian and hence $X(t)$ is a Gaussian process.

The kth-order joint distribution of $X(t_1), X(t_2), \ldots, X(t_k)$ will involve the parameters $E\{X(t_i)\} = 0$, and $E\{X(t_i)X(t_j)\} = R_{XX}(|t_i - t_j|)$, which depends only on the time difference $t_i - t_j$. Hence, the joint distribution of $X(t_1)$, $X(t_2), \ldots, X(t_k)$, and the joint distribution of $X(t_1 + \tau), X(t_2 + \tau), \ldots, X(t_k + \tau)$ will be the same for all values of τ and $t_i \in \Gamma$, which proves that $X(t)$ is SSS.

A Gaussian random process provides one of the few examples where WSS implies SSS.

3.5.4 Other Forms of Stationarity

A process $X(t)$ is *asymptotically* stationary if the distribution of $X(t_1 + \tau)$, $X(t_2 + \tau), \ldots, X(t_n + \tau)$ does not depend on τ when τ is large.

A process $X(t)$ is *stationary in an interval* if Equation 3.23 holds for all τ for which $t_1 + \tau, t_2 + \tau, \ldots, t_k + \tau$ lie in an interval that is a subset of Γ.

A process $X(t)$ is said to have *stationary increments* if its increments $Y(t) = X(t + \tau) - X(t)$ form a stationary process for every τ. The Poisson and Wiener processes are examples of processes with stationary increments.

Finally, a process is *cyclostationary* or periodically stationary if it is stationary under a shift of the time origin by integer multiples of a constant T_0 (which is the period of the process).

3.5.5 Tests for Stationarity

If a fairly detailed description of a random process is available, then it is easy to verify the stationarity of the process as illustrated by the examples given in Section 3.5.3. When a complete description is not available, then the stationarity of the process has to be established by collecting and analyzing a few sample functions of the process. The general approach is to divide the interval of observation into N nonoverlapping subintervals where the data in each interval may be considered independent; estimate the parameters of the process using the data from nonoverlapping intervals; and test these values for time dependency. If the process is stationary, then we would not expect these estimates from the different intervals to be significantly different. Excessive variation in the estimated values from different time intervals would indicate that the process is nonstationary.

Details of the estimation and testing procedures are presented in Chapters 8 and 9.

3.6 AUTOCORRELATION AND POWER SPECTRAL DENSITY FUNCTIONS OF REAL WSS RANDOM PROCESSES

Frequency domain descriptions of deterministic signals are obtained via their Fourier transforms, and this technique plays an important role in the characterization of random waveforms. However, direct transformation usually is not applicable for random waveforms since a transform of each member function of the ensemble is often impossible. Thus, spectral analysis of random processes differs from that of deterministic signals.

For stationary random processes, the autocorrelation function $R_{XX}(\tau)$ tells us something about how rapidly we can expect the random signal to change as a function of time. If the autocorrelation function decays rapidly to zero it indicates that the process can be expected to change rapidly with time. And a slowly changing process will have an autocorrelation function that decays slowly. Furthermore if the autocorrelation function has periodic components, then the underlying process will also have periodic components. Hence we conclude, correctly, that the autocorrelation function contains information about the expected frequency content of the random process. The relationship between the autocorrelation function and the frequency content of a random process is the main topic of discussion in this section.

Throughout this section we will assume the process to be real-valued. The concepts developed in this section can be extended to complex-valued random processes. These concepts rely heavily on the theory of Fourier transforms.

3.6.1 Autocorrelation Function of a Real WSS Random Process and Its Properties

The autocorrelation function of a real-valued WSS random process is defined as

$$R_{XX}(\tau) = E\{X(t)X(t + \tau)\}$$

There are some general properties that are common to all autocorrelation functions of stationary random processes, and we discuss these properties briefly before proceeding to the development of power spectral densities.

1. If we assume that $X(t)$ is a voltage waveform across a 1-Ω resistance, then the ensemble average value of $X^2(t)$ is the average value of power delivered to the 1-Ω resistance by $X(t)$:

$$E\{X^2(t)\} = \text{Average power}$$
$$= R_{XX}(0) \geq 0 \quad (3.31)$$

2. $R_{XX}(\tau)$ is an even function of τ

$$R_{XX}(\tau) = R_{XX}(-\tau) \quad (3.32)$$

3. $R_{XX}(\tau)$ is bounded by $R_{XX}(0)$

$$|R_{XX}(\tau)| \leq R_{XX}(0)$$

This can be verified by starting from the inequalities

$$E\{[X(t + \tau) - X(t)]^2\} \geq 0$$
$$E\{[X(t + \tau) + X(t)]^2\} \geq 0$$

which yield

$$E\{X^2(t + \tau)\} + E\{X^2(t)\} - 2R_{XX}(\tau) \geq 0$$
$$E\{X^2(t + \tau)\} + E\{X^2(t)\} + 2R_{XX}(\tau) \geq 0$$

Since $E\{X^2(t + \tau)\} = E\{X^2(t)\} = R_{XX}(0)$, we have

$$2R_{XX}(0) - 2R_{XX}(\tau) \geq 0$$
$$2R_{XX}(0) + 2R_{XX}(\tau) \geq 0$$

Hence, $-R_{XX}(0) \leq R_{XX}(\tau) \leq R_{XX}(0)$ or $|R_{XX}(\tau)| \leq R_{XX}(0)$

4. If $X(t)$ contains a periodic component, then $R_{XX}(\tau)$ will also contain a periodic component.
5. If $\lim_{\tau \to \infty} R_{XX}(\tau) = C$, then $C = \mu_X^2$.
6. If $R_{XX}(T_0) = R_{XX}(0)$ for some $T_0 \neq 0$, then R_{XX} is periodic with a period T_0. Proof of this follows from the cosine inequality (Problem 2.22a)

$$[E\{[X(t + \tau + T_0) - X(t + \tau)]X(t)\}]^2 \leq E\{[X(t + \tau + T_0) - X(t + \tau)]^2\}E\{X^2(t)\}$$

Hence

$$[R_{XX}(\tau + T_0) - R_{XX}(\tau)]^2 \leq 2[R_{XX}(0) - R_{XX}(T_0)]R_{XX}(0)$$

for every τ and T_0. If $R_{XX}(T_0) = R_{XX}(0)$, then $R_{XX}(\tau + T_0) = R_{XX}(\tau)$ for every τ and $R_{XX}(\tau)$ is periodic with period T_0.
7. If $R_{XX}(0) < \infty$ and $R_{XX}(\tau)$ is continuous at $\tau = 0$, then it is continuous for every τ.

Properties 2 through 7 say that any arbitrary function cannot be an autocorrelation function.

3.6.2 Cross-correlation Function and Its Properties

The cross-correlation function of two real random processes $X(t)$ and $Y(t)$ that are jointly WSS will be independent of t, and we can write it as

$$R_{XY}(\tau) = E\{X(t)Y(t + \tau)\}$$

The cross-correlation function has the following properties:

1. $R_{XY}(\tau) = R_{YX}(-\tau)$ (3.33)
2. $|R_{XY}(\tau)| \leq \sqrt{R_{XX}(0)R_{YY}(0)}$ (3.34)
3. $|R_{XY}(\tau)| \leq \dfrac{1}{2}[R_{XX}(0) + R_{YY}(0)]$ (3.35)
4. $R_{XY}(\tau) = 0$ if the processes are orthogonal, and

   $R_{XY}(\tau) = \mu_X \mu_Y$ if the processes are independent.

Proofs of these properties are left as exercises for the reader.

3.6.3 Power Spectral Density Function of a WSS Random Process and Its Properties

For a deterministic power signal, $x(t)$, the average power in the signal is defined as

$$P_x = \lim_{T \to \infty} \frac{1}{2T} \int_{-T}^{T} x^2(t) \, dt \tag{3.36}$$

If the deterministic signal is periodic with period T_0, then we can define a time-averaged autocorrelation function $\langle R_{xx}(\tau) \rangle_{T_0}$ as*

$$\langle R_{xx}(\tau) \rangle_{T_0} = \frac{1}{T_0} \int_0^{T_0} x(t) x(t + \tau) \, dt \tag{3.37}$$

and show that the Fourier transform $S_{xx}(f)$ of $\langle R_{xx}(\tau) \rangle_{T_0}$ yields

$$P_x = \int_{-\infty}^{\infty} S_{xx}(f) \, df \tag{3.38}$$

In Equation 3.38, the left-hand side represents the total average power in the signal, f is the frequency variable expressed usually in Hertz (Hz), and $S_{xx}(f)$ has the units of power (watts) per Hertz. The function $S_{xx}(f)$ thus describes the power distribution in the frequency domain, and it is called the power spectral density function of the deterministic signal $x(t)$.

The concept of power spectral density function also applies to stationary random processes and the power spectral density function of a WSS random process $X(t)$ is defined as the Fourier transform of the autocorrelation function

$$S_{XX}(f) = F\{R_{XX}(\tau)\} = \int_{-\infty}^{\infty} R_{XX}(\tau) \exp(-j2\pi f \tau) \, d\tau \tag{3.39}$$

Equation 3.39 is called the Wiener–Khinchine relation. Given the power spectral density function, the autocorrelation function is obtained as

$$R_{XX}(\tau) = F^{-1}\{S_{XX}(f)\} = \int_{-\infty}^{\infty} S_{XX}(f) \exp(j2\pi f \tau) \, df \tag{3.40}$$

---

*The notation $\langle \ \rangle_{T_0}$ denotes integration or averaging in the time domain for a duration of T_0 seconds whereas $E\{\ \}$ denotes ensemble averaging.

Properties of the Power Spectral Density Function. The power spectral density (psd) function, which is also called the spectrum of $X(t)$, possesses a number of important properties:

1. $S_{XX}(f)$ is real and nonnegative.
2. The average power in $X(t)$ is given by

$$E\{X^2(t)\} = R_{XX}(0) = \int_{-\infty}^{\infty} S_{XX}(f)\, df \qquad (3.41)$$

Note that if $X(t)$ is a current or voltage waveform then $E\{X^2(t)\}$ is the average power delivered to a one-ohm load. Thus, the left-hand side of the equation represents power and the integrand $S_{XX}(f)$ on the right-hand side has the units of power per Hertz. That is, $S_{XX}(f)$ gives the distribution of power as a function of frequency and hence is called the power spectral density function of the stationary random process $X(t)$.

3. For $X(t)$ real, $R_{XX}(\tau)$ is an even function and hence $S_{XX}(f)$ is also even. That is

$$S_{XX}(-f) = S_{XX}(f) \qquad (3.42)$$

4. If $X(t)$ has periodic components, then $S_{XX}(f)$ will have impulses.

Lowpass and Bandpass Processes. A random process is said to be lowpass if its psd is zero for $|f| > B$, and B is called the bandwidth of the process. On the other hand, a process is said to be bandpass if its psd is zero outside the band

$$f_c - \frac{B}{2} \le |f| \le f_c + \frac{B}{2}$$

f_c is usually referred to as the center frequency and B is the bandwidth of the process. Examples of lowpass and bandpass spectra are shown in Figure 3.11.

Notice that we are using positive and negative values of frequencies and the psd is shown on both sides of $f = 0$. Such a spectral characterization is called a *two-sided psd*.

Power and Bandwidth Calculations. As stated in Equation 3.41, the area under the psd function gives the total power in $X(t)$. The power in a finite band of frequencies, f_1 to f_2, $0 < f_1 < f_2$ is the area under the psd from $-f_2$ to $-f_1$ plus the area between f_1 to f_2, and for real $X(t)$

$$P_X[f_1, f_2] = 2 \int_{f_1}^{f_2} S_{XX}(f)\, df \qquad (3.43)$$

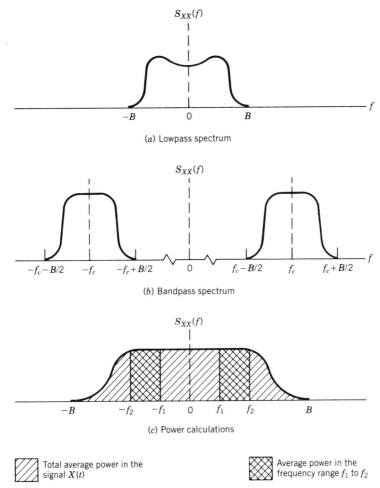

Figure 3.11 Examples of power spectral densities.

The proof of this equation is given in the next chapter. Figure 3.11.c makes it seem reasonable. The factor 2 appears in Equation 3.43 since we are using a two-sided psd and $S_{XX}(f)$ is an even function (see Figure 3.11.c and Equation 3.42).

Some processes may have psd functions with nonzero values for all finite values of f, for example, $S_{XX}(f) = \exp(-f^2/2)$. For such processes, several indicators are used as measures of the spread of the psd in the frequency domain. One popular measure is the *effective (or equivalent) bandwidth* B_{eff}. For zero mean random processes with continuous psd, B_{eff} is defined as

$$B_{\text{eff}} = \frac{1}{2} \frac{\int_{-\infty}^{\infty} S_{XX}(f)\, df}{\max[S_{XX}(f)]} \tag{3.44}$$

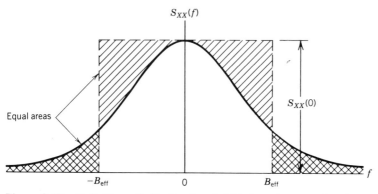

Figure 3.12 Definition of effective bandwidth for a lowpass signal.

(See Figure 3.12.) The effective bandwidth is related to a measure of the spread of the autocorrelation function called the *correlation time* τ_c, where

$$\tau_c = \frac{\int_{-\infty}^{\infty} R_{XX}(\tau)\, d\tau}{R_{XX}(0)} \tag{3.45}$$

If $S_{XX}(f)$ is continuous and has a maximum at $f = 0$, then it can be shown that

$$B_{\text{eff}} = \frac{1}{2\tau_c} \tag{3.46}$$

Other measures of spectral spread include the rms bandwidth defined as the standard deviation of the psd and the half-power bandwidth (see Problems 3.23 and 3.24).

3.6.4 Cross-power Spectral Density Function and Its Properties

The relationship between two real-valued random processes $X(t)$ and $Y(t)$ is expressed in the frequency domain via the cross-power spectral density (cpsd) function $S_{XY}(f)$, which is defined as the Fourier transform of the cross-correlation function $R_{XY}(\tau)$,

$$S_{XY}(f) = \int_{-\infty}^{\infty} R_{XY}(\tau)\exp(-j2\pi f\tau)\, d\tau \tag{3.47}$$

and

$$R_{XY}(\tau) = \int_{-\infty}^{\infty} S_{XY}(f)\exp(j2\pi f\tau)\, df \qquad (3.48)$$

Unlike the psd, which is a real-valued function of f, the cpsd will, in general, be a complex-valued function. Some of the properties of cpsd are as follows:

1. $S_{XY}(f) = S_{YX}^*(f)$
2. The real part of $S_{XY}(f)$ is an even function of f, and the imaginary part of f is an odd function of f.
3. $S_{XY}(f) = 0$ if $X(t)$ and $Y(t)$ are orthogonal and $S_{XY}(f) = \mu_X \mu_Y \delta(f)$ if $X(t)$ and $Y(t)$ are independent.

In many applications involving the cpsd, a real-valued function

$$\rho_{XY}^2(f) = \frac{|S_{XY}(f)|^2}{S_{XX}(f)S_{YY}(f)} \leq 1 \qquad (3.49)$$

called the *coherence function* is used as an indicator of the dependence between two random processes $X(t)$ and $Y(t)$. When $\rho_{XY}^2(f_0) = 0$ at a particular frequency, f_0, then $X(t)$ and $Y(t)$ are said to be incoherent at that frequency, and the two processes are said to be fully coherent at a particular frequency, f_0, when $\rho_{XY}^2(f_0) = 1$. If $X(t)$ and $Y(t)$ are statistically independent, then $\rho_{XY}^2(f) = 0$ at all frequencies except at $f = 0$.

3.6.5 Power Spectral Density Function of Random Sequences

The psd of a random sequence $X(nT_s)$ with a uniform sampling time of one second ($T_s = 1$) is defined by the Fourier Transform of the sequence as

$$S_{XX}(f) = \sum_{n=-\infty}^{\infty} \exp(-j2\pi fn) R_{XX}(n), \qquad -\frac{1}{2} < f < \frac{1}{2} \qquad (3.50.\text{a})$$

The definition implies that $S_{XX}(f)$ is periodic in f with period 1. We will only consider the principal part, $-1/2 < f < 1/2$. Then it follows that

$$R_{XX}(n) = \int_{-1/2}^{1/2} S_{XX}(f) \exp(j2\pi fn)\, df \qquad (3.50.\text{b})$$

150 RANDOM PROCESSES AND SEQUENCES

It is important to observe that if the uniform sampling time (T_s) is not one second (i.e., if nT_s is the time index instead of n) then the actual frequency range is not 1, but is $1/T_s$.

If $X(n)$ is real, then $R_{XX}(n)$ will be even and

$$S_{XX}(f) = \sum_{n=-\infty}^{\infty} \cos 2\pi fn \, R_{XX}(n), \quad |f| < \frac{1}{2} \quad (3.50.c)$$

which implies that $S_{XX}(f)$ is real and even. It is also nonnegative. In fact, $S_{XX}(f)$ of a sequence has the same properties as $S_{XX}(f)$ of a continuous process except of course, as defined, $S_{XX}(f)$ of a sequence is periodic.

Although the psd of a random sequence can be defined as the Fourier transform of the autocorrelation function $R_{XX}(n)$ as in Equation 3.50.a, we present a slightly modified version here that will prove quite useful later on. To simplify the derivation, let us assume that $E\{X(n)\} = 0$.

We start with the assumption that the observation times of the random sequence are uniformly spaced in the time domain and that the index n denotes $t = nT$. From the random sequence $X(n)$, we create a random process $X_p(t)$ of the form

$$X_p(t) = \sum_{n=-\infty}^{\infty} X(n)p(t - nT - D)$$

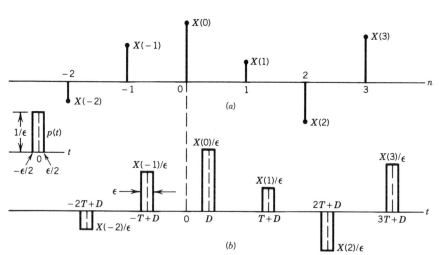

Figure 3.13a Random sequence $X(n)$. **Figure 3.13b** Random process $X_p(t)$.

AUTOCORRELATION AND POWER SPECTRAL DENSITY

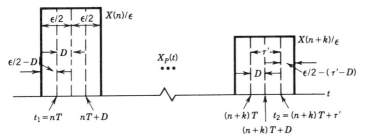

Figure 3.14 Details of calculations for $R_{X_p X_p}(kT + \tau')$.

where $p(t)$ is a pulse of height $1/\epsilon$ and duration $\epsilon \ll T$, and D is a random delay that has a uniform probability density function in the interval $[-T/2, T/2]$ (see Figure 3.13). Except for its width and varying height, $X_p(t)$ is similar in structure to the random binary waveform discussed earlier. It is fairly easy to verify that $X_p(t)$ will be WSS if $X(n)$ is WSS.

To find the autocorrelation function of $X_p(t)$, let us arbitrarily choose $t_1 = nT$, and $t_2 = nT + kT + \tau'$, $0 < \tau' < \epsilon$ (see Figure 3.14). Following the line of reasoning used in the derivation of the autocorrelation function of the random binary waveform, we start with

$$E\{X_p(t_1)X_p(t_2)\} = E\{X_p(nT)X_p(nT + kT + \tau')\}$$
$$= R_{X_p X_p}(kT + \tau')$$

From Figure 3.14, we see that the value of the product $X_p(t_1)X_p(t_2)$ will depend on the value of D according to

$$X_p(t_1)X_p(t_2) = \begin{cases} \dfrac{X(n)X(n+k)}{\epsilon^2}, & -\left(\dfrac{\epsilon}{2} - \tau'\right) \leq D \leq \dfrac{\epsilon}{2}; \quad 0 < \tau' < \epsilon \\ 0 & \text{otherwise} \end{cases}$$

and $R_{X_p X_p}(kT + \tau')$ is given by

$$R_{X_p X_p}(kT + \tau') = E\left\{X_p(t_1)X_p(t_2) \Big| -\left(\dfrac{\epsilon}{2} - \tau'\right) \leq D \leq \dfrac{\epsilon}{2}\right\}$$
$$\cdot P\left[-\left(\dfrac{\epsilon}{2} - \tau'\right) \leq D \leq \dfrac{\epsilon}{2}\right]$$
$$= E\{X(n)X(n+k)\}\dfrac{\epsilon - \tau'}{T\epsilon^2}, \quad 0 < \tau' < \epsilon$$

When $\tau' > \epsilon$, then irrespective of the value of D, t_2 will fall outside of the pulse at $t = kT$ and hence $X(t_2)$ and the product $X(t_1)X(t_2)$ will be zero. Since $X_p(t)$ is stationary, we can generalize the result to arbitrary values of τ' and k and write $R_{X_p X_p}$ as

$$R_{X_p X_p}(kT + \tau') = \begin{cases} R_{XX}(k) \dfrac{\epsilon - |\tau'|}{T\epsilon^2}, & |\tau'| < \epsilon \\ 0 & \epsilon < |\tau'| < T - \epsilon \end{cases}$$

or

$$\begin{aligned} R_{X_p X_p}(\tau) &= \begin{cases} R_{XX}(k) \dfrac{\epsilon - |(\tau - kT)|}{T\epsilon^2}, & |kT - \tau| < \epsilon \\ 0 & \text{elsewhere} \end{cases} \\ &= \frac{1}{T} \sum_k R_{XX}(k) q(\tau - kT) \end{aligned} \qquad (3.51)$$

where $q(t)$ is a triangular pulse of width 2ϵ and height $1/\epsilon$. An example of $R_{X_p X_p}(\tau)$ is shown in Figure 3.15. Now if we let $\epsilon \to 0$, then both $p(t)$ and $q(t) \to \delta(t)$ and we have

$$X_p(t) = \sum_{n=-\infty}^{\infty} X(n) \delta(t - nT - D) \qquad (3.52.a)$$

and

$$R_{X_p X_p}(\tau) = \frac{1}{T} \sum_{k=-\infty}^{\infty} R_{XX}(k) \delta(\tau - kT) \qquad (3.52.b)$$

The psd of the random sequence $X(n)$ is defined as the Fourier transform of $R_{X_p X_p}(\tau)$, and we have

$$\begin{aligned} S_{X_p X_p}(f) &= F\{R_{X_p X_p}(\tau)\} \\ &= \frac{1}{T} \left[R_{XX}(0) + 2 \sum_{k=1}^{\infty} R_{XX}(k) \cos 2\pi k f T \right] \end{aligned} \qquad (3.53)$$

Note that if $T = 1$, this is the Fourier transform of an even sequence as defined in Equation 3.50.a, except the spectral density given in Equation 3.53 is valid for $-\infty < f < \infty$.

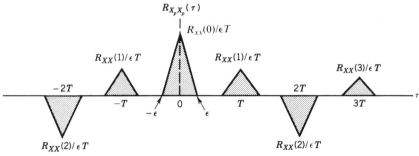

Figure 3.15 Autocorrelation function of $X_p(t)$.

If the random sequence $X(n)$ has a nonzero mean, then $S_{XX}(f)$ will have discrete frequency components at multiples of $1/T$ (see Problem 3.35). Otherwise, $S_{XX}(f)$ will be continuous in f.

The derivation leading to Equation 3.53 seems to be a convoluted way of obtaining the psd of a random sequence. The advantage of this formulation will be explained in the next chapter.

EXAMPLE 3.10.

Find the power spectral density function of the random process $X(t) = 10\cos(2000\pi t + \Theta)$ where Θ is a random variable with a uniform pdf in the interval $[-\pi, \pi]$.

SOLUTION:

$$R_{XX}(\tau) = 50\cos(2000\pi\tau)$$

and hence

$$S_{XX}(f) = 25[\delta(f - 1000) + \delta(f + 1000)]$$

The psd of $S_{XX}(f)$ shown in Figure 3.16 has two discrete components in the frequency domain at $f = \pm 1000$ Hz. Note that

$$R_{XX}(0) = \text{average power in the signal}$$
$$= \frac{(10)^2}{2} = \int_{-\infty}^{\infty} S_{XX}(f)\,df$$

154 RANDOM PROCESSES AND SEQUENCES

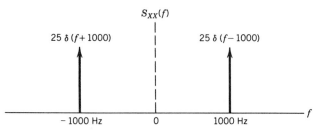

Figure 3.16 Psd of $10\cos(2000\pi t + \Theta)$ and $10\sin(2000\pi t + \Theta)$.

Also, the reader can verify that $Y(t) = 10\sin(2000\pi t + \Theta)$ has the same psd as $X(t)$, which illustrates that the psd does not contain any phase information.

EXAMPLE 3.11.

A WSS random sequence $X(n)$ has the following autocorrelation function:

$$R_{XX}(k) = 4 + 6\exp(-0.5|k|)$$

Find the psd of $X_p(t)$ and of $X(n)$.

SOLUTION: We assume that as $k \to \infty$, the sequence is uncorrelated. Thus $R_{XX}(k) = [E\{X(n)\}]^2 = 4$. Hence $E\{X(n)\} = \pm 2$. If we define $X(n) = Z(n) + Y(n)$, with $Y(n) = \pm 2$, then $Z(n)$ is a zero mean stationary sequence with $R_{ZZ}(k) = R_{XX}(k) - 4 = 6\exp(-0.5|k|)$, and $R_{YY}(k) = 4$.

The autocorrelation functions of the continuous-time versions of $Z(n)$ and

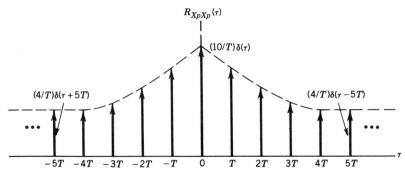

Figure 3.17 Autocorrelation function of the random sequence $X(n)$.

$Y(n)$ are given by

$$R_{Z_pZ_p}(\tau) = \frac{1}{T} \sum_{k=-\infty}^{\infty} 6 \exp(-0.5|k|)\delta(\tau - kT)$$

$$R_{Y_pY_p}(\tau) = \frac{1}{T} \sum_{k=-\infty}^{\infty} 4\delta(\tau - kT)$$

and $R_{X_pX_p}(\tau) = R_{Z_pZ_p}(\tau) + R_{Y_pY_p}(\tau)$ (see Figure 3.17). Taking the Fourier transform, we obtain the psd's as

$$S_{Z_pZ_p}(f) = \frac{1}{T}\left[6 + \sum_{k=1}^{\infty} 12 \exp(-0.5k) \cos 2\pi kfT\right]$$

$$= \frac{6}{T}\left[-1 + \sum_{k=0}^{\infty} \exp\{-(.5 + j2\pi fT)k\} + \sum_{k=0}^{\infty} \exp\{-(.5 - j2\pi fT)k\}\right]$$

$$= \frac{6}{T}[-1 + 1/(1 - \exp\{-(.5 + j2\pi fT)\}) + 1/(1 - \exp\{-(.5 - j2\pi fT)\})]$$

$$= \frac{6}{T}[(1-e^{-1})/(1 - 2e^{-.5}\cos 2\pi fT + e^{-1})]$$

$$S_{Y_pY_p}(f) = \frac{1}{T^2} \sum_{k=-\infty}^{\infty} 4\delta\left(f - \frac{k}{T}\right)$$

and

$$S_{X_pX_p}(f) = S_{Z_pZ_p}(f) + S_{Y_pY_p}(f)$$

The psd of $X_p(t)$ has a continuous part $S_{Z_pZ_p}(f)$ and a discrete sequence of impulses at multiples of $1/T$.

The psd of $X(n)$ is the Fourier transform of $R_{ZZ}(k)$ plus the Fourier transform of $R_{YY}(k)$ where

$$S_{YY}(f) = 4\delta(f), \quad |f| < \frac{1}{2}$$

and

$$S_{ZZ}(f) = 6\left[\sum_{k=-\infty}^{\infty} \exp(-.5|k|)\exp(-j2\pi fk)\right]$$

$$= 6[(1 - e^{-1})/(1 - 2e^{-.5}\cos 2\pi f + e^{-1})], \quad |f| < \frac{1}{2}$$

Thus

$$S_{xx}(f) = 4\delta(f) + 6[(1 - e^{-1})/(1 - 2e^{-.5}\cos 2\pi f + e^{-1})], \quad |f| < \frac{1}{2}$$

Note the similarities and the differences between $S_{X_p X_p}$ and S_{xx}. Essentially $S_{xx}(f)$ is the principal part of $S_{X_p X_p}$ (i.e. the value of $S_{X_p X_p}(f)$ for $-\frac{1}{2} < f < \frac{1}{2}$) and it assumes that T is 1.

EXAMPLE 3.12.

Find the psd of the random binary waveform discussed in Section 3.4.4.

SOLUTION: The autocorrelation function of $X(t)$ is

$$R_{xx}(\tau) = \begin{cases} 1 - \frac{|\tau|}{T}, & |\tau| < T \\ 0, & \text{elsewhere} \end{cases}$$

The psd of $X(t)$ is obtained (see the table of Fourier transform pairs in Appendix A) as

$$S_{xx}(f) = T\left[\frac{\sin \pi f T}{\pi f T}\right]^2$$

A sketch of $S_{xx}(f)$ is shown in Figure 3.18b. The main "lobe" of the psd extends from $-1/T$ to $1/T$ Hz, and 90% of the signal power is contained in the main

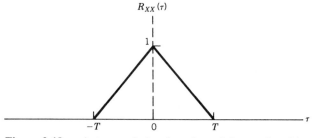

Figure 3.18a Autocorrelation function of the random binary waveform.

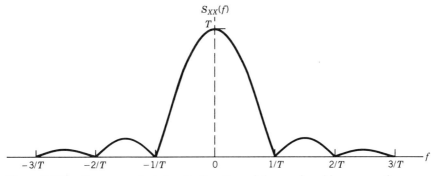
Figure 3.18b Power spectral density function of the random binary waveform.

lobe. For many applications, the "bandwidth" of the random binary waveform is defined to be $1/T$.

EXAMPLE 3.13.

The autocorrelation function $R_{XX}(\tau)$ of a WSS random process is given by

$$R_{XX}(\tau) = A \exp(-\alpha|\tau|); \quad A, \alpha > 0$$

Find the psd and the effective bandwidth of $X(t)$.

SOLUTION:

$$S_{XX}(f) = \int_{-\infty}^{\infty} A \exp(-\alpha|\tau|)\exp(-j2\pi f\tau)\, d\tau$$

$$= \frac{2A\alpha}{\alpha^2 + (2\pi f)^2}$$

The effective bandwidth of $X(t)$ is calculated from Equation 3.44 as

$$B_{\text{eff}} = \frac{1}{2}\frac{\int_{-\infty}^{\infty} S_{XX}(f)\, df}{\max[S_{XX}(f)]} = \frac{1}{2}\frac{R_{XX}(0)}{S_{XX}(0)}$$

$$= \frac{1}{2} \cdot \frac{A}{2A/\alpha} = \frac{\alpha}{4}\,\text{Hz}$$

EXAMPLE 3.14.

The power spectral density function of a zero mean Gaussian random process is given by (Figure 3.19)

$$S_{XX}(f) = \begin{cases} 1, & |f| < 500 \text{ Hz} \\ 0 & \text{elsewhere} \end{cases}$$

Find $R_{XX}(\tau)$ and show that $X(t)$ and $X(t + 1 \text{ ms})$ are uncorrelated and, hence, independent.

SOLUTION:

$$R_{XX}(\tau) = \int_{-500}^{500} \exp(j2\pi f\tau) \, df = \left. \frac{\exp(j2\pi f\tau)}{j2\pi\tau} \right|_{-500}^{500}$$

$$= (2B) \frac{\sin 2\pi B\tau}{2\pi B\tau}, \quad B = 500 \text{ Hz}$$

To show that $X(t)$ and $X(t + 1 \text{ ms})$ are uncorrelated we need to show that $E\{X(t)X(t + 1 \text{ ms})\} = 0$.

$$E\{X(t)X(t + 1 \text{ ms})\} = R_{XX}(1 \text{ ms})$$

$$= 2B \frac{\sin \pi}{\pi} = 0$$

Hence, $X(t)$ and $X(t + 1 \text{ ms})$ are uncorrelated. Since $X(t)$ and $X(t + 1 \text{ ms})$ have a joint Gaussian distribution, being uncorrelated implies their independence.

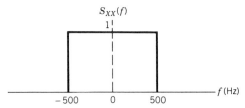

Figure 3.19a Psd of a lowpass random process $X(t)$.

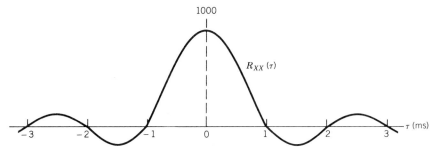

Figure 3.19b Autocorrelation function of $X(t)$.

EXAMPLE 3.15.

$X(t)$ is a stationary random process with a psd

$$S_{XX}(f) = \begin{cases} 1, & |f| < B \\ 0 & \text{elsewhere} \end{cases}$$

$X(t)$ is multiplied by a random process $Y(t)$ of the form $Y(t) = A \cos(2\pi f_c t + \Theta)$, $f_c \gg B$, where Θ is a random variable with a uniform distribution in the interval $[-\pi, \pi]$. Assume that $X(t)$ and $Y(t)$ are independent and find the psd of $Z(t) = X(t)Y(t)$.

SOLUTION:

$$R_{YY}(\tau) = \frac{A^2}{2} \cos(2\pi f_c \tau)$$

and

$$\begin{aligned}
R_{ZZ}(\tau) &= E\{X(t)Y(t)X(t + \tau)Y(t + \tau)\} \\
&= E\{X(t)X(t + \tau)\}E\{Y(t)Y(t + \tau)\} \\
&= R_{XX}(\tau)R_{YY}(\tau) \\
&= R_{XX}(\tau) \cdot \frac{A^2}{2} \cos(2\pi f_c \tau) \\
&= R_{XX}(\tau) \frac{A^2}{4} [\exp(2\pi j f_c \tau) + \exp(-2\pi j f_c \tau)]
\end{aligned}$$

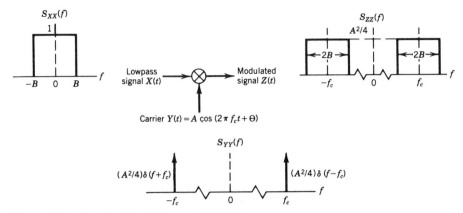

Figure 3.20 Psd of $X(t)$, $Y(t)$, and $X(t)Y(t)$.

$$S_{ZZ}(f) = F\{R_{ZZ}(\tau)\}$$

$$= \frac{A^2}{4}\left[\int_{-\infty}^{\infty} R_{XX}(\tau)\exp(j2\pi f_c\tau)\exp(-j2\pi f\tau)\,d\tau \right.$$

$$\left. + \int_{-\infty}^{\infty} R_{XX}(\tau)\exp(-j2\pi f_c\tau)\exp(-j2\pi f\tau)\,d\tau\right]$$

$$= \frac{A^2}{4}\left[\int_{-\infty}^{\infty} R_{XX}(\tau)\exp[-j2\pi(f - f_c)\tau]\,d\tau \right.$$

$$\left. + \int_{-\infty}^{\infty} R_{XX}(\tau)\exp[-j2\pi(f + f_c)\tau]\,d\tau\right]$$

$$= \frac{A^2}{4}[S_{XX}(f - f_c) + S_{XX}(f + f_c)]$$

The preceding equations shows that the spectrum of $Z(t)$ is a translated version of the spectrum of $X(t)$ (Figure 3.20). The operation of multiplying a "message" signal $X(t)$ by a "carrier" $Y(t)$ is called "modulation" and it is a fundamental operation in communication systems. Modulation is used primarily to alter the frequency content of a message signal so that it is suitable for transmission over a given communication channel.

3.7 CONTINUITY, DIFFERENTIATION, AND INTEGRATION

Many dynamic electrical systems can be considered linear as a first approximation and their dynamic behavior can be described by linear differential or difference

CONTINUITY, DIFFERENTIATION, AND INTEGRATION

equations. In analyzing the response of these systems to deterministic input signals, we make use of rules of calculus as they apply to continuity, differentiation, and integration. These concepts can be applied to random signals also, either on a sample-function-by-sample-function basis or to the ensemble as a whole. When we discuss any of these concepts or properties as applying to the whole ensemble, this will be done in terms of probabilities.

Consider, for example, the continuity property. A real (deterministic) function $x(t)$ is said to be continuous at $t = t_0$ if

$$\lim_{t \to t_0} x(t) = x(t_0)$$

We can define continuity of a random process $X(t)$ at t_0 by requiring every member function of the process to be continuous at t_0 (sample continuity) or by requiring continuity in probability,

$$P[X(t) \text{ is continuous at } t_0] = 1 \tag{3.54}$$

or in a mean square (MS) sense by requiring

$$\operatorname*{l.i.m.}_{t \to t_0} X(t) = X(t_0)$$

where l.i.m. denotes mean square (MS) convergence, which stands for

$$\lim_{t \to t_0} E\{[X(t) - X(t_0)]^2\} = 0 \tag{3.55}$$

While sample continuity is the strongest requirement, MS continuity is most useful since it involves only the first two moments of the process and much of the analysis in electrical engineering is based on the first two moments.

In the following sections we will define continuity, differentiation, and integration operations in a MS sense as they apply to real *stationary* random processes, and derive conditions for the existence of derivatives and integrals of random processes.

3.7.1 Continuity

A stationary, finite variance real random process $X(t)$, $t \in \Gamma$, is said to be continuous in a mean square sense at $t_0 \in \Gamma$ if

$$\lim_{t \to t_0} E\{[X(t) - X(t_0)]^2\} = 0$$

Continuity of the autocorrelation function $R_{XX}(\tau)$ at $\tau = 0$ is a sufficient condition for the MS continuity of the process.

The sufficient condition for MS continuity can be shown by writing $E\{[X(t) - X(t_0)]^2\}$ as

$$E\{[X(t) - X(t_0)]^2\} = E\{X^2(t)\} + E\{X^2(t_0)\} - 2E\{X(t)X(t_0)\}$$
$$= R_{XX}(0) + R_{XX}(0) - 2R_{XX}(t - t_0)$$

and taking the ordinary limit

$$\lim_{t \to t_0} E\{[X(t) - X(t_0)]^2\} = R_{XX}(0) + R_{XX}(0) - 2 \lim_{t \to t_0} R_{XX}(t - t_0)$$

Now, since $R_{XX}(0) < \infty$, and if we assume $R_{XX}(\tau)$ to be continuous at $\tau = 0$, then

$$\lim_{t \to t_0} R_{XX}(t - t_0) = R_{XX}(t_0 - t_0) = R_{XX}(0)$$

and hence

$$\lim_{t \to t_0} E\{[X(t) - X(t_0)]^2\} = 0$$

Thus, continuity of the autocorrelation function at $\tau = 0$ is a sufficient condition for MS continuity of the process.

MS continuity and finite variance guarantee that we can interchange limiting and expected value operations, for example

$$\lim_{t \to t_0} E\{g(X(t))\} = E\{g(X(t_0))\}$$

when $g(\cdot)$ is any ordinary, continuous function.

3.7.2 Differentiation

The derivative of a finite variance stationary process $X(t)$ is said to exist in a mean square sense if there exists a random process $X'(t)$ such that

$$\operatorname*{l.i.m.}_{\epsilon \to 0} \frac{X(t + \epsilon) - X(t)}{\epsilon} = X'(t) \qquad (3.56)$$

CONTINUITY, DIFFERENTIATION, AND INTEGRATION 163

Note that the definition does not explicitly define the derivative random process $X'(t)$. To establish a sufficient condition for the existence of the MS derivative, we make use of the Cauchy criteria (see Equation 2.97) for MS convergence which when applied to Equation 3.56 requires that

$$\lim_{\epsilon_1,\epsilon_2 \to 0} E\left\{\left[\frac{X(t+\epsilon_1)-X(t)}{\epsilon_1} - \frac{X(t+\epsilon_2)-X(t)}{\epsilon_2}\right]^2\right\} = 0 \quad (3.57)$$

Completing the square and taking expected values, we have for the first term

$$E\left\{\left[\frac{X(t+\epsilon_1)-X(t)}{\epsilon_1}\right]^2\right\} = \frac{2[R_{XX}(0) - R_{XX}(\epsilon_1)]}{\epsilon_1^2}$$

Now, suppose that the first two derivatives of $R_{XX}(\tau)$ exist at $\tau = 0$. Then, since $R_{XX}(\tau)$ is even in τ, we must have

$$R'_{XX}(0) = 0$$

and

$$R''_{XX}(0) = \lim_{\epsilon \to 0} \frac{2[R_{XX}(\epsilon) - R_{XX}(0)]}{\epsilon^2}$$

Hence

$$\lim_{\epsilon_1 \to 0} E\left\{\left[\frac{X(t+\epsilon_1)-X(t)}{\epsilon_1}\right]^2\right\} = -R''_{XX}(0)$$

Proceeding along similar lines, we can show that the cross-product term in Equation 3.57 is equal to $2R''_{XX}(0)$, and the last term is equal to $-R''_{XX}(0)$. Thus,

$$\lim_{\epsilon_1,\epsilon_2 \to 0} E\left\{\left[\frac{X(t+\epsilon_1)-X(t)}{\epsilon_1} - \frac{X(t+\epsilon_2)-X(t)}{\epsilon_2}\right]^2\right\}$$
$$= 2[-R''_{XX}(0) + R''_{XX}(0)] = 0$$

if the first two derivatives of $R_{XX}(\tau)$ exist at $\tau = 0$, which guarantees the existence of the MS derivative of $X(t)$. This development is summarized by:

A finite variance stationary real random process $X(t)$ has a MS derivative, $X'(t)$, if $R_{XX}(\tau)$ has derivatives of order up to two at $\tau = 0$.

The mean and autocorrelation function of $X'(t)$ can be obtained easily as follows. The mean of $X'(t)$ is given by

$$E\{X'(t)\} = E\left\{\lim_{\epsilon \to 0}\left[\frac{X(t+\epsilon) - X(t)}{\epsilon}\right]\right\}$$

$$= \lim_{\epsilon \to 0} \frac{E\{X(t+\epsilon)\} - E\{X(t)\}}{\epsilon}$$

$$= \mu'_X(t)$$

For a stationary process, $\mu_X(t)$ is constant and hence

$$E\{X'(t)\} = 0$$

To find the autocorrelation function of $X'(t)$, let us start with

$$E\{X(t_1)X'(t_2)\} = R_{XX'}(t_1, t_2) = E\left\{X(t_1)\lim_{\epsilon \to 0}\frac{X(t_2+\epsilon) - X(t_2)}{\epsilon}\right\}$$

which yields

$$R_{XX'}(t_1, t_2) = \lim_{\epsilon \to 0}\left\{\frac{R_{XX}(t_1, t_2+\epsilon) - R_{XX}(t_1, t_2)}{\epsilon}\right\}$$

The functions on the right-hand side of the preceding equation are deterministic and the limiting operation yields the partial derivative of $R_{XX}(t_1, t_2)$ with respect to t_2. Thus,

$$R_{XX'}(t_1, t_2) = \frac{\partial R_{XX}(t_1, t_2)}{\partial t_2}$$

Proceeding along the same lines, we can show that

$$R_{X'X'}(t_1, t_2) = \frac{\partial R_{XX'}(t_1, t_2)}{\partial t_1}$$

$$= \frac{\partial^2 R_{XX}(t_1, t_2)}{\partial t_1 \partial t_2}$$

For a stationary process $X(t)$, $\mu_X(t) = $ constant, and $R_{XX}(t_1, t_2) = R_{XX}(t_2 - $

$t_1) = R_{XX}(\tau)$, and we have

$$E\{X'(t)\} = 0 \tag{3.58}$$

$$R_{XX'}(\tau) = \frac{dR_{XX}(\tau)}{d\tau} \tag{3.59}$$

$$R_{X'X'}(\tau) = -\frac{d^2 R_{XX}(\tau)}{d\tau^2} \tag{3.60}$$

3.7.3 Integration

The Riemann integral of an ordinary function is defined as the limit of a summing operation

$$\int_{t_0}^{t} x(\tau)\, d\tau = \lim_{n \to \infty} \sum_{i=0}^{n-1} x(\tau_i)\, \Delta t_i$$

where $t_0 < t_1 < t_2 < \cdots < t_n = t$ is an equally spaced partition of the interval, $[t_0, t]$, $\Delta t_i = t_{i+1} - t_i$, and τ_i is a point in the ith interval, $[t_i, t_{i+1}]$. For a random process $X(t)$, the MS integral is defined as the process $Y(t)$

$$Y(t) = \int_{t_0}^{t} X(\tau)\, d\tau = \lim_{n \to \infty} \sum_{i=0}^{n-1} X(\tau_i)\, \Delta t_i \tag{3.61}$$

It can be shown that a sufficient condition for the existence of the MS integral $Y(t)$ of a stationary finite variance process $X(t)$ is the existence of the integral

$$\int_{t_0}^{t} \int_{t_0}^{t} R_{XX}(t_1 - t_2)\, dt_1\, dt_2$$

Note that finite variance implies that $R_{XX}(0) < \infty$ and MS continuity implies continuity of $R_{XX}(\tau)$ at $\tau = 0$, which also implies continuity for all values of τ. These two conditions guarantee the existence of the preceding integral and, hence, the existence of the MS integral.

When the MS integral exists, we can show that

$$E\{Y(t)\} = (t - t_0)\mu_X \tag{3.62}$$

and

$$R_{YY}(t_1, t_2) = \int_{t_0}^{t_1} \int_{t_0}^{t_2} R_{XX}(\tau_1 - \tau_2)\, d\tau_1\, d\tau_2 \qquad (3.63)$$

EXAMPLE 3.16.

Discuss whether the random binary waveform is MS continuous, and whether the MS derivative and integral exist.

SOLUTION: For the random binary waveform $X(t)$, the autocorrelation function is

$$R_{XX}(\tau) = \begin{cases} 1 - \dfrac{|\tau|}{T}, & |\tau| < T \\ 0 & \text{elsewhere} \end{cases}$$

(a) Since $R_{XX}(\tau)$ is continuous at $\tau = 0$, $X(t)$ is MS continuous for all t.
(b) The derivative of $X(t)$ does not exist on a sample-function-by-sample-function basis and $R'_{XX}(0)$ and $R''_{XX}(0)$ do not exist. However, since their existence is only a sufficient condition for the existence of the MS derivative of $X(t)$, we cannot conclude whether or not $X(t)$ has a MS derivative.
(c) Finite variance plus MS continuity guarantees the existence of the MS integral over any finite interval $[t_0, t]$.

The MS integral of a random process is used to define the moving average of a random process $X(t)$ as

$$\langle X(t) \rangle_T = \frac{1}{T} \int_{t-T}^{t} X(\tau)\, d\tau$$

$\langle X(t) \rangle_T$ is also referred to as the time average of $X(t)$ and has many important applications. Properties of $\langle X(t) \rangle_T$ and its applications are discussed in the following section.

3.8 TIME AVERAGING AND ERGODICITY

When taking laboratory measurements, it is a common practice to obtain multiple measurements of a variable and "average" them to "reduce measurement

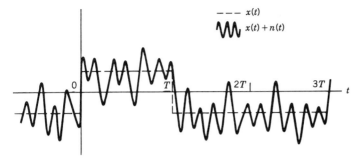

Figure 3.21 A member function of signal + noise.

errors." If the value of the variable being measured is constant, and errors are due to "noise" or due to the instability of the measuring instrument, then averaging is indeed a valid and useful technique. Time averaging is an extension of this concept and is used to reduce the variance associated with the estimation of the value of a random signal or the parameters of a random process.

As an example, let us consider the problem of estimating the amplitudes of the pulses in a random binary waveform that is corrupted by additive noise. That is, we observe $Y(t) = X(t) + N(t)$ where $X(t)$ is a random binary waveform, $N(t)$ is the independent noise, and we want to estimate the pulse amplitudes by processing $Y(t)$. A sample function of $Y(t)$ is shown in Figure 3.21. Suppose we observe a sample function $y(t)$ with $D = 0$ over the time interval $(0, T)$, or from $(k - 1)T$ to kT in general, and estimate the amplitude of $x(t)$ in the interval $(0, T)$. A simple way to estimate the amplitude of the pulse is to take one sample of $y(t)$ at some point in time, say $t_1 \in (0, T)$, and estimate the value of $x(t)$ as

$$\hat{x}(t) = \begin{cases} +1 & \text{for } 0 < t < T \text{ if } y(t_1) > 0; \quad t_1 \in (0, T) \\ -1 & \text{for } 0 < t < T \text{ if } y(t_1) \leq 0; \quad t_1 \in (0, T) \end{cases}$$

The ^ on $x(t)$ denotes that $\hat{x}(t)$ is an estimate of $x(t)$.

Because of noise, $y(t)$ has positive and negative values in the interval $(0, T)$ even though the pulse amplitude $x(t)$ is positive, and whether we estimate the pulse amplitude correctly will depend on the instantaneous value of the noise.

Instead of basing our decision on a single sample of $y(t)$, we can take m samples of $y(t)$ in the interval $(0, T)$, average the values, and decide

$$\hat{x}(t) = \begin{cases} +1 & \text{for } 0 < t < T \text{ if } \frac{1}{m}\sum_{i=1}^{m} y(t_i) > 0; \quad t_i \in (0, T) \\ -1 & \text{for } 0 < t < T \text{ if } \frac{1}{m}\sum_{i=1}^{m} y(t_i) \leq 0; \quad t_i \in (0, T) \end{cases}$$

168 RANDOM PROCESSES AND SEQUENCES

If the distribution of noise is assumed to be symmetrical about 0, then $y(t)$ is more likely to have positive values than negative values when $x(t) = 1$, and hence, the average value is more likely to be >0 than a single sample of $y(t)$. And we can conclude correctly that a decision based on averaging a large number of samples is more likely to be correct than a decision based on a single sample.

We can extend this concept one step further and use continuous time averaging to estimate the value of $x(t)$ as

$$\hat{x}(t) = \begin{cases} +1 & \text{if } \frac{1}{T}\int_0^T y(t)\,dt > 0 \\ -1 & \text{if } \frac{1}{T}\int_0^T y(t)\,dt \le 0 \end{cases}$$

The decision rule given above, which is based on time averaging, is extensively used in communication systems. The relationship between the duration of the integration and the variance of the estimator is a fundamental one in the design of communication and control systems. Derivation of this relationship is one of the topics covered in this section.

We have used ensemble averages such as the mean and autocorrelation function for characterizing random processes. To estimate ensemble averages one has to perform a weighted average over *all* the member functions of the random process. An alternate practical approach, which is often misused, involves estimation via time averaging over a *single* member function of the process. Laboratory instruments such as spectrum analyzers and integrating voltmeters routinely use time-averaging techniques. The relationship between integration time and estimation accuracy, and whether time averages will converge to ensemble averages (i.e., the concept of *ergodicity*) are important issues addressed in this section.

3.8.1 Time Averages

Definitions. The time average of a function of a random process is defined as

$$\langle g[X(t)]\rangle_T = \frac{1}{T}\int_{-T/2}^{T/2} g[X(t)]\,dt \quad \text{(continuous case)} \qquad (3.64.\text{a})$$

$$= \frac{1}{m}\sum_{i=1}^{m} g[X(i)] \quad \text{(discrete case)} \qquad (3.64.\text{b})$$

The corresponding ensemble average is given by

$$E\{g[X(t)]\} = \int_{-\infty}^{\infty} g(\alpha)f_X(\alpha)\,d\alpha \quad \text{(continuous case)} \qquad (3.65.\text{a})$$

or

$$= \sum_i g(x_i) P(X = x_i) \quad \text{(discrete case)} \quad (3.65.b)$$

Some time averages that are of interest include the following.

Time-averaged Mean.

$$\langle X(t) \rangle_T \triangleq \langle \mu_X \rangle_T \triangleq \frac{1}{T} \int_{-T/2}^{T/2} X(t)\, dt \quad (3.66)$$

Time-averaged Autocorrelation Function.

$$\langle X(t)X(t+\tau) \rangle_T \triangleq \langle R_{XX}(\tau) \rangle_T \triangleq \frac{1}{T} \int_{-T/2}^{T/2} X(t)X(t+\tau)\, dt \quad (3.67)$$

Time-averaged Power Spectral Density Function or Periodogram.

$$\frac{1}{T}|\langle X(t)\exp(-2\pi jft) \rangle_T|^2 \triangleq \langle S_{XX}(f) \rangle_T$$

$$\triangleq \frac{1}{T} \left| \int_{-T/2}^{T/2} X(t)\exp(-j2\pi ft)\, dt \right|^2 \quad (3.68)$$

Interpretation of Time Averages. Although the ensemble average has a unique numerical value, the time average of a function of a random process is, in general, a random variable. For any one sample function of the random process, time averaging produces a number. However, when all sample functions of a random process are considered, time averaging produces a random variable. For example, the time averaged mean of the random process shown in Figure 3.9 produces a discrete random variable.

$$\langle \mu_X \rangle_T = \frac{1}{T}\int_{-T/2}^{T/2} X(t)\, dt = \begin{cases} 5 & \text{when } X(t) = x_1(t) \\ 3 & \text{when } X(t) = x_2(t) \\ 1 & \text{when } X(t) = x_3(t) \\ -1 & \text{when } X(t) = x_4(t) \\ -3 & \text{when } X(t) = x_5(t) \\ -5 & \text{when } X(t) = x_6(t) \end{cases}$$

Notice that in this example, none of the values of $\langle \mu_X \rangle_T$ equals the true ensemble mean of $X(t)$, which is zero.

The determination of the probability distribution function of the random variable $\langle g[X(t)] \rangle_T$ is in general very complicated. For this reason, we will focus our attention only on the mean and variance of $\langle g[X(t)] \rangle_T$ and use them to analyze the asymptotic distribution of $\langle g[X(t)] \rangle_T$ as $T \to \infty$. In the following derivation, we will assume the process to be stationary so that the ensemble

170 RANDOM PROCESSES AND SEQUENCES

averages do not depend on time. Finite variance and MS continuity will also be assumed so that the existence of the time averages is guaranteed.

Mean and Variance of Time Averages. If we define a random variable Y as the average of m values of a real-valued stationary random process $X(t)$

$$Y = \frac{1}{m} \sum_{i=1}^{m} X(i\Delta) \qquad (3.69)$$

where Δ is the time between samples, then we can calculate $E\{Y\}$ and σ_Y^2 as

$$E\{Y\} = E\left\{\frac{1}{m} \sum_{i=1}^{m} X(i\Delta)\right\} = \frac{1}{m} \sum_{i=1}^{m} E\{X(i\Delta)\}$$
$$= \mu_X \qquad (3.70)$$

and

$$\sigma_Y^2 = E\{(Y - \mu_X)^2\}$$
$$= E\left\{\frac{1}{m^2} \sum_i \sum_j [X(i\Delta) - \mu_X][X(j\Delta) - \mu_X]\right\}$$
$$= \frac{1}{m^2} \sum_i \sum_j C_{XX}(|i - j|\Delta) \qquad (3.71)$$

If the samples of $X(t)$, taken Δ seconds apart, are uncorrelated, then

$$E\{Y\} = \mu_X \quad \text{and} \quad \sigma_Y^2 = \frac{\sigma_X^2}{m} \qquad (3.72)$$

which shows that averaging of m uncorrelated samples of a stationary random process leads to a reduction in the variance by a factor of m.

We can extend this development to continuous time averages as follows. To simplify the notation, let us define

$$Z(t) = g[X(t)]$$

and

$$Y = \frac{1}{T} \int_{-T/2}^{T/2} Z(t)\, dt$$

Then

$$E\{Y\} = E\left\{\frac{1}{T}\int_{-T/2}^{T/2} Z(t)\,dt\right\}$$

$$= \frac{1}{T}\int_{-T/2}^{T/2} E\{Z(t)\}\,dt$$

$$= \frac{1}{T}\int_{-T/2}^{T/2} \mu_Z\,dt = \mu_Z \qquad (3.73)$$

To calculate the variance, we need to find $E\{Y^2\}$. By writing Y^2 as a double integral and taking the expected value, we have

$$E\{Y^2\} = E\left\{\frac{1}{T}\int_{-T/2}^{T/2} Z(t_1)\,dt_1\,\frac{1}{T}\int_{-T/2}^{T/2} Z(t_2)\,dt_2\right\}$$

$$= \frac{1}{T^2}\iint_{-T/2}^{T/2} E\{Z(t_1)Z(t_2)\}\,dt_1\,dt_2$$

$$= \frac{1}{T^2}\iint_{-T/2}^{T/2} R_{ZZ}(t_1 - t_2)\,dt_1\,dt_2$$

and

$$\sigma_Y^2 = \frac{1}{T^2}\iint_{-T/2}^{T/2} C_{ZZ}(t_1 - t_2)\,dt_1\,dt_2 \qquad (3.74)$$

With reference to Figure 3.22, if we evaluate the integral over the shaded strip centered on the line $t_1 - t_2 = \tau$, the integrand $C_{ZZ}(t_1 - t_2)$ is constant and equal to $C_{ZZ}(\tau)$, and the area of the shaded strip is $[T - |\tau|]\,d\tau$. Hence, we can write the double integral in Equation 3.74 as

$$\frac{1}{T^2}\iint_{-T/2}^{T/2} C_{ZZ}(t_1 - t_2)\,dt_1\,dt_2 = \frac{1}{T^2}\int_{-T}^{T} [T - |\tau|]C_{ZZ}(\tau)\,d\tau$$

or

$$\sigma_Y^2 = \frac{1}{T}\int_{-T}^{T}\left[1 - \frac{|\tau|}{T}\right]C_{ZZ}(\tau)\,d\tau \qquad (3.75.a)$$

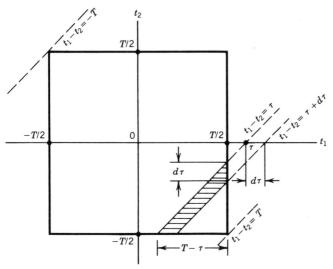

Figure 3.22 Evaluation of the double integral given in Equation 3.74.

It is left as an exercise for the reader to show that σ_Y^2 can be expressed as the following integral in the frequency domain:

$$\sigma_Y^2 = \int_{-\infty}^{\infty} S_{ZZ}^{\#}(f) \left(\frac{\sin \pi fT}{\pi fT} \right)^2 df \qquad (3.75.b)$$

where

$$S_{ZZ}^{\#}(f) = F\{C_{ZZ}(\tau)\} = \int_{-\infty}^{\infty} \exp(-j2\pi f\tau) C_{ZZ}(\tau) \, d\tau$$

The advantages of time averaging and the use of Equations 3.71, 3.75.a, and 3.75.b to compute the variances of time averages are illustrated in the following examples.

EXAMPLE 3.17.

$X(t)$ is a stationary, zero-mean, Gaussian random process whose power spectral density is shown in Figure 3.23. Let $Y = 1/10\{X(\Delta) + X(2\Delta) + \cdots + X(10\Delta)\}$, $\Delta = 1$ μs. Find the mean and variance of Y.

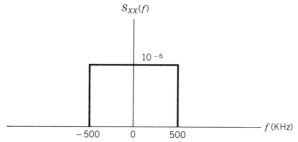

Figure 3.23 Psd of $X(t)$ for Example 3.17.

SOLUTION:

$$E\{Y\} = \frac{1}{10} E\{X(\Delta) + X(2\Delta) + \cdots + X(10\Delta)\}$$
$$= \frac{1}{10} [E\{X(\Delta)\} + E\{X(2\Delta)\} + \cdots + E\{X(10\Delta)\}]$$
$$= 0$$
$$E\{Y^2\} = E\left\{\left[\frac{1}{10} \sum_{i=1}^{10} X(i\Delta)\right]\left[\frac{1}{10} \sum_{j=1}^{10} X(j\Delta)\right]\right\}$$
$$= \frac{1}{100} \sum_i \sum_j E\{X(i\Delta)X(j\Delta)\}$$
$$= \frac{1}{100} \sum_i \sum_j R_{XX}(|i-j|\Delta)$$

Since $R_{XX}(k) = 0$ for $k \neq 0$ (why?), and $R_{XX}(0) = \sigma_X^2 = 1$, we obtain

$$E\{Y^2\} = \frac{1}{100} \sum_{i=1}^{10} R_{XX}(0) = \frac{1}{10}$$

or

$$\sigma_Y^2 = \frac{1}{10} = \frac{\sigma_X^2}{10}$$

EXAMPLE 3.18.

A lowpass, zero-mean, stationary Gaussian random process $X(t)$ has a power spectral density of

$$S_{XX}(f) = \begin{cases} A & \text{for } |f| < B \\ 0 & \text{for } |f| \geq B \end{cases}$$

Let

$$Y = \frac{1}{T} \int_{-T/2}^{T/2} X(t)\, dt$$

Assuming that $T \gg 1/B$, calculate σ_Y^2 and compare it with σ_X^2.

SOLUTION:

$$\sigma_X^2 = E\{X^2\} = R_{XX}(0) = \int_{-\infty}^{\infty} S_{XX}(f)\, df$$
$$= 2AB$$
$$E\{Y\} = \frac{1}{T} \int_{-T/2}^{T/2} E\{X(t)\}\, dt = 0$$
$$\sigma_Y^2 = E\{Y^2\} = \int_{-\infty}^{\infty} S_{XX}(f) \left(\frac{\sin \pi f T}{\pi f T}\right)^2 df$$

From Figure 3.24, we see that the bandwidth or the duration of $(\sin \pi f T/\pi f T)^2$ is very small compared to the bandwidth of $S_{XX}(f)$ and hence the integral of the product can be approximated as

$$\int_{-\infty}^{\infty} S_{XX}(f) \left(\frac{\sin \pi f T}{\pi f T}\right)^2 df \approx S_{XX}(0) \left[\text{area under } \left(\frac{\sin \pi f T}{\pi f T}\right)^2\right]$$

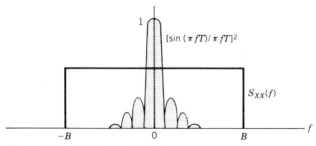

Figure 3.24 Variance calculations in the frequency domain.

or

$$\sigma_Y^2 = S_{XX}(0) \cdot \frac{1}{T} = \frac{A}{T}$$

and

$$\frac{\sigma_X^2}{\sigma_Y^2} \approx \frac{2AB}{(A/T)} = 2BT, \quad BT \gg 1$$

The result derived in this example is important and states that time averaging of a lowpass random process over a long interval results in a reduction in variance by a factor of $2BT$ (when $BT \gg 1$). Since this is equivalent to a reduction in variance that results from averaging $2BT$ uncorrelated samples of a random sequence, it is often stated that there are $2BT$ uncorrelated samples in a T second interval or there are $2B$ uncorrelated samples per second in a lowpass random process with a bandwith B.

EXAMPLE 3.19.

Consider the problem of estimating the pulse amplitudes in a random binary waveform $X(t)$, which is corrupted by additive Gaussian noise $N(t)$ with $\mu_N = 0$ and $R_{NN}(\tau) = \exp(-|\tau|/\alpha)$. Assume that the unknown amplitude of $X(t)$ in the interval $(0, T)$ is 1, $T = 1$ ms, $\alpha = 1$ μs, and compare the accuracy of the following two estimators for the unknown amplitude:

(a) $\hat{S}_1 = Y(t_1), \quad t_1 \in (0, T)$

(b) $\hat{S}_2 = \frac{1}{T} \int_0^T Y(t)\, dt$

where $Y(t) = X(t) + N(t)$.

SOLUTION:

$$\hat{S}_1 = X(t_1) + N(t_1) = 1 + N(t_1)$$
$$E\{\hat{S}_1\} = 1$$

and

$$\text{var}(\hat{S}_1) = R_{NN}(0) = 1$$

$$E\{\hat{S}_2\} = \frac{1}{T}\int_0^T E\{1 + N(t)\}\,dt = 1$$

and

$$\text{var}\{\hat{S}_2\} = \text{var}\left\{\frac{1}{T}\int_0^T N(t)\,dt\right\}$$

$$= \frac{1}{T}\int_{-T}^T \left[1 - \frac{|\tau|}{T}\right] C_{NN}(\tau)\,d\tau$$

$$= \frac{2}{T}\int_0^T \left[1 - \frac{\tau}{T}\right] \exp(-\tau/\alpha)\,d\tau$$

$$= \frac{2\alpha}{T} - \frac{2\alpha^2}{T^2}[1 - \exp(-T/\alpha)]$$

Since $\alpha/T \ll 1$, the second term in the preceding equation can be neglected and we have

$$\text{var}\{\hat{S}_2\} \approx \frac{2\alpha}{T} = \frac{1}{500}$$

and standard deviation of $\hat{S}_2 \approx 1/\sqrt{500} \approx 0.0447$.

Comparing the standard deviation (σ) of the estimators, we find that σ of $\hat{S}_1$ is 1, which is the same order of magnitude of the unknown signal amplitude being estimated. On the other hand, σ of $\hat{S}_2$ is 0.0447, which is quite small compared to the signal amplitude. Hence, the fluctuations in the estimated value due to noise will be very small for $\hat{S}_2$ and quite large for $\hat{S}_1$. Thus, we can expect $\hat{S}_2$ to be a much more accurate estimator.

3.8.2 Ergodicity

In the analysis and design of systems that process random signals, we often assume that we have prior knowledge of such quantities as the means, autocorrelation functions, and power spectral densities of the random processes involved. In many applications, such prior knowledge will not be available, and a central problem in the theory of random processes is the estimation of the parameters of random processes (see Chapter 9). If the theory of random processes is to be useful, then we have to be able to estimate such quantities as the mean and autocorrelation from data. From a practical point of view, it would be very attractive if we can do this estimation from an actual recording of one sample function of the random process.

Suppose we want to estimate the mean $\mu_X(t)$ of the random process $X(t)$. The mean is defined as an ensemble average, and if we observe the values of $X(t)$ over several member functions, then we can use their average as an ensemble estimate of $\mu_X(t)$. On the other hand, if we have access to only a single member function of $X(t)$, say $x(t)$, then we can form a time average $\langle x(t) \rangle_T$

$$\langle x(t) \rangle_T = \frac{1}{T} \int_{-T/2}^{T/2} x(t) \, dt$$

and attempt to use the time average as an estimate of the ensemble average, $\mu_X(t)$.

Whereas the time average $\langle x(t) \rangle_T$ is a constant for a particular member function, the set of values taken over all member functions is a random variable. That is, $\langle X(t) \rangle_T$ is a random variable and $\langle x(t) \rangle_T$ is a particular value of this random variable. Now, if $\mu_X(t)$ is a constant (i.e., independent of t), then the "quality" of the time-averaged estimator will depend on whether $E\{\langle X(t) \rangle_T\} \to \mu_X$ and the variance of $\{\langle X(t) \rangle_T\} \to 0$ as $T \to \infty$. If

$$\lim_{T \to \infty} E\{\langle X(t) \rangle_T\} = \mu_X$$

and

$$\lim_{T \to \infty} \text{var}\{\langle X(t) \rangle_T\} = 0$$

then we can conclude that the time-averaged mean converges to the ensemble mean and that they are equal. In general, ensemble averages and time averages are not equal except for a very special class of random processes called ergodic processes. The concept of ergodicity deals with the equality of time averages and ensemble averages.

The problem of determining the properties of a random process by time averaging over a single member function of finite duration belongs to statistics and is covered in detail in Chapter 9. In the following sections, we will derive the conditions for time averages to be equal to ensemble averages. We will focus our attention on the mean, autocorrelation, and power spectral density functions of stationary random processes.

178   RANDOM PROCESSES AND SEQUENCES

General Definition of Ergodicity. A stationary random process $X(t)$ is called ergodic if its ensemble averages equal (in a mean-square sense) appropriate time averages. This definition implies that, with probability one, any ensemble average of $X(t)$ can be determined from a single member function of $X(t)$. In most applications we are usually interested in only certain ensemble averages such as the mean and autocorrelation function, and we can define ergodicity with respect to these averages. In presenting these definitions, we will focus our attention on time averages over a finite interval $(-T/2, T/2)$ and the conditions under which the variances of the time averages tend to zero as $T \to \infty$.

It must be pointed out here that ergodicity is a stronger condition than stationarity and that not all processes that are stationary are ergodic. Furthermore, ergodicity is usually defined with respect to one or more specific ensemble averages, and a process may be ergodic with respect to some ensemble averages but not others.

Ergodicity of the Mean. A stationary random process $X(t)$ is said to be ergodic in the mean if

$$\operatorname*{l.i.m.}_{T \to \infty} \langle \mu_X \rangle_T = \mu_X$$

where l.i.m. stands for equality in the mean square sense, which requires

$$\lim_{T \to \infty} E\{\langle \mu_X \rangle_T\} = \mu_X$$

and

$$\lim_{T \to \infty} \operatorname{var}\{\langle \mu_X \rangle_T\} = 0$$

Now, the expected value of $\langle \mu_X \rangle_T$ for a finite value of T is given by

$$\begin{aligned} E\{\langle \mu_X \rangle_T\} &= \frac{1}{T} E\left\{ \int_{-T/2}^{T/2} X(t)\, dt \right\} \\ &= \frac{1}{T} \int_{-T/2}^{T/2} E\{X(t)\}\, dt = \frac{1}{T} \int_{-T/2}^{T/2} \mu_X\, dt \\ &= \mu_X \end{aligned} \qquad (3.76)$$

and the variance of $\langle \mu_X \rangle_T$ can be obtained from Equation 3.75.a as

$$\text{var}\{\langle \mu_X \rangle_T\} = \frac{1}{T}\int_{-T}^{T}\left[1 - \frac{|\tau|}{T}\right] C_{XX}(\tau)\, d\tau$$

If the variance given in the preceding equation approaches zero, then $X(t)$ is ergodic in the mean. Note that $E\{\langle \mu_X \rangle_T\}$ is always equal to μ_X for a stationary random process. Thus, a stationary process $X(t)$ is ergodic in the mean if

$$\lim_{T\to\infty} \frac{1}{T}\int_{-T}^{T}\left(1 - \frac{|\tau|}{T}\right) C_{XX}(\tau)\, d\tau = 0 \tag{3.77}$$

Although Equation 3.77 states the condition for ergodicity of the mean of $X(t)$, it does not have much use in applications involving testing for ergodicity of the mean. In order to use Equation 3.77 to justify time averaging, we need prior knowledge of $C_{XX}(\tau)$. However, Equation 3.77 might be of use in some situations if only partial knowledge of $C_{XX}(\tau)$ is available. For example, if we know that $|C_{XX}(\tau)|$ decreases exponentially for large values of $|\tau|$, then we can show that Equation 3.77 is satisfied and hence the process is ergodic in the mean.

Ergodicity of the Autocorrelation Function. A stationary random process $X(t)$ is said to be ergodic in the autocorrelation function if

$$\underset{T\to\infty}{\text{l.i.m.}}\ \langle R_{XX}(\alpha)\rangle_T = R_{XX}(\alpha) \tag{3.78}$$

The reader can show using Equations 3.73 and 3.75.a that

$$E\{\langle R_{XX}(\alpha)\rangle_T\} = R_{XX}(\alpha) \tag{3.79}$$

and

$$\text{var}\{\langle R_{XX}(\alpha)\rangle_T\} = \frac{1}{T}\int_{-T}^{T}\left(1 - \frac{|\tau|}{T}\right) C_{ZZ}(\tau)\, d\tau \tag{3.80}$$

where $Z(t) = X(t)X(t + \alpha)$.

As in the case of the time-averaged mean, the expected value of the time-averaged autocorrelation function is equal to $R_{XX}(\tau)$ irrespective of the length of averaging (T). If the right-hand side of Equation 3.80 approaches zero as $T \to \infty$, then the time-averaged autocorrelation function equals the true autocorrelation function. Hence, for any given α

$$\underset{T\to\infty}{\text{l.i.m.}} \langle R_{XX}(\alpha)\rangle_T = R_{XX}(\alpha)$$

if

$$\lim_{T\to\infty} \frac{1}{T} \int_{-T}^{T} \left(1 - \frac{|\tau|}{T}\right) [E\{Z(t)Z(t+\tau)\} - R_{XX}^2(\alpha)] \, d\tau = 0 \quad (3.81)$$

where $Z(t) = X(t)X(t+\alpha)$.

Note that to verify ergodicity of the autocorrelation function we need to have knowledge of the fourth-order moments of the process.

Ergodicity of the Power Spectral Density Function. The psd of a stationary random process plays a very important role in the frequency domain analysis and design of signal-processing systems, and the determination of the spectral characteristics of a random process from experimental data is a common engineering problem. The psd may be estimated by taking the Fourier transform of the time-averaged autocorrelation function.

A faster method of estimating the psd function involves the use of the time average

$$\langle S_{XX}(f)\rangle_T = \frac{1}{T} \left| \int_{-T/2}^{T/2} X(t)\exp(-j2\pi ft) \, dt \right|^2 \quad (3.82)$$

which is also called the *periodogram* of the process. Note that the integral represents the finite Fourier transform; the magnitude of the Fourier transform squared is the energy spectral density function (Parseval's theorem); and $1/T$ is the conversion factor for going from energy spectrum to power spectrum.

Unfortunately, the time average $\langle S_{XX}(f)\rangle_T$ does not converge to the ensemble average $S_{XX}(f)$ as $T \to \infty$. We will show in Chapter 9 that while

$$\lim_{T\to\infty} E\{\langle S_{XX}(f)\rangle_T\} = S_{XX}(f)$$

the variance of $\langle S_{XX}(f)\rangle_T$ does not go to zero as $T \to \infty$. Further averaging of the estimator $\langle S_{XX}(f)\rangle_T$ in the frequency domain is a technique that is commonly used to reduce the variance of $\langle S_{XX}(f)\rangle_T$. Although we will deal with the problem of estimating psd functions in some detail in Chapter 9, we want to point out here that estimation of psd is one important application in which a direct substitution of the time-averaged estimate $\langle S_{XX}(f)\rangle_T$ for the ensemble average $S_{XX}(f)$ is incorrect.

TIME AVERAGING AND ERGODICITY

EXAMPLE 3.20.

For the stationary random process shown in Figure 3.9, find $E\{\langle\mu_X\rangle_T\}$ and $\text{var}\{\langle\mu_X\rangle_T\}$. Is the process ergodic in the mean?

SOLUTION: $\langle\mu_X\rangle_T$ has six values: 5, 3, 1, −1, −3, −5, and

$$E\{\langle\mu_X\rangle_T\} = \frac{1}{6}\{5 + 3 + 1 - 1 - 3 - 5\} = 0$$

Variance of $\langle\mu_X\rangle_T$ can be obtained as

$$\text{var}\{\langle\mu_X\rangle_T\} = \frac{1}{6}\{5^2 + 3^2 + 1^2 + (-1)^2 + (-3)^2 + (-5)^2\}$$
$$= 70/6$$

Note that the variance of $\langle\mu_X\rangle_T$ does not depend on T and it does not decrease as we increase T. Thus, the condition stated in Equation 3.77 is not met and the process is not ergodic in the mean. This is to be expected since a single member function of this process has only one amplitude, and it does not contain any of the other five amplitudes that $X(t)$ can have.

EXAMPLE 3.21.

Consider the stationary random process

$$X(t) = 10\cos(100t + \Theta)$$

where Θ is a random variable with a uniform probability distribution in the interval $[-\pi, \pi]$. Show that $X(t)$ is ergodic in the autocorrelation function.

SOLUTION:

$$R_{XX}(\tau) = E\{100\cos(100t + \Theta)\cos(100t + 100\tau + \Theta)\}$$
$$= 50\cos(100\tau)$$
$$\langle R_{XX}(\tau)\rangle_T = \frac{1}{T}\int_{-T/2}^{T/2} X(t)X(t + \tau)\, dt$$

$$= \frac{1}{T} \int_{-T/2}^{T/2} 100 \cos(100t + \Theta) \cos(100t + 100\tau + \Theta) \, dt$$

$$= \frac{1}{T} \int_{-T/2}^{T/2} 50 \cos(100\tau) \, dt$$

$$+ \frac{1}{T} \int_{-T/2}^{T/2} 50 \cos(200t + 100\tau + 2\Theta) \, dt$$

Irrespective of which member function we choose to form the time-averaged correlation function (i.e., irrespective of the value of Θ), as $T \to \infty$, we have

$$\langle R_{XX}(\tau) \rangle_T = 50 \cos(100\tau)$$
$$= R_{XX}(\tau)$$

Hence, $E\{\langle R_{XX}(\tau) \rangle_T\} = R_{XX}(\tau)$ and $\text{var}\{\langle R_{XX}(\tau) \rangle_T\} = 0$. Thus, the process is ergodic in autocorrelation function.

Other Forms of Ergodicity. There are several other forms of ergodicity and some of the important ones include the following:

Wide Sense Ergodic Processes. A random process is said to be wide-sense ergodic (WSE) if it is ergodic in the mean and the autocorrelation function. WSE processes are also called *weakly ergodic.*

Distribution Ergodic Processes. A random process is said to be distribution ergodic if time-averaged estimates of distribution functions are equal to the appropriate (ensemble) distribution functions.

Jointly Ergodic Processes. Two random processes are jointly (wide-sense) ergodic if they are ergodic in their means and autocorrelation functions and also have a time-averaged cross-correlation function that equals the ensemble averaged cross-correlation functions.

Tests for Ergodicity. Conditions for ergodicity derived in the preceding sections are in general of limited use in practical applications since they require prior knowledge of parameters that are often not available. Except for certain simple cases, it is usually very difficult to establish whether a random process meets the conditions for the ergodicity of a particular parameter. In practice, we are usually forced to consider the physical origin of the random process to make an intuitive judgment of ergodicity.

For a process to be ergodic, each member function should "look" random, even though we view each member function to be an ordinary time signal. For example, if we consider the member functions of a random binary waveform, randomness is evident in *each member function* and it might be reasonable to

TIME AVERAGING AND ERGODICITY

expect the process to be at least weakly ergodic. On the other hand, each of the member functions of the random process shown in Figure 3.9 is a constant and by observing one member function we learn nothing about other member functions of the process. Hence, for this process, time averaging will tell us nothing about the ensemble averages. Thus, intuitive justification of ergodicity boils down to deciding whether a single member function is a "truly random signal" whose variations along the time axis can be assumed to represent typical variations over the ensemble.

The comments given in the previous paragraph may seem somewhat circular, and the reader may feel that the concept of ergodicity is on shaky ground. However, we would like to point out that in many practical situations we are forced to use models that are often hard to justify under rigorous examination.

Fortunately, for Gaussian random processes, which are extensively used in a variety of applications, the test for ergodicity is very simple and is given below.

EXAMPLE 3.22.

Show that a stationary, zero-mean, finite variance Gaussian random process is ergodic in the general sense if

$$\int_{-\infty}^{\infty} |R_{XX}(\tau)| \, d\tau < \infty$$

SOLUTION: Since a stationary Gaussian random process is completely specified by its mean and autocorrelation function, we need to be concerned only with the mean and autocorrelation function (i.e., weakly ergodic implies ergodicity in the general sense for a stationary Gaussian random process). For the process to be ergodic in mean, we need to show that

$$\lim_{T \to \infty} \frac{1}{T} \int_{-T}^{T} \left(1 - \frac{|\tau|}{T}\right) C_{XX}(\tau) \, d\tau = 0$$

The preceding integral can be written as

$$0 \leq \frac{1}{T} \int_{-T}^{T} \left(1 - \frac{|\tau|}{T}\right) C_{XX}(\tau) \, d\tau \leq \frac{1}{T} \int_{-T}^{T} |C_{XX}(\tau)| \, d\tau$$

Hence,

$$\lim_{T \to \infty} \frac{1}{T} \int_{-T}^{T} \left(1 - \frac{|\tau|}{T}\right) C_{XX}(\tau) \, d\tau = 0$$

since

$$\int_{-\infty}^{\infty} |R_{XX}(\tau)|\, d\tau < \infty$$

To prove ergodicity of the autocorrelation function, we need to show that, for every α, the integral

$$V = \frac{1}{T}\int_{-T}^{T}\left(1 - \frac{|\tau|}{T}\right) C_{ZZ}(\tau)\, d\tau; \qquad Z(t) = X(t)X(t+\alpha)$$

approaches zero as $T \to \infty$.
The integral V can be bounded as

$$0 \le V \le \frac{1}{T}\int_{-T}^{T} |C_{ZZ}(\tau)|\, d\tau$$

where

$$C_{ZZ}(\tau) = E\{X(t)X(t+\alpha)X(t+\tau)X(t+\alpha+\tau)\} - R_{XX}^2(\alpha)$$

Now, making use of the following relationship for a four-dimensional Gaussian distribution (Equation 2.69)

$$E\{X_1 X_2 X_3 X_4\} = E\{X_1 X_2\}E\{X_3 X_4\} + E\{X_1 X_3\}E\{X_2 X_4\} \\ + E\{X_1 X_4\}E\{X_2 X_3\}$$

we have

$$C_{ZZ}(\tau) = R_{XX}^2(\alpha) + R_{XX}^2(\tau) + R_{XX}(\tau+\alpha)R_{XX}(\tau-\alpha) - R_{XX}^2(\alpha)$$

and

$$0 \le V \le \frac{1}{T}\int_{-T}^{T} |R_{XX}^2(\tau)|\, d\tau + \frac{1}{T}\int_{-T}^{T} |R_{XX}(\tau+\alpha)R_{XX}(\tau-\alpha)|\, d\tau$$
$$\le R_{XX}(0)\left\{\frac{1}{T}\int_{-T}^{T} |R_{XX}(\tau)|\, d\tau + \frac{1}{T}\int_{-T}^{T} |R_{XX}(\tau+\alpha)|\, d\tau\right\}$$

Since $\int_{-\infty}^{\infty} |R_{XX}(\tau)| \, d\tau < \infty$, the upper bound approaches 0 as $T \to \infty$, and hence the variance (V) of the time-averaged autocorrelation function $\to 0$ as $\tau \to \infty$. Thus, if the autocorrelation function is absolutely integrable, then the stationary Gaussian process is ergodic. Note that this is a sufficient (but not a necessary) condition for ergodicity. Also note that $\int_{-\infty}^{\infty} |R_{XX}(\tau)| \, d\tau < \infty$ requires that $\mu_X = 0$.

3.9 SPECTRAL DECOMPOSITION AND SERIES EXPANSION OF RANDOM PROCESSES

We have seen that a stationary random process can be described in the frequency domain by its power spectral density function which is defined as the Fourier transform of the autocorrelation function of the process. In the case of deterministic signals, the expansion of a signal as a superposition of complex exponentials plays an important role in the study of linear systems. In the following discussion, we will examine the possibility of expressing a random process $X(t)$ by a sum of exponentials or other orthogonal functions. Before we start our discussion, we would like to point out that each member function of a stationary random process has infinite energy and hence its ordinary Fourier transform does not exist.

We present three forms for expressing random processes in a series form, starting with the simple Fourier series expansion.

3.9.1 Ordinary Fourier Series Expansion

A stationary random process that is MS periodic and MS continuous can be expanded in a Fourier series of the form

$$\tilde{X}(t) = \sum_{n=-N}^{N} C_X(nf_0) \exp(j2\pi n f_0 t); \qquad (3.83.a)$$

where

$$C_X(nf_0) = \frac{1}{T} \int_{-T/2}^{T/2} X(t) \exp(-j2\pi n f_0 t) \, dt \qquad (3.83.b)$$

186 RANDOM PROCESSES AND SEQUENCES

and T is the period of the process and $f_0 = 1/T$. $\tilde{X}(t)$ converges to $X(t)$ in a MS sense, that is,

$$\lim_{N \to \infty} E\{|X(t) - \tilde{X}(t)|^2\} = 0$$

for all values of $t \in (-\infty, \infty)$.

Note that the coefficients $C_X(nf_0)$ of the Fourier series are complex-valued random variables. For each member function of the random process these random variables have a particular set of values.

The reader can easily verify the following:

1. $E\{C_X(nf_0)C_X^*(mf_0)\} = 0, \quad n \neq m;$

   that is, the coefficients are orthogonal

2. $R_{XX}(\tau) = \sum_{n=-\infty}^{\infty} \alpha_n \exp(j2\pi nf_0\tau),$ where $\alpha_n = E\{|C_X(nf_0)|^2\}$

3. $E\{|X(t)|^2\} = \sum_{n=-\infty}^{\infty} E\{|C_X(nf_0)|^2\}$ (Parseval's theorem)

The rate of convergence or how many terms should be included in the series expansion in order to provide an "accurate" representation of $X(t)$ can be determined as follows: The MS difference between $X(t)$ and the series $\tilde{X}(t)$ is given by

$$E\{|X(t) - \tilde{X}(t)|^2\} = E\left\{\left|X(t) - \sum_{n=-N}^{N} C_X(nf_0)\exp(j2\pi nf_0 t)\right|^2\right\}$$

$$= E\{|X(t)|^2\} - \sum_{n=-N}^{N} E\{|C_X(nf_0)|^2\}$$

and the normalized MS error, which is defined as

$$\epsilon_N^2 = \frac{E\{|X(t) - \tilde{X}(t)|^2\}}{E\{|X(t)|^2\}} \tag{3.84}$$

can be used to measure the rate of convergence and the accuracy of the series representation as a function of N. As a rule of thumb, ϵ_N^2 is chosen to have a value less than 0.05, which implies that the series representation accounts for 95% of the normalized MS variation of $X(t)$.

3.9.2 Modified Fourier Series for Aperiodic Random Signals

A stationary MS continuous *aperiodic* random process $X(t)$ can be expanded in a series form as

$$\tilde{X}(t) = \sum_{n=-N}^{N} C_X(nf_0)\exp(j2\pi nf_0 t), \qquad |t| \ll \frac{1}{f_0} \qquad (3.85)$$

where

$$C_X(nf_0) = \int_{-\infty}^{\infty} X(t) \frac{\sin(\pi f_0 t)}{\pi t} \exp(-j2\pi nf_0 t)\, dt \qquad (3.86)$$

The constants N and f_0 are chosen to yield an acceptable level of normalized MS error defined in Equation 3.84. As $N \to \infty$ and $f_0 \to 0$, $\tilde{X}(t)$ converges in MS sense to $X(t)$ for all values of $|t| \ll 1/f_0$. It can be shown that this series representation has the following properties:

1. $E\{C_X(nf_0)C_X^*(mf_0)\} = 0, \qquad m \neq n$
2. $E\{|\tilde{X}(t)|^2\} = E\{|X(t)|^2\} \qquad \text{as } n \to \infty$
3. $E\{|C_X(nf_0)|^2\} = \int_{(n-1/2)f_0}^{(n+1/2)f_0} S_{XX}(f)\, df$
4. $S_{\tilde{X}\tilde{X}}(f) = \sum_{n=-N}^{N} E\{|C_X(nf_0)|^2\}\delta(f - nf_0)$

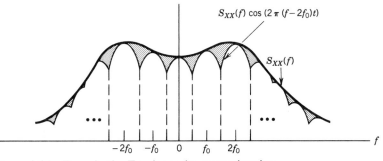

Figure 3.25 Error in the Fourier series approximation.

188 RANDOM PROCESSES AND SEQUENCES

5. $\lim\limits_{N \to \infty} E\{|X(t) - \tilde{X}(t)|^2\}$

$$= 2 \sum_{n=-\infty}^{\infty} \int_{(n-1/2)f_0}^{(n+1/2)f_0} S_{XX}(f)[1 - \cos 2\pi t(f - nf_0)] \, df$$

$$\leq 4 \, E\{X^2(t)\} \sin^2\left(\frac{\pi f_0 t}{2}\right) \qquad \text{for } |t| < \frac{1}{2f_0}$$

6. For any finite value of N, the MS error is the shaded area shown in Figure 3.25.

The proofs of some of these statements are rather lengthy and the reader is referred to Section 13-2 of the first edition of [9] for details.

3.9.3 Karhunen–Loeve (K-L) Series Expansion

The normalized mean squared error $E\{|X(t) - \tilde{X}(t)|^2\}$ between $X(t)$ and its series representation $\tilde{X}(t)$ depends on the number of terms in the series and the (basis) functions used in the series expansion. A series expansion is said to be optimum in a MS sense if it yields the smallest MS error for a given number of terms. The K-L expansion is optimum in a MS sense for expanding a stationary random process $X(t)$ over any finite time interval $[-T/2, T/2]$.

The orthonormal basis function, $\phi_i(t)$, used in the K-L expansion are obtained from the solutions of the integral equation

$$\int_{-T/2}^{T/2} R_{XX}(t - \tau)\phi(\tau) \, d\tau = \lambda \phi(t), \qquad |t| < \frac{T}{2} \qquad (3.87)$$

The solution yields a set of eigenvalues $\lambda_1 > \lambda_2 > \lambda_3, \ldots$, and eigenfunctions $\phi_1(t), \phi_2(t), \phi_3(t), \ldots$, and the K-L expansion is written in terms of the eigenfunctions as

$$\tilde{X}(t) = \sum_{n=1}^{N} A_n \phi_n(t), \qquad |t| < \frac{T}{2} \qquad (3.88)$$

where

$$A_n = \int_{-T/2}^{T/2} X(t) \phi_n^*(t) \, dt, \qquad n = 1, 2, \cdots N \qquad (3.89)$$

The K-L series expansion has the following properties:

1. $\text{l.i.m. } \tilde{X}(t) = X(t)$

2. $\int_{-T/2}^{T/2} \phi_n(t)\phi_m^*(t)\, dt = \begin{cases} 1, & m = n \\ 0 & m \neq n \end{cases}$

3. $E\{A_n A_m^*\} = \begin{cases} \lambda_n, & m = n \\ 0 & m \neq n \end{cases}$

4. $E\{X^2(t)\} = R_{XX}(0) = \sum_{n=1}^{\infty} \lambda_n$

5. Normalized MSE $= \dfrac{\sum_{n=N+1}^{\infty} \lambda_n}{\sum_{n=1}^{\infty} \lambda_n}$

The main difficulty in the use of Karhunen–Loeve expansion lies in finding the eigenfunctions of the random process. While much progress has been made in developing computational algorithms for solving integral equations of the type given in Equation 3.87, the computational burden is still a limiting factor in the application of the K-L series expansion.

3.10 SAMPLING AND QUANTIZATION OF RANDOM SIGNALS

Information-bearing random signals such as the output of a microphone, a TV camera, or a pressure or temperature sensor are predominantly analog (continuous-time, continuous-amplitude) in nature. These signals are often transmitted over digital transmission facilities and are also processed digitally. To make these analog signals suitable for digital transmission and processing, we make use of two operations: sampling and quantization. The sampling operation is used to convert a continuous-time signal to a discrete-time sequence. The quantizing operation converts a continuous-amplitude signal to a discrete-amplitude signal.

In this section, we will discuss techniques for sampling and quantizing a continuous-amplitude, continuous-time signal $X(t)$. We will first show that, given the values of $X(t)$ at $t = kT_s$, $k = \cdots -3, -2, -1, 0, 1, 2, 3, \cdots$, we can reconstruct the signal $X(t)$ for all values of t if $X(t)$ is a stationary random process with a bandwidth of B and T_s is chosen to be smaller than $1/2B$. Then we will develop procedures for representing the analog amplitude of $X(kT_s)$ by a finite set of precomputed values. This operation amounts to approximating a continuous random variable X by a discrete random variable X_q, which can take on one of Q possible values such that $E\{(X - X_q)^2\} \to 0$ as $Q \to \infty$.

190 RANDOM PROCESSES AND SEQUENCES

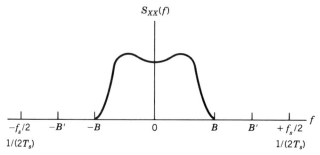

Figure 3.26 Power spectral density of the signal being sampled.

3.10.1 Sampling of Lowpass Random Signals

Let $X(t)$ be a real-valued stationary random process with a continuous power spectral density function, $S_{XX}(f)$, that is zero for $|f| > B$ (Figure 3.26). Since $S_{XX}(f)$ is a real function of f, we can use an ordinary Fourier series to represent $S_{XX}(f)$ as

$$S_{XX}(f) = \sum_{n=-\infty}^{\infty} C_X(nT_s)\exp(j2\pi n f T_s), \quad |f| < B_0 \qquad (3.90)$$

where

$$B_0 = \frac{1}{2T_s}, \quad B_0 > B$$

and

$$C_X(nT_s) = \frac{1}{2B_0} \int_{-B_0}^{B_0} S_{XX}(f)\exp(-j2\pi n f T_s)\, df \qquad (3.91)$$

Taking the inverse Fourier transform of $S_{XX}(f)$ as given in Equation 3.90, we have

$$R_{XX}(\tau) = F^{-1}\left\{ \sum_{n=-\infty}^{\infty} C_X(nT_s)\exp(j2\pi n f T_s) \right\}$$

$$= \int_{-B'}^{B'} \sum_{n=-\infty}^{\infty} C_X(nT_s)\exp(j2\pi n f_1 T_s)\exp(j2\pi f_1 \tau)\, df_1, \quad B \leq B' \leq B_0$$

$$= \sum_{n=-\infty}^{\infty} \int_{-B'}^{B'} \frac{1}{2B_0} \int_{-B_0}^{B_0} S_{XX}(f_2)\exp(-j2\pi n f_2 T_s)\, df_2$$
$$\times \exp[j2\pi f_1(\tau + nT_s)]\, df_1$$
$$= \sum_{n=-\infty}^{\infty} R_{XX}(-nT_s) \frac{1}{2B_0} \left[\frac{\sin 2\pi B'(\tau + nT_s)}{\pi(\tau + nT_s)} \right]$$

If we choose the limits of integration B' to be equal to B, we have

$$R_{XX}(\tau) = 2BT_s \sum_{n=-\infty}^{\infty} R_{XX}(nT_s) \frac{\sin 2\pi B(\tau - nT_s)}{2\pi B(\tau - nT_s)} \qquad (3.92)$$

It is convenient to state two other versions of Equation 3.92 for use in deriving the sampling theorem for lowpass random signals. With a an arbitrary constant, the transform of $R_{XX}(\tau - a)$ is equal to $S_{XX}(f)\exp(-j2\pi fa)$. This function is also lowpass, and hence Equation 3.92 can be applied to $R_{XX}(\tau - a)$ as

$$R_{XX}(\tau - a) = 2BT_s \sum_{n=-\infty}^{\infty} R_{XX}(nT_s - a)\operatorname{sinc} 2B(\tau - nT_s) \qquad (3.93)$$

where

$$\operatorname{sinc} x = \frac{\sin \pi x}{\pi x}$$

Changing $(\tau - a)$ to τ in Equation 3.93, we have

$$R_{XX}(\tau) = 2BT_s \sum_{n=-\infty}^{\infty} R_{XX}(nT_s - a)\operatorname{sinc} 2B(\tau + a - nT_s) \qquad (3.94)$$

We now state and prove the sampling theorem for band-limited random processes.

The Uniform Sampling Theorem for Band-limited Random Signals. If a real random process $X(t)$ is band-limited to B Hz, then $X(t)$ can be represented using the instantaneous values $X(kT_s)$ as

$$\tilde{X}_N(t) = 2BT_s \sum_{n=-N}^{N} X(nT_s)\operatorname{sinc}[2B(t - nT_s)], \qquad T_s < 1/(2B) \qquad (3.95)$$

192 RANDOM PROCESSES AND SEQUENCES

and $\tilde{X}_N(t)$ converges to $X(t)$ in a MS sense. That is $E\{[X(t) - \tilde{X}_N(t)]^2\} = 0$, as $N \to \infty$.

To prove MS convergence of $\tilde{X}_N(t)$ to $X(t)$, we need to show that

$$E\{[X(t) - \tilde{X}_N(t)]^2\} = 0 \quad \text{as} \quad N \to \infty \tag{3.96}$$

Let $N \to \infty$, then

$$\tilde{X}(t) = 2BT_s \sum_{n=-\infty}^{\infty} X(nT_s)\operatorname{sinc}[2B(t - nT_s)], \quad T_s < \frac{1}{2B}$$

Now

$$E\{[X(t) - \tilde{X}(t)]^2\} = E\{[X(t) - \tilde{X}(t)]X(t)\} \\ - E\{[X(t) - \tilde{X}(t)]\tilde{X}(t)\} \tag{3.97}$$

The first term on the right-hand side of the previous equation may be written as

$$E\{[X(t) - \tilde{X}(t)]X(t)\} \\ = R_{XX}(0) - 2BT_s \sum_{n=-\infty}^{\infty} R_{XX}(nT_s - t)\operatorname{sinc}[2B(t - nT_s)]$$

From Equation 3.94 with $\tau = 0$ and $a = t$, we have

$$2BT_s \sum_{n=-\infty}^{\infty} R_{XX}(nT_s - t)\operatorname{sinc}[2B(t - nT_s)] = R_{XX}(0)$$

and hence

$$E\{[X(t) - \tilde{X}(t)]X(t)\} = 0 \tag{3.98}$$

The second term in Equation 3.97 can be written as

$$E\{[X(t) - \tilde{X}(t)]\tilde{X}(t)\} \\ = \sum_{m=-\infty}^{\infty} E\{[X(t) - \tilde{X}(t)]X(mT_s)\}2BT_s \operatorname{sinc}[2B(t - mT_s)]$$

Now

$$E\{[X(t) - \tilde{X}(t)]\tilde{X}(mT_s)\}$$
$$= R_{XX}(t - mT_s) - \sum_{n=-\infty}^{\infty} 2BT_s R_{XX}(nT_s - mT_s)\text{sinc}[2B(t - nT_s)]$$

and from Equation 3.93 with $\tau = t$ and $a = mT_s$, we have

$$R_{XX}(t - mT_s) = 2BT_s \sum_{n=-\infty}^{\infty} R_{XX}(nT_s - mT_s)\text{sinc}[2B(t - nT_s)]$$

Hence

$$E\{[X(t) - \tilde{X}(t)]\tilde{X}(t)\} = 0 \qquad (3.99)$$

Substitution of Equations 3.98 and 3.99 in Equation 3.97 completes the proof of the uniform sampling theorem.

The sampling theorem permits us to store, transmit, and process the sequence $X(nT_s)$ rather than the continuous time signal $X(t)$, as long as the samples are taken at intervals less than $1/(2B)$. The minimum sampling rate is $2B$ and is called the *Nyquist rate*. If $X(t)$ is sampled at rates lower than $2B$ samples/second, then we cannot reconstruct $X(t)$ from $X(nT_s)$ due to "aliasing," which is explained next.

Aliasing Effect. To examine the aliasing effect, let us define the sampling operation as

$$X_s(t) = X(t) \cdot S(t)$$

where $X_s(t)$ is the sampled version of a band-limited process $X(t)$ and $S(t)$ is the sampling waveform. Assume that the sampling waveform $S(t)$ is an impulse sequence (see Figure 3.27) of the form

$$S(t) = \sum_{k=-\infty}^{\infty} \delta(t - kT_s - D)$$

where D is a random variable with a uniform distribution in the interval $[0, T_s]$, and D is independent of $X(t)$. The product $X_s(t) = X(t) \cdot S(t)$ as shown in

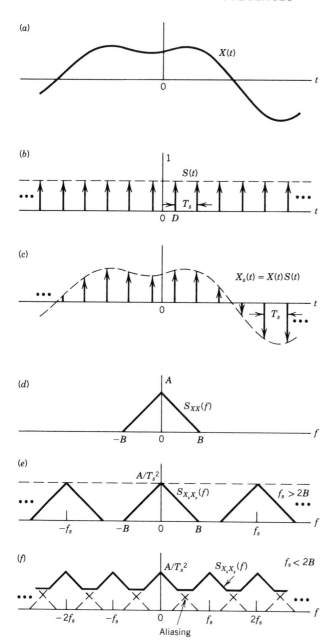

Figure 3.27 Sampling operation.

Figure 3.27c can be written as

$$X_s(t) = \sum_{k=-\infty}^{\infty} X(t - kT_s - D)\delta(t - kT_s - D)$$

Following the derivation in Section 3.6.5, the reader can show that the autocorrelation function of $X_s(t)$ is given by

$$R_{X_sX_s}(\tau) = \sum_{k=-\infty}^{\infty} \frac{1}{T_s} R_{XX}(kT_s)\delta(\tau - kT_s)$$

$$= \frac{1}{T_s} R_{XX}(\tau) \sum_{k=-\infty}^{\infty} \delta(\tau - kT_s)$$

The last step results from one of the properties of delta functions. Taking the Fourier transform of $R_{X_sX_s}(\tau)$ we obtain

$$S_{X_sX_s}(f) = \frac{1}{T_s} S_{XX}(f) * F\left\{\sum_{k=-\infty}^{\infty} \delta(\tau - kT_s)\right\}$$

The reader can show that

$$F\left\{\sum_{k=-\infty}^{\infty} \delta(t - kT_s)\right\} = \frac{1}{T_s} \sum_{k=-\infty}^{\infty} \delta(f - kf_s)$$

where $f_s = 1/T_s$ is the sampling rate, and hence

$$S_{X_sX_s}(f) = \frac{1}{T_s^2}\left\{\sum_{k=-\infty}^{\infty} S_{XX}(f - kf_s)\right\}$$

$$= \frac{1}{T_s^2}\{S_{XX}(f) + S_{XX}(f - f_s) + S_{XX}(f + f_s)$$

$$+ S_{XX}(f - 2f_s) + S_{XX}(f + 2f_s) + \cdots\} \qquad (3.100)$$

The preceding equation shows that the psd of the sampled version $X_s(t)$ of $X(t)$ consists of replicates of the original spectrum $S_{XX}(f)$ with a replication rate of f_s. For a band-limited process $X(t)$, the psd of $X(t)$ and $X_s(t)$ are shown in Figure 3.27 for two sampling rates $f_s > 2B$ and $f_s < 2B$.

When $f_s > 2B$ or $T_s < 1/(2B)$, $S_{X_sX_s}(f)$ contains the original spectrum of $X(t)$ intact and recovery of $X(t)$ from $X_s(t)$ is possible. But when $f_s < 2B$, replicates of $S_{XX}(f)$ overlap and the psd of $X_s(t)$ does not bear much resemblance to the psd of $X(t)$. This is called the *aliasing effect,* and it often prevents us from reconstructing $X(t)$ from $X_s(t)$ with the required accuracy.

When $f_s > (2B)$, we have shown that $X(t)$ can be reconstructed in the time domain from samples of $X(t)$ according to Equation 3.95. Examination of Figure 3.27e shows that if we select only that portion of $S_{X_sX_s}(f)$ that lies in the interval $[-B, B]$, we can recover the psd of $X(t)$. This selection can be accomplished in the frequency domain by an operation known as "lowpass filtering," which will be discussed in Chapter 4. Indeed, Equation 3.95 is the time domain equivalent of lowpass filtering in the frequency domain.

3.10.2 Quantization

The instantaneous value of a continuous amplitude (analog) random process $X(t)$ is a continuous random variable. If the instantaneous values are to be processed digitally, then the continuous random variable X, which can have an uncountably infinite number of possible values, has to be represented by a discrete random variable with a finite number of values. For example, if the instantaneous value is sampled by a 4-bit analog-to-digital converter, then X is approximated at the output by a discrete random variable with one of 2^4 possible values. We now develop procedures for quantizing or approximating a continuous random variable X by a discrete random variable X_q. The device that performs this operation is referred to as a quantizer or analog-to-digital converter.

An example of the quantizing operation is shown in Figure 3.28. The input to the quantizer is a random process $X(t)$, and we will assume that the random signal $X(t)$ is sampled at an appropriate rate and the sample values $X(kT_s)$ are converted to one of Q allowable levels, $m_1, m_2, \ldots, m_Q$, according to some predetermined rule:

$$X_q(kT_s) = m_i \quad \text{if} \quad x_{i-1} < X(kT_s) \le x_i$$
$$x_0 = -\infty, \quad x_Q = +\infty \tag{3.101}$$

The output of the quantizer is a sequence of levels, shown in Figure 3.28 as a waveform $X_q(t)$, where

$$X_q(t) = X_q(kT_s), \quad kT_s \le t < (k+1)T_s$$

We see from Figure 3.28 that the quantized signal is an approximation to the original signal. The quality of the approximation can be improved by increasing the number of quantizing levels Q and for fixed Q by a careful choice of x_i's and m_i's such that some measure of performance is optimized. The measure of performance that is most commonly used for evaluating the performance of a quantizing scheme is the normalized MS error

$$\epsilon_Q^2 = \frac{E\{[X(kT_s) - X_q(kT_s)]^2\}}{E\{X_q^2(kT_s)\}}$$

We will now consider several methods of quantizing the sampled values of a random process $X(t)$. For convenience, we will assume $X(t)$ to be a zero-

SAMPLING AND QUANTIZATION OF RANDOM SIGNALS 197

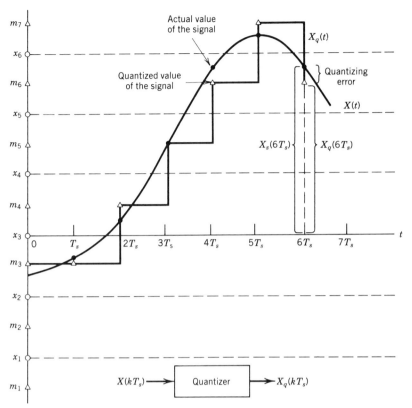

Figure 3.28 Quantizing operation; $m_1, m_2, \ldots, m_7$ are the seven output levels of the quantizer.

mean, stationary random process with a pdf $f_X(x)$. We will use the abbreviated notation, X to denote $X(kT_s)$ and X_q to denote $X_q(kT_s)$. The problem of quantizing consists of approximating the continuous random variable X by a discrete random variable X_q such that $E\{(X - X_q)^2\}$ is minimized.

3.10.3 Uniform Quantizing

In this method of quantizing, the range of the continuous random variable X is divided into Q intervals of equal length, say Δ. If the value of X falls in the ith quantizing interval, then the quantized value of X is taken to be the midpoint of the interval (see Figure 3.29). If a and b are the minimum and maximum values of X, respectively, then the step size or interval length Δ is given by

$$\Delta = \frac{(b - a)}{Q} \tag{3.102.a}$$

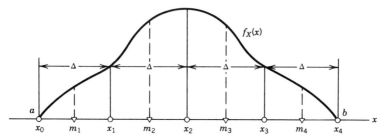

Figure 3.29 Example of uniform quantizing. Step size = Δ, $Q = 4$.

The quantized output X_q is generated according to

$$X_q = m_i \quad \text{if} \quad x_{i-1} < X \le x_i, \quad i = 1, 2, \ldots, Q \quad (3.102.\text{b})$$

where

$$x_i = a + i\Delta \quad (3.102.\text{c})$$

and

$$m_i = \frac{x_{i-1} + x_i}{2} \quad (3.102.\text{d})$$

The quantizing "noise power" N_Q for the uniform quantizer is given by

$$\begin{aligned} N_Q &= E\{(X - X_q)^2\} \\ &= \int_a^b (x - x_q)^2 f_X(x) \, dx \\ &= \sum_{i=1}^{Q} \int_{x_{i-1}}^{x_i} (x - m_i)^2 f_X(x) \, dx \end{aligned} \quad (3.103.\text{a})$$

where $x_i = a + i\Delta$ and $m_i = a + i\Delta - \Delta/2$. The "signal power" S_Q at the output of the quantizer can be obtained from

$$\begin{aligned} S_Q &= E\{(X_q)^2\} \\ &= \sum_{i=1}^{Q} (m_i)^2 \int_{x_{i-1}}^{x_i} f_X(x) \, dx \end{aligned} \quad (3.103.\text{b})$$

SAMPLING AND QUANTIZATION OF RANDOM SIGNALS

The ratio N_Q/S_Q is ϵ_Q^2 and it gives us a measure of the MS error of the uniform quantizer. This ratio can be computed if the pdf of X is known.

EXAMPLE 3.23.

The input to a Q-step uniform quantizer has a uniform pdf over the interval $[-a, a]$. Calculate the normalized MS error as a function of the number of quantizer levels.

SOLUTION: From Equation 3.103.a we have

$$
\begin{aligned}
N_Q &= \sum_{i=1}^{Q} \int_{x_{i-1}}^{x_i} (x - m_i)^2 \left(\frac{1}{2a}\right) dx \\
&= \sum_{i=1}^{Q} \int_{-a+(i-1)\Delta}^{-a+i\Delta} \left(x + a - i\Delta + \frac{\Delta}{2}\right)^2 \frac{1}{2a} dx \\
&= \sum_{i=1}^{Q} \left(\frac{1}{2a}\right)\left(\frac{\Delta^3}{12}\right) \\
&= \frac{Q\Delta^3}{(2a)12} = \frac{\Delta^2}{12} \qquad \text{since } Q\Delta = 2a
\end{aligned}
$$

Now, the output signal power S_Q can be obtained using Equation 3.103.b as

$$
\begin{aligned}
S_Q &= \sum_{i=1}^{Q} (m_i)^2 \left(\frac{\Delta}{2a}\right) \\
&= \frac{Q^2 - 1}{12} (\Delta)^2
\end{aligned}
$$

and hence the normalized MS error is given by

$$
\frac{N_Q}{S_Q} = \frac{1}{Q^2 - 1} \approx \frac{1}{Q^2} \qquad \text{when } Q \gg 1 \qquad (3.104)
$$

Equation 3.104 can be used to determine the number of quantizer levels needed for a given application. In quantizing audio and video signals the ratio N_Q/S_Q is kept lower than 10^{-4}, which requires that Q be greater than 100. It is a common practice to use 7-bit A/D converters (128 levels) to quantize voice and video signals.

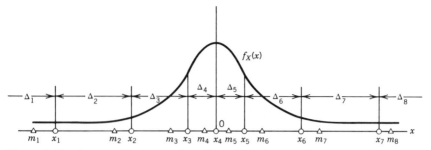

Figure 3.30 A nonuniform quantizer for a Gaussian variable. $X_0 = -\infty$, $X_Q = \infty$, $Q = 8$, and $\Delta_i = \Delta_{Q+1-i}$, $(i = 1, 2, 3, 4)$.

3.10.4 Nonuniform Quantizing

The uniform quantizer is optimum (yields the lowest N_Q/S_Q for a given value of Q) if the random process $X(t)$ has uniform amplitude distribution. If the pdf is nonuniform, then the quantizer step size should be variable, with smaller step sizes near the mode of the pdf and larger step sizes near the tails of the pdf. An example of nonuniform quantizing is shown in Figure 3.30. The input to the quantizer is a Gaussian random variable and the quantizer output is determined according to

$$X_q = m_i \quad \text{if} \quad x_{i-1} < X \le x_i, \quad i = 1, 2, \ldots, Q$$
$$x_0 = -\infty, \quad x_Q = \infty \tag{3.105}$$

The step size $\Delta_i = x_i - x_{i-1}$ is variable. The quantizer end points x_i's and the output levels m_i's are chosen to minimize N_Q/S_Q.

The design of an optimum nonuniform quantizer can be approached as follows. We are given a continuous random variable X with a pdf $f_X(x)$. We want to approximate X by a discrete random variable X_q according to Equation 3.105. The quantizing intervals and the levels are to be chosen such that N_Q is minimized. This minimizing can be done as follows. We start with

$$N_Q = \sum_{i=1}^{Q} \int_{x_{i-1}}^{x_i} (x - m_i)^2 f_X(x) \, dx, \quad x_0 = -\infty \quad \text{and} \quad x_Q = \infty$$

Since we wish to minimize N_Q for a fixed Q, we get the necessary* conditions by differentiating N_Q with respect to the x_j's and m_j's and setting the derivatives

---

*After finding all the x_i's and m_i's that satisfy the necessary conditions, we may evaluate N_Q at these points to find a set of x_i's and m_i's that yield the absolute minimum value of N_Q. In most practical cases we will get a unique solution for Equations 3.106.a and 3.106.b.

equal to zero:

$$\frac{\partial N_Q}{\partial x_j} = (x_j - m_j)^2 f_X(x_j) - (x_j - m_{j+1})^2 f_X(x_j) = 0$$

$$j = 1, 2, \ldots, Q - 1 \quad (3.106.\text{a})$$

$$\frac{\partial N_Q}{\partial m_j} = -2 \int_{x_{j-1}}^{x_j} (x - m_j) f_X(x)\, dx = 0, \quad j = 1, 2, \ldots, Q \quad (3.106.\text{b})$$

From Equation 3.106.a we obtain

$$x_j = \frac{1}{2}(m_j + m_{j+1})$$

or

$$m_j = 2x_{j-1} - m_{j-1}, \quad j = 2, 3, \ldots, Q \quad (3.107.\text{a})$$

Equation 3.106.b reduces to

$$\int_{x_{j-1}}^{x_j} (x - m_j) f_X(x)\, dx = 0, \quad j = 1, 2, \ldots, Q \quad (3.107.\text{b})$$

which implies that m_j is the centroid (or mean) of the jth quantizer interval. The foregoing set of simultaneous equations cannot be solved in closed form for an arbitrary $f_X(x)$. For a specific $f_X(x)$, a method of solving Equations 3.107.a and 3.107.b is to pick m_1 and calculate the succeeding x_i's and m_i's using Equations 3.107.a and 3.107.b. If m_1 is chosen correctly, then at the end of the iteration, m_Q will be the mean of the interval $[x_{Q-1}, \infty]$. If m_Q is not the centroid or the mean of the Qth interval, then a different choice of m_1 is made and the procedure is repeated until a suitable set of x_i's and m_i's is reached. A computer program to solve for the quantizing intervals and the means by this iterative method can be written.

Quantizer for a Gaussian Random Variable. The end points of the quantizer intervals and the output levels for a Gaussian random variable have been computed by J. Max [15]. Attempts have also been made to determine the functional dependence of N_Q on the number of levels Q. For a Gaussian random variable with a variance of 1, Max has found that N_Q is related to Q by

$$N_Q \approx (2.2) Q^{-1.96}, \quad \text{when} \quad Q \gg 1$$

If the variance is σ_X^2, then the preceding expression becomes

$$N_Q \approx (2.2)\sigma_X^2 Q^{-1.96} \qquad (3.108)$$

Now if we assume X to have zero mean, then $S_Q \approx E\{X^2\} = \sigma_X^2$, and hence

$$\epsilon_Q^2 = \frac{N_Q}{S_Q} \approx 2.2 Q^{-1.96} \qquad (3.109)$$

Equation 3.109 can be used to determine the number of quantizer levels needed to achieve a given normalized mean-squared error for a zero-mean Gaussian random process.

3.11 SUMMARY

In this chapter, we introduced the concept of random processes, which may be viewed as an extension of the concept of random variables. A random process maps outcomes of a random experiment to functions of time and is a useful model for both signals and noise. For many engineering applications, a random process can be characterized by first-order and second-order probability distribution functions, or perhaps just the mean and variance and autocorrelation function. For stationary random processes, the mean and autocorrelation functions are often used to describe the time domain structure of the process in an average or ensemble sense. The Fourier transform of the autocorrelation function, called the power spectral density function, provides a frequency domain description of the random processes.

Markov, independent increments, Martingale, and Gaussian random processes were defined. The random walk; its limiting version, the Wiener process; the Poisson process; and the random binary waveform were introduced as important examples of random processes, and their mean and autocorrelation functions were found.

Different types of stationarity were defined and wide-sense stationarity (weak stationarity) was emphasized because of its importance in applications. The properties of the autocorrelation and the cross-correlation functions of real wide-sense stationary (WSS) processes were presented. The Fourier transforms of these functions are called the power spectral density function and cross-power density function, respectively. The Fourier transform was used to define the spectral density function of random sequences. Power and bandwidth calculations, which are patterned after deterministic signal definitions, were introduced.

The concepts of continuity, differentiation, and integration were introduced for random processes. If all member functions of the ensemble have one of these three properties, then the random process has that property. In addition, these properties were defined in the mean-square sense as they apply to stationary (WSS) processes. It was shown that this extends these important operations to a wider class of random signals.

The time average of a random process or a function, for example $(X(t) - \mu)^2$, of a random process is a random variable. This time average will have a mean and a variance. For stationary processes, it was shown that the mean of the time average equals the ensemble mean. In order for the time average to equal the ensemble average, it was shown that it is necessary for the variance of the time average to be zero. When this is the case, the stationary process is called ergodic. Various definitions of ergodicity were given.

Series expansions of random processes were introduced. Fourier series and a modified Fourier series were presented, and the Karhunen–Loeve series expansion, which is optimum in the MS sense for a specified number of terms, was introduced.

The sampling theorem for a random process band-limited to B Hz was proved. It shows that if the sampling rate f_s is greater than $2B$, then samples $X(nT_s)$ can be used to reproduce, in the MS sense, the original process. Such sampling often requires quantization, which was introduced and analyzed in Section 3.10. The mean-square error and normalized mean-square error were suggested as measures of performance for quantizers.

3.12 REFERENCES

A number of texts are available to the interested reader who needs additional material on the topics discussed in this chapter. Background material on deterministic signal processing may be found in References [7] and [10]. Introductory treatment of the material of this chapter may be found in Cooper and McGillem [1], Gardner [4], Helstrom [5], Peebles [11], O'Flynn [12], and Schwartz and Shaw [13], and a slightly higher level treatment is contained in Papoulis [9]. Davenport and Root [3] is the classical book in this area, whereas Doob [2] is a primary reference in this field from the mathematical perspective. Advanced material on random processes may be found in texts by Larson and Shubert [6], and Wong and Hajek [14], and Mohanty [8].

[1] G. R. Cooper and C. D. McGillem, *Probabilistic Methods of Signal and System Analysis,* 2nd ed., Holt, Rinehart, and Winston, New York, 1986.

[2] J. L. Doob, *Stochastic Processes,* John Wiley & Sons, New York, 1953.

[3] W. B. Davenport, Jr. and W. L. Root, *Introduction to Random Signals and Noise*, McGraw-Hill, New York, 1958.

[4] W. A. Gardner, *Introduction to Random Processes: With Applications to Signals and Systems*, Macmillan, New York, 1986.

[5] C. W. Helstrom, *Probability and Stochastic Processes for Engineers*, Macmillan, New York, 1984.

[6] H. J. Larson and B. O. Shubert, *Probabilistic Models in Engineering and Science*, Vols. I and II, John Wiley & Sons, New York, 1979.

[7] C. D. McGillem and G. R. Cooper, *Continuous and Discrete Signal and System Analysis*, 2nd ed., Holt, Rinehart, and Winston, New York, 1984.

[8] N. Mohanty, *Random Signals, Estimation and Identification*, Van Nostrand, New York, 1986.

[9] A. Papoulis, *Probability, Random Variables and Stochastic Processes*, McGraw-Hill, New York, 1965, 1984.

[10] A. Papoulis, *Signal Analysis*, McGraw-Hill, New York, 1977.

[11] P. J. Peebles, *Probability, Random Variables and Random Signal Principles*, 2nd ed., McGraw-Hill, New York, 1986.

[12] M. O'Flynn, *Probabilities, Random Variables and Random Processes*, Harper and Row, New York, 1982.

[13] M. Schwartz and L. Shaw, *Signal Processing: Discrete Spectral Analysis, Detection and Estimation*, McGraw-Hill, New York, 1975.

[14] E. Wong and B. Hajek, *Stochastic Processes in Engineering Systems*, Springer-Verlag, New York, 1971, 1985.

[15] Max, J., "Quantizing for Minimum Distortion," *IRE Transaction on Information Theory*, Vol. IT-6, 1960, pp. 7–12.

3.14 PROBLEMS

3.1 Define a random process $X(t)$ based on the outcome k of tossing a die as

$$X(t) = \begin{cases} -2 & k = 1 \\ -1 & k = 2 \\ 1 & k = 3 \\ 2 & k = 4 \\ t & k = 5 \\ -t & k = 6 \end{cases}$$

  a. Find the joint probability mass function of $X(0)$ and $X(2)$.

  b. Find the marginal probability mass functions of $X(0)$ and $X(2)$.

  c. Find $E\{X(0)\}$, $E\{X(2)\}$, and $E\{X(0)X(2)\}$.

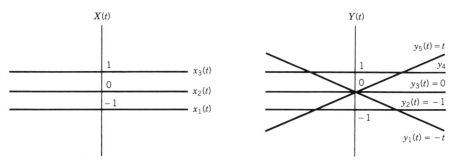

Figure 3.31 Member functions of $X(t)$ and $Y(t)$.

3.2 The member functions of two random processes $X(t)$ and $Y(t)$ are shown in Figure 3.31. Assume that the member functions have equal probabilities of occurrence.

   a. Find $\mu_X(t)$, and $R_{XX}(t, t + \tau)$. Is $X(t)$ WSS?

   b. Find $\mu_Y(t)$, and $R_{YY}(t, t + \tau)$. Is $Y(t)$ WSS?

   c. Find $R_{XY}(0, 1)$ assuming that the underlying random experiments are independent.

3.3 $X(t)$ is a Gaussian random process with mean $\mu_X(t)$ and autocorrelation function $R_{XX}(t_1, t_2)$. Find $E\{X(t_2)|X(t_1)\}$, $t_1 < t_2$.

3.4 Using the Markov property, show that if $X(n)$, $n = 1, 2, 3, \ldots$, is Markov, then

$$E\{X(n + 1)|X(1), X(2), \ldots, X(n)\} = E\{X(n + 1)|X(n)\}$$

3.5 For a Markov process $X(t)$ show that, for $t > t_1 > t_0$,

$$f_{X(t)|X(t_0)}(x|x_0) = \int_{-\infty}^{\infty} f_{X(t)|X(t_1)}(x|x_1) f_{X(t_1)|X(t_0)}(x_1|x_0) \, dx_1$$

(The preceding equation is called the Chapman–Kolmogoroff equation.)

3.6 Show that the Wiener process is a Martingale.

3.7 Consider the random walk discussed in Section 3.4.2. Assuming $d = 1$, and $T = 1$, find

   a. $P[X(2) = 0]$

   b. $P[X(8) = 0|X(6) = 2]$

   c. $E\{X(10)\}$

   d. $E\{X(10)|X(4) = 4\}$

206 RANDOM PROCESSES AND SEQUENCES

3.8 A symmetric Bernoulli random walk is defined by the sequence $S(n)$ as

$$S(n) = \sum_{k=1}^{n} X(k), \quad X(0) = 0, \quad n = 1, 2, 3, \ldots$$

where $X(n)$, $n = 1, 2, 3, \ldots$ is a sequence of independent and identically distributed (i.i.d) Bernoulli random variables with

$$P[X(n) = 1] = P[X(n) = -1] = \frac{1}{2}$$

    a. Show that $S(n)$ is a Martingale sequence.

    b. Show that $Z(n) = S^2(n) - n$ is also a Martingale.

3.9 Let $X(1), X(2), \ldots, X(n), \ldots$ be a sequence of zero-mean i.i.d. random variables with a pdf $f_X(x)$. Define $Y(n)$ as

$$Y(n) = \sum_{k=1}^{n} X(k), \quad n = 1, 2, 3, \ldots$$

    a. Show that $Y(n)$ is a Markov sequence and a Martingale.

    b. Show that

$$f_{Y_1,Y_2,\ldots,Y_n}(y_1, y_2, \ldots, y_n) = f_X(y_1)f_X(y_2 - y_1) \cdots f_X(y_n - y_{n-1})$$

    c. Find the conditional pdf $f_{Y_n|Y_{n-1}}$.

3.10 Let $N(t)$, $t \geq 0$ be the Poisson process with parameter λ, and define

$$X(t) = \begin{cases} 1 & \text{if } N(t) \text{ is odd} \\ -1 & \text{if } N(t) \text{ is even} \end{cases}$$

$X(t)$ is called a random telegraph signal.

    a. Show that $X(t)$ has the Markov property.

    b. Find $\mu_X(t)$ and $R_{XX}(t_1, t_2)$.

3.11 $X(t)$ is a real WSS random process with an autocorrelation function $R_{XX}(\tau)$. Prove the following:

    a. If $X(t)$ has periodic components, then $R_{XX}(\tau)$ will also have periodic components.

    b. If $R_{XX}(0) < \infty$, and if $R_{XX}(\tau)$ is continuous at $\tau = 0$, then it is continuous for every τ.

3.12 $X(t)$ and $Y(t)$ are real random processes that are jointly WSS. Prove the following:

    a. $|R_{XY}(\tau)| \leq \sqrt{R_{XX}(0)R_{YY}(0)}$

    b. $R_{XY}(\tau) \leq \frac{1}{2}[R_{XX}(0) + R_{YY}(0)]$

3.13 $X(t)$ and $Y(t)$ are independent WSS random processes with zero means. Find the autocorrelation function of $Z(t)$ when

   a. $Z(t) = a + bX(t) + cY(t)$
   b. $Z(t) = aX(t)Y(t)$

3.14 $X(t)$ is a WSS process and let $Y(t) = X(t+a) - X(t-a)$. Show that

   a. $R_{YY}(\tau) = 2R_{XX}(\tau) - R_{XX}(\tau + 2a) - R_{XX}(\tau - 2a)$
   b. $S_{YY}(f) = 4S_{XX}(f)\sin^2(2\pi a f)$

3.15 Determine whether the following functions can be the autocorrelation functions of real-valued WSS random processes:

   a. $(1 + 2\tau^2)^{-1}$
   b. $2 \sin 2\pi(1000)\tau$
   c. $\dfrac{\sin 2\pi f_0 \tau}{f_0 \tau}$, $f_0 > 0$
   d. $\delta(\tau) + \cos 2\pi f_0 \tau$

3.16 Determine whether the following functions can be power spectral density functions of real-valued WSS random processes.

   a. $(1 + 10f^2)^{-1/2}$
   b. $\dfrac{\sin 1000f}{1000f}$
   c. $50 + 20\delta(f - 1000)$
   d. $10\delta(f) + 5\delta(f + 500) + 5\delta(f - 500)$
   e. $\exp(-200\pi f^2)$
   f. $\dfrac{f}{(f^2 + 100)}$

3.17 For each of the autocorrelation functions below, find the power spectral density function.

   a. $\exp(-\alpha|\tau|)$, $\alpha > 0$
   b. $\dfrac{\sin 1000 \tau}{1000 \tau}$
   c. $\dfrac{1}{4}\exp(-|\tau|)[\cos \tau + \sin |\tau|]$
   d. $\exp(-10^{-2}f_0^2\tau^2)$
   e. $\cos(1000\tau)$

3.18 For each of the power spectral density functions given below, find the autocorrelation function.

 a. $(40\pi^2 f^2 + 35)/[(4\pi^2 f^2 + 9)(4\pi^2 f^2 + 4)]$
 b. $1/(1 + 4\pi^2 f^2)^2$
 c. $100\delta(f) + 2\alpha/(\alpha^2 + 4\pi^2 f^2)$

3.19 $X(n)$, $n = \cdots, -1, 0, 1, \ldots$ is a real discrete-time, zero-mean, WSS sequence. Find the power spectral density function for each of the following cases.

 a. $X(n)$ is a sequence of i.i.d. random variables with unit variance.
 b. $X(n)$ is a discrete time Markov sequence with $R_{XX}(m) = \exp(-\alpha|m|)$.
 c. $X(n)$ is a sequence with $R_{XX}(0) = 1$, $R_{XX}(\pm 1) = -1/2$ and $R_{XX}(k) = 0$ for $|k| > 1$.

3.20 The psd of a WSS random process $X(t)$ is shown in Figure 3.32.

 a. Find the power in the DC term.
 b. Find $E\{X^2(t)\}$.
 c. Find the power in the frequency range $[0, 100 \text{ Hz}]$.

3.21 Let X and Y be independent Gaussian random variables with zero-mean and unit variance. Define

$$Z(t) = X \cos 2\pi(1000)t + Y \sin 2\pi(1000)t$$

 a. Show that $Z(t)$ is a Gaussian random process.
 b. Find the joint pdf of $Z(t_1)$ and $Z(t_2)$.
 c. Is the process WSS?
 d. Is the process SSS?
 e. Find $E\{Z(t_2)|Z(t_1)\}$, $t_2 > t_1$.

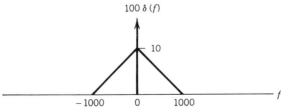

Figure 3.32 Psd of $X(t)$ for Problem 3.20.

$S_{XX}(f) = 10 \exp(-f^2/10000)$ $S_{XX}(f) = 100/[1 + (2\pi f/100)^2]^2$

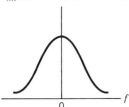 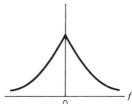

Figure 3.33 Psd functions for Problem 3.23 and 3.44.

3.22 For a wide-sense stationary random process, show that

  a. $R_{XX}(0)$ = area under $S_{XX}(f)$.

  b. $S_{XX}(0)$ = area under $R_{XX}(\tau)$.

3.23 For the random process $X(t)$ with the psd's shown in Figure 3.33, determine

  a. The effective bandwidth, and

  b. The rms bandwidth which is defined as

$$B_{rms}^2 = \frac{\int_{-\infty}^{\infty} f^2 S_{XX}(f)\, df}{\int_{-\infty}^{\infty} S_{XX}(f)\, df}$$

[*Note:* The rms bandwidth exists only if $S_{XX}(f)$ decays faster than $1/f^2$]

3.24 For bandpass processes, the rms bandwidth is defined as

$$B_{rms}^2 = \frac{4 \int_0^\infty (f - f_0)^2 S_{XX}(f)\, df}{\int_0^\infty S_{XX}(f)\, df}$$

where the mean or center frequency f_0 is defined as

$$f_0 = \frac{\int_0^\infty f\, S_{XX}(f)\, df}{\int_0^\infty S_{XX}(f)\, df}$$

Find the rms bandwidth of

$$S_{XX}(f) = \frac{A}{\left[1 + \left(\frac{f - f_0}{B}\right)^2\right]^2} + \frac{A}{\left[1 + \left(\frac{f + f_0}{B}\right)^2\right]^2}; \quad A, B, f_0 > 0$$

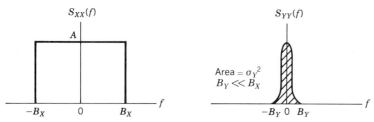

Figure 3.34 Psd functions for Problem 3.26.

3.25 $X(t)$ is a complex-valued WSS random process defined as

$$X(t) = A \exp(2\pi jYt + j\Theta)$$

where A, Y and Θ are independent random variables with the following pdfs:

$$f_A(a) = a \exp(-a^2/2), \quad a > 0$$
$$\quad\quad\quad = 0 \quad\quad\quad\quad\quad\quad \text{elsewhere}$$

$$f_Y(y) = \begin{cases} 1/1000 & \text{for } 10{,}000 < y < 11{,}000 \\ 0 & \text{elsewhere} \end{cases}$$

$$f_\Theta(\theta) = \begin{cases} \dfrac{1}{2\pi} & \text{for } -\pi < \theta < \pi \\ 0 & \text{elsewhere} \end{cases}$$

Find the psd of $X(t)$.

3.26 $X(t)$ and $Y(t)$ are two independent WSS random processes with the power spectral density functions shown in Figure 3.34. Let $Z(t) = X(t)Y(t)$. Sketch the psd of $Z(t)$, and find $S_{ZZ}(0)$.

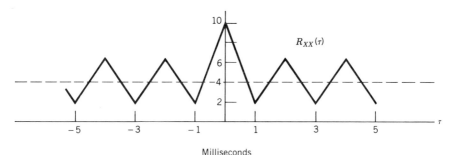

Figure 3.35 Autocorrelation function for Problem 3.27.

3.27 A WSS random process $X(t)$ has a mean of 2 volts, a periodic component $X_p(t)$, and a random component $X_r(t)$; that is, $X(t) = 2 + X_p(t) + X_r(t)$. The autocorrelation function of $X(t)$ is given in Figure 3.35.

  a. What is the average power in the periodic component?

  b. What is the average power in the random component?

3.28 A stationary zero-mean random process $X(t)$ has an autocorrelation function

$$R_{XX}(\tau) = 10 \exp(-0.1\tau^2)$$

  a. Find the autocorrelation function of $X'(t)$ if $X'(t)$ exists.

  b. Find the mean and variance of

$$Y = \frac{1}{5} \int_0^5 X(t) \, dt$$

3.29 Show that if a finite variance process is MS differentiable, then it is necessarily MS continuous.

3.30 Show that for a lowpass process with a bandwidth B, the amount of change from t to $t + \tau$ is bounded by

  a. $E\{[X(t + \tau) - X(t)]^2\} \le (2\pi B\tau)^2 [\sigma_X^2 + \mu_X^2]$

  b. $R_{XX}(0) - R_{XX}(\tau) \le (2\pi B\tau)^2 R_{XX}(0)/2$

3.31 $X(t)$ and $Y(t)$ are two independent WSS processes that are MS continuous.

  a. Show that the sum $X(t) + Y(t)$ is MS continuous.

  b. Show that the product $X(t)Y(t)$ is also MS continuous.

3.32 Show that both MS differentiation and integration obey the following rules of calculus:

  a. Differentiating and integrating linear combinations.

  b. Differentiating and integrating products of independent random processes.

3.33 Show that the sufficient condition for the existence of the MS integral of a stationary finite variance process $X(t)$ is the existence of the integral

$$\int_{t_0}^{t} \int_{t_0}^{t} R_{XX}(t_1 - t_2) \, dt_1 dt_2$$

3.34 $X(t)$ is WSS with $E\{X(t)\} = 2$ and $R_{XX}(\tau) = 4 + \exp(-|\tau|/10)$.

  a. Find the mean and variance of $S = \int_0^1 X(\tau) \, d\tau$

b. How large should T be chosen so that
$$P\{|\langle\mu_X\rangle_T - 2| < 0.1\} > 0.95$$

3.35 Let $Z(t) = x(t) + Y(t)$ where $x(t)$ is a deterministic, periodic power signal with a period T and $Y(t)$ is a zero mean *ergodic* random process. Find the autocorrelation function and also the psd function of $Z(t)$ using time averages.

3.36 $X(t)$ is a random binary waveform with a bit rate of $1/T$, and let
$$Y(t) = X(t)X(t - T/2)$$

a. Show that $Y(t)$ can be written as $Y(t) = v(t) + W(t)$ where $v(t)$ is a periodic deterministic signal and $W(t)$ is a random binary waveform of the form
$$\sum_k A_k p(t - kT - D); \quad p(t) = \begin{cases} 1 & \text{for } |t| < T/2 \\ 0 & \text{elsewhere} \end{cases}$$

b. Find the psd of $Y(t)$ and show that it has discrete frequency spectral components.

3.37 Consider the problem of estimating the unknown value of a constant signal by observing and processing a noisy version of the signal for T seconds. Let $X(t) = c + N(t)$ where c is the unknown signal value (which is assumed to remain constant), and $N(t)$ is a zero-mean stationary Gaussian random process with a psd $S_{NN}(f) = N_0$ for $|f| < B$ and zero elsewhere ($B \gg 1/T$). The estimate of c is the time-averaged value
$$\hat{c} = \frac{1}{T}\int_0^T X(t)\, dt$$

a. Show that $E\{\hat{c}\} = c$.

b. Find the value of T such that $P\{|\hat{c} - c| < 0.1c\} \geq 0.999$. (Express T in terms of c, B, and N_0.)

3.38 Give an example of a random process that is WSS but not ergodic in mean.

3.39 A stationary zero-mean Gaussian random process $X(t)$ has an autocorrelation function
$$R_{XX}(\tau) = 10\exp(-|\tau|)$$
Show that $X(t)$ is ergodic in the mean and autocorrelation function.

3.40 $X(t)$ is a stationary zero-mean Gaussian random process.

a. Show that
$$\text{Var}\{\langle R_{XX}(\tau)\rangle_T\} \leq \frac{4}{T}\int_0^\infty R_{XX}^2(\tau)\, d\tau$$

b. Show that

$$E\{\langle S_{XX}(f)\rangle_T\} = S_{XX}(f), \quad \text{as } T \to \infty$$

and

$$\text{Var}\{\langle S_{XX}(f)\rangle_T\} \geq [E\{\langle S_{XX}(f)\rangle_T\}]^2$$

3.41 Define the time-averaged mean and autocorrelation function of a real-valued stationary random sequence as

$$\langle \mu_X \rangle_N = \frac{1}{N} \sum_i^N X(i)$$

and

$$\langle R_{XX}(k) \rangle_N = \frac{1}{N} \sum_{i=1}^N X(i) X(i+k)$$

a. Find the mean and variance of $\langle \mu_X \rangle_N$ and $\langle R_{XX}(k) \rangle_N$

b. Derive the condition for the ergodicity of the mean.

3.42 Prove the properties of the Fourier series expansion given in section 3.9.1 and 3.9.2.

3.43 Let $\mathbf{X} = [X_1, X_2, \ldots, X_n]^T$ be a random vector with a covariance matrix $\Sigma_{\mathbf{X}}$. Let $\lambda_1 > \lambda_2 > \cdots > \lambda_n$ be the eigenvalues of $\Sigma_{\mathbf{X}}$. Suppose we want to approximate $\mathbf{X}$ as

$$\tilde{\mathbf{X}} = A_1 \mathbf{v}_1 + A_2 \mathbf{v}_2 + \cdots + A_m \mathbf{v}_m, \quad m < n$$

such that $E\{[\tilde{\mathbf{X}} - \mathbf{X}]^T [\tilde{\mathbf{X}} - \mathbf{X}]\}$ is minimized.

a. Show that the basis vectors $\mathbf{v}_1, \mathbf{v}_2, \ldots, \mathbf{v}_m$ are the eigenvectors of $\Sigma_{\mathbf{X}}$ corresponding to $\lambda_1, \lambda_2, \ldots, \lambda_m$, respectively.

b. Show that the coefficients A_i are random variables and that $A_i = \mathbf{X}^T \mathbf{v}_i$.

c. Find the mean squared error.

3.44 Suppose we want to sample the random processes whose power spectral densities are shown in Figure 3.33. Find a suitable sampling rate using the constraint that the ratio of $S_{XX}(0)$ to the aliased spectral component at $f = 0$ has to be greater than 100.

3.45 Show that a WSS bandpass random process can also be represented by sampled values. Establish a relationship between the bandwidth B and the minimum sampling rate.

3.46 The probability density function of a random variable X is shown in Figure 3.36.

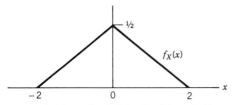
Figure 3.36 Pdf of X for Problem 3.46.

a. If X is quantized into four levels using a uniform quantizing rule, find the MSE.

b. If X is quantized into four levels using a minimum MSE nonuniform quantizer, find the quantizer end points and output levels as well as the MSE.

CHAPTER FOUR

Response of Linear Systems to Random Inputs

In many cases, physical systems are modeled as lumped, linear, time invariant (LLTIV), and causal, and their dynamic behavior is described by linear differential or difference equations with constant coefficients. The response (i.e., the output) of a LTIV (lumped is not a requirement if the impulse response is known) system driven by a deterministic input signal can be computed in the time domain via the convolution integral or in the transform domain via Fourier, Laplace, or Z transforms. Although the analysis of LTIV systems follows a rather standard and unified approach, such is not the case when the system is nonlinear or time varying. Here, a variety of numerical techniques are used and the specific approach used will be highly problem dependent.

In this chapter, we develop techniques for calculating the response of linear systems driven by random input signals. Regardless of whether or not the system is linear, for each member function $x(t)$ of the input process $X(t)$, the system produces an output $y(t)$ and the ensemble of output functions form a random process $Y(t)$, which is the response of the system to the random input signal $X(t)$. Given a description of the input process $X(t)$ and a description of the system, we want to obtain the properties of $Y(t)$ such as the mean, autocorrelation function, and at least some of the lower order probability distribution functions of $Y(t)$. In most cases we will obtain just the mean and autocorrelation function. Only in some special cases will we want to (and be able to) determine the probability distribution functions.

We will show that the determination of the response of a LTIV system responding to a random input is rather straightforward. However, the problem of determining the output of a nonlinear system responding to a random input signal is very difficult except in some special cases. No general tractable analytical techniques are available to handle nonlinear systems. However, the analysis of nonlinear systems can be carried out using Monte Carlo simulation techniques, which were introduced in Chapter 2.

In the remaining sections of this chapter, we will assume that the functional relationship between the input and output is given and that the system parameters are constants. Occasionally, there arises a need for using system models in which some of the parameters are modeled as random variables. For example, the gain of a certain lot of IC amplifiers or the resistance of a ±10% resistor can be modeled as a random variable. In this chapter, we will consider only fixed-parameter systems.

4.1 CLASSIFICATION OF SYSTEMS

Mathematically, a "system" is a functional relationship between the input $x(t)$ and the output $y(t)$. We can write this input–output relationship as:

$$y(t_0) = f[x(t); -\infty < t < \infty], \quad -\infty < t_0 < \infty \quad (4.1)$$

Based on the properties of the functional relationship given in Equation 4.1, we can classify systems into various categories. Rather than listing all possible classifications, we list only those classes of systems that we will study in this chapter in some detail.

4.1.1 Lumped Linear Time-invariant Causal (LLTIVC) System

A system is said to be LLTIVC if it has all of the following properties:

1. **Lumped.** A dynamic system is called lumped if it can be modeled by a set of ordinary differential or difference equations.
2. **Linear.** If

$$y_1(t) = f[x_1(t); -\infty < t < \infty]$$

and

$$y_2(t) = f[x_2(t); -\infty < t < \infty]$$

then

$$f[a_1 x_1(t) + a_2 x_2(t)] = a_1 f[x_1(t)] + a_2 f[x_2(t)]$$

(i.e., superposition applies).

3. **Time Invariant.** If $y(t) = f[x(t)]$, then

$$y(t - t_0) = f[x(t - t_0)], \quad -\infty < t, t_0 < \infty$$

(i.e., a time shift in the input results in a corresponding time shift in the output).

4. *Causal.* The value of the output at $t = t_0$ depends only on the past values of the input $x(t)$, $t \leq t_0$, that is

$$y(t_0) = f[x(t); \quad t \leq t_0], \quad -\infty < t, t_0 < \infty$$

Almost all of the systems analyzed in this chapter will be linear, time invariant, and causal (LTIVC). An exception is the memoryless systems discussed in the next subsection.

4.1.2 Memoryless Nonlinear Systems

Any system in which superposition does not apply is called a nonlinear system. A system is said to be memoryless if the output at $t = t_0$ depends only on the instantaneous value of the input at $t = t_0$. A commonly used model for memoryless nonlinear systems is the power series model in which

$$y(t) = \sum_{i=0}^{n} a_i x^i(t), \quad n \geq 2$$

where the a_i's are known constants. Such systems can be analyzed using the techniques of Section 2.6, as illustrated by the following example.

EXAMPLE 4.1.

Let $X(t)$ be a stationary Gaussian process with

$$\mu_X(t) = 0$$
$$R_{XX}(\tau) = \exp(-|\tau|)$$
$$Y(t) = X^2(t)$$

Find $\mu_Y(t)$ and $R_{YY}(t_1, t_2)$.

SOLUTION:

$$E\{Y(t)\} = E\{X^2(t)\} = \int_{-\infty}^{\infty} x^2 \frac{1}{\sqrt{2\pi}} \exp[-(x)^2/2] \, dx$$

Because $E\{X^2(t)\} = R_{XX}(0)$

$$\mu_Y(t) = R_{XX}(0) = 1$$

$$E[Y(t_1)Y(t_2)] = \int\int_{-\infty}^{\infty} \frac{x_1^2 x_2^2}{2\pi\sqrt{1 - \exp\{-2|t_1 - t_2|\}}}$$

$$\times \exp\left[-\frac{(x_1)^2 - 2x_1 x_2 \exp\{-|t_1 - t_2|\} + (x_2)^2}{2[1 - \exp\{-2|t_1 - t_2|\}]}\right] dx_1\, dx_2$$

The evaluation of the integral is given in Equation 2.70 as

$$E\{X_1^2 X_2^2\} = E\{X_1^2\}E\{X_2^2\} + 2E^2\{X_1 X_2\}$$

Thus

$$\begin{aligned}
R_{YY}(t_1, t_2) &= E\{Y(t_1)Y(t_2)\} = E\{X^2(t_1)X^2(t_2)\} \\
&= E\{X^2(t_1)\}E\{X^2(t_2)\} + 2E^2\{X(t_1)X(t_2)\} \\
&= R_{XX}(0)R_{XX}(0) + 2R_{XX}^2(|t_1 - t_2|) \\
&= 1 + 2[\exp(-|t_1 - t_2|)]^2 = 1 + 2\exp(-2|\tau|)
\end{aligned}$$

where $\tau = t_2 - t_1$.

4.2 RESPONSE OF LTIVC DISCRETE TIME SYSTEMS

4.2.1 Review of Deterministic System Analysis

The input–output relationship of a LTIVC system with a deterministic input can be described by an Nth order difference equation,

$$\sum_{m=0}^{N} a_m y[(m + n)T_s] = \sum_{m=0}^{N} b_m x[(m + n)T_s] \qquad (4.2)$$

where $x(kT_s)$ and $y(kT_s)$ are the input and output sequences, T_s is the time between samples, and the a_i's and b_i's are known constants. We will assume x, y, a_i's, and b_i's to be real-valued and set $T_s = 1$. The last assumption is equivalent to describing the system in terms of normalized frequency. Given a set of initial conditions and the input sequence, the output sequence can be obtained using a variety of techniques (see, for example, Reference [1]). If we assume zero initial conditions, or that we are observing the output after the transients have died out, then we can write the input–output response in the form of a convolution

$$y(n) = x(n) * h(n)$$
$$= \sum_{m=-\infty}^{\infty} h(n-m)x(m) \quad (4.3.a)$$

or

$$= \sum_{m=-\infty}^{\infty} h(m)x(n-m) \quad (4.3.b)$$

where * represents convolution.

The sequence $h(k)$ in Equation 4.3.a and 4.3.b is the *unit pulse (impulse) response* of the system, defined as the output $y(k)$ at time k when the input is a sequence of zeros except for a unit input at $t = 0$. Since we assume the system to be causal, $h(k) = 0$ for $k < 0$, and for a *stable* system (which yields a bounded output sequence when the input sequence is bounded)

$$\sum_{k=0}^{\infty} |h(k)| < \infty$$

The Fourier transform of the unit pulse response is called the *transfer function*, $H(f)$, and is

$$H(f) = F\{h(n)\} = \sum_{n=-\infty}^{\infty} h(n)\exp(-j2\pi nf), \quad |f| < \frac{1}{2} \quad (4.4.a)$$

where f is the frequency variable. The unit pulse response can be obtained from $H(f)$ by taking the inverse transform, which is defined as

$$h(n) = F^{-1}\{H(f)\} = \int_{-1/2}^{1/2} H(f)\exp(j2\pi nf)\, df \quad (4.4.b)$$

If we assume that the Fourier transforms of $x(n)$ and $y(n)$ exist and are called $X_F(f)$ and $Y_F(f)$, respectively, then the input–output relationship can be expressed in the transform domain as

$$Y_F(f) = \sum_{n=-\infty}^{\infty} \left[\sum_{m=-\infty}^{\infty} h(m)x(n-m) \right] \exp(-j2\pi nf)$$

If the system is stable, then the order of summation can be interchanged, and we have

$$Y_F(f) = \sum_{m=-\infty}^{\infty} h(m) \left[\sum_{n=-\infty}^{\infty} x(n-m)\exp(-j2\pi(n-m)f) \right] \exp(-j2\pi mf)$$

$$= \sum_{m=-\infty}^{\infty} h(m) \left[\sum_{n'=-\infty}^{\infty} x(n')\exp(-j2\pi n'f) \right] \exp(-j2\pi mf)$$

$$= \sum_{m=-\infty}^{\infty} h(m) X_F(f)\exp(-j2\pi mf)$$

Now, since $X_F(f)$ is not a function of m, we take it outside the summation and write

$$Y_F(f) = X_F(f)H(f) \qquad (4.5.a)$$

Equation 4.5.a is an important result, namely, the Fourier transform of the convolution of two time sequences is equal to the product of their transforms. From $Y_F(f)$ we obtain $y(n)$ as

$$y(n) = F^{-1}\{Y_F(f)\} = \int_{-1/2}^{1/2} X_F(f)H(f)\exp(j2\pi nf)\, df \qquad (4.5.b)$$

The Z transform of a discrete sequence is also useful and is defined by

$$X_Z(z) = \sum_{n=0}^{\infty} x(n)z^{-n}$$

$$H_Z(z) = \sum_{n=0}^{\infty} h(n)z^{-n}$$

Note that there are two significant differences between the Z transform and the Fourier transform. The Z transform is applicable when the sequence is defined on the nonnegative integers, and with this restriction the two transforms are equal if $z = \exp(j2\pi f)$. Also the Z transform of a stable system will exist if $|z| > 1$.

It is easy to show that an expression equivalent to Equation 4.5.a is

$$Y_Z(z) = X_Z(z)H_Z(z) \tag{4.6}$$

A brief table of the Z transform is given in Appendix C.

With a random input sequence, the response of the system to each sample sequence can be computed via Equation 4.3. However, Equation 4.5.a cannot be used in general since the Fourier transform of the input sequence $x(n)$ may or may not exist. Note that in a stable system, the output always exists and will be bounded when the input is bounded. It is just that the direct Fourier technique for computing the output sequences may not be applicable.

Rather than trying to compute the response of the system to each member sequence of the input and obtain the properties of the ensemble of the output sequences, we may compute the properties of the output directly as follows.

4.2.2 Mean and Autocorrelation of the Output

With a random input sequence $X(n)$, the output of the system may be written as

$$Y(n) = \sum_{m=-\infty}^{\infty} h(m)X(n-m) \tag{4.7.a}$$

Note that $Y(n)$ represents a random sequence, where each member function is subject to Equation 4.3.

The mean and the autocorrelation of the output can be calculated by taking the expected values

$$E\{Y(n)\} = \mu_Y(n) = \sum_{m=-\infty}^{\infty} h(m)E\{X(n-m)\} \tag{4.7.b}$$

and

$$R_{YY}(n_1, n_2) = E\{Y(n_1)Y(n_2)\}$$

$$= \sum_{m_1=-\infty}^{\infty}\sum_{m_2=-\infty}^{\infty} h(m_1)h(m_2)R_{XX}(n_1 - m_1, n_2 - m_2) \tag{4.7.c}$$

4.2.3 Distribution Functions

The distribution functions of the output sequence are in general very difficult to obtain. Even in the simplest case when the impulse response has a finite number of nonzero entries, Equation 4.7.a represents a linear transformation of the input sequence and the joint distribution functions cannot in general be expressed in a closed functional form. One important exception here is the Gaussian case. If the input is a discrete-time Gaussian process with finite variance, then $Y(n)$ is a linear combination of Gaussian variables and hence is Gaussian (see Section 2.5). All joint distributions of the output will be Gaussian. The mean vector and the covariance matrix of the joint Gaussian distributions can be obtained from Equations 2.65.a and 2.65.b. The Central Limit Theorem (Section 2.8.2) also suggests that $Y(n)$ will tend to be Gaussian for a number of other distributions of $X(n)$.

4.2.4 Stationarity of the Output

If $X(n)$ is wide-sense stationary (WSS), then from Equations 4.7.b and 4.7.c we obtain

$$\mu_Y = E\{Y(n)\} = \sum_{m=-\infty}^{\infty} h(m)\mu_X = \mu_X \sum_{m=-\infty}^{\infty} h(m)$$
$$= \mu_X H(0) \qquad (4.8.\text{a})$$

and

$$R_{YY}(n_1, n_2) = \sum_{m_1=-\infty}^{\infty} \sum_{m_2=-\infty}^{\infty} h(m_1)h(m_2) \\ \times R_{XX}[(n_2 - n_1) - (m_2 - m_1)] \qquad (4.8.\text{b})$$

Equation 4.8.a shows that the mean of Y does not depend on the time index n. The right-hand side of Equation 4.8.b depends only on the difference of n_1 and n_2 and hence $R_{YY}(n_1, n_2)$ will be a function of $n_2 - n_1$. Thus, the output $Y(n)$ of a LTIVC system is WSS when the input $X(n)$ is WSS.

It can also be shown that if the input to a LTIVC system is strict-sense stationary (SSS), then the output will also be SSS.

The assumption that we made earlier about zero initial conditions has an important bearing on the stationarity of the output. If we have nonzero initial conditions or if the input to the system is applied at t (or time index n) equal to 0, then the output will not be stationary. However, in either case, $Y(n)$ will be asymptotically stationary if the system is stable and the input is stationary.

4.2.5 Correlation and Power Spectral Density of the Output

Suppose the input to a LTIVC system is a real WSS sequence $X(n)$. To find the psd of the output $Y(n)$, let us start with the crosscorrelation function $R_{YX}(k)$

$$
\begin{aligned}
R_{YX}(k) &= E\{Y(n)X(n+k)\} \\
&= E\left\{\left[\sum_{m=-\infty}^{\infty} h(m)X(n-m)\right]X(n+k)\right\} \\
&= \sum_{m=-\infty}^{\infty} h(m)E\{X(n-m)X(n+k)\} \\
&= \sum_{m=-\infty}^{\infty} h(m)R_{XX}(k+m) = \sum_{n=-\infty}^{\infty} h(-n)R_{XX}(k-n)
\end{aligned}
$$

or

$$R_{YX}(k) = h(-k) * R_{XX}(k) \qquad (4.9)$$

It also follows from Equation 3.33 that

$$R_{XY}(k) = h(k) * R_{XX}(k) \qquad (4.10)$$

Similarly, we can show that

$$R_{YY}(k) = R_{YX}(k) * h(k)$$

and hence

$$R_{YY}(k) = R_{XX}(k) * h(k) * h(-k) \qquad (4.11.a)$$

Defining the psd of $Y(n)$ as

$$S_{YY}(f) = \sum_{n=-\infty}^{\infty} R_{YY}(n)\exp(-j2\pi nf), \qquad |f| < \frac{1}{2}$$

we have

$$
\begin{aligned}
S_{YY}(f) &= F\{R_{YY}(k)\} \\
&= F\{R_{XX}(k) * h(k) * h(-k)\}
\end{aligned}
$$

Since the Fourier transform of the convolution of two time sequences is the product of their transforms, we have

$$S_{YY}(f) = F\{R_{XX}(k)\}F\{h(k)\}F\{h(-k)\}$$
$$= S_{XX}(f)H(f)H(-f)$$
$$= S_{XX}(f)H(f)H^*(f)$$
$$= S_{XX}(f)|H(f)|^2 \qquad (4.11.b)$$

Equation 4.11.b is the basis of frequency domain techniques for the design of LTIVC systems. It shows that the spectral properties of a signal can be modified by passing it through a LTIVC system with the appropriate transfer function. By carefully choosing $H(f)$ we can remove or *filter out* certain spectral components in the input. For example, suppose we have $X(n) = S(n) + N(n)$, where $S(n)$ is a signal of interest and $N(n)$ is an unwanted noise process. Then, if the psd of $S(n)$ and $N(n)$ are nonoverlapping in the frequency domain, the noise $N(n)$ can be removed by passing $X(n)$ through a filter $H(f)$ that has a response of 1 for the range of frequencies occupied by the signal and a response of 0 for the range of frequencies occupied by the noise. Unfortunately, in most practical situations there is spectral overlap and the design of optimum filters to separate signal and noise is somewhat difficult. We will discuss this problem in some detail in Chapter 7. Also note that if $X(n)$ is a zero-mean white noise sequence, then $S_{YY}(f) = \sigma^2|H(f)|^2$, and $R_{XY}(k) = \sigma^2 h(k)$. Thus, white noise might be used to determine $h(k)$ for a linear time-invariant system.

From the definition of the Z transform, it follows that

$$H_Z[\exp(j2\pi f)] = H(f)$$

Defining

$$S_{XX}^{\#}(z) = \sum_{n=0}^{\infty} z^{-n} R_{XX}(n)$$

Then it follows that

$$S_{XX}^{\#}[\exp(j2\pi f)] = S_{XX}(f)$$

And we can show that

$$S_{YY}^{\#}(z) = S_{XX}^{\#}(z)|H(z)|^2 \qquad (4.12.a)$$
$$= S_{XX}^{\#}(z)H(z)H(z^{-1}) \qquad (4.12.b)$$

EXAMPLE 4.2.

The input to a LLTIVC system is a stationary random sequence $X(n)$ with

$$\mu_X = 0$$

and

$$R_{XX}(k) = \begin{cases} 1 & \text{for } k = 0 \\ 0 & \text{for } k \neq 0 \end{cases}$$

The impulse response of the system is

$$h(k) = \begin{cases} 1 & \text{for } k = 0, 1 \\ 0 & \text{for } k > 1 \end{cases}$$

Find the mean, the autocorrelation function, and the power spectral density function of the output $Y(n)$.

SOLUTION:

$$\mu_Y = 0 \quad \text{since} \quad \mu_X = 0$$

To find $R_{YY}(k)$, let us first find $S_{YY}(f)$ from Equation 4.11.b. We are given that

$$H(f) = \sum_{k=0}^{\infty} h(k)\exp(-j2\pi kf)$$
$$= 1 + \exp(-j2\pi f)$$

and

$$S_{XX}(f) = F\{R_{XX}(k)\}$$
$$= 1, \quad |f| < \frac{1}{2}$$

Hence

$$S_{YY}(f) = (1)|1 + \exp(-j2\pi f)|^2$$
$$= 2 + 2\cos 2\pi f, \quad |f| < \frac{1}{2}$$

Taking the inverse transform, we obtain

$$R_{YY}(0) = 2$$
$$R_{YY}(\pm 1) = 1$$
$$R_{YY}(k) = 0, \quad |k| > 1$$

EXAMPLE 4.3.

The input $X(n)$ to a certain digital filter is a zero-mean white noise sequence with variance σ^2. The digital filter is described by

$$H_z(z) = \frac{a_0 + a_1 z^{-1} + a_2 z^{-2}}{1 + b_1 z^{-1}}$$

If the filter output is $Y(n)$, find μ_Y, $S^{\#}_{YY}(z)$, and the power spectral density of Y in the normalized frequency domain.

SOLUTION: From the problem statement

$$\mu_X = 0; \quad R_{XX}(n) = \begin{cases} \sigma^2, & n = 0 \\ 0 & \text{elsewhere} \end{cases}$$

Using Equation 4.8.a

$$\mu_Y = 0(H(0)) = 0$$

Using Equation 4.12

$$S^{\#}_{YY}(z) = \left(\frac{a_0 + a_1 z^{-1} + a_2 z^{-2}}{1 + b_1 z^{-1}}\right)\left(\frac{a_0 + a_1 z + a_2 z^2}{1 + b_1 z}\right)\sigma^2$$

$$= \sigma^2 \frac{a_0^2 + a_1^2 + a_2^2 + a_1(a_0 + a_2)(z + z^{-1}) + a_0 a_2(z^2 + z^{-2})}{1 + b_1^2 + b_1(z + z^{-1})}$$

Thus, substituting $z = \exp(j2\pi f)$

$$S_{YY}(f) = \sigma^2 \frac{a_0^2 + a_1^2 + a_2^2 + 2a_1(a_0 + a_2)\cos 2\pi f + 2a_0 a_2 \cos 4\pi f}{1 + b_1^2 + 2b_1 \cos 2\pi f}$$

$$|f| < \frac{1}{2}$$

4.3 RESPONSE OF LTIVC CONTINUOUS-TIME SYSTEMS

The input–output relationship of a linear, time-invariant, and causal system driven by a deterministic input signal $x(t)$ can be represented by the convolution integral

$$y(t) = \int_{-\infty}^{\infty} x(\tau) h(t - \tau) \, d\tau \qquad (4.13.\text{a})$$

$$= \int_{-\infty}^{\infty} h(\tau) x(t - \tau) \, d\tau \qquad (4.13.\text{b})$$

where $h(t)$ is the impulse response of the system and we assume zero initial conditions. For a stable causal system

$$\int_{-\infty}^{\infty} |h(\tau)| \, d\tau < \infty$$

and

$$h(\tau) = 0, \quad \tau < 0$$

In the frequency domain, the input–output relationship can be expressed as

$$Y_F(f) = H(f) X_F(f) \qquad (4.14)$$

and $y(t)$ is obtained by taking the inverse Fourier transform of $Y_F(f)$. The forward and inverse transforms are defined as

$$Y_F(f) = \int_{-\infty}^{\infty} y(t)\exp(-j2\pi ft)\, dt$$

$$y(t) = F^{-1}\{Y_F(f)\} = \int_{-\infty}^{\infty} Y_F(f)\exp(j2\pi ft)\, df$$

Note that the frequency variable f ranges from $-\infty$ to ∞ in the continuous time case.

When the input to the system is a random process $X(t)$, the resulting output process $Y(t)$ is given by

$$Y(t) = \int_{-\infty}^{\infty} X(t-\tau)h(\tau)\, d\tau \qquad (4.15.\text{a})$$

$$= \int_{-\infty}^{\infty} X(\tau)h(t-\tau)\, d\tau \qquad (4.15.\text{b})$$

Note that Equation 4.15 implies that each member function of $X(t)$ produces a member function of $Y(t)$ according to Equation 4.13.

As with discrete time inputs, distribution functions of the process $Y(t)$ are very difficult to obtain except for the Gaussian case in which $Y(t)$ is Gaussian when $X(t)$ is Gaussian. Rather than attempting to obtain a complete description of $Y(t)$, we settle for a less complete description of the output than we have for deterministic problems. In most cases with random inputs, we find the mean, autocorrelation function, spectral density function, and mean-square value of the output process.

4.3.1 Mean and Autocorrelation Function

Assuming that $h(t)$ and $X(t)$ are real-valued and that the expectation and integration order can be interchanged because integration is a linear operator, we can calculate the mean and autocorrelation function of the output as

$$E\{Y(t)\} = E\left\{\int_{-\infty}^{\infty} X(t-\tau)h(\tau)\, d\tau\right\}$$

$$= \int_{-\infty}^{\infty} E\{X(t-\tau)\}h(\tau)\, d\tau$$

$$= \int_{-\infty}^{\infty} \mu_X(t-\tau)h(\tau)\, d\tau \qquad (4.16)$$

and

$$R_{YY}(t_1, t_2) = E\{Y(t_1) Y(t_2)\}$$
$$= E\left\{\int_{-\infty}^{\infty}\int_{-\infty}^{\infty} X(t_1 - \tau_1)h(\tau_1)X(t_2 - \tau_2)h(\tau_2)\, d\tau_1\, d\tau_2\right\} \quad (4.17)$$
$$= \int_{-\infty}^{\infty}\int_{-\infty}^{\infty} h(\tau_1)h(\tau_2)R_{XX}(t_1 - \tau_1, t_2 - \tau_2)\, d\tau_1\, d\tau_2$$

4.3.2 Stationarity of the Output

From Equation 4.15.a we have

$$Y(t) = \int_{-\infty}^{\infty} X(t - \tau)h(\tau)\, d\tau$$

and

$$Y(t + \epsilon) = \int_{-\infty}^{\infty} X(t + \epsilon - \tau)h(\tau)\, d\tau$$

Now, if the processes $X(t)$ and $X(t + \epsilon)$ have the same distributions [i.e., $X(t)$ is strict-sense stationary] then the same is true for $Y(t)$ and $Y(t + \epsilon)$ and hence $Y(t)$ is strict-sense stationary.

If $X(t)$ is WSS, then $\mu_X(t)$ does not depend on t and we have from Equation 4.16

$$E\{Y(t)\} = \int_{-\infty}^{\infty} \mu_X h(\tau)\, d\tau$$
$$= \mu_X \int_{-\infty}^{\infty} h(\tau)\, d\tau = \mu_X H(0) \quad (4.18)$$

Thus, the mean of the output does not depend on time. The autocorrelation function of the output given in Equation 4.17 becomes

$$R_{YY}(t_1, t_2) = \int_{-\infty}^{\infty}\int_{-\infty}^{\infty} h(\tau_1)h(\tau_2)R_{XX}[(t_2 - t_1) - (\tau_2 - \tau_1)]\, d\tau_1\, d\tau_2 \quad (4.19)$$

Since the integral depends only on the time difference $t_2 - t_1$, $R_{YY}(t_1, t_2)$ will also be a function of the difference $t_2 - t_1$. This coupled with the fact that μ_Y

is a constant establishes that the output process $Y(t)$ is WSS if the input process $X(t)$ is WSS.

4.3.3 Power Spectral Density of the Output

When $X(t)$ is WSS it can be shown that

$$R_{YX}(\tau) = R_{XX}(\tau) * h(-\tau) \tag{4.20.a}$$

$$R_{XY}(\tau) = R_{XX}(\tau) * h(\tau) \tag{4.20.b}$$

and

$$R_{YY}(\tau) = R_{YX}(\tau) * h(\tau) \tag{4.21}$$

$$= R_{XX}(\tau) * h(\tau) * h(-\tau) \tag{4.22}$$

where $*$ denotes convolution. Taking the Fourier transform of both sides of Equation 4.22, we obtain the power spectral density of the output as

$$S_{YY}(f) = S_{XX}(f)|H(f)|^2 \tag{4.23}$$

Equation 4.23, which is of the same form as Equation 4.11.b, is a very important relationship in the frequency domain analysis of systems that are driven by random input signals. This equation shows that an input spectral component at frequency f is modified according to $|H(f)|^2$, which is sometimes referred to as the *power transfer function*. By choosing $H(f)$ appropriately, we can emphasize or reject selected spectral components of the input signal. Such operations are referred to as "filtering."

Note that in the sinusoidal steady-state analysis of electrical circuits we use an input voltage (or current) of the form

$$x(t) = A\sin(2\pi ft)$$

as the input to the system and write the output voltage (or current) as

$$y(t) = A|H(f)|\sin[2\pi ft + \text{angle of } H(f)]$$

Note that the preceding equation is a *voltage* to *voltage* relationship and it involves the magnitude and phase of $H(f)$. In contrast, Equation 4.23 is a power to power relationship defined by $|H(f)|^2$.

RESPONSE OF LTIVC CONTINUOUS-TIME SYSTEMS

Power Spectral Density Function. The definition of psd given in Equation 3.43 can now be justified using Equation 4.23. If we have an ideal bandpass filter which is defined by

$$H(f) = 1, \quad f_1 \leq |f| \leq f_2$$
$$= 0 \quad \text{elsewhere}$$

then because (Equation 3.41)

$$E[Y^2(t)] = \int_{-\infty}^{\infty} S_{YY}(f)\, df$$

Using the definition of $H(f)$ and the fact that $S_{XX}(f)$ is even

$$E[Y^2(t)] = 2 \int_{f_1}^{f_2} S_{XX}(f)\, df$$

Because the average power of the output $Y(t)$ of the ideal bandpass filter is the integral of the power spectral density between $-f_2$ and $-f_1$ and between f_1 and f_2, we say that the power of $X(t)$ between the frequencies f_1 and f_2 is given by Equation 3.43. Thus, we naturally call $S_{XX}(f)$ the power spectral density function.

The foregoing development also shows, because $E[Y^2(t)] \geq 0$, that $S_{XX}(f) \geq 0$ for all f.

EXAMPLE 4.4.

$X(t)$ is the input voltage to the system shown in Figure 4.1, and $Y(t)$ is the output voltage. $X(t)$ is a stationary random process with $\mu_X = 0$ and $R_{XX}(\tau) = \exp(-\alpha|\tau|)$. Find μ_Y, $S_{YY}(f)$, and $R_{YY}(\tau)$.

SOLUTION: From the circuit in Figure 4.1

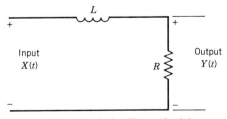

Figure 4.1 Circuit for Example 4.1.

232 RESPONSE OF LINEAR SYSTEMS TO RANDOM INPUTS

$$H(f) = \frac{R}{R + j2\pi f L}$$

Also

$$S_{XX}(f) = \int_{-\infty}^{0} \exp(\alpha\tau)\exp(-j2\pi f\tau)\, d\tau + \int_{0}^{\infty} \exp(-\alpha\tau)\exp(-j2\pi f\tau)\, d\tau$$

$$= \frac{2\alpha}{\alpha^2 + (2\pi f)^2}$$

Using Equation 4.18

$$\mu_Y = 0$$

Using Equation 4.23

$$S_{YY}(f) = \left[\frac{2\alpha}{\alpha^2 + (2\pi f)^2}\right] \frac{R^2}{R^2 + (2\pi f L)^2}$$

Taking the inverse Fourier transform

$$R_{YY}(\tau) = \frac{\left(\dfrac{R}{L}\right)^2}{\left(\dfrac{R}{L}\right)^2 - \alpha^2} \exp(-\alpha|\tau|) + \frac{\left(\dfrac{R}{L}\right)\alpha}{\alpha^2 - \left(\dfrac{R}{L}\right)^2} \exp\left[-\left(\dfrac{R}{L}\right)|\tau|\right]$$

EXAMPLE 4.5.

A differentiator is a system for which

$$H(f) = j2\pi f$$

If a stationary random process $X(t)$ is differentiated, find the power density spectrum and autocorrelation function of the derivative random process $X'(t)$.

SOLUTION:

$$S_{X'X'}(f) = (2\pi f)^2 S_{XX}(f)$$

and

$$R_{X'X'}(\tau) = -\frac{d^2 R_{XX}(\tau)}{d\tau^2}$$

which agrees with Equation 3.60. Note that differentiation results in the multiplication of the spectral power density by f^2. If $X(t)$ is noise, then differentiation greatly magnifies the noise at higher frequencies and provides a theoretical explanation for the practical result that it is impossible to build a differentiator that is not "noisy."

EXAMPLE 4.6.

An averaging circuit with an integration period T has an impulse response

$$h(t) = \frac{1}{T}, \quad 0 \leq t \leq T$$
$$= 0 \quad \text{elsewhere}$$

Indeed it is called averaging because

$$Y(t) = X(t) * h(t)$$
$$Y(t) = \frac{1}{T} \int_{t-T}^{t} X(\tau) \, d\tau$$

Find $S_{YY}(f)$ in terms of the input spectral density $S_{XX}(f)$.

SOLUTION:

$$H(f) = \frac{1}{T} \int_0^T \exp(-j2\pi f \tau) \, d\tau = \exp(-j\pi f T) \frac{\sin(\pi f T)}{\pi f T}$$

thus

$$S_{YY}(f) = \frac{\sin^2(\pi f T)}{(\pi f T)^2} S_{XX}(f)$$

which agrees with the result implied in Equation 3.75.b.

This result demonstrates, if $X(t)$ is noise, how the higher frequency noise is reduced by integration.

4.3.4 Mean-square Value of the Output

The mean square value of the output, which is a measure of the average value of the output "power," is of interest in many applications. The mean-square value is given by

$$E\{Y^2(t)\} = R_{YY}(0) = \int_{-\infty}^{\infty} S_{YY}(f) \, df$$

$$= \int_{-\infty}^{\infty} S_{XX}(f) |H(f)|^2 \, df \tag{4.24}$$

Except in some simple cases, the evaluation of the preceding integral is somewhat difficult.

If we make the assumption that $R_{XX}(\tau)$ can be expressed as a sum of complex exponentials [i.e., $S_{XX}(f)$ is a rational function of f], then the evaluation of the integral can be simplified. Since $S_{XX}(f)$ is an even function, we can make a transformation $s = 2\pi j f$ and factor $S_{XX}(f)$ as

$$S_{XX}\left(\frac{s}{2\pi j}\right) = \frac{a(s)a(-s)}{b(s)b(-s)}$$

where $a(s)/b(s)$ has all of its poles and zeros (roots) in the left-half of the s-plane and $a(-s)/b(-s)$ has all of its roots in the right half plane. No roots of $b(s)$ are permitted on the imaginary axis. We can factor $|H(f)|^2$ in a similar fashion and write Equation 4.24 as

$$E\{Y^2(t)\} = \frac{1}{2\pi j} \int_{-j\infty}^{j\infty} \frac{c(s)c(-s)}{d(s)d(-s)} \, ds \tag{4.25}$$

where

$$S_{XX}(f)H(f)H^*(f)|_{f=s/2\pi j} = \frac{c(s)c(-s)}{d(s)d(-s)}$$

TABLE 4.1 TABLE OF INTEGRALS

$$I_n = \frac{1}{2\pi j} \int_{-j\infty}^{j\infty} \frac{c(s)c(-s)}{d(s)d(-s)} ds$$

$$c(s) = c_{n-1} s^{n-1} + c_{n-2} s^{n-2} + \cdots + c_0$$
$$d(s) = d_n s^n + d_{n-1} s^{n-1} + \cdots + d_0$$

$$I_1 = \frac{c_0^2}{2 d_0 d_1}$$

$$I_2 = \frac{c_1^2 d_0 + c_0^2 d_2}{2 d_0 d_1 d_2}$$

$$I_3 = \frac{c_2^2 d_0 d_1 + (c_1^2 - 2 c_0 c_2) d_0 d_3 + c_0^2 d_2 d_3}{2 d_0 d_3 (d_1 d_2 - d_0 d_3)}$$

$$I_4 = \frac{c_3^2(-d_0^2 d_3 + d_0 d_1 d_2) + (c_2^2 - 2 c_1 c_3) d_0 d_1 d_4 + (c_1^2 - 2 c_0 c_2) d_0 d_3 d_4 + c_0^2(-d_1 d_4^2 + d_2 d_3 d_4)}{2 d_0 d_4 (-d_0 d_3^2 - d_1^2 d_4 + d_1 d_2 d_3)}$$

and $c(s)$ and $d(s)$ contain the left-half plane roots of S_{YY}. Values of integrals of the form given in Equation 4.25 have been tabulated in many books and an abbreviated table is given in Table 4.1.

We now present an example on the use of these tabulated values.

EXAMPLE 4.7.

The input to an R-C lowpass filter with

$$H(f) = \frac{1}{1 + j(f/1000)}$$

is a zero-mean stationary random process with

$$S_{XX}(f) = 10^{-12} \text{ watt/Hz}$$

Find $E\{Y^2(t)\}$ where $Y(t)$ is the output.

SOLUTION:

$$S_{YY}(f) = 10^{-12} \frac{1}{1 + j(f/1000)} \cdot \frac{1}{1 - j(f/1000)}$$

Transforming to the s-domain with $s = 2\pi jf$, we can write the integral for $E\{Y^2(t)\}$ using Equation 4.25 as

$$E\{Y^2(t)\} = \frac{1}{2\pi j} \int_{-j\infty}^{j\infty} \frac{10^{-6}}{1 + (s/2000\pi)} \frac{10^{-6}}{1 - (s/2000\pi)} ds$$

With $n = 1$, $c_0 = 10^{-6}$, $d_0 = 1$, and $d_1 = 1/(2000\pi)$, we find from Table 4.1

$$E\{Y^2(t)\} = c_0^2/(2 d_0 d_1) = 10^{-12}(1000\pi)$$

EXAMPLE 4.8.

A random (pulse) process $Y(t)$ has the form

$$Y(t) = \sum_{k=-\infty}^{\infty} A(k)p(t - kT_s - D)$$

where $p(t)$ is an arbitrary deterministic pulse of known shape and duration less than T_s, D is a random variable with a uniform distribution in the interval $[0, T_s]$, and $A(k)$ is a stationary random sequence (see Figure 4.2 for an example where $A(k)$ is binary). Find the psd of $Y(t)$ in terms of the autocorrelation (and psd) function of $A(k)$ and $P_F(f)$.

SOLUTION: Now, suppose we define a new process

$$X(t) = \sum_{k=-\infty}^{\infty} A(k)\delta(t - kT_s - D)$$

The only difference between $Y(t)$ and $X(t)$ is the "pulse" shape and it is easy to see that if $X(t)$ is passed through a linear time invariant system that converts each impulse $\delta(t)$ into a pulse $p(t)$, the resulting output will be $Y(t)$. Such a system will have an impulse response of $p(t)$ and we can write

$$Y(t) = X(t) * p(t)$$

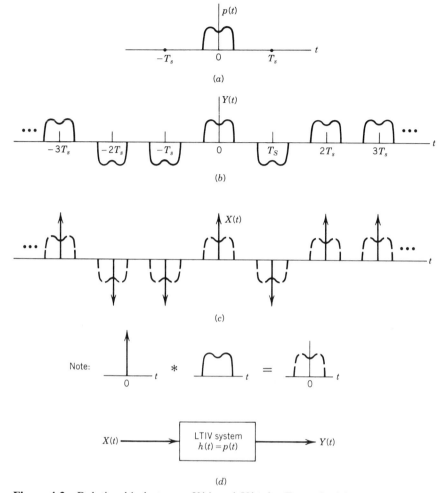

Figure 4.2 Relationship between $X(t)$ and $Y(t)$ for Example 4.8.

and hence from Equation 4.23 we have

$$S_{YY}(f) = S_{XX}(f)|P_F(f)|^2$$

From Equation 3.53

$$S_{XX}(f) = \frac{1}{T_s}\left\{R_{AA}(0) + 2\sum_{k=1}^{\infty} R_{AA}(k)\cos 2\pi k f T_s\right\}$$

and hence

$$S_{YY}(f) = \frac{|P_F(f)|^2}{T_s}\left\{R_{AA}(0) + 2\sum_{k=1}^{\infty} R_{AA}(k)\cos 2\pi k f T_s\right\}$$

Note that the preceding equation, which gives the psd of an arbitrary pulse process, has two parts. The first part $|P_F(f)|^2$ shows the influence of the pulse shape on the shape of the spectral density and the second part, in brackets, shows the effect of the correlation properties of the amplitude sequence. The factor $1/T_s$ converts energy distribution (or density) to power distribution.

4.3.5 Multiple Input–Output Systems

Occasionally we will have to analyze systems with multiple inputs and outputs. Analysis of these systems can be reduced to the study of several single input–single output systems (see Figure 4.3). Consider two such linear systems with two inputs $X_1(t)$ and $X_2(t)$ and impulse responses $h_1(t)$ and $h_2(t)$ as shown in Figure 4.3.

Assuming the systems to be LTIVC, and the inputs to be jointly stationary, we have

$$Y_1(t) = \int_{-\infty}^{\infty} X_1(t-\alpha)h_1(\alpha)\,d\alpha$$

$$Y_2(t) = \int_{-\infty}^{\infty} X_2(t-\beta)h_2(\beta)\,d\beta$$

$$Y_1(t)Y_2(t+\tau) = \int_{-\infty}^{\infty} X_1(t-\alpha)\,Y_2(t+\tau)h_1(\alpha)\,d\alpha$$

$$X_1(t)Y_2(t+\tau) = \int_{-\infty}^{\infty} X_1(t)\,X_2(t+\tau-\beta)h_2(\beta)\,d\beta$$

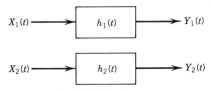

Figure 4.3 Multiple input–output systems.

Taking the expected values on both sides, we conclude

$$R_{Y_1Y_2}(\tau) = R_{X_1Y_2}(\tau) * h_1(-\tau) \quad (4.26.a)$$

$$R_{X_1Y_2}(\tau) = R_{X_1X_2}(\tau) * h_2(\tau) \quad (4.26.b)$$

The Fourier transforms of these equations yield

$$S_{Y_1Y_2}(f) = S_{X_1Y_2}(f)H_1^*(f) \quad (4.27.a)$$

$$S_{X_1Y_2}(f) = S_{X_1X_2}(f)H_2(f) \quad (4.27.b)$$

Hence

$$S_{Y_1Y_2}(f) = S_{X_1X_2}(f)H_1^*(f)H_2(f) \quad (4.27.c)$$

Equations 4.26 and 4.27 describe the input–output relationship for multiple input–output systems in terms of the joint properties of the input signals and the system impulse responses (or transfer functions).

4.3.6 Filters

Filtering is commonly used in electrical systems to reject undesirable signals and noise and to select the desired signal. A simple example of filtering occurs when we "tune" in a particular radio or TV station to "select" one of many signals. Filters are also used extensively to remove noise in communication links.

A filter has a transfer function $H(f)$ that is selected carefully to modify the spectral components of the input signal. Ideal versions of three types of filters are shown in Figure 4.4. In every case, the idealized system has a transfer function whose magnitude is flat within its "passband" and zero outside of this band; its midband gain is unity and its phase is a linear function of frequency.

The transfer function of practical filters will deviate from their corresponding ideal versions. The Butterworth lowpass filter, for example, has a magnitude response of the form

$$|H(f)|^2 = \frac{1}{1 + (f/B)^{2n}}$$

where n is the order of the filter and B is a parameter that determines the bandwidth of the filter. For a detailed discussion of filters, see Reference [1].

240   RESPONSE OF LINEAR SYSTEMS TO RANDOM INPUTS

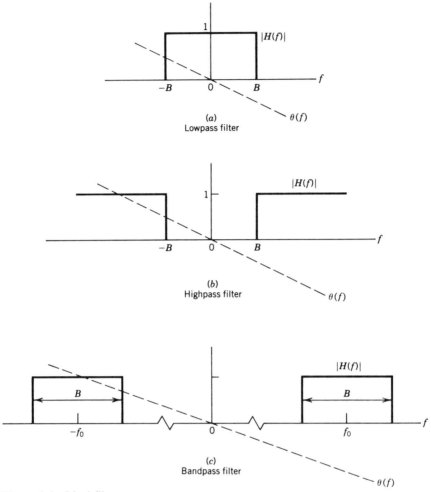

Figure 4.4 Ideal filters.

To simplify analysis, it is often convenient to approximate the transfer function of a practical filter $H(f)$ by an ideal version $\tilde{H}(f)$ as shown in Figure 4.5. In replacing an actual system with an ideal one, the later would be assigned a "midband" gain and phase slope that approximate the actual values. The bandwidth B_N of the ideal approximation (in lowpass and bandpass cases) is chosen according to some convenient basis. For example, the bandwidth of the ideal filter can be set equal to the 3-dB (or half-power) bandwidth of the actual filter or it can be chosen to satisfy a specific requirement. An example of the latter case is to choose B_N such that the actual and ideal systems produce the same output power when each is excited by the same source.

RESPONSE OF LTIVC CONTINUOUS-TIME SYSTEMS   241

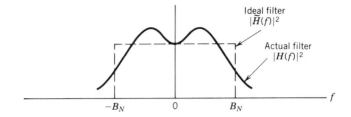

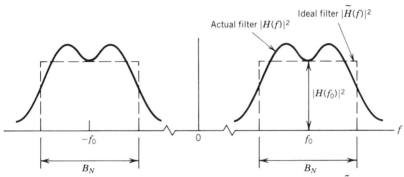

Figure 4.5 Noise bandwidth of filter; areas under $|H(f)|^2$ and $|\tilde{H}(f)|^2$ are equal.

Consider an ideal and actual lowpass filter whose input is white noise, that is, a noise process whose power spectral density has a constant value, say $\eta/2$, for all frequencies. The average output powers of the two filters are given by

$$E\{Y^2(t)\} = \frac{\eta}{2} \int_{-\infty}^{\infty} |H(f)|^2 \, df$$

for the actual filter and

$$E\{Y^2(t)\} = \left(\frac{\eta}{2}\right) |H(0)|^2 \, 2B_N$$

for the ideal version. By equating the output powers, we obtain

$$B_N = \frac{\int_{-\infty}^{\infty} |H(f)|^2 \, df}{2|H(0)|^2} \qquad (4.28)$$

This value of B_N is called the *noise-equivalent bandwidth* of the actual filter. Extension of this definition to the bandpass case is obvious (see Figure 4.5).

EXAMPLE 4.9.

Find the noise-equivalent bandwidth of a first-order Butterworth filter with

$$|H(f)|^2 = 1/[1 + (f/B)^2]$$

SOLUTION: Using Equation 4.28

$$B_N = \int_0^\infty 1/[1 + (f/B)^2] \, df$$

Using Table 4.1

$$= B(\pi/2)$$

The reader can verify (Problem 4.18) that the noise equivalent bandwidth of an nth order Butterworth filter is

$$B_N = B[(\pi/2n)/\sin(\pi/2n)]$$

As $n \to \infty$, the Butterworth filter approaches the transfer function of an ideal lowpass filter with a bandwidth B.

4.4 SUMMARY

After reviewing deterministic system analysis for linear-time invariant causal systems, we considered these systems when the input is a random process.

It was shown that when the input to a LTIVC system is a SSS (or WSS) random process, then the output is a SSS (or WSS) random process. While the distribution functions of the output process are difficult to find except in the Gaussian case, simple relations were developed for the mean, autocorrelation function, and power spectral density functions of the output process. These are as follows:

$$\mu_Y = H(0)\mu_X$$
$$R_{YY}(\tau) = R_{XX}(\tau) * h(\tau) * h(-\tau)$$
$$S_{YY}(f) = |H(f)|^2 S_{XX}(f)$$

where $X(t)$ is the input process, $Y(t)$ is the output process, $h(t)$ is the impulse response, and $H(f)$ is the transfer function of the system. The average power in the output $Y(t)$ can be obtained by integrating $S_{YY}(f)$ using the table of integrals provided in this chapter.

In the case of random sequences, the relation for power spectral density function was found using the Fourier transform and the Z transform and an application to digital filters was shown. The relations for the mean, correlation functions, and power spectral density functions for continuous random processes were found to be of the same form as those for sequences.

The only nonlinear systems considered in this chapter were instantaneous systems. Such systems with single inputs can be handled relatively easily as illustrated by Example 4.1.

4.5 REFERENCES

A large number of textbooks treat the subject of deterministic signals and systems. References [1], [3], [4], [6] and [8] are typical undergraduate-level textbooks that provide excellent treatment of discrete and continuous time signals and systems.

Response of systems to random inputs is treated in References [2], [5] and [7] with [2] providing an introductory-level treatment and [5] providing in-depth coverage.

[1] N. Ahmed and T. Natarajan, *Discrete-Time Signals and Systems,* Reston Publishing Co., Reston, Va., 1983.

[2] R. G. Brown, *Random Signal Analysis and Kalman Filtering,* John Wiley & Sons, New York, 1983.

[3] R. A. Gable and R. A. Roberts, *Signals and Linear Systems,* 2nd ed., John Wiley & Sons, New York, 1981.

[4] M. T. Jong, *Discrete-Time Signals and Systems,* McGraw-Hill, New York, 1982.

[5] H. J. Larson and B. O. Schubert, *Probabilistic Models in Engineering Sciences,* Vol. 2, John Wiley & Sons, New York, 1979.

[6] C. D. McGillem and G. R. Cooper, *Continuous and Discrete Signal and System Analysis,* 2nd ed., Holt, Rinehart and Winston, New York, 1984.

[7] A. Papoulis, *Probability, Random Variables and Stochastic Processes*, 2nd ed., McGraw-Hill, New York, 1984.

[8] R. E. Ziemer, W. H. Tranter, and D. R. Fanin, *Signals and Systems*, Macmillan, New York, 1983.

4.6 PROBLEMS

4.1 $X(t)$ is a zero-mean stationary Gaussian random process with a power spectral density function $S_{XX}(f)$. Find the power spectral density function of

$$Y(t) = X^2(t)$$

4.2 Show that the output of the LTIVC system is SSS if the input is SSS.

4.3 Show that in a LTIVC system

$$R_{YY}(k) = R_{YX}(k) * h(k)$$

4.4 The output of a discrete-time system is related to the input by

$$Y(n) = \frac{1}{k} \sum_{i=1}^{k} X(n-i)$$

    a. Find the transfer function of the system.

    b. If the input $X(n)$ is stationary with

$$E\{X(n)\} = 0$$

$$R_{XX}(k) = \begin{cases} 1, & \text{for } k = 0 \\ 0 & \text{for } k \neq 0 \end{cases}$$

    find $S_{YY}(f)$ and $E\{Y^2(n)\}$.

4.5 Repeat Problem 4.4 with $Y(n) = X(n) - X(n-1)$.

4.6 The input–output relationship of a discrete-time LTIVC system is given by

$$Y(n) = h(0) X(n) + h(1) X(n-1) + \cdots + h(k) X(n-k)$$

The input sequence $X(n)$ is stationary, zero mean, Gaussian with

$$E\{X(n) X(n+j)\} = \begin{cases} 1, & j = 0 \\ 0 & j \neq 0 \end{cases}$$

    a. Find the pdf of $Y(n)$.

    b. Find $R_{YY}(n)$ and $S_{YY}(f)$.

4.7 Consider the difference equation
$$X(n+1) = \sqrt{n}\, X(n) + U(n), \quad n = 0, 1, 2, \ldots$$
with $X(0) = 1$, and $U(n)$, $n = 0, 1, \ldots$, a sequence of zero-mean, uncorrelated, Gaussian variables.

    a. Find $\mu_X(n)$.

    b. Find $R_{XX}(0, n)$, $R_{XX}(1, 1)$, $R_{XX}(1, 2)$, and $R_{XX}(3, 1)$.

4.8 An autoregressive moving average process (ARMA) is described by
$$Y(t) = a_1 Y(t - T) + a_2 Y(t - 2T) + \cdots + a_m Y(t - mT)$$
$$+ X(t) + b_1 X(t - T) + \cdots + b_n X(t - nT)$$
Find $S_{YY}(f)$ in terms of $S_{XX}(f)$ and the coefficients of the model.

4.9 With reference to the model defined in Problem 4.8, find $S_{YY}(f)$ for the following two special cases:

    a. $X(t)$ is Gaussian with $S_{XX}(f) = \eta/2$ for all f, and
$$b_2 = b_3 = \cdots = b_n = 0$$
$$a_1 = a_2 = \cdots = a_m = 0$$
(first-order moving average process)

    b. Same $X(t)$ as in (a), with
$$a_2 = a_3 = \cdots = a_m = 0$$
$$b_1 = b_2 = \cdots = b_n = 0$$
(first-order autoregressive process)

4.10 Establish the modified versions of Equations 4.20.a and 4.20.b when both $X(t)$ and $h(t)$ are complex-valued functions of time.

4.11 Repeat Problem 4.10 for Equations 4.26.a, 4.26.b, and 4.27.c.

4.12 Consider an ideal integrator
$$Y(t) = \frac{1}{T} \int_{t-T}^{t} X(\alpha)\, d\alpha$$

    a. Find the transfer function of the integrator.

    b. If the integrator input is a stationary, zero-mean white Gaussian noise with
$$S_{XX}(f) = \eta/2$$
find $E\{Y^2(t)\}$.

4.13 Using the spectral factorization method, find $E\{Y^2(t)\}$ where $Y(t)$ is a stationary random process with

$$S_{YY}(f) = \frac{(2\pi f)^2 + 1}{(2\pi f)^4 + 8(2\pi f)^2 + 16}$$

4.14 Assume that the input to a linear time-invariant system is a zero-mean Gaussian random process with

$$S_{XX}(f) = \eta/2$$

and that the impulse response of the system is

$$h(t) = \begin{cases} \exp(-t), & t \geq 0 \\ 0 & \text{elsewhere} \end{cases}$$

a. Find $S_{YY}(f)$, where $Y(t)$ is the output.
b. Find $E\{Y^2(t)\}$.

4.15 Let $X(t)$ be a random binary waveform of the form

$$X(t) = \sum_{k=-\infty}^{\infty} A(k)p(t - kT_s - D)$$

where $A(k)$ is a sequence of independent amplitudes, $A(k) = \pm 1$ with equal probability, $1/T_s$ is the pulse rate, $p(t)$ is a unit amplitude rectangular pulse with a duration T_s, and D is a random delay with a uniform distribution in the interval $[0, T_s]$. Let

$$Y(t) = \sum_{k=-\infty}^{\infty} B(k)p(t - kT_s - D)$$

where $B(k) = 0$ if $A(k) = -1$, otherwise $B(k)$ takes on alternating values of $+1$ and -1 [i.e., the negative amplitude pulses in $X(t)$ appear with 0 amplitude in $Y(t)$, and the positive amplitude pulses in $X(t)$ appear with alternating polarities in $Y(t)$]. $Y(t)$ is called a bipolar random binary waveform.

a. Sketch a member function of $X(t)$ and the corresponding member function of $Y(t)$.
b. Find the psd of $Y(t)$ and compare it with the psd of $X(t)$.

4.16 Consider a pulse waveform

$$Y(t) = \sum_{k=-\infty}^{\infty} A(k)p(t - kT_s - D)$$

where $p(t)$ is a rectangular pulse of height 1 and width $T_s/2$, D is a random variable uniformly distributed in the interval $[0, T_s]$, and $A(k)$ is a stationary sequence with

$$E\{A(k)\} = \frac{3}{4} \quad E\{A(k)^2\} = \frac{5}{8}$$

$$E\{A(k) A(k+j)\} = \frac{9}{16} \quad \text{for all } j \geq 1$$

Find $R_{YY}(\tau)$ and $S_{YY}(f)$ and sketch $S_{YY}(f)$.

4.17 Find the transfer function of a shaping filter that will produce an output spectrum

$$S_{YY}(f) = \frac{(2\pi f)^2 + 1}{(2\pi f)^4 + 13(2\pi f)^2 + 36}$$

from an input spectrum

$$S_{XX}(f) = \eta/2$$

4.18 a. Find the noise bandwidth of the nth order Butterworth filter with the magnitude response

$$|H(f)|^2 = 1/[1 + (f/B)^{2n}]$$

for $n = 1, 2, 3, 4,$ and 8.

b. From a noise-rejection point of view, is there much to be gained by using anything higher than a third-order Butterworth filter?

4.19 Find the noise bandwidth of the filters shown in Figure 4.6.

4.20 The input to a lowpass filter with a transfer function

$$H(f) = \frac{1}{1 + j(f/f_0)}$$

is $X(t) = S(t) + N(t)$. The signal $S(t)$ has the form

$$S(t) = A \sin(2\pi f_c t + \Theta)$$

where A and f_c are real constants and Θ is a random variable uniformly distributed in the interval $[-\pi, \pi]$. The noise $N(t)$ is white Gaussian noise with $S_{NN}(f) = \eta/2$.

248   RESPONSE OF LINEAR SYSTEMS TO RANDOM INPUTS

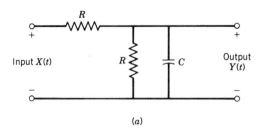

(a)

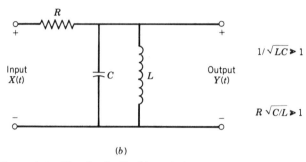

$1/\sqrt{LC} \gg 1$

$R\sqrt{C/L} \gg 1$

(b)

Figure 4.6   Circuits for Problem 4.19.

a. Find the power spectral density function of the output signal and noise.

b. Find the ratio of the average output signal power to output noise power.

c. What value of f_0 will maximize the ratio of part (b)?

CHAPTER FIVE

Special Classes of Random Processes

5.1 INTRODUCTION

In deterministic signal theory, classes of special signals such as impulses and complex exponentials play an important role. There are several classes of random processes that play a similar role in the theory and application of random processes. In this chapter, we discuss four important classes of random processes: autoregressive and moving average processes, Markov processes, Poisson processes, and Gaussian processes. The basic properties of these processes are derived and their applications are illustrated with a number of examples.

We start with two discrete time processes that are generated by linear time-invariant difference equations. These two models, the autoregressive and moving average models are widely used in data analysis. A very useful application of these two processes lies in fitting models to data and for model-based estimation of autocorrelation and power spectral densities. We derive the properties of autoregressive and moving average processes in Section 5.2. Detailed discussion of their statistical application is contained in Chapter 9.

Markov sequences and processes are discussed next. Markov processes have the property that the value of the process depends only on the most recent value, and given that value, the random process is independent of all values in the more distant past. Models in which output is independent of past input values and past output values given the present output are common in electrical engineering (for example, the output of a linear time-invariant causal system). Properties and applications of Markov processes are discussed in detail in Section 5.3.

The next class of model that is developed in this chapter is the point-process model with an emphasis on the Poisson process. Point processes are very useful

250   SPECIAL CLASSES OF RANDOM PROCESSES

for modeling and analyzing queues, and for describing "shot noise" in communication systems. In Section 5.4, we develop several point-process models and illustrate their usefulness in several interesting applications.

By virtue of the central limit theorem, many random phenomena are well approximated by Gaussian random processes. One of the most important uses of the Gaussian process is to model and analyze the effects of "thermal" noise in electronic circuits. Properties of the Gaussian process are derived and the use of Gaussian process models to analyze the effects of noise in communication systems is illustrated in Section 5.5.

5.2   DISCRETE LINEAR MODELS

In this section, we introduce two stationary linear models that are often used to model random sequences. These models can be "derived" from data as is shown in Chapter 9. Combinations of these two models describe the output of a LLTIVC system, and they are the most used empirical models of random sequences.

5.2.1   Autoregressive Processes

An autoregressive process is one represented by a difference equation of the form:

$$X(n) = \sum_{i=1}^{p} \phi_{p,i} X(n - i) + e(n) \tag{5.1}$$

where $X(n)$ is the real random sequence, $\phi_{p,i}$, $i = 1, \ldots, p$, $\phi_{p,p} \neq 0$ are parameters, and $e(n)$ is a sequence of independent and identically distributed zero-mean Gaussian random variables, that is,

$$E\{e(n)\} = 0$$

$$E\{e(n)e(j)\} = \begin{cases} \sigma_N^2, & \text{for } n = j \\ 0 & \text{for } n \neq j \end{cases}$$

$$f_{e(n)}(\lambda) = \frac{1}{\sqrt{2\pi}\,\sigma_N} \exp\left\{-\frac{\lambda^2}{2\sigma_N^2}\right\}$$

The sequence $e(n)$ is called white Gaussian noise. (See Section 5.5.2.) Thus, an autoregressive process is simply another name for a linear difference equation model when the input or forcing function is white Gaussian noise. Further, if the difference equation is of order p (i.e., $\phi_{p,p} \neq 0$), then the sequence is called

DISCRETE LINEAR MODELS 251

a pth order autoregressive model. We now study such models in some detail because of their importance in applications, primarily due to their use in creating models of random processes from data. Autoregressive models are also called state models, recursive digital filters, and all-pole models as explained later.

Equation 5.1 can be easily reduced to a state model (see Problem 5.1) of the form

$$\mathbf{X}(n) = \mathbf{\Phi}\mathbf{X}(n-1) + \mathbf{E}(n) \tag{5.2}$$

In addition, models of the form of Equation 5.1 are often called recursive digital filters. In this case, the $\phi_{p,i}$'s are usually called h_i's, which are terms of the unit pulse response, and Equation 5.1 is usually written as

$$X(n) = \sum_{i=1}^{p} h_i X(n-i) + e(n)$$

and the typical block diagram for this model is shown in Figure 5.1.

Using the results derived in Chapter 4, we can show that the transfer function of the system represented in Equation 5.1 and Figure 5.1 is

$$H(f) = \frac{1}{1 - \sum_{i=1}^{p} \phi_{p,i} \exp(-j2\pi f i)}, \quad |f| < \frac{1}{2} \tag{5.3}$$

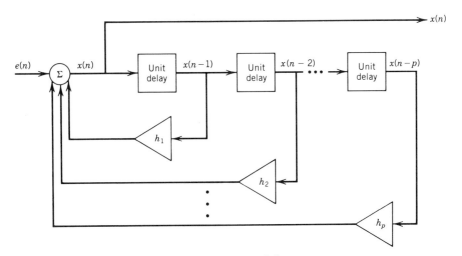

Figure 5.1 Recursive filter (autoregressive model).

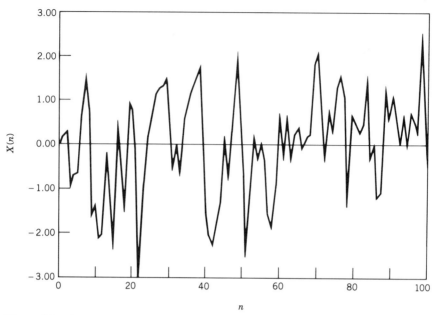

Figure 5.2 Sample function of the first-order autoregressive process: $X(n) = .48X(n-1) + e(n)$.

and the autoregressive process $X(n)$ is the output of this system when the input is $e(n)$. Note that there are no zeros in the transfer function given in Equation 5.3.

First-order Autoregressive Model. Consider the model

$$X(n) = \phi_{1,1} X(n-1) + e(n) \tag{5.4}$$

where $e(n)$ is zero-mean stationary white Gaussian noise. Note that Equation 5.4 defines $X(n)$ as a Markov process. Also note that Equation 5.4 is a first-order regression equation with $X(n-1)$ as the "controlled" variable. (Regression is discussed in Chapter 8.) A sample function for $X(n)$ when $\phi_{1,1} = .48$ is shown in Figure 5.2.

We now find the mean, variance, autocorrelation function, which is also the autocovariance, correlation coefficient, and the power spectral density of this process.

Since we wish the model to be stationary, this requirement imposes certain conditions on the parameters of the model. The mean of $X(n)$ can be obtained as

$$\mu_X = E\{X(n)\} = E\{\phi_{1,1} X(n-1) + e(n)\}$$
$$= \phi_{1,1}\mu_X + 0$$

Thus, except for the case where $\phi_{1,1} = 1$

$$\mu_X = 0$$

These models are sometimes used for $n \geq 0$. In such cases, a starting or initial condition for the difference equation at time 0 is required. In these cases we require that $X(0)$ is Gaussian and

$$E\{X(0)\} = 0 \tag{5.5}$$

The variance of $X(n)$ is

$$\begin{aligned}\sigma_X^2 &= E\{X(n)^2\} \\ &= E\{\phi_{1,1}^2 X(n-1)^2 + e(n)^2 + 2\phi_{1,1} X(n-1)e(n)\}\end{aligned} \tag{5.6}$$

Because $X(n-1)$ consists of a linear combination of $e(n-1)$, $e(n-2)$, ..., it follows that $X(n-1)$ and $e(n)$ are independent. If a starting condition $X(0)$ is considered, then we also assume that $e(n)$ is independent of $X(0)$. Returning to Equation 5.6 and using independence of $X(n-1)$ and $e(n)$ plus stationarity, we obtain

$$\sigma_X^2 = \phi_{1,1}^2 \sigma_X^2 + \sigma_N^2$$

and hence

$$\sigma_X^2 = \frac{\sigma_N^2}{1 - \phi_{1,1}^2} \tag{5.7}$$

In order for σ_X^2 to be finite and nonnegative, $\phi_{1,1}$ must satisfy

$$-1 < \phi_{1,1} < 1 \tag{5.8}$$

The autocorrelation function of the first-order autoregressive process is given by

$$\begin{aligned}R_{XX}(m) &= E\{X(n)X(n-m)\}, \quad m \geq 1 \\ &= E\{[\phi_{1,1}X(n-1) + e(n)][X(n-m)]\} \\ &= \phi_{1,1} R_{XX}(m-1), \quad m \geq 1\end{aligned}$$

254   SPECIAL CLASSES OF RANDOM PROCESSES

Thus, $R_{XX}(m)$ is the solution to a first-order linear homogeneous difference equation, that is,

$$R_{XX}(m) = \phi_{1,1}^m R_{XX}(0) = \phi_{1,1}^m \sigma_X^2, \quad m \geq 0 \tag{5.9}$$

The autocorrelation coefficient of the process is

$$r_{XX}(m) = \frac{R_{XX}(m)}{R_{XX}(0)} = \phi_{1,1}^m, \quad m \geq 0 \tag{5.10}$$

This autocorrelation coefficient, for $\phi_{1,1} = 0.48$ is pictured as a histogram in Figure 5.3. When $\phi_{1,1}$ is negative, $r_{XX}(m)$ will decay in magnitude but oscillate in sign.

Finally, the power spectral density is found using Equation 4.11b as

$$S_{XX}(f) = |H(f)|^2 S_{ee}(f)$$

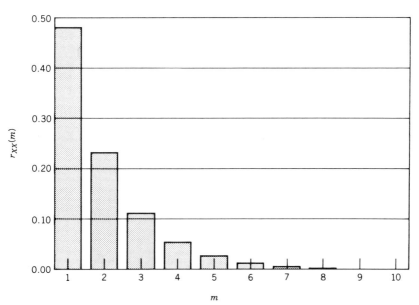

Figure 5.3   Correlation coefficient of the first-order autoregressive process: $X(n) = .48X(n - 1) + e(n)$.

The definition of $e(n)$ implies that

$$S_{ee}(f) = \sigma_N^2, \quad |f| < \frac{1}{2}$$

A special case of Equation 5.3 is

$$H(f) = \frac{1}{1 - \phi_{1,1}\exp(-j2\pi f)}, \quad |f| < \frac{1}{2}$$

and hence

$$|H(f)|^2 = \frac{1}{1 - 2\phi_{1,1}\cos 2\pi f + \phi_{1,1}^2}, \quad |f| < \frac{1}{2}$$

Thus

$$S_{XX}(f) = \frac{\sigma_N^2}{1 - 2\phi_{1,1}\cos 2\pi f + \phi_{1,1}^2}, \quad |f| < \frac{1}{2} \qquad (5.11)$$

Finally, using Equation 5.7 in Equation 5.11

$$S_{XX}(f) = \frac{\sigma_X^2(1 - \phi_{1,1}^2)}{1 - 2\phi_{1,1}\cos 2\pi f + \phi_{1,1}^2}, \quad |f| < \frac{1}{2} \qquad (5.12)$$

Equation 5.12 also can be found by taking the Fourier transform of Equation 5.9 (see Problem 5.5).

If we define z^{-1} to be the backshift or the delay operator, that is,

$$z^{-1}[X(n)] = X(n-1); \quad z^{-1}[e(n)] = e(n-1)$$
$$z^{-k}[X(n)] = X(n-k); \quad z^{-k}[e(n)] = e(n-k)$$

then Equation 5.4 becomes

$$X(n) = \phi_{1,1}z^{-1}[X(n)] + e(n) \qquad (5.13)$$

or

$$X(n) = \frac{e(n)}{1 - \phi_{1,1}z^{-1}} \tag{5.14}$$

And recognizing that if $|\phi_{1,1}| < 1$ as required by Equation 5.8, then

$$X(n) = \left[\sum_{i=0}^{\infty} \phi_{1,1}^i z^{-i}\right] e(n) = \sum_{i=0}^{\infty} \phi_{1,1}^i e(n-i) \tag{5.15}$$

(See Problem 5.20.) Thus, this first-order autoregressive model can be viewed as a weighted infinite sum of white noise terms.

Second-order Autoregressive Model. The second-order autoregressive process is given by

$$X(n) = \phi_{2,1} X(n-1) + \phi_{2,2} X(n-2) + e(n) \tag{5.16}$$

A typical sample function of a second-order autoregressive process is shown in Figure 5.4.

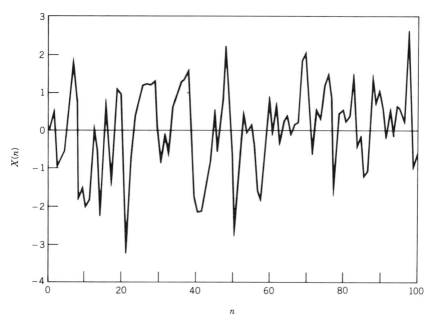

Figure 5.4 Sample function of the second-order autoregressive model.

DISCRETE LINEAR MODELS 257

We now seek μ_X, σ_X^2, R_{XX}, r_{XX}, and S_{XX} and sufficient conditions on $\phi_{2,1}$ and $\phi_{2,2}$ in order to ensure stationarity. Taking the expected value of Equation 5.16

$$\mu_X = \phi_{2,1}\mu_X + \phi_{2,2}\mu_X$$

and hence

$$\mu_X = 0$$

if $\phi_{2,1} + \phi_{2,2} \neq 1$, a required condition, as will be seen later. The variance can be calculated as

$$\begin{aligned}\sigma_X^2 &= E\{X(n)X(n)]\} = E\{\phi_{2,1}X(n)X(n-1) \\ &\quad + \phi_{2,2}X(n)X(n-2) + X(n)e(n)\} \\ &= \phi_{2,1}R_{XX}(1) + \phi_{2,2}R_{XX}(2) + \sigma_N^2\end{aligned}$$

Substituting $R_{XX}(k) = \sigma_X^2 r_{XX}(k)$ into the previous equation and solving for σ_X^2, we have

$$\sigma_X^2 = \frac{\sigma_N^2}{1 - \phi_{2,1}r_{XX}(1) - \phi_{2,2}r_{XX}(2)} \tag{5.17}$$

In order for σ_X^2 to be finite and positive

$$\phi_{2,1}r_{XX}(1) + \phi_{2,2}r_{XX}(2) < 1 \tag{5.18}$$

We now find $R_{XX}(m)$ for $m \geq 1$:

$$\begin{aligned}R_{XX}(m) &= E\{X(n-m)X(n)\} \\ &= E\{\phi_{2,1}X(n-1)X(n-m) + \phi_{2,2}X(n-2)X(n-m) \\ &\quad + X(n-m)e(n)\}\end{aligned}$$

or

$$R_{XX}(m) = \phi_{2,1}R_{XX}(m-1) + \phi_{2,2}R_{XX}(m-2) \tag{5.19}$$

258   SPECIAL CLASSES OF RANDOM PROCESSES

This is a second-order linear homogeneous difference equation, which has the solution

$$R_{XX}(m) = A_1\lambda_1^m + A_2\lambda_2^m \quad \text{if } \lambda_1 \neq \lambda_2 \tag{5.20.a}$$
$$= B_1\lambda^m + B_2 m\lambda^m \quad \text{if } \lambda_1 = \lambda_2 \tag{5.20.b}$$

where λ_1 and λ_2 are the roots of the characteristic equation obtained by assuming $R_{XX}(m) = \lambda^m$, for $m \geq 1$, in Equation 5.19. This produces

$$\lambda^2 = \phi_{2,1}\lambda + \phi_{2,2}$$

or

$$\lambda = \frac{\phi_{2,1} \pm \sqrt{\phi_{2,1}^2 + 4\phi_{2,2}}}{2} \tag{5.21}$$

Thus, $R_{XX}(m)$ can be a linear combination of geometric decays (λ_1 and λ_2 real) or decaying sinusoids (λ_1 and λ_2 complex conjugates) or of the form, $B_1\lambda^m + B_2 m\lambda^m$, where $\lambda_1 = \lambda_2 = \lambda$.

The coefficients A_1 and A_2 (or B_1 and B_2) must satisfy the initial conditions

$$R_{XX}(0) = \sigma_X^2$$

and

$$R_{XX}(1) = \phi_{2,1}R_{XX}(0) + \phi_{2,2}R_{XX}(-1)$$

Furthermore, stationarity implies $R_{XX}(-1) = R_{XX}(1)$, thus

$$R_{XX}(1) = \frac{\phi_{2,1}}{1 - \phi_{2,2}} \sigma_X^2 \tag{5.22}$$

If λ_1 and λ_2 are distinct, then

$$R_{XX}(0) = \sigma_X^2 = A_1 + A_2 \tag{5.23.a}$$

$$R_{XX}(1) = A_1\lambda_1 + A_2\lambda_2 = \frac{\phi_{2,1}}{1 - \phi_{2,2}} \sigma_X^2 \tag{5.23.b}$$

Equations 5.23.a and 5.23.b can be solved simultaneously for A_1 and A_2. Thus, $R_{XX}(m)$ is known in terms of σ_X^2 and $\phi_{2,1}$ and $\phi_{2,2}$. Also

$$r_{XX}(m) = \frac{R_{XX}(m)}{\sigma_X^2} = a_1\lambda_1^m + a_2\lambda_2^m \tag{5.24.a}$$

where

$$a_i = \frac{A_i}{\sigma_X^2}, \quad i = 1, 2 \tag{5.24.b}$$

We now find $r_{XX}(1)$ and $r_{XX}(2)$ directly in order to find an expression for σ_X^2 in terms of only the constants, $\phi_{2,1}$ and $\phi_{2,2}$. Using Equations 5.22 and 5.24.a, we have

$$r_{XX}(1) = \frac{\phi_{2,1}}{1 - \phi_{2,2}} \tag{5.25}$$

Now, using this in Equation 5.19 with $m = 2$ produces

$$r_{XX}(2) = \frac{\phi_{2,1}^2}{1 - \phi_{2,2}} + \phi_{2,2} \tag{5.26}$$

Substitution of Equations 5.25 and 5.26 in Equation 5.17 results in

$$\sigma_X^2 = \frac{\sigma_N^2(1 - \phi_{2,2})}{(1 + \phi_{2,2})(1 - \phi_{2,1} - \phi_{2,2})(1 + \phi_{2,1} - \phi_{2,2})} \tag{5.27}$$

This will be finite if

$$\phi_{2,2} \neq -1$$
$$\phi_{2,1} + \phi_{2,2} \neq 1$$
$$\phi_{2,2} - \phi_{2,1} \neq 1$$

Furthermore, σ_X^2 will be positive if

$$-1 < \phi_{2,2} < 1$$
$$\phi_{2,1} + \phi_{2,2} < 1$$
$$-\phi_{2,1} + \phi_{2,2} < 1$$

The power spectral density of the second-order autoregressive process is given by

$$S_{XX}(f) = |H(f)|^2 \sigma_N^2, \qquad |f| < \frac{1}{2}$$

where

$$H(f) = \frac{1}{1 - \phi_{2,1}\exp(-j2\pi f) - \phi_{2,2}\exp(-j4\pi f)}, \qquad |f| < \frac{1}{2}$$

Thus

$$S_{XX}(f) = \frac{\sigma_N^2}{|1 - \phi_{2,1}\exp(-j2\pi f) - \phi_{2,2}\exp(-j4\pi f)|^2}, \qquad |f| < \frac{1}{2}$$

which can also be found by taking the Fourier transform of $R_{XX}(m)$ as given by Equation 5.20.a. In this case it can be seen that

$$S_{XX}(f) = \frac{A_1(1 - \lambda_1^2)}{1 - 2\lambda_1 \cos 2\pi f + \lambda_1^2} + \frac{A_2(1 - \lambda_2^2)}{1 - 2\lambda_2 \cos 2\pi f + \lambda_2^2}, \qquad |f| < \frac{1}{2}$$

Using Equations 5.21 and 5.23 one can show that the two expressions for $S_{XX}(f)$ are equivalent (see Problem 5.17).

General Autoregressive Model. Returning to Equation 5.1,

$$X(n) = \sum_{i=1}^{p} \phi_{p,i} X(n - i) + e(n)$$

or again using z^{-1} as the backshift operator

$$X(n) = \sum_{i=1}^{p} \phi_{p,i} z^{-i}[X(n)] + e(n) \qquad (5.28)$$

We now find the mean, variance, autocorrelation function, correlation coefficient, and power spectral density of the general autoregressive process.

We have, taking expected values

$$\mu_X = 0 \qquad (5.29.\text{a})$$

and

$$\sigma_X^2 = E\{X(n)X(n)\} = E\left\{X(n)\sum_{i=1}^{p}\phi_{p,i}X(n-i) + X(n)e(n)\right\}$$

$$= \sum_{i=1}^{p}\phi_{p,i}R_{XX}(i) + \sigma_N^2 \qquad (5.29.\text{b})$$

The autocorrelation coefficient is obtained from

$$r_{XX}(k) = \frac{R_{XX}(k)}{\sigma_X^2} = \frac{E\{X(n-k)X(n)\}}{\sigma_X^2}$$

Using Equation 5.28 for $X(n)$, we obtain

$$r_{XX}(k) = \sum_{i=1}^{p}\phi_{p,i} r_{XX}(k-i), \qquad k \geq 1 \qquad (5.30)$$

This is a pth order difference equation. Equation 5.30 for $k = 1, 2, \ldots, p$, can be expressed in matrix form as

$$\begin{bmatrix} r_{XX}(1) \\ r_{XX}(2) \\ \vdots \\ r_{XX}(p) \end{bmatrix} = \begin{bmatrix} 1 & r_{XX}(1) & r_{XX}(2) & \cdots & r_{XX}(p-1) \\ r_{XX}(1) & 1 & r_{XX}(1) & r_{XX}(2) & \cdots & r_{XX}(p-2) \\ \vdots & & & & \vdots \\ r_{XX}(p-1) & r_{XX}(p-2) & \cdot & \cdot & \cdot & 1 \end{bmatrix} \begin{bmatrix} \phi_{p,1} \\ \phi_{p,2} \\ \vdots \\ \phi_{p,p} \end{bmatrix}$$

$$(5.31.\text{a})$$

or

$$\mathbf{r}_{XX} = \mathbf{R}\,\boldsymbol{\Phi} \qquad (5.31.b)$$

where $\mathbf{R}$ is the correlation coefficient matrix, $\mathbf{r}_{XX}$ is the correlation coefficient vector, and $\boldsymbol{\Phi}$ is the autoregressive coefficient vector.

This matrix equation is called the Yule–Walker equation. Because $\mathbf{R}$ is invertible, we can obtain

$$\boldsymbol{\Phi} = \mathbf{R}^{-1}\,\mathbf{r}_{XX} \qquad (5.32)$$

Equation 5.32 can be used to estimate the parameters $\phi_{p,i}$ of the model from the estimated values of the correlation coefficient $r_{XX}(k)$, and this is of considerable importance in data analysis.

The power spectral density of $X(n)$ can be shown to be

$$S_{XX}(f) = S_{ee}(f)|H(f)|^2 = \frac{\sigma_N^2}{\left|1 - \sum_{i=1}^{p} \phi_{p,i}\exp(-j2\pi fi)\right|^2}, \quad |f| < \frac{1}{2} \qquad (5.33)$$

This power spectral density is sometimes called the all-pole model.

5.2.2 Partial Autocorrelation Coefficient

Consider the first-order autoregressive model

$$X(n) = \frac{1}{2}X(n-1) + e(n)$$

It is clear that this is a Markov process, that is given $X(n-1)$, the previous X's, that is, $X(n-2), X(n-3), \ldots$, are of no use in determining or predicting $X(n)$. But as we see from Equation 5.10, the correlation between $X(n)$ and $X(n-2)$ is not zero, indeed:

$$r_{XX}(2) = \left(\frac{1}{2}\right)^2 = \frac{1}{4}$$

DISCRETE LINEAR MODELS 263

Similarly $X(n)$ and $X(n-3)$ are correlated:

$$r_{XX}(3) = \left(\frac{1}{2}\right)^3 = \frac{1}{8}$$

We now suggest that the partial autocorrelation between $X(n)$ and $X(n-2)$ after the effect of $X(n-1)$ has been eliminated might be of some interest. In fact, it turns out to be of considerable interest when estimating models from data.

In order to define the partial autocorrelation function in general, we return to the Yule–Walker equation, Equation 5.31. When $p = 1$, Equation 5.31 reduces to

$$r_{XX}(1) = \phi_{1,1}$$

When $p = 2$, Equation 5.31 becomes

$$\begin{bmatrix} r_{XX}(1) \\ r_{XX}(2) \end{bmatrix} = \begin{bmatrix} 1 & r_{XX}(1) \\ r_{XX}(1) & 1 \end{bmatrix} \begin{bmatrix} \phi_{2,1} \\ \phi_{2,2} \end{bmatrix}$$

and in general

$$\begin{bmatrix} r_{XX}(1) \\ r_{XX}(2) \\ \vdots \\ r_{XX}(p) \end{bmatrix} = \begin{bmatrix} 1 & r_{XX}(1) & r_{XX}(2) & \cdots & r_{XX}(p-1) \\ r_{XX}(1) & 1 & r_{XX}(1) & \cdots & r_{XX}(p-2) \\ \vdots & \vdots & \vdots & & \vdots \\ r_{XX}(p-1) & & r_{XX}(p-3) & \cdots & 1 \end{bmatrix} \begin{bmatrix} \phi_{p,1} \\ \phi_{p,2} \\ \vdots \\ \phi_{p,p} \end{bmatrix}$$

(5.34)

The coefficient $\phi_{k,k}$, found from the Yule-Walker equation when $p = k$, is defined as the kth *partial autocorrelation coefficient*. It is a measure of the effect of $X(n-k)$ on $X(n)$. For example if $p = 3$, then $r_{XX}(3) = \phi_{3,1} r_{XX}(2) + \phi_{3,2} r_{XX}(1) + \phi_{3,3}$. The first two terms describe the effects of $r_{XX}(2)$ and $r_{XX}(1)$ on $r_{XX}(3)$. The last term $\phi_{3,3}$ describes that part of the correlation $r_{XX}(3)$ after these two effects are accounted for; that is, $\phi_{3,3}$ is the partial correlation of $X(n)$ and $X(n-3)$ after the intervening correlation associated with lag 1 and lag 2 have been subtracted.

In the case of $k = 2$

$$\begin{bmatrix} \phi_{2,1} \\ \phi_{2,2} \end{bmatrix} = \begin{bmatrix} 1 & r_{XX}(1) \\ r_{XX}(1) & 1 \end{bmatrix}^{-1} \begin{bmatrix} r_{XX}(1) \\ r_{XX}(2) \end{bmatrix}$$

264   SPECIAL CLASSES OF RANDOM PROCESSES

or

$$\begin{bmatrix} \phi_{2,1} \\ \phi_{2,2} \end{bmatrix} = \frac{1}{1 - [r_{XX}(1)]^2} \begin{bmatrix} 1 & -r_{XX}(1) \\ -r_{XX}(1) & 1 \end{bmatrix} \begin{bmatrix} r_{XX}(1) \\ r_{XX}(2) \end{bmatrix}$$

Thus the second partial autocorrelation coefficient is

$$\phi_{2,2} = \frac{r_{XX}(2) - r_{XX}^2(1)}{1 - r_{XX}^2(1)} \tag{5.35}$$

For a first-order autoregressive process (Markov process)

$$r_{XX}(m) = \phi_{1,1}^m$$

Thus, Equation 5.35 produces

$$\phi_{2,2} = \frac{\phi_{1,1}^2 - \phi_{1,1}^2}{1 - \phi_{1,1}^2} = 0$$

showing that for a first-order autoregressive process the partial correlation between $X(n)$ and $X(n-2)$ is zero.

The partial autocorrelation function of a second-order autoregressive process

$$X(n) = \phi_{2,1} X(n-1) + \phi_{2,2} X(n-2) + e(n)$$

can be obtained as follows. Using Equations 5.25 and 5.34 with $k = 1$, the first partial correlation coefficient is

$$\phi_{1,1} = r_{XX}(1) = \frac{\phi_{2,1}}{1 - \phi_{2,2}}$$

Also using Equations 5.35, 5.25, and 5.26, we find the second partial correlation coefficient for a second-order autoregressive model as

$$\phi_{2,2} = \frac{\left(\dfrac{\phi_{2,1}^2}{1 - \phi_{2,2}} + \phi_{2,2}\right) - \dfrac{\phi_{2,1}^2}{(1 - \phi_{2,2})^2}}{1 - \dfrac{\phi_{2,1}^2}{(1 - \phi_{2,2})^2}}$$

$$= \frac{\phi_{2,1}^2 - \phi_{2,1}^2\phi_{2,2} + \phi_{2,2}(1-\phi_{2,2})^2 - \phi_{2,1}^2}{(1-\phi_{2,2})^2 - \phi_{2,1}^2}$$

$$= \phi_{2,2}$$

This justifies the notation for the partial correlation coefficient agreeing with the parameter in the autoregressive model. It can be shown that for a second-order autoregressive process (see Problem 5.18)

$$\phi_{k,k} = 0, \quad k > 2$$

In general, for a pth order autoregressive process,

$$\phi_{k,k} = 0, \quad k > p$$

In Chapter 9, this fact will be used to estimate the order of the model from data.

5.2.3 Moving Average Models

A moving average process is one represented by a difference equation

$$X(n) = \theta_0 e(n) + \theta_1 e(n-1) + \theta_2 e(n-2) + \cdots + \theta_k e(n-k)$$

Note that if $\Sigma \theta_i = 1$ and $\theta_i \geq 0$, then this is the usual moving average of the inputs $e(n)$. We change the parameter limits slightly, and rewrite the preceding equation as

$$X(n) = \sum_{i=1}^{q} \theta_{q,i} e(n-i) + e(n) = \sum_{i=0}^{q} \theta_{q,i} z^{-i}(e(n)) \qquad (5.36)$$

where $\theta_{q,0} = 1$ and $\theta_{q,q} \neq 0$. The model given in Equation 5.36 can be represented in block diagram form as shown in Figure 5.5.

The reader can show that the transfer function of the system shown in Figure 5.5 is

$$H(f) = 1 + \sum_{i=1}^{q} \theta_{q,i} \exp(-j2\pi f i)$$

266 SPECIAL CLASSES OF RANDOM PROCESSES

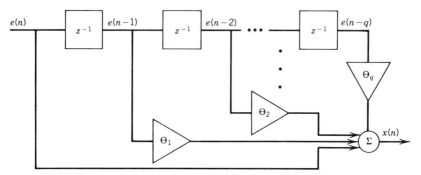

Figure 5.5 Moving average filter.

Note that this transfer function does not have any poles and hence is called an all-zero model.

First-order Moving Average Models. Consider the model

$$X(n) = \theta_{1,1} e(n-1) + e(n) \tag{5.37}$$

A sample sequence is plotted in Figure 5.6. A different form of this model can be obtained using the backshift operator

$$X(n) = (\theta_{1,1} z^{-1} + 1) e(n)$$

or

$$(1 + \theta_{1,1} z^{-1})^{-1} X(n) = e(n)$$

And if

$$-1 < \theta_{1,1} < 1$$

then

$$e(n) = (1 + \theta_{1,1} z^{-1})^{-1} X(n) = \left(\sum_{i=0}^{\infty} (-\theta_{1,1})^i (z)^{-i} \right) X(n)$$

$$= \sum_{i=0}^{\infty} (-\theta_{1,1})^i X(n-i)$$

DISCRETE LINEAR MODELS 267

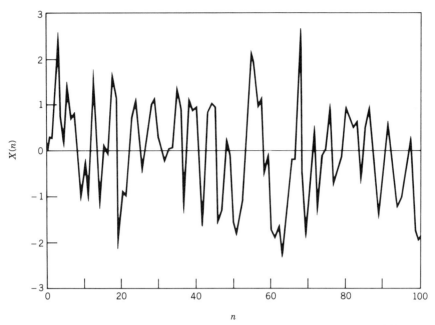

Figure 5.6 Sample function of the first-order moving average model: $X(n) = .45e(n - 1) + e(n)$.

Rearranging the preceding equation, we have

$$X(n) = -\sum_{i=1}^{\infty} (-\theta_{1,1})^i X(n - i) + e(n) \qquad (5.38)$$

Thus, the first-order moving average model can be *inverted* to an infinite autoregressive model. In order to be invertible, it is required that $-1 < \theta_{1,1} < 1$.

Returning to Equation 5.37, we find $\mu_X, \sigma_X^2, R_{XX}, r_{XX}$, the partial correlation coefficients, and $S_{XX}(f)$ as

$$\mu_X = E\{\theta_{1,1} e(n - 1) + e(n)\} = 0 \qquad (5.39.\text{a})$$

$$\sigma_X^2 = (\theta_{1,1}^2 + 1)\sigma_N^2 \qquad (5.39.\text{b})$$

$$\begin{aligned}
R_{XX}(k) &= E\{X(n)X(n - k)\} \\
&= E\{[\theta_{1,1} e(n - 1) + e(n)] \\
&\quad \times [\theta_{1,1} e(n - k - 1) + e(n - k)]\} \\
&= \begin{cases} \theta_{1,1}\sigma_N^2, & k = 1 \\ 0 & k > 1 \end{cases}
\end{aligned} \qquad (5.40.\text{a})$$

and hence

$$r_{xx}(1) = \frac{\theta_{1,1}}{1 + \theta_{1,1}^2}$$

and

$$r_{xx}(k) = 0, \quad k > 1 \tag{5.40.b}$$

Note the important result that the autocorrelation function is zero for k greater than one for the first-order moving average sequence.

The partial autocorrelation coefficients can be obtained from Equation 5.34 as

$$\phi_{1,1} = r_{xx}(1) = \frac{\theta_{1,1}}{1 + \theta_{1,1}^2} \tag{5.41.a}$$

$$\phi_{2,2} = \frac{r_{xx}(2) - r_{xx}^2(1)}{1 - r_{xx}^2(1)} = \frac{\dfrac{-\theta_{1,1}^2}{(1 + \theta_{1,1}^2)^2}}{1 - \dfrac{\theta_{1,1}^2}{(1 + \theta_{1,1}^2)^2}}$$

$$= \frac{-\theta_{1,1}^2}{1 + \theta_{1,1}^2 + \theta_{1,1}^4} \tag{5.41.b}$$

It can be shown that (see Problem 5.22)

$$\phi_{k,k} = \frac{(-1)^{k-1}\theta_{1,1}^k(1 - \theta_{1,1}^2)}{1 - \theta_{1,1}^{2(k+1)}} \tag{5.42}$$

Thus, the partial autocorrelation coefficients do not become zero as the correlation coefficients do for this moving average process.

The spectral density function $S_{xx}(f)$ is

$$S_{xx}(f) = \sum_{k=-1}^{+1} \exp(-j2\pi f k) R_{xx}(k)$$

$$= \exp(j2\pi f)\sigma_N^2 \theta_{1,1} + (\theta_{1,1}^2 + 1)\sigma_N^2 + \sigma_N^2 \theta_{1,1}\exp(-j2\pi f)$$

$$S_{xx}(f) = \sigma_N^2[\theta_{1,1}^2 + 2\theta_{1,1}\cos 2\pi f + 1], \quad |f| < \frac{1}{2} \tag{5.43}$$

Second-order Moving Average Models. The second-order moving average process described by

$$X(n) = \theta_{2,1}e(n-1) + \theta_{2,2}e(n-2) + e(n)$$

has a mean

$$E\{X(n)\} = 0 \tag{5.44.a}$$

and an autocorrelation function

$$\begin{aligned}
R_{XX}(k) &= E\{[\theta_{2,1}e(n-1) + \theta_{2,2}e(n-2) + e(n)][\theta_{2,1}e(n-k-1) \\
&\quad + \theta_{2,2}e(n-k-2)) + e(n-k)]\} \\
&= (\theta_{2,1}^2 + \theta_{2,2}^2 + 1)\sigma_N^2, \quad k = 0 \\
&= (\theta_{2,1} + \theta_{2,1}\theta_{2,2})\sigma_N^2, \quad k = 1 \\
&= \theta_{2,2}\sigma_N^2, \quad k = 2 \\
&= 0, \quad k > 2
\end{aligned} \tag{5.44.b}$$

Thus

$$\sigma_X^2 = R_{XX}(0) = (\theta_{2,1}^2 + \theta_{2,2}^2 + 1)\sigma_N^2 \tag{5.45.a}$$

$$r_{XX}(1) = \frac{R_{XX}(1)}{\sigma_X^2} = \frac{\theta_{2,1} + \theta_{2,1}\theta_{2,2}}{1 + \theta_{2,1}^2 + \theta_{2,2}^2} \tag{5.45.b}$$

$$r_{XX}(2) = \frac{\theta_{2,2}}{1 + \theta_{2,1}^2 + \theta_{2,2}^2} \tag{5.45.c}$$

$$r_{XX}(k) = 0, \quad k > 2 \tag{5.45.d}$$

The last result, that is, Equation 5.45.d, is particularly important in identifying the order of models, as discussed in Chapter 9.

The power spectral density function of the second-order moving average process is given by

$$\begin{aligned}
S_{XX}(f) &= \sigma_N^2 |1 + \theta_{2,1}\exp(-j2\pi f) + \theta_{2,2}\exp(-j4\pi f)|^2, \quad |f| < \frac{1}{2} \\
&= \sigma_N^2 [1 + \theta_{2,1}^2 + \theta_{2,2}^2 + 2\theta_{2,1}(1 + \theta_{2,2})\cos 2\pi f \\
&\quad + 2\theta_{2,2}\cos 4\pi f]
\end{aligned} \tag{5.46}$$

270 SPECIAL CLASSES OF RANDOM PROCESSES

General Moving Average Model. We now find the mean, autocorrelation function, and spectral density function of a qth-order moving average process which is modeled as

$$X(n) = \sum_{i=1}^{q} \theta_{q,i} e(n-i) + e(n)$$

The mean and variance can be calculated as

$$\mu_X = E\{X(n)\} = 0 \tag{5.47.a}$$

and

$$\sigma_X^2 = E\{X(n)X(n)\} = E\left\{\left[\sum_{i=0}^{q} \theta_{q,i} e(n-i)\right]\left[\sum_{j=0}^{q} \theta_{q,j} e(n-j)\right]\right\}$$

$$= \sum_{i=0}^{q} \theta_{q,i}^2 \sigma_N^2$$

$$= \sigma_N^2 \left[1 + \sum_{i=1}^{q} \theta_{q,i}^2\right] \tag{5.47.b}$$

The autocorrelation function is given by

$$R_{XX}(m) = E\left\{\left[\sum_{i=0}^{q} \theta_{q,i} e(n-i)\right]\left[\sum_{j=0}^{q} \theta_{q,j} e(n-m-j)\right]\right\}$$

$$= \sigma_N^2 \left[1 + \sum_{j=1}^{q} \theta_{q,j}^2\right], \quad m = 0$$

$$= \sigma_N^2 \left[\theta_{q,1} + \sum_{j=2}^{q} \theta_{q,j} \theta_{q,j-1}\right], \quad m = 1$$

$$= \sigma_N^2 \left[\theta_{q,2} + \sum_{j=3}^{q} \theta_{q,j} \theta_{q,j-2}\right], \quad m = 2$$

In general

$$R_{XX}(m) = \sigma_N^2 \left[\theta_{q,m} + \sum_{j=m+1}^{q} \theta_{q,j} \theta_{q,j-m}\right], \quad m < q$$

$$= \sigma_N^2 \theta_{q,q}, \quad m = q$$

$$= 0, \quad m > q \tag{5.47.c}$$

Because Equation 5.47.c is a finite series, no restrictions on the $\theta_{q,i}$'s are necessary in order to ensure stationarity. However, some restrictions are necessary on the $\theta_{q,i}$'s in order to be able to invert this model into an infinite autoregressive model.

Taking the transform of $R_{XX}(k)$ we obtain the spectral density function of the moving average process as

$$S_{XX}(f) = \sigma_N^2 \left| 1 + \sum_{i=1}^{q} \theta_{q,i} \exp(-j2\pi i f) \right|^2, \quad |f| < \frac{1}{2} \quad (5.48)$$

Equation 5.48 justifies calling a moving average model an all-zero model.

5.2.4 Autoregressive Moving Average Models

An autoregressive moving average (ARMA) model is of the form

$$X(n) = \sum_{i=1}^{p} \phi_{p,i} X(n-i) + \sum_{k=1}^{q} \theta_{q,k} e(n-k) + e(n) \quad (5.49)$$

A block diagram representation of an ARMA (p, q) model is shown in Figure 5.7. This model can also be described using the backshift operator as

$$\left(1 - \sum_{i=1}^{p} \phi_{p,i} z^{-i}\right) X(n) = \left(1 + \sum_{k=1}^{q} \theta_{q,k} z^{-k}\right) e(n) \quad (5.50)$$

Using Equation 5.50 to suggest the transfer function and using

$$S_{XX}(f) = |H(f)|^2 \sigma_N^2, \quad |f| < \frac{1}{2}$$

we obtain

$$S_{XX}(f) = \frac{\sigma_N^2 \left| 1 + \sum_{k=1}^{q} \theta_{q,k} \exp(-j2\pi f k) \right|^2}{\left| 1 - \sum_{i=1}^{p} \phi_{p,i} \exp(-j2\pi f i) \right|^2}, \quad |f| < \frac{1}{2} \quad (5.51)$$

272 SPECIAL CLASSES OF RANDOM PROCESSES

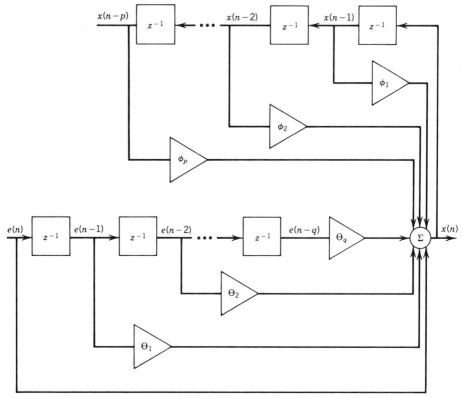

Figure 5.7 An autoregressive moving average ARMA (p, q) filter.

Note that the transfer function $H(f)$ and the power spectral density $S_{XX}(f)$ have both poles and zeros. The autocorrelation function $R_{XX}(m)$ of the ARMA process is

$$\begin{aligned}
R_{XX}(m) &= E\{X(n - m)X(n)\} \\
&= E\left\{[X(n - m)]\left[\sum_{i=1}^{p} \phi_{p,i} X(n - i)\right.\right. \\
&\quad \left.\left. + \sum_{k=1}^{q} \theta_{q,k} e(n - k) + e(n)\right]\right\} \\
&= \sum_{i=1}^{p} \phi_{p,i} R_{XX}(m - i) + E\{X(n - m)e(n)\} \\
&\quad + \sum_{k=1}^{q} \theta_{q,k} E\{X(n - m)e(n - k)\}
\end{aligned}$$

DISCRETE LINEAR MODELS

Because

$$E\{X(n-m)e(n)\} = 0, \quad m \geq 1$$

the preceding equation reduces to

$$R_{XX}(m) = \sum_{i=1}^{p} \phi_{p,i} R_{XX}(m-i), \quad m \geq q+1 \tag{5.52}$$

Thus, for an ARMA (p, q) model $R_{XX}(0), R_{XX}(1), \ldots, R_{XX}(q)$ will depend upon both the autoregressive and the moving average parameters. The remainder of the autocorrelation function, that is, $R_{XX}(k), k > q$ is determined by the pth order difference equation given in Equation 5.52.

The ARMA random process described by Equation 5.49 can also be written as

$$X(n) = \left(1 + \sum_{k=1}^{q} \theta_{q,k} z^{-k}\right)\left(1 - \sum_{i=1}^{p} \phi_{p,i} z^{-i}\right)^{-1} e(n) \tag{5.53}$$

The expansion of the middle term in an infinite series shows that $X(n)$ is an infinite series in z^{-1}. Thus, $X(n)$ depends upon the infinite past and the partial autocorrelation function will be nonzero for an infinite number of values.

The ARMA (1, 1) Process. The ARMA (1, 1) process is described by

$$X(n) = \phi_{1,1} X(n-1) + \theta_{1,1} e(n-1) + e(n)$$

A sample sequence with $\phi_{1,1} = .5, \theta_{1,1} = .5$ is shown in Figure 5.8.
The mean of the process is

$$\mu_X = \phi_{1,1}\mu_X + 0$$

and for stationarity it is required that $\phi_{1,1} \neq 1$. Thus

$$\mu_X = 0 \tag{5.54.a}$$

The variance of $X(n)$ is obtained from

$$\sigma_X^2 = E\{X(n)^2\} = \phi_{1,1}^2 \sigma_X^2 + (1 + \theta_{1,1}^2)\sigma_N^2 + 2\phi_{1,1}\theta_{1,1} E\{X(n-1)e(n-1)\}$$

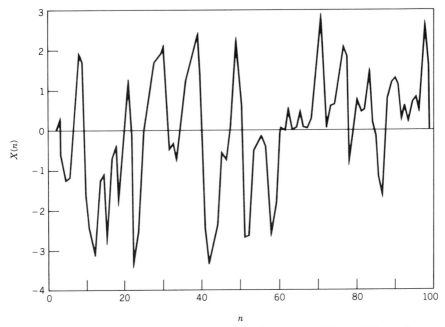

Figure 5.8 Sample function of the ARMA (1, 1) model: $X(n) = .5X(n-1) + .5e(n-1) + e(n)$.

which leads to

$$\sigma_X^2 = \frac{[1 + \theta_{1,1}^2 + 2\phi_{1,1}\theta_{1,1}]\sigma_N^2}{(1 - \phi_{1,1}^2)} \tag{5.54.b}$$

Since $\sigma_X^2 > 0$, $\phi_{1,1}$ should satisfy

$$-1 < \phi_{1,1} < +1 \tag{5.55}$$

The autocorrelation function of the first order ARMA process can be obtained from

$$\begin{aligned} R_{XX}(1) &= E\{X(n-1)X(n)\} \\ &= E\{X(n-1)\phi_{1,1}X(n-1) + X(n-1)\Theta_{1,1}e(n-1) \\ &\quad + X(n-1)e(n)\} \\ &= \phi_{1,1}R_{XX}(0) + \theta_{1,1}\sigma_N^2 \end{aligned}$$

Thus using Equation 5.54.b

$$R_{XX}(1) = \frac{\phi_{1,1}(1 + \theta_{1,1}^2) + 2\phi_{1,1}^2\theta_{1,1}}{1 - \phi_{1,1}^2}\sigma_N^2 + \theta_{1,1}\frac{1 - \phi_{1,1}^2}{1 - \phi_{1,1}^2}\sigma_N^2$$

$$= \frac{(1 + \phi_{1,1}\theta_{1,1})(\phi_{1,1} + \theta_{1,1})\sigma_N^2}{1 - \phi_{1,1}^2} \quad (5.56.a)$$

and

$$R_{XX}(2) = E\{X(n - 2)X(n)\} = \phi_{1,1}R_{XX}(1) \quad (5.56.b)$$

$$R_{XX}(k) = \phi_{1,1}R_{XX}(k - 1), \quad k \geq 2 \quad (5.56.c)$$

Note that this autocorrelation function decays exponentially from $R_{XX}(1)$, that is

$$R_{XX}(k) = R_{XX}(1)\phi_{1,1}^{k-1}, \quad k \geq 2 \quad (5.57)$$

The decay may be either monotonic or alternating depending upon whether $\phi_{1,1}$ is positive or negative. Because stationarity requires that $|\phi_{1,1}| < 1$, the sign of $R_{XX}(1)$ depends upon the sign of $(\phi_{1,1} + \theta_{1,1})$.

The power spectral density of the first-order ARMA process can be shown to be

$$S_{XX}(f) = \frac{\sigma_N^2|1 + \theta_{1,1}\exp(-j2\pi f)|^2}{|1 - \phi_{1,1}\exp(-j2\pi f)|^2}, \quad |f| < \frac{1}{2} \quad (5.58)$$

5.2.5 Summary of Discrete Linear Models

The primary application of autoregressive moving average (ARMA) models is their use as random process models that can be derived from data. Briefly, the order of the model is identified (or estimated) from data using the sample autocorrelation function. For example, if the autocorrelation function, $R_{XX}(k)$ is zero for $k > 2$, then Equation 5.45 suggests that an ARMA (0, 2) model is appropriate. Similarly, if the partial autocorrelation coefficients, $\phi_{i,i}$ are zero for $i > 2$, then, as shown in Section 5.2.2, an ARMA (2, 0) model is suggested.

If the autocorrelation coefficients were known (as opposed to estimated from data), then equations such as Equation 5.32 and Equation 5.45 could be used to determine the parameters in the model from the autocorrelation coefficients.

An extensive introduction to estimating these models is contained in Chapter 9. The purpose of this section was to introduce the models themselves and explore some of their properties.

5.3 MARKOV SEQUENCES AND PROCESSES

The least complicated model of a random process is the trivial one in which the value of the process at any given time is independent of the values at all other times. In this case, a random process model is not needed; a single random variable model will suffice with no loss of generality.

A more complicated model is one in which the value of the random process depends only upon the one most recent previous value and given that value the random process is independent of all values in the more distant past. Such a model is called a Markov model and is often described by saying that a Markov process is one in which the future value is independent of the past values given the present value.

Models in which the future depends only upon the present are common among electrical engineering models. Indeed a first-order linear differential equation or a first-order linear difference equation is such a model. For example, the solution for $i(t)$ that satisfies

$$\frac{di}{dt} + a_0 i(t) = f(t)$$

for $t > t_0$ requires only $i(t_0)$ and the solution cannot use knowledge of $i(t)$ for $t < t_0$ when $i(t_0)$ is given. Even if $f(t)$ is random, values of $f(t)$ or $i(t)$ for $t < t_0$ are of no use in predicting $i(t)$ for $t > t_0$ given $i(t_0)$.

Higher order difference equations require more past values (an nth order equation requires the present and $n - 1$ past values) for a solution. Similarly, an nth order differential equation requires an initial value and $n - 1$ derivatives at the initial time. An nth order difference equation can be transformed to n first-order difference equations (a state variable formulation) and thus the dependence on initial conditions at n different times is transformed to n values at one time. Such models are analogous to an nth order Markov processes.

We have argued that Markov processes are simple and analogous to familiar models. We present later several examples that have proved to be useful. Before presenting these examples and discussing methods for analyzing Markov proc-

TABLE 5.1 CLASSIFICATION OF MARKOV PROCESSES

$X(t)$ \ t	Continuous	Discrete
Continuous	Continuous random process	Continuous random sequence
Discrete	Discrete random process	Discrete random sequence

Markov Chains

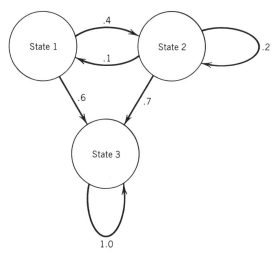

Figure 5.9 State diagram of a Markov chain.

esses, we classify Markov processes and present a diagram called a state diagram which will be useful for describing Markov processes.

The classification of Markov process is given in Table 5.1. Note that if the values of $X(t)$ are discrete, then Markov processes are called Markov chains. In this section only Markov chains, including both sequences and continuous-time processes, are discussed.

Markov chains, that is, Markov processes with discrete $X(t)$, are usually described by referring to their states. There are a finite or at most a countable number of such states, and $X(t)$ maps each state to a discrete value or number. With the Markov concept, the next state is dependent only upon the present state. Thus, a diagram like Figure 5.9 is often used to describe a Markov chain that is a sequence and a similar diagram is used to describe a Markov chain in which time is continuous. In Figure 5.9, each number adjacent to an arrow represents the conditional probability of the Markov chain making the state transition in the direction of the arrow, given that it is in the state from which the arrow emanates. For example, given that the Markov chain of Figure 5.9 is in state 1, the probability is .4 that its next transition will be to state 2.

EXAMPLE 5.1 (MESSAGE SOURCES).

For many communication systems it is desirable to "code" messages into equiprobable symbols in order to fully utilize available bandwidth. Such coding requires knowledge of the probability of the various messages, and in particular it may be desirable to know the probabilities of the letters of the English alphabet (26 letters plus a space). The probability of a letter obviously depends upon at

278   SPECIAL CLASSES OF RANDOM PROCESSES

least the preceding letter (e.g., the probability of a "u" is 1 given the preceding letter is a "q"). If the probability of a letter depended only upon the preceding letter, then the sequence of letters could be modeled as a Markov chain. However the dependence in English text usually extends considerably beyond simply the previous letter. Thus, a Markov model with a state space of the 26 letters plus a space would not be adequate. However, if instead of a single symbol, the states were to represent blocks of say 5 consecutive symbols, the resulting Markov model might be adequate. In this case there are approximately $(27)^5$ states, but this complexity is often compensated by the fact that a Markov chain model may be used. With the expanded state model the message:

"This-book-is-easy-for . . ."

would be transformed into the states:

This-;book-;is-ea;sy-fo; . . .

and we model each state as being dependent only upon the previous state.

EXAMPLE 5.2 (EQUIPMENT FAILURE).

This example differs from the preceding one in the sense that time is continuous, while the preceding example consisted of a sequence. A piece of equipment, for example, a communication receiver, can have two states, operable and nonoperable. The transitions from the operable to the nonoperable state occur at a prescribed rate called the failure rate. The transitions from the nonoperable to the operable state occur at the repair rate. If the rates of transition depend only on the present state and not on the repair history, then a Markov model can be used.

5.3.1 Analysis of Discrete-time Markov Chains

We model $X(n)$ to be a random sequence that represents the state of a system at time n (time is discrete) and we assume that $X(n)$ can take on only a finite or perhaps a countably infinite number of states. Thus, the general Markov

MARKOV SEQUENCES AND PROCESSES 279

property can be described by the transition probabilities:

$$P[X(m) = x_m | X(m-1) = x_{m-1}, X(m-2) = x_{m-2},$$
$$\ldots, X(0) = x_0]$$
$$= P[X(m) = x_m | X(m-1) = x_{m-1}] \tag{5.59}$$

In this section, we will develop, in matrix notation, a method for finding the probability that a finite Markov chain is in a specified state at a specified time. That is, we want to find the *state probabilities*

$$p_j(n) \triangleq P[X(n) = j], \quad j = 1, 2, \ldots \tag{5.60}$$

To find these probabilities we use the single-step (conditional) *transition probabilities* defined by

$$P_{i,j}(m-1, m) \triangleq P[X(m) = j | X(m-1) = i] \tag{5.61}$$

Now from Chapter 2, the joint probability is given by the product of the marginal and the conditional probability, that is

$$P\{[X(m-1) = i], [X(m) = j]\} \tag{5.62}$$
$$= P[X(m-1) = i] P[X(m) = j | X(m-1) = i]$$

Using the notation of Equations 5.60 and 5.61 in the preceding equation, we have

$$P[(X(m-1) = i), (X(m) = j)] = p_i(m-1) P_{i,j}(m-1, m) \tag{5.63}$$

The state probabilities $p_j(m), j = 1, 2, \ldots$, may be found using the probability laws given in Chapter 2 as

$$p_j(m) = \sum_{\text{all } i} p_i(m-1) P_{i,j}(m-1, m) \tag{5.64}$$

To illustrate the use of Equation 5.64 consider the following example.

280 SPECIAL CLASSES OF RANDOM PROCESSES

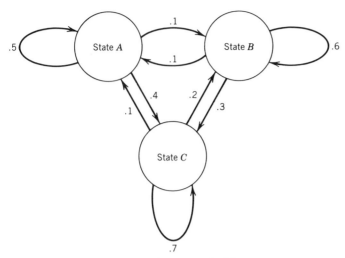

Figure 5.10 State diagram for Example 5.3.

EXAMPLE 5.3.

Three possible messages, A, B, and C, can be transmitted, and sequences of messages are Markov. The transition probabilities from the current message to the next message are independent of when the transition occurs and are as follows:

Current Message	Next Message		
	A	B	C
A	.5	.1	.4
B	.1	.6	.3
C	.1	.2	.7

Note that the sum across each row is one, as it must be. If we assume that A corresponds with message one, B corresponds with message two, and C corresponds with message three, then the conditional probability in row i and column j is $P_{i,j}(m-1, m)$, $i = 1, 2, 3$, $j = 1, 2, 3$, for all m. For instance $P_{2,3}(m-1, m) = .3$. This example is displayed in the state diagram of Figure 5.10.

Assume that the probabilities of the three starting states are given as

$$p_1(0) = .5, \quad p_2(0) = .3, \quad p_3(0) = .2$$

We now want the probabilities of the next message. These are found using Equation 5.64 as follows:

$$p_1(1) = p_1(0)P_{1,1}(0, 1) + p_2(0)P_{2,1}(0, 1) + p_3(0)P_{3,1}(0, 1)$$
$$= (.5)(.5) + (.3)(.1) + (.2)(.1) = .30$$
$$p_2(1) = (.5)(.1) + (.3)(.6) + (.2)(.2) = .27$$
$$p_3(1) = (.5)(.4) + (.3)(.3) + (.2)(.7) = .43$$

The state probabilities and the transition probabilities can be conveniently expressed in matrix form (for a finite chain) with the following definitions:

$$\mathbf{P}(m, n) \triangleq [P_{i,j}(m, n)] \tag{5.65}$$

where $\mathbf{P}(m, n)$ is a matrix, and

$$\mathbf{p}^T(n) \triangleq [p_1(n), p_2(n), \ldots, p_k(n)] \tag{5.66}$$

where $\mathbf{p}^T(n)$ is a row vector, and k is the number of states.
Using this notation, Equation 5.64 can be expressed

$$\mathbf{p}^T(m) = \mathbf{p}^T(m - 1)\mathbf{P}(m - 1, m) \tag{5.67}$$

EXAMPLE 5.4.

We return to Example 5.3 to illustrate the use of Equation 5.67.

$$\mathbf{p}^T(1) = [.5 \quad .3 \quad .2] \begin{bmatrix} .5 & .1 & .4 \\ .1 & .6 & .3 \\ .1 & .2 & .7 \end{bmatrix}$$

$$= [.3 \quad .27 \quad .43]$$

as found earlier.

Equation 5.67 can be used to find $\mathbf{p}(n)$ from $\mathbf{p}(0)$ as follows:

$$\mathbf{p}^T(1) = \mathbf{p}^T(0)\mathbf{P}(0, 1)$$
$$\mathbf{p}^T(2) = \mathbf{p}^T(1)\mathbf{P}(1, 2) = \mathbf{p}^T(0)\mathbf{P}(0, 1)\mathbf{P}(1, 2)$$
$$\mathbf{p}^T(3) = \mathbf{p}^T(0)\mathbf{P}(0, 1)\mathbf{P}(1, 2)\mathbf{P}(2, 3)$$

This procedure can be continued for values of $n = 4, 5, 6 \ldots$.

Homogeneous Markov Chains. In many models of Markov chains the transition probabilities are independent of when the transition occurs, that is, $P_{i,j}(m - 1, m) = P_{i,j}(n - 1, n)$ for all $i, j, m,$ and n. If this is the case then the chain is called *homogeneous* and the state transition matrix is called stationary. (Note that a *stationary transition matrix* does not imply a stationary random sequence).

If the transition probabilities are homogeneous, then Equation 5.67 becomes

$$\mathbf{p}^T(m) = \mathbf{p}^T(m - 1)\mathbf{P}(1) \tag{5.68}$$

where

$$\mathbf{P}(1) \triangleq \mathbf{P}(n - 1, n) = \mathbf{P}(m - 1, m) \tag{5.69}$$

That is, because $\mathbf{P}(n - 1, n) = \mathbf{P}(m - 1, m)$, the argument of the $\mathbf{P}$ matrix may be reduced to the time difference between steps. In this case it follows that

$$\mathbf{p}^T(1) = \mathbf{p}^T(0)\mathbf{P}(1)$$
$$\mathbf{p}^T(2) = \mathbf{p}^T(0)\mathbf{P}(1)\mathbf{P}(1) = \mathbf{p}^T(0)\mathbf{P}(1)^2 = \mathbf{p}^T(0)\mathbf{P}(2)$$
$$\vdots$$
$$\mathbf{p}^T(n) = \mathbf{p}^T(0)\mathbf{P}(1)^n = \mathbf{p}^T(0)\mathbf{P}(n) \tag{5.70}$$

For this homogeneous case

$$\mathbf{P}(n) \triangleq \mathbf{P}(1)^n \tag{5.71}$$

$\mathbf{P}(1)^n$ is an n-stage transition matrix, that is, the i, jth element of $\mathbf{P}(1)^n$ represents the probability of transferring, in n time intervals, from state i to state j.

EXAMPLE 5.5.

Find $\mathbf{P}(2)$, $\mathbf{P}(3)$, ..., $\mathbf{P}(10)$ for the homogeneous Markov chain represented by the matrix

$$\mathbf{P}(1) = \begin{bmatrix} .5 & .1 & .4 \\ .1 & .6 & .3 \\ .1 & .2 & .7 \end{bmatrix}$$

SOLUTION: By matrix multiplication we note that

$$\mathbf{P}(2) = \mathbf{P}(1)^2 = \begin{bmatrix} .3 & .19 & .51 \\ .14 & .43 & .43 \\ .14 & .27 & .59 \end{bmatrix}$$

$$\mathbf{P}(3) = \mathbf{P}(1)^3 = \begin{bmatrix} .220 & .246 & .534 \\ .156 & .358 & .486 \\ .156 & .294 & .550 \end{bmatrix}$$

$$\mathbf{P}(4) = \mathbf{P}(1)^4 = \begin{bmatrix} .1880 & .2764 & .5356 \\ .1624 & .3276 & .5100 \\ .1624 & .3020 & .5356 \end{bmatrix}$$

$$\mathbf{P}(5) = \mathbf{P}(1)^5 = \begin{bmatrix} .17520 & .29176 & .53304 \\ .16496 & .31480 & .52024 \\ .16496 & .30456 & .53048 \end{bmatrix}$$

$$\mathbf{P}(10) = \mathbf{P}(1)^{10} \approx \begin{bmatrix} .1668 & .3053 & .5279 \\ .1666 & .3057 & .5277 \\ .1666 & .3056 & .5278 \end{bmatrix}$$

Note that the elements in each row seem to be approaching a constant value. Thus

$$\mathbf{p}^T(10) = \mathbf{p}^T(0)\mathbf{P}(1)^{10} \approx [.1667 \quad .3055 \quad .5278]$$

independent of $\mathbf{p}(0)$, which indicates steady-state behavior.

The state probability vectors may be found from Equation 5.70. If $\mathbf{p}^T(0) = [.5 \quad .3 \quad .2]$

$$\mathbf{p}^T(1) = \mathbf{p}^T(0)\mathbf{P}(1) = [.3 \quad .27 \quad .43]$$
$$\mathbf{p}^T(2) = \mathbf{p}^T(0)\mathbf{P}(1)^2 = [.22 \quad .278 \quad .502]$$
$$\mathbf{p}^T(5) = \mathbf{p}^T(0)\mathbf{P}(1)^5 \approx [.170 \quad .301 \quad .529]$$

Chapman–Kolmogorov Equation. We now show that for a homogeneous discrete-time Markov chain with $n_1 < n_2 < n_3$

$$P_{i,j}(n_3 - n_1) = \sum_k P_{i,k}(n_2 - n_1) P_{k,j}(n_3 - n_2) \tag{5.72}$$

where $P_{i,j}(n) \triangleq P[X(m + n) = j | X(m) = i]$.

Proof: A two-dimensional marginal probability can be obtained by summing the joint probabilities (see Equation 2.12). Thus

$$P[(X(n_1) = i), (X(n_3) = j)]
= \sum_{\text{all } k} P[(X(n_1) = i), (X(n_2) = k), (X(n_3) = j)] \tag{5.73}$$

Since the $X(n)$ are from a homogeneous Markov process, then

$$P[(X(n_1) = i), (X(n_3) = j)] = p_i(n_1) P_{i,j}(n_3 - n_1) \tag{5.74}$$

and

$$P[(X(n_1) = i), (X(n_2) = k), (X(n_3) = j)]
= p_i(n_1) P_{i,k}(n_2 - n_1) P_{k,j}(n_3 - n_2) \tag{5.75}$$

Using Equations 5.74 and 5.75 in Equation 5.73 results in

$$p_i(n_1) P_{i,j}(n_3 - n_1) = \sum_{\text{all } k} p_i(n_1) P_{i,k}(n_2 - n_1) P_{k,j}(n_3 - n_2)$$

Dividing both sides by $p_i(n_1)$ produces the desired result. Equation 5.72 is called the Chapman–Kolmogorov equation and can be rewritten in matrix form for finite chains as

$$\mathbf{P}(n_3 - n_1) = \mathbf{P}(n_2 - n_1) \mathbf{P}(n_3 - n_2)$$

Long-run (Asymptotic) Behavior of Homogeneous Chains. Example 5.5 suggests, at least for the example, that a homogeneous Markov chain will reach steady-state probability after many transitions. That is,

$$\lim_{n \to \infty} \mathbf{P}(n) = \lim_{n \to \infty} \mathbf{P}(n - 1) = \mathbf{P} \tag{5.76}$$

Then, using Equation 5.76 we have

$$\lim_{n\to\infty} \mathbf{p}^T(n) \stackrel{\Delta}{=} \boldsymbol{\pi}^T = \mathbf{p}^T(0) \lim_{n\to\infty} \mathbf{P}(n) = \mathbf{p}^T(0)\mathbf{P} \tag{5.77}$$

where $\boldsymbol{\pi}$ is called the *limiting state probabilities* and

$$\pi_j \stackrel{\Delta}{=} \lim_{n\to\infty} p_j(n)$$

Now if Equation 5.76 holds, then

$$\boldsymbol{\pi}^T = \mathbf{p}^T(0) \lim_{n\to\infty} \mathbf{P}(1)^n = \mathbf{p}^T(0) \lim_{n\to\infty} \mathbf{P}(1)^{n-1}\mathbf{P}(1) \tag{5.78}$$

or

$$\boldsymbol{\pi}^T = \boldsymbol{\pi}^T \mathbf{P}(1) \tag{5.79}$$

Equation 5.79 can be used to find the steady-state probabilities if they exist. The solution to Equation 5.79 is not unique because $\mathbf{P}$ is singular. However, a unique solution can be obtained by using

$$\sum_{\text{all } j} \pi_j = 1 \tag{5.80}$$

EXAMPLE 5.6.

Find the steady-state probabilities for Example 5.5.

SOLUTION: The steady-state probabilities may be found using Equation 5.79 as follows:

$$\boldsymbol{\pi}^T = \boldsymbol{\pi}^T \begin{bmatrix} .5 & .1 & .4 \\ .1 & .6 & .3 \\ .1 & .2 & .7 \end{bmatrix}$$

or

$$\pi_1 = .5\pi_1 + .1\pi_2 + .1\pi_3$$
$$\pi_2 = .1\pi_1 + .6\pi_2 + .2\pi_3$$
$$\pi_3 = .4\pi_1 + .3\pi_2 + .7\pi_3$$

These equations are linearly dependent (the sum of the first two equations is equivalent to the last equation). However, any two of them plus Equation 5.80, that is,

$$\pi_1 + \pi_2 + \pi_3 = 1$$

can be used to find the steady-state probabilities, which are

$$\pi = [6/36 \quad 11/36 \quad 19/36] \approx [0.1667 \quad 0.3056 \quad 0.5278]$$

Limiting Behavior of a Two-state Discrete-time Homogeneous Markov Chain.
We now investigate the limiting-state probabilities of a general two-state discrete-time homogeneous Markov chain. This chain can be described by the state diagram of Figure 5.11. Because of homogeneity, we use Equation 5.71, that is

$$\mathbf{P}(n) = \mathbf{P}(1)^n$$

where for $0 < a < 1$ and $0 < b < 1$

$$\mathbf{P}(1) = \begin{bmatrix} 1 - a & a \\ b & 1 - b \end{bmatrix}$$

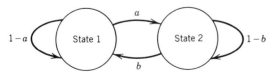

Figure 5.11 Markov chain with two states.

Now we show, using induction, that

$$\mathbf{P}(1)^n = \begin{bmatrix} \dfrac{b + a(1-a-b)^n}{a+b} & \dfrac{a - a(1-a-b)^n}{a+b} \\ \dfrac{b - b(1-a-b)^n}{a+b} & \dfrac{a + b(1-a-b)^n}{a+b} \end{bmatrix} \qquad (5.81)$$

First the root of the induction follows by letting $n = 1$ in Equation 5.81; this shows that

$$\mathbf{P}(1) = \begin{bmatrix} \dfrac{b + a - a^2 - ab}{a+b} & \dfrac{a - a + a^2 + ab}{a+b} \\ \dfrac{b - b + ab + b^2}{a+b} & \dfrac{a + b - ab - b^2}{a+b} \end{bmatrix} = \begin{bmatrix} 1-a & a \\ b & 1-b \end{bmatrix}$$

We now assume $\mathbf{P}(n)$ is correct and show that $\mathbf{P}(n+1)$ is consistent with Equation 5.81;

$$\mathbf{P}(n+1) = \begin{bmatrix} 1-a & a \\ b & 1-b \end{bmatrix}$$
$$\times \begin{bmatrix} \dfrac{b + a(1-a-b)^n}{a+b} & \dfrac{a - a(1-a-b)^n}{a+b} \\ \dfrac{b - b(1-a-b)^n}{a+b} & \dfrac{a + b(1-a-b)^n}{a+b} \end{bmatrix}$$

Letting $r = (1 - a - b)$

$$= \dfrac{1}{a+b} \begin{bmatrix} b + ar^n - ab - a^2 r^n + ab - abr^n & a - ar^n - a^2 + a^2 r^n + a^2 + abr^n \\ b^2 + abr^n + b - br^n - b^2 + b^2 r^n & ab - abr^n + a + br^n - ab - b^2 r^n \end{bmatrix}$$

$$= \dfrac{1}{a+b} \begin{bmatrix} b + ar^n(1-a-b) & a - ar^n(1-a-b) \\ b - br^n(1-a-b) & a + br^n(1-a-b) \end{bmatrix}$$

$$= \dfrac{1}{a+b} \begin{bmatrix} b + ar^{n+1} & a - ar^{n+1} \\ b - br^{n+1} & a + br^{n+1} \end{bmatrix}$$

This completes the inductive proof of Equation 5.81. Note that if $0 < a < 1$ and $0 < b < 1$, then $|r| < 1$. Thus

$$\lim_{n \to \infty} |r|^n = \lim_{n \to \infty} |1 - a - b|^n = 0$$

and the limiting transition matrix is

$$\lim_{n \to \infty} \mathbf{P}(n) = \begin{bmatrix} \dfrac{b}{a+b} & \dfrac{a}{a+b} \\ \dfrac{b}{a+b} & \dfrac{a}{a+b} \end{bmatrix}$$

Note that

$$\boldsymbol{\pi}^T = \lim_{n \to \infty} \mathbf{p}^T(n) = \mathbf{p}^T(0) \lim_{n \to \infty} \mathbf{P}(n) = \left[\dfrac{b}{a+b}, \dfrac{a}{a+b} \right]$$

is independent of $\mathbf{p}(0)$.

Once again, steady-state or stationary-state probabilities have been achieved. Note that if $a = b = 0$, then

$$\mathbf{p}(n) = \mathbf{p}(0) \begin{bmatrix} 1 & 0 \\ 0 & 1 \end{bmatrix}^n = \mathbf{p}(0)$$

Also if $a = b = 1$, then

$$\mathbf{p}(n) = \mathbf{p}(0) \begin{bmatrix} 0 & 1 \\ 1 & 0 \end{bmatrix}^n = \begin{cases} \mathbf{p}(0), & \text{if } n \text{ is even} \\ \mathbf{p}(0) \begin{bmatrix} 0 & 1 \\ 1 & 0 \end{bmatrix}, & \text{if } n \text{ is odd} \end{cases}$$

Note that in this last case, that is, $a = b = 1$, true limiting-state probabilities do not exist.

We have observed in both Examples 5.5 and 5.6 that as $n \to \infty$, the n-step transition probabilities $P_{i,j}(n)$ approach a limit that is independent of i (all elements in a given column of the $\mathbf{P}$ matrix are the same). Many, but not all, Markov chains have such a limiting steady-state or long-term behavior. The investigation and analysis of the limiting conditions of different types of Markov chains is beyond the scope of this book. See the references listed in Section 5.7.

5.3.2 Continuous-time Markov Chains

Let $X(t)$, $t \geq 0$ be a continuous-time Markov chain with homogeneous (transitions depend only on time difference) transition probability function

$$P_{i,j}(\tau) \triangleq P[X(t + \tau) = j | X(t) = i] \tag{5.82}$$

We assume $P_{i,j}(\tau)$ is continuous at $\tau = 0$; thus

$$\lim_{\epsilon \to 0} P_{i,j}(\epsilon) = \delta_{i,j} = \begin{cases} 1 & \text{if } i = j \\ 0 & \text{if } i \neq j \end{cases} \tag{5.83}$$

Note that the Markov property implies that for $\tau > 0$

$$\begin{aligned} P_{i,j}(\tau) = P[X(t_n + \tau) &= j | X(t_n) = i, \\ X(t_{n-1}) &= k_{n-1}, \ldots, \quad X(t_1) = k_1] \\ \text{for } \tau \geq 0, \quad t_n &> t_{n-1} > \cdots > t_1 \end{aligned} \tag{5.84}$$

Chapman–Kolmogorov Equation. The transition probabilities of a Markov chain satisfy the Chapman–Kolmogorov equation for all $0 < t < \tau$, that is

$$P_{i,j}(\tau) = \sum_{\text{all } k} P_{i,k}(t) P_{k,j}(\tau - t) \tag{5.85}$$

This is proved by the same steps used in proving Equation 5.72. (See Problem 5.36.)

Analysis of Continuous-time Markov Chains. In order to apply the Chapman–Kolmogorov equations for analysis of continuous-time Markov chains we use the following approach.

Transition intensities are defined in terms of the derivatives of the transition probability functions evaluated at 0,

$$\lambda_{i,j} \triangleq \frac{\partial}{\partial \tau} [P_{i,j}(\tau)]|_{\tau=0} \quad i \neq j \tag{5.86}$$

These derivatives can be interpreted as intensities of passage; $(\lambda_{ij}\epsilon)$ is approximately the probability of transition from state i to a different state j in a very small time interval ϵ.

SPECIAL CLASSES OF RANDOM PROCESSES

We now find the approximate probability of no transition from state i in a very small increment of time ϵ. Taking derivatives of both sides of

$$\sum_{\text{all } j} P_{i,j}(\tau) = 1 \tag{5.87}$$

shows that

$$\sum_{\text{all } j} \lambda_{i,j} = 0 \tag{5.88}$$

or

$$\lambda_{i,i} = -\sum_{j \neq i} \lambda_{i,j} \tag{5.89}$$

For small ϵ, the probability, $P_{i,i}(\epsilon)$ of no transition to a different state is

$$P_{i,i}(\epsilon) = 1 - P \text{ (transition to another state)}$$

$$\approx 1 - \sum_{j \neq i} \lambda_{i,j}\epsilon = 1 + \lambda_{i,i}\epsilon \tag{5.90}$$

Note that $\lambda_{i,j}$, $i \neq j$ will be positive or zero and thus by Equation 5.89, $\lambda_{i,i}$ will be nonpositive.

Now we use the transition intensities to find the state probabilities. Using the basic Markov property, that is

$$p_i(\tau) = \sum_{\text{all } j} p_j(0) P_{j,i}(\tau) \tag{5.91}$$

we have

$$p_i(t + \epsilon) - p_i(t) = \sum_{j \neq i} p_j(t) P_{j,i}(\epsilon)$$

$$- \sum_{k \neq i} p_i(t) P_{i,k}(\epsilon) \tag{5.92}$$

because the first sum is the probability of changing from another state to state i in time ϵ whereas the second sum is the probability of changing from state i to a different state in time ϵ.

Dividing both sides of Equation 5.92 by ϵ and then taking the limit as ϵ approaches zero, we have, using Equation 5.86

$$\frac{dp_i(t)}{dt} = \sum_{j \neq i} \lambda_{j,i} p_j(t) - \sum_{k \neq i} \lambda_{i,k} p_i(t) \qquad (5.93)$$

Equation 5.93 is true for all i; thus using Equation 5.89 we have for finite chains

$$\left[\frac{dp_1(t)}{dt} \cdots \frac{dp_m(t)}{dt}\right] = [p_1(t) \cdots p_m(t)] \begin{bmatrix} \lambda_{1,1} & \lambda_{1,2} & \cdots & \lambda_{1,m} \\ \vdots & & & \vdots \\ \lambda_{m,1} & \lambda_{m,2} & \cdots & \lambda_{m,m} \end{bmatrix} \qquad (5.94)$$

where m is the number of states.

A Two-state Continuous-time Homogeneous Markov Chain. Example 5.7 illustrates a general two-state continuous-time homogeneous Markov chain.

EXAMPLE 5.7.

A communication receiver can have two states: operable (state 1) and inoperable (state 2). The intensity of transition from 1 to 2, $\lambda_{1,2}$, is called the failure rate and the intensity of transition from 2 to 1, $\lambda_{2,1}$ is called the repair rate. To simplify and to correspond with the notation commonly used in this reliability model, we will call $\lambda_{1,2} = \lambda$ and $\lambda_{2,1} = \mu$. Find

(a) Expressions for $p_1(t)$ and $p_2(t)$.
(b) $p_1(t)$ and $p_2(t)$ if $p_1(0) = 1$.
(c) Steady-state values of $p_1(t)$ and $p_2(t)$.

SOLUTION: With the notation introduced in this example, Equation 5.94 becomes

$$[p_1'(t) \quad p_2'(t)] = [p_1(t) \quad p_2(t)] \begin{bmatrix} -\lambda & \lambda \\ \mu & -\mu \end{bmatrix}$$

or

$$p_1'(t) = -\lambda p_1(t) + \mu p_2(t)$$
$$= -\lambda p_1(t) + \mu[1 - p_1(t)]$$

We can solve the preceding differential equation using Laplace transforms as

$$sP_1(s) - p_1(0) = -\lambda P_1(s) + \frac{\mu}{s} - \mu P_1(s)$$

Solving for $P_1(s)$ and using partial fractions

$$P_1(s) = \frac{\frac{\mu}{(\lambda + \mu)}}{s} + \frac{\frac{[\lambda p_1(0) - \mu p_2(0)]}{(\lambda + \mu)}}{s + \lambda + \mu}$$

Taking inverse transforms, we solve for $p_1(t)$ and $p_2(t)$

$$p_1(t) = \frac{\mu}{\lambda + \mu} + \exp[-(\lambda + \mu)t]\frac{[\lambda p_1(0) - \mu p_2(0)]}{(\lambda + \mu)}$$

$$p_2(t) = 1 - p_1(t) = \frac{\lambda}{\lambda + \mu} + \exp[-(\lambda + \mu)t]\frac{[\mu p_2(0) - \lambda p_1(0)]}{(\lambda + \mu)}$$

If $p_1(0) = 1$, then these equations reduce to the time-dependent probabilities of a repairable piece of equipment being operable given that it started in the operable state. Letting $p_1(0) = 1$, and $p_2(0) = 0$, we have

$$p_1(t) = \frac{\mu}{\lambda + \mu} + \frac{\lambda}{\lambda + \mu}\exp[-(\lambda + \mu)t]$$

$$p_2(t) = \frac{\lambda}{\lambda + \mu} - \frac{\lambda}{\lambda + \mu}\exp[-(\lambda + \mu)t]$$

As t becomes large, the steady-state values are

$$\lim_{t \to \infty} p_1(t) = \frac{\mu}{\lambda + \mu}$$

$$\lim_{t \to \infty} p_2(t) = \frac{\lambda}{\lambda + \mu}$$

Birth and Death Process. Suppose that a continuous-time Markov chain takes on the values 0, 1, 2, . . . and that its changes equal +1 or −1. We say that this Markov chain is a *Birth–Death process*, and

Birthrate (arrival rate) $\lambda_{i,i+1} = \lambda_{a,i},$ $i = 0, 1, \ldots$

Deathrate (departure rate) $\lambda_{i,i-1} = \lambda_{d,i},$ $i = 1, 2, \ldots$

$\lambda_{i,i} = -\lambda_{a,i} - \lambda_{d,i},$ $i = 1, 2, \ldots$

$\lambda_{i,j} = 0,$ $j \neq i-1, i, i+1$

(The state diagram for this process is shown in Figure 5.14.c (page 305) for the case when the rates are independent of i.) Then for small ϵ

$$P[X(t + \epsilon) = n | X(t) = n - 1] = \lambda_{a,n-1}\epsilon$$
$$P[X(t + \epsilon) = n | X(t) = n + 1] = \lambda_{d,n+1}\epsilon$$
$$P[X(t + \epsilon) = n | X(t) = n] = 1 - (\lambda_{a,n} + \lambda_{d,n})\epsilon$$

Using the notation $P[X(t + \epsilon) = n] = p_n(t + \epsilon)$ we have

$$p_n(t + \epsilon) = p_{n-1}(t)[\lambda_{a,n-1}\epsilon] + p_n(t)[1 - (\lambda_{a,n} + \lambda_{d,n})\epsilon]$$
$$+ p_{n+1}(t)[\lambda_{d,n+1}\epsilon] \tag{5.95}$$

Then for $n \geq 1$

$$p_n'(t) = \lim_{\epsilon \to 0} \frac{p_n(t + \epsilon) - p_n(t)}{\epsilon}$$
$$= \lambda_{a,n-1} p_{n-1}(t) - (\lambda_{a,n} + \lambda_{d,n}) p_n(t) + \lambda_{d,n+1} p_{n+1}(t) \tag{5.96.a}$$

and for $n = 0$

$$p_0'(t) = -\lambda_{a,0} p_0(t) + \lambda_{d,1} p_1(t) \tag{5.96.b}$$

which is deduced from the general equation by recognizing that $p_{-1}(t) = 0$ and $\lambda_{d,0} = 0$.

The solution of this system of difference-differential equations is beyond the scope of this book. However, in order to find the steady-state solution, we assume $p_n'(t) = 0$ and arrive at the equations

$$0 = \lambda_{a,n-1} p_{n-1} - (\lambda_{a,n} + \lambda_{d,n}) p_n + \lambda_{d,n+1} p_{n+1} \tag{5.97.a}$$

$$0 = -\lambda_{a,0} p_0 + \lambda_{d,1} p_1 \tag{5.97.b}$$

Equation 5.97 is called the balance equation, which we now solve. From Equation 5.97.b we have

$$p_1 = \frac{\lambda_{a,0}}{\lambda_{d,1}} p_0$$

From Equation 5.97.a with $n = 1$

$$0 = \lambda_{a,0} p_0 - (\lambda_{a,1} + \lambda_{d,1}) p_1 + \lambda_{d,2} p_2 \qquad (5.98)$$

Rearranging Equation 5.98 we obtain

$$p_2 = \frac{1}{\lambda_{d,2}} [(\lambda_{a,1} + \lambda_{d,1}) p_1 - \lambda_{a,0} p_0]$$

Now using Equation 5.97.b in the preceding equation

$$p_2 = \frac{1}{\lambda_{d,2}} [(\lambda_{a,1} + \lambda_{d,1}) p_1 - \lambda_{d,1} p_1]$$

$$p_2 = \frac{\lambda_{a,1}}{\lambda_{d,2}} p_1 = \frac{\lambda_{a,1}}{\lambda_{d,2}} \frac{\lambda_{a,0}}{\lambda_{d,1}} p_0$$

It is then easy to show that

$$p_n = \frac{\lambda_{a,n-1}}{\lambda_{d,n}} p_{n-1} = \frac{\lambda_{a,0} \cdot \lambda_{a,1} \cdots \lambda_{a,n-1}}{\lambda_{d,1} \cdot \lambda_{d,2} \cdots \lambda_{d,n}} p_0 \qquad (5.99)$$

Since $\sum_{k=0}^{\infty} p_k = 1$

$$p_0 = \frac{1}{1 + \sum_{k=1}^{\infty} \prod_{i=0}^{k-1} \left(\frac{\lambda_{a,i}}{\lambda_{d,i+1}} \right)} \qquad (5.100)$$

The birth–death process models are used in a number of important applications. In fact, this same process is analyzed in Section 5.4.2 for the case where both $\lambda_{a,i}$ and $\lambda_{d,i}$ are independent of i. Such a model is useful in analyzing queues, as illustrated in Section 5.4.2 and in Example 5.8 of that section.

5.3.3 Summary of Markov Models

We restricted our introduction of Markov models to Markov chains (the state space is discrete). The reader should note that the matrix analysis we presented is valid only for finite state spaces, and furthermore the limiting state behavior is valid only for certain types of chains. For other types of chains one might be interested in other analysis such as the mean time to absorption (in a certain state) or the mean recurrence time (of a certain state). Classification of chains with different characteristics is discussed in References [5] through [10] listed in Section 5.7.

We discussed both discrete- and continuous-time Markov chains. The emphasis was placed upon the Markov (future independent of the past given the present) concept and analysis of certain chains using elementary matrix representation of the difference equations (sequences) and the differential equations (continuous-time). The birth–death process will be revisited in the next section.

5.4 POINT PROCESSES

Previously we have presented random processes as an ensemble of sample functions (of time) together with a probability distribution specifying probabilities of events associated with the sample paths. In the case of random sequences, the random variables $X(n)$ represent successive values at times $\ldots t_{-1}, t_0, t_1, \ldots$, with the index n in $X(n)$ representing t_n. The time index set $\{t_n\}$ is *fixed* (e.g., $t_{n+1} = t_n + \Delta t$ for all n) and we studied models of $X(t)$ or $X(n)$.

There are, however, many practical situations where the random times of occurrences of some specific events are of primary interest. For example, we may want to study the times at which components fail in a large system or analyze the times at which jobs enter the queue in a computer system. Other examples include the arrival of phone calls at an exchange, or the emission of electrons from the cathode of a vacuum tube. In these examples, our main interest, at least for the initial phase of analysis, is not the phenomenon itself, but the sequence of random time instants at which the phenomena occur. An ensemble of collections of discrete sets of points from the time domain, called a *point process,* is used as a model to analyze phenomena such as the ones just mentioned.

A point process is a rather special kind of random process that may be described as an ensemble of sample paths, where each sample path is obtained by performing the underlying random experiment once. Whereas the sample path of an ordinary random process is a deterministic function of time $x(t)$, the sample path of a point process is a list of time instants at which a specific phenomenon such as a component failure occurred. A graphic example is shown in Figure 5.12. If the point process is defined over the time domain $\Gamma = (-\infty, \infty)$, the ith member function is defined by a sequence of times of occurrence

$$\cdots t^i(-2) < t^i(-1) \leq 0 < t^i(1) < t^i(2) < \cdots \quad (5.101)$$

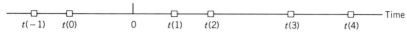

Figure 5.12 Sample path of a point process.

Note that i denotes a particular member function and the argument 1 denotes the first occurrence time after $t = 0$. Now, if $\Gamma = (0, \infty)$, then we specify the ith member function by

$$0 < t^i(1) < t^i(2) < \cdots \qquad (5.102)$$

We can represent the ensemble or the collection of sample paths of a point process by a random sequence

$$T(n), n = 1, 2, \ldots, \quad \text{or} \quad n = \cdots, -1, 0, 1, \ldots \qquad (5.103)$$

For example, $T(5)$ is a random variable that represents the time of the fifth occurrence after $t = 0$. If the point process has an uncountably infinite number of sample paths, then we will use $t(n)$ to denote a particular sample path. If there are only a countable number of sample paths, then we will use $t^i(n)$ to denote the ith sample path.

A point process may also be described by a *waiting* (or interarrival) *time sequence* $W(n)$ where $W(n)$ is defined as

$$\begin{aligned} W(n) &= T(n+1) - T(n) \\ n &= \cdots, -1, 0, 1, \ldots \quad \text{if} \quad \Gamma = (-\infty, \infty) \end{aligned} \qquad (5.104)$$

or

$$\begin{aligned} W(n) &= T(n+1) - T(n), \quad n = 1, 2, \ldots \quad \text{if } \Gamma = (0, \infty) \\ W(0) &= T(1) \end{aligned} \qquad (5.105)$$

The random variable $W(n)$ represents the waiting time between the nth and $(n + 1)$st occurrences.

A third method of characterizing a point process is through the use of a counting procedure. For example, we can define a process, $X(t)$, $t \in \Gamma$, such that

For $t > 0$, $X(t)$ = number of occurrences in the interval $(0, t)$
For $t < 0$, $X(t)$ = $-$[number of occurrences in the interval $(t, 0)$] (5.106)

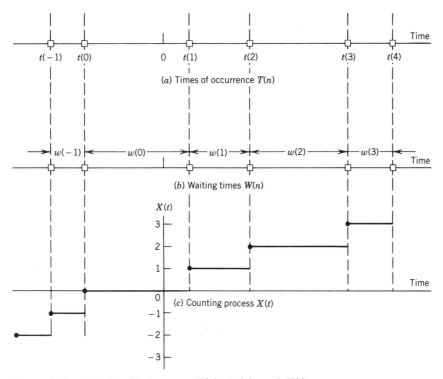

Figure 5.13 Relationship between $T(n)$, $W(n)$, and $X(t)$.

The relationship between $T(n)$, $W(n)$, and $X(t)$ is shown in Figure 5.13. It is clear from Figure 5.13 that the times-of-occurrence sequence $T(n)$, the waiting-time sequence $W(n)$, and the counting process $X(t)$, are equivalent in the sense that each by itself provides a complete specification of the probability law of the underlying point process and thus also of the other two processes. However, the choice of $T(n)$, $W(n)$, or $X(t)$ will be dictated by the specific application that is of interest. For example, if one is interested in analyzing the waiting times in queues, then $W(n)$ is the appropriate model, whereas if queue length is of interest, then $X(t)$ may be more appropriate.

The theory of point processes is built around a set of reasonable assumptions that we state here:

1. Times of occurrences are distinct.
2. Any finite interval contains only a finite number of occurrences.
3. Any infinite time interval contains an infinite number of occurrences.
4. Events do not occur at predetermined times.

298 SPECIAL CLASSES OF RANDOM PROCESSES

Although models that do not meet all the foregoing requirements have been developed, they are beyond the scope of this text, and we will focus our attention only on models that satisfy all four assumptions.

In the following sections we develop two point-process models and illustrate their use in several interesting applications.

5.4.1 Poisson Process

A point process with the additional property that the numbers of occurrences in any finite collection of nonoverlapping time intervals are *independent* random variables (i.e., an independent increment process) leads to a Poisson process. The Poisson process is a *counting* process that arises in many applied problems such as the emission of charged particles from a radioactive material, times of failures of components of a system, and times of demand for services (queueing problems). The Poisson process also serves as a basic building block for more complicated point processes. In the following paragraphs, we derive the Poisson process and its properties.

Suppose we have an independent increments point process with $\Gamma = (0, \infty)$ in which the occurrence of events is governed by a rate function $\lambda(t)$ with the following properties in the limit as $\Delta t \to 0$:

1. $P[1 \text{ occurrence in the interval } (t, t + \Delta t)] = \lambda(t) \Delta t$ \hfill (5.107.a)
2. $P[0 \text{ occurrence in the interval } (t, t + \Delta t)] = 1 - \lambda(t) \Delta t$ \hfill (5.107.b)
3. $P[2 \text{ or more occurrences in the interval } (t, t + \Delta t)] = 0$ \hfill (5.107.c)

If $\lambda(t)$ does not depend on t, then the process is called a *homogeneous* process. The counting process associated with this point process is called the Poisson process, and we now show that for a homogeneous Poisson (counting) process, the number of occurrences in the interval $(0, t)$ is governed by a Poisson distribution.

Let us denote the probability that there are k occurrences in the time interval $(0, t)$ by $P(k, t)$. Then the probability of no occurrences in the time interval $(0, t)$ is $P(0, t)$, and $P(0, t + \Delta t)$ is related to $P(0, t)$ by

$$P(0, t + \Delta t) = P[0 \text{ occurrences in the interval } (0, t) \text{ and}$$
$$0 \text{ occurrences in the interval } (t, t + \Delta t)]$$

Using the independent increments assumption and Equation 5.107.b, we can write $P(0, t + \Delta t)$ as

$$P(0, t + \Delta t) = P(0, t)(1 - \lambda \Delta t)$$

where λ is the rate. We can rewrite the preceding equation as

$$\frac{P(0, t + \Delta t) - P(0, t)}{\Delta t} = -\lambda P(0, t)$$

and taking the limit $\Delta t \to 0$, we have

$$\frac{dP(0, t)}{dt} = -\lambda P(0, t)$$

Solving this first-order differential equation with the initial condition $P(0, 0) = 1$, we have

$$P(0, t) = \exp(-\lambda t) \qquad (5.108)$$

Proceeding in a similar manner, we can show that for any $k > 0$

$$P(k, t + \Delta t) = P(k, t)(1 - \lambda \Delta t) + P(k - 1, t)\lambda \Delta t$$

Rearranging and taking the limit

$$\frac{dP(k, t)}{dt} + \lambda P(k, t) = \lambda P(k - 1, t) \qquad (5.109)$$

This linear differential equation has the solution

$$P(k, t) = \lambda \exp(-\lambda t) \int_0^t \exp(\lambda \tau) P(k - 1, \tau) \, d\tau \qquad (5.110)$$

Starting with $P(0, t) = \exp(-\lambda t)$, we can recursively solve Equation 5.110 for $P(1, t)$, $P(2, t)$, and so on, as

$$P(1, t) = \lambda t \exp(-\lambda t)$$
$$P(2, t) = \frac{(\lambda t)^2}{2} \exp(-\lambda t)$$
$$\vdots \qquad \vdots$$
$$P(k, t) = \frac{(\lambda t)^k}{k!} \exp(-\lambda t)$$
$$\vdots \qquad \vdots \qquad (5.111)$$

Since we define the counting process to have $X(0) = 0$, Equation 5.111 also gives the probability that $X(t) - X(0) = k$. That is

$$P[X(t) - X(0) = k] = \frac{(\lambda t)^k}{k!} \exp(-\lambda t), \quad k = 0, 1, 2, \ldots \quad (5.112)$$

Since λ is constant, we can show that for any $t_1, t_2 \in \Gamma$, $t_2 > t_1$

$$P[(X(t_2) - X(t_1)) = k] = \frac{(\lambda[t_2 - t_1])^k}{k!} \exp(-\lambda[t_2 - t_1]) \quad$$

When λ is time dependent, then the distribution has the form

$$P[(X(t_2) - X(t_1)) = k] = \frac{[\mu(t_2) - \mu(t_1)]^k}{k!} \exp(-[\mu(t_2) - \mu(t_1)]) \quad (5.113)$$

where

$$\mu(t_2) - \mu(t_1) = \int_{t_1}^{t_2} \lambda(\tau) \, d\tau \quad (5.114)$$

Thus, the increments of a Poisson process have Poisson probability distributions.

Properties of the Homogeneous Poisson Process. The Poisson counting process $X(t)$ is a discrete amplitude process with the first-order probability function

$$P[X(t) = k] = \frac{(\lambda t)^k \exp(-\lambda t)}{k!}, \quad k = 0, 1, \ldots \quad (5.115)$$

Higher-order probability functions of the process can be obtained as follows, starting with the second-order probability function:

$$P[X(t_1) = k_1, X(t_2) = k_2]$$
$$= P[X(t_2) = k_2 | X(t_1) = k_1] P[X(t_1) = k_1], \quad t_2 > t_1 \quad (5.116)$$

The conditional probability that $X(t_2) = k_2$ given $X(t_1) = k_1$ is the probability that there are $k_2 - k_1$ occurrences between t_1 and t_2. Hence

$$P[X(t_2) = k_2 | X(t_1) = k_1]$$
$$= P[X(t_2) - X(t_1) = k_2 - k_1]$$
$$= \frac{(\lambda[t_2 - t_1])^{k_2-k_1}}{(k_2 - k_1)!} \exp(-\lambda[t_2 - t_1]), \quad k_2 \geq k_1$$
$$= 0 \qquad\qquad k_2 < k_1 \quad (5.117)$$

Combining Equations 5.115, 5.116, and 5.117, we obtain the second-order probability function as

$$P[X(t_1) = k_1, X(t_2) = k_2]$$
$$= \begin{cases} \dfrac{(\lambda t_1)^{k_1}(\lambda[t_2 - t_1])^{k_2-k_1} \exp(-\lambda t_2)}{k_1!(k_2 - k_1)!}, & k_2 \geq k_1 \\ 0 & k_2 < k_1 \end{cases} \quad (5.118)$$

Proceeding in a similar fashion we can obtain, for example, the third-order joint probability function as

$$P[X(t_1) = k_1, \ X(t_2) = k_2, X(t_3) = k_3]$$

$$= \begin{cases} \dfrac{(\lambda t_1)^{k_1}(\lambda[t_2 - t_1])^{k_2-k_1}(\lambda[t_3 - t_2])^{k_3-k_2} \exp(-\lambda t_3)}{k_1!(k_2 - k_1)!(k_3 - k_2)!}, \\ \qquad\qquad\qquad\qquad\qquad k_1 \leq k_2 \leq k_3 \\ 0 \qquad\qquad\qquad\qquad \text{otherwise} \end{cases} \quad (5.119)$$

The reader can verify that

$$P[X(t_3) = k_3 | X(t_2) = k_2, X(t_1) = k_1]$$
$$= P[X(t_3) = k_3 | X(t_2) = k_2]$$

and thus show that the Poisson process has the Markov property. Recall also that the Poisson process is an independent increments process.

In Chapter 3, the mean, autocorrelation, and autocovariance functions of the Poisson process were derived as

$$E\{X(t)\} = \mu(t) = \lambda t, \qquad t \in \Gamma \tag{5.120.a}$$

$$R_{XX}(t_1, t_2) = \lambda^2 t_1 t_2 + \lambda \min(t_1, t_2) \tag{5.120.b}$$

$$C_{XX}(t_1, t_2) = \lambda \min(t_1, t_2), \qquad t_1, t_2, \in \Gamma \tag{5.120.c}$$

Note that the Poisson process is nonstationary. The distribution of the waiting or interarrival time of a Poisson process is of interest in the analysis of queueing problems. The probability that the interarrival time W is greater than w is equivalent to the probability that no events occur in the time interval $(0, w)$. That is

$$P(W > w) = P(0, w) = \exp(-\lambda w), \qquad w \geq 0$$

or

$$P(W \leq w) = 1 - \exp(-\lambda w), \qquad w \geq 0$$
$$= 0 \qquad w < 0$$

Thus, W is an exponential random variable, and the probability density function of W has the form

$$\begin{aligned} f_W(w) &= \frac{d}{dw}[P(W \leq w)] \\ &= \begin{cases} \lambda \exp(-\lambda w), & w \geq 0 \\ 0 & \text{elsewhere} \end{cases} \end{aligned} \tag{5.121}$$

Note that the expected value of the waiting time has the value

$$E\{W\} = \int_0^\infty \lambda w \exp(-\lambda w) \, dw$$
$$= 1/\lambda \tag{5.122}$$

where λ is the rate of the process.

Superposition of Poisson Processes. The Poisson process has the very interesting property that the sum of n independent Poisson processes with rate functions $\lambda_1(t), \lambda_2(t), \ldots, \lambda_n(t)$ is a Poisson process with rate function $\lambda(t)$ where

$$\lambda(t) = \lambda_1(t) + \lambda_2(t) + \cdots + \lambda_n(t) \tag{5.123}$$

(See Problem 5.44.)

5.4.2 Application of Poisson Processes—Analysis of Queues

Most of us have spent a considerable amount of time waiting for service at checkout lines at supermarkets, banks, and airline counters. Arrival of customers (who demand services) at these service locations is a random phenomenon that can be modeled by a point process. The time it takes to serve a customer is also a random quantity that can be modeled as a random variable. Several models for arrival and service times have been developed and used to analyze average queue lengths and the average time a customer has to wait for service. Such models are used in analyzing time-shared computer and communication systems. One of the most widely used model is the M/M/1* model.

In the M/M/1 model, the arrival process is assumed to be a homogeneous Poisson process with an arrival rate of λ_a. Customers wait in a queue for service, and after they reach the top of the queue they depart from the line at a departure rate of λ_d. It is assumed that the service time S has an exponential probability density function

$$\begin{aligned} f_S(s) &= \lambda_d \exp(-\lambda_d s), & s &\geq 0 \\ &= 0 & s &< 0 \end{aligned} \tag{5.124}$$

Note that the average service time for a customer (after the customer reaches the server) is

$$E\{S\} = \frac{1}{\lambda_d} \tag{5.125}$$

and the service times for different customers are assumed to be independent random variables.

In the analysis of queues, one of the important quantities of interest is the average number of customers waiting for service, that is, the average queue length. This will be a function of the arrival (or "birth") rate and the departure

---

*Queueing models are designated by three letters: A/R/S where the first letter denotes arrival model (M for Markov), the second letter denotes the service time distribution (M for Markov), and the last letter represents the number of servers.

304 SPECIAL CLASSES OF RANDOM PROCESSES

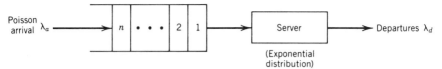

Figure 5.14a M/M/1 queue.

(or "death") rate. We now find $E\{N\}$ where N is the number of customers waiting in a queue. (This analysis is similar to that given for the birth and death process in Section 5.3.2.)

Because of the Poisson assumptions involved, a simple way of analyzing the queue is to focus on two successive time intervals t and $t + \Delta t$, $\Delta t \to 0$. With reference to Figures 5.14a and 5.14b, suppose that there are n customers in the queue at time $t + \Delta t$. With the Poisson arrival assumption, it is apparent that the queue could have been at only one of three states, $n + 1$, n, or $n - 1$, at time t, Δt seconds prior. For in a Δt-second interval no more than one customer could have arrived and no more than one customer could have been served. If we let $p_n(t)$ denote the probability that there are n customers in the queue at time t, then we can derive the following relationship between $p_n(t)$ and $p_n(t + \Delta t)$.

$$\begin{aligned}
p_n(t + \Delta t) &= p_n(t) \cdot P[\text{0 arrivals and 0 departures or} \\
&\qquad \text{1 arrival and 1 departure in } (t, t + \Delta t)] \\
&+ p_{n+1}(t) \cdot P[\text{0 arrivals and 1 departures in } (t, t + \Delta t)] \\
&+ p_{n-1}(t) \cdot P[\text{1 arrival and 0 departures in } (t, t + \Delta t)] \\
&= p_n(t) \cdot [(1 - \lambda_a \Delta t)(1 - \lambda_d \Delta t) + \lambda_a \Delta t \lambda_d \Delta t] \\
&+ p_{n+1}(t)[1 - \lambda_a \Delta t]\lambda_d \Delta t] \\
&+ p_{n-1}(t)[\lambda_a \Delta t(1 - \lambda_d \Delta t)], \qquad n > 0 \qquad (5.126)
\end{aligned}$$

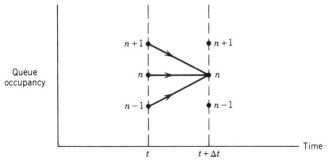

Figure 5.14b Analysis of M/M/1 queue.

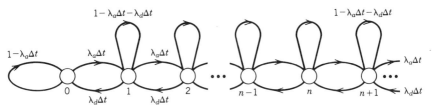

Figure 5.14c State diagram of an M/M/1 queue.

Subtracting $p_n(t)$ from both sides, dividing by Δt, and taking the limit as $\Delta t \to 0$ produces the derivative $p'_n(t)$. Now if we focus on the stationary behavior of the queue after the queue has been operating for a while, we can assume that $p'_n(t) = 0$. Furthermore, if we assume that $\lambda_a \Delta t$ and $\lambda_d \Delta t$ are both $\ll 1$, then as $\Delta t \to 0$ we can ignore all terms involving Δt^2 and rewrite Equation 5.126 as

$$(\lambda_a + \lambda_d) p_n = \lambda_d p_{n+1} + \lambda_a p_{n-1}, \qquad n \geq 1 \qquad (5.127)$$

Equation 5.126 is shown diagrammed in Figure 5.14c. Note that this defines a continuous time Markov chain, and that

$$\sum_{n=0}^{\infty} p_n = 1 \qquad (5.128)$$

When $n = 0$, Equation 5.127 reduces to

$$p_0 \lambda_a = \lambda_d p_1 \qquad (5.129)$$

We can recursively solve for $p_1, p_2, \ldots$, starting with

$$p_1 = p_0 \left(\frac{\lambda_a}{\lambda_d}\right)$$
$$= p_0 \rho$$

where

$$\rho = \left(\frac{\lambda_a}{\lambda_d}\right) < 1 \quad \text{(for a stable queue)}$$

306 SPECIAL CLASSES OF RANDOM PROCESSES

Substituting p_1 in Equation 5.127, we obtain p_2 as

$$p_2 = (\rho + 1)p_1 - \rho p_0 = \rho^2 p_0$$

Similarly, we can show that

$$p_n = \rho^n p_0$$

Finally, substituting p_n in Equation 5.128 we can solve for p_0 and show that

$$p_n = (1 - \rho)\rho^n \qquad (5.130)$$

where p_n is the probability that there are n customers in the queue.

The average number of customers waiting in the queue can be obtained as

$$E\{N\} = \sum_{n=0}^{\infty} n p_n$$

$$= \frac{\rho}{1 - \rho} \qquad (5.131)$$

Using the result given in Equation 5.131, it can be shown [11] that the average waiting time W is

$$E\{W\} = \text{Average queue length} \times \text{average service time per customer}$$

$$= \left[\frac{\rho}{1 - \rho}\right] E\{S\}, \qquad \rho = \frac{\lambda_a}{\lambda_d} \qquad (5.132)$$

Equations 5.131 and 5.132 are used extensively in the design of queueing systems.

EXAMPLE 5.8.

Customers arrive at an airline ticketing counter at the rate of 30/hour. Assume Poisson arrivals, exponential service-time distribution, and a single server queue model (M|M|1).

(a) If the average service time is 100 seconds, find the average waiting time (before being served).
(b) If the waiting time is to be less than 3 minutes, what should be the average service time?

SOLUTION:

(a) With $\lambda_a = 30$ and $\lambda_d = 36$, $\rho = 30/36$ and using Equation 5.132

$$E\{W\} = \frac{\frac{30}{36}}{1 - \frac{30}{36}} \left(\frac{100}{60}\right) \approx 8.33 \text{ minutes}$$

(b) $E\{W\} \leq 3$ requires

$$\frac{\frac{\lambda_a}{\lambda_d}}{1 - \frac{\lambda_a}{\lambda_d}} \left(\frac{60}{\lambda_d}\right) \leq 3$$

Or $\lambda_d \geq 44$ departures per hour; thus a service time of approximately 84 seconds is needed.

5.4.3 Shot Noise

Emission of charged particles is a fundamental phenomenon that occurs in electronic systems. In lightwave communication systems, for example, the information transmitted is represented by intensities of light and is detected by a photosensitive device in the receiver that produces current pulses in proportion to the incident light. The current pulses result from photoelectrons emitted at random times τ_k with an average rate λ proportional to the incident light energy. The resulting waveform can be represented by a random process of the form

$$X(t) = \sum_{k=-\infty}^{\infty} h(t - \tau_k) \tag{5.133}$$

where $h(t)$ is the electrical pulse produced by a single photoelectron emitted at $t = 0$ (see Figure 5.15). The emission times τ_k are assumed to form a Poisson point process with a fixed rate λ, and the Poisson process is assumed to be continuous for all time, that is, $\Gamma = (-\infty, \infty)$. Because λ is constant and there is no distinguishing origin of time, the process $X(t)$ is stationary.

$X(t)$ is called a shot noise process and has been used extensively to model randomly fluctuating components of currents in electronic circuits containing devices such as photodiodes and vacuum tubes. This process can also be used to model phenomenon such as the background return (clutter) seen by radar. Consider for example an airplane flying over a terrain and sending out radar pulses. These pulses will be reflected by various objects in the terrain and the

308   SPECIAL CLASSES OF RANDOM PROCESSES

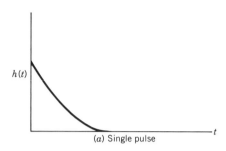

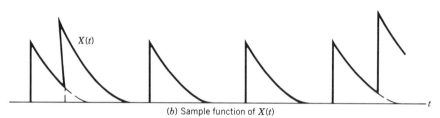

Figure 5.15 Shot-noise process. (a) Single pulse. (b) Sample function of $X(t)$.

radar receiver will receive a multitude of echoes of random amplitudes and delays. The received signal can be modeled as a random process of the form

$$Y(t) = \sum_{k=-\infty}^{\infty} A_k h(t - \tau_k) \qquad (5.134)$$

where $h(t)$ is the shape of the reflected pulse, τ_k is the time at which the kth pulse reaches the receiver, and A_k its amplitude which is independent of τ_k. A_k and A_j, for $j \neq k$, can be assumed to be independent and identically distributed random variables with the probability density function $f_A(a)$.

We now derive the properties of $X(t)$ and $Y(t)$. The derivation of the probability density functions of $X(t)$ and $Y(t)$ is in general a very difficult problem, and we refer the interested reader to References [7] and [13] of Section 5.7. Only the mean and autocorrelation function are derived here.

Mean and Autocorrelation of Shot Noise. Suppose we divide the time axis into nonoverlapping intervals of length Δt, where Δt is so short that $\lambda \, \Delta t \ll 1$. Let I_m be a random variable such that for all integer values of m

$$I_m = \begin{cases} 0 & \text{if no new pulse is emitted in} \quad m \, \Delta t < t < (m+1) \, \Delta t \\ 1 & \text{if one new pulse is emitted in} \quad m \, \Delta t < t < (m+1) \, \Delta t \end{cases}$$

Since $\lambda \, \Delta t \ll 1$, we can neglect the probability of more than one new pulse in any interval of length Δt. Then, because of the Poisson assumption,

$$P(I_m = 0) = \exp(-\lambda \, \Delta t) \approx 1 - \lambda \, \Delta t$$
$$P(I_m = 1) = \lambda \, \Delta t \exp(-\lambda \, \Delta t) \approx \lambda \, \Delta t$$

and

$$E\{I_m\} = \lambda \, \Delta t$$
$$E\{I_m^2\} = \lambda \, \Delta t$$

The process $X(t)$ can be approximated by $\tilde{X}(t)$, where

$$\tilde{X}(t) = \sum_{m=-\infty}^{\infty} I_m h(t - m \, \Delta t)$$

This new process $\tilde{X}(t)$ in which at most one new pulse can start in each interval Δt approaches $X(t)$ as $\Delta t \to 0$, and we use $\tilde{X}(t)$ to derive the mean and autocorrelation of $X(t)$.

The mean of $X(t)$ is obtained as

$$E\{X(t)\} \approx E\{\tilde{X}(t)\}$$
$$= \sum_{m=-\infty}^{\infty} E\{I_m\} h(t - m \, \Delta t)$$
$$= \sum_{m=-\infty}^{\infty} \lambda \, \Delta t \, h(t - m \, \Delta t)$$
$$= \lambda \left[\sum_{m=-\infty}^{\infty} h(t - m \, \Delta t) \, \Delta t \right]$$

As $\Delta t \to 0$, the summation in the previous equation [which is the area under $h(t)$] becomes an integral and we have

$$E\{X(t)\} = \lambda \int_{-\infty}^{\infty} h(u) \, du \qquad (5.135.a)$$

The autocorrelation function of $X(t)$ can be approximated the same way:

$$R_{XX}(t_1, t_2) = E\{X(t_1)X(t_2)\}$$
$$\approx E\{\tilde{X}(t_1)\tilde{X}(t_2)\} = R_{\tilde{X}\tilde{X}}(t_1, t_2)$$
$$R_{\tilde{X}\tilde{X}}(t_1, t_2) = \sum_{m_1}\sum_{m_2} E\{I_{m_1}I_{m_2}\}h(t_1 - m_1\,\Delta t)h(t_2 - m_2\,\Delta t)$$

Now, using independent increments

$$E\{I_{m_1}I_{m_2}\} = \begin{cases} E\{I_m^2\} = \lambda\,\Delta t, & m_1 = m_2 = m \\ E\{I_{m_1}\}E\{I_{m_2}\} = [\lambda\,\Delta t]^2; & m_1 \neq m_2 \end{cases}$$

Thus

$$R_{\tilde{X}\tilde{X}}(t_1, t_2) = \sum_m \lambda\,\Delta t\, h(t_1 - m\,\Delta t)h(t_2 - m\,\Delta t)$$
$$+ \left[\sum_{m_1} \lambda\,\Delta t\, h(t_1 - m_1\,\Delta t)\right]\left[\sum_{m_2 \neq m_1} \lambda\,\Delta t\, h(t_2 - m_2\,\Delta t)\right]$$

and when $\Delta t \to 0$,

$$R_{XX}(t_1, t_2) = \lambda \int_{-\infty}^{\infty} h(t_1 - v)h(t_2 - v)\,dv + \left[\lambda \int_{-\infty}^{\infty} h(t - u)\,du\right]^2$$

Since the process is stationary, the autocorrelation function can be written as

$$R_{XX}(\tau) = \lambda \int_{-\infty}^{\infty} h(u)h(\tau + u)\,du + \left[\lambda \int_{-\infty}^{\infty} h(u)\,du\right]^2 \quad (5.135.b)$$

Equation 5.135.b is known as *Cambell's theorem*. Recognizing the second term in Equation 5.135.b as $[E\{X(t)\}]^2$, we can write the autocovariance function of $X(t)$ as

$$C_{XX}(\tau) = \lambda \int_{-\infty}^{\infty} h(u)h(\tau + u)\,du \quad (5.135.c)$$

Using the Fourier transform $H(f)$ of $h(\tau)$, we obtain

$$E\{X(t)\} = \lambda H(0) \qquad (5.136.\text{a})$$

and

$$S_{XX}(f) = [\lambda H(0)]^2 \delta(f) + \lambda |H(f)|^2 \qquad (5.136.\text{b})$$

The reader can show that the process $Y(t)$ given in Equation 5.134 has

$$E\{Y(t)\} = \lambda E\{A\} \int_{-\infty}^{\infty} h(u) \, du \qquad (5.137.\text{a})$$

and

$$C_{YY}(\tau) = \lambda E\{A^2\} \int_{-\infty}^{\infty} h(u) h(\tau + u) \, du \qquad (5.137.\text{b})$$

EXAMPLE 5.9.

The pulses in a shot noise process are rectangles of height A_k and duration T. Assume that A_k's are independent and identically distributed with $P[A_k = 1] = 1/2$ and $P[A_k = -1] = 1/2$ and that the rate, $\lambda = 1/T$. Find the mean, autocorrelation, and the power spectral density of the process.

SOLUTION:

$$Y(t) = \sum_k A_k h(t - \tau_k)$$

$$E\{Y(t)\} = \frac{1}{T} E\{A_k\} \int_{-\infty}^{\infty} h(u) \, du$$

$$= 0$$

The autocorrelation function is given by

$$R_{YY}(\tau) = C_{YY}(\tau) = \frac{1}{T} \cdot 1 \int_{-\infty}^{\infty} h(u) h(u + \tau) \, du$$

$$= \begin{cases} 1 - \frac{|\tau|}{T}, & 0 < |\tau| < T \\ 0 & \text{elsewhere} \end{cases}$$

and the power spectral density function has the form

$$S_{YY}(f) = \frac{1}{T}|H(f)|^2$$

where

$$H(f) = \sin(\pi f T)/\pi f$$

The reader should note that both the autocorrelation function and the spectral density function of this shot noise process is the same as those of the random binary waveform discussed in Section 3.4.4. These two examples illustrate that even though the two processes have entirely different models in the time domain, their frequency domain descriptions are identical. This is because the spectral density function, which is the frequency domain description, and autocorrelation function are average (second moment) descriptors and they do not uniquely describe a random process.

5.4.4 Summary of Point Processes

Processes where the times (points) of occurrences of events are of primary interest are called point processes. Such processes were defined by the times of occurrence, waiting time, and the count (of the number of occurrences).

The Poisson process was introduced as an independent increments point process, where the count was shown to be a Poisson random variable, and the waiting time was shown to be an exponential random variable. The Poisson process was applied to find the average queue length and the average waiting time in a queue.

Finally, a model of shot noise was developed using a Poisson process as a model for emission times. The mean and autocorrelation function were found and used in an example.

5.5 GAUSSIAN PROCESSES

In this section we introduce the most common models of *noise* and of signals used in analysis of communication systems. Many random phenomena in physical problems including noise are well approximated by Gaussian random processes. By virtue of the central limit theorem, a number of processes such as the Wiener process as well as the shot-noise process can be approximated by a Gaussian process. Furthermore, the output of a linear system made up of a weighted sum of a large number of independent samples of the input random process tends to approach a Gaussian process. Gaussian processes play a central role in the theory and analysis of random phenomena both because they are good approximations to the observations and because multivariate Gaussian distributions are analytically simple.

One of the most important uses of the Gaussian process is to model and analyze the effects of "thermal" noise in electronic circuits used in communication systems. Individual circuits contain resistors, inductors, and capacitors as well as semiconductor devices. The resistors and semiconductor elements contain charged particles subjected to random motion due to thermal agitation. The random motion of charged particles causes fluctuations in the current waveforms (or information-bearing signals) that flow through these components. These, fluctuations are called thermal noise and are often of sufficient strength to mask a weak signal and make the recognition of signals a difficult task. Models of thermal noise are used to understand and minimize the effects of noise on signal detection (or recognition).

5.5.1 Definition of a Gaussian Process

A real-valued random process $X(t)$, $t \in \Gamma$, is called a Gaussian process if all its nth-order density functions are n-variate Gaussian. If we denote $X(t_i)$ by X_i, then for any n

$$f_\mathbf{X}(\mathbf{x}) = [(2\pi)^{n/2} |\Sigma_\mathbf{X}|^{1/2}]^{-1} \exp\left[-\frac{1}{2} (\mathbf{x} - \boldsymbol{\mu}_\mathbf{X})^T \Sigma_\mathbf{X}^{-1} (\mathbf{x} - \boldsymbol{\mu}_\mathbf{X})\right] \quad (5.138)$$

where

$$\mathbf{X} = \begin{bmatrix} X_1 \\ X_2 \\ \vdots \\ X_n \end{bmatrix} = \begin{bmatrix} X(t_1) \\ X(t_2) \\ \vdots \\ X(t_n) \end{bmatrix} \qquad \mathbf{x} = \begin{bmatrix} x_1 \\ x_2 \\ \vdots \\ x_n \end{bmatrix}$$

$$\boldsymbol{\mu}_\mathbf{X} = \begin{bmatrix} E\{X(t_1)\} \\ E\{X(t_2)\} \\ \vdots \\ E\{X(t_n)\} \end{bmatrix} = \begin{bmatrix} \mu_X(t_1) \\ \mu_X(t_2) \\ \vdots \\ \mu_X(t_n) \end{bmatrix}$$

$$\Sigma_\mathbf{X} = \begin{bmatrix} C_{XX}(t_1, t_1) & \cdots & C_{XX}(t_1, t_j) & \cdots & C_{XX}(t_1, t_n) \\ \vdots & & & & \vdots \\ \vdots & & C_{XX}(t_i, t_j) & & \vdots \\ & & \vdots & & \\ C_{XX}(t_n, t_1) & \cdots & C_{XX}(t_n, t_j) & \cdots & C_{XX}(t_n, t_n) \end{bmatrix}$$

and

$$C_{XX}(t_i, t_j) = E\{X(t_i)X(t_j)\} - \mu_X(t_i)\mu_X(t_j)$$
$$= R_{XX}(t_i, t_j) - \mu_X(t_i)\mu_X(t_j)$$

The joint density function given in Equation 5.138 is a multivariate Gaussian density function; its properties are given in Section 2.5.

If the process is WSS, then we have

$$E\{X(t_i)\} = \mu_X$$

and

$$C_{XX}(t_i, t_j) = R_{XX}(|t_i - t_j|) - \mu_X^2$$

The nth-order distributions of a Gaussian process depend on the two functions $\mu_X(t)$, and $C_{XX}(t, t + \tau)$. When the process is WSS, then $\mu_X(t)$, and $C_{XX}(t, t + \tau)$ do not depend on t, which implies that

$$f_\mathbf{X}[x(t_1), x(t_2), \ldots, x(t_n)] = f_\mathbf{X}[x(t_1 + \tau), x(t_2 + \tau), \ldots, x(t_n + \tau)]$$

That is, the process is strict-sense stationary. In this case $\mu_X(t) = \mu_X$, $C_{XX}(t, t + \tau) = C_{XX}(\tau)$, and $R_{XX}(t, t + \tau) = R_{XX}(\tau)$.

5.5.2 Models of White and Band-limited White Noise

Thermal noise generated in resistors and semiconductors is modeled as a zero-mean, stationary Gaussian random process $N(t)$ with a power spectral density that is flat over a very wide range of frequencies. Such a process is called *white (Gaussian) noise* in analogy to white light, whose spectral density is broad and uniform over a wide frequency range. The power spectral density of thermal noise has been shown to have the value $kT/2$ Joules, where k is Boltzman's constant (1.38×10^{-23} Joules/°Kelvin) and T is the equivalent temperature in °Kelvin of the noise source.

It is customary to denote the uniform spectral density of white noise by $\eta/2$, (or $N_0/2$).

$$S_{NN}(f) = \frac{\eta}{2} \tag{5.139}$$

Note that the spectral density given Equation 5.139 yields

$$R_{NN}(\tau) = \frac{\eta}{2}\delta(\tau) \qquad (5.140)$$

which implies that $N(t)$ and $N(t + \tau)$ are independent for any value of $\tau \neq 0$.

The spectral density given in Equation 5.139 is not physically realizable since it implies infinite average power, that is

$$\int_{-\infty}^{\infty} S_{NN}(f)\, df \to \infty$$

However, since the bandwidths of real systems are always finite, and since

$$\int_{-B}^{B} S_{NN}(f)\, df = \eta B < \infty$$

for any finite bandwidth B, the spectral density given in Equation 5.139 can be used over finite bandwidths.

Noise having a nonzero and constant spectral density over a finite frequency band and zero elsewhere is called *band-limited white noise*. Figure 5.16 shows such a spectrum where

$$S_{NN}(f) = \begin{cases} \dfrac{\eta}{2}, & |f| < B \\ 0 & \text{elsewhere} \end{cases}$$

The reader can verify that this process has the following properties:

1. $E\{N^2(t)\} = \eta B$
2. $R_{NN}(\tau) = \eta B \dfrac{\sin 2\pi B\tau}{2\pi B\tau}$
3. $N(t)$ and $N(t + k\tau_0)$, where k is an integer (nonzero) and $\tau_0 = \dfrac{1}{2B}$ are independent.

It should be pointed out here that terms "white" and "band-limited white" refer to the spectral shape of the process. These terms by themselves do not imply that the distributions associated with $X(t)$ are Gaussian. A process that is not a Gaussian process may also have a flat, that is, white, power spectral density.

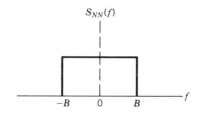

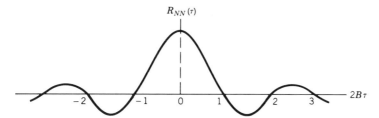

Figure 5.16 Bandlimited white noise.

For random sequences, white noise $N(k)$ is a stationary sequence that has a mean of zero and

$$E\{N(n)N(m)\} = 0, \quad n \neq m$$
$$= \sigma_N^2, \quad n = m$$

If it is a stationary (zero-mean) Gaussian white noise sequence, then $N(n)$ is Gaussian:

$$f_{N(n)}(x) = \frac{1}{\sqrt{2\pi}\,\sigma_N} \exp(-x^2/2\sigma_N^2)$$

Also

$$R_{NN}(n) = \begin{cases} \sigma_N^2, & n = 0 \\ 0, & n \neq 0 \end{cases}$$

and thus

$$S_{NN}(f) = \sigma_N^2, \quad |f| < \frac{1}{2}$$

5.5.3 Response of Linear Time Invariant Systems to White Gaussian Noise

The response of an LTIV system driven by a random process was derived in Chapter 4. If the input is a zero-mean, white Gaussian process $N(t)$, then the output process $Y(t)$ has the following properties:

1. $E\{Y(t)\} = \mu_N H(0) = 0$ (5.140.a)

2. $R_{YY}(\tau) = R_{NN}(\tau) * h(\tau) * h(-\tau) = \dfrac{\eta}{2}\delta(\tau) * h(\tau) * h(-\tau)$ (5.140.b)

3. $S_{YY}(f) = S_{NN}(f)|H(f)|^2 = \dfrac{\eta}{2}|H(f)|^2$ (5.140.c)

In Equations 5.140.a–c, $h(t)$ and $H(f)$ are the impulse response and transfer function of the system, respectively.

Since the convolution integral given by

$$Y(t) = \int_{-\infty}^{\infty} h(t - \alpha) N(\alpha)\, d\alpha$$

is a linear operation on the Gaussian process $N(t)$, the output $Y(t)$ will have Gaussian densities and hence is a Gaussian process.

5.5.4 Quadrature Representation of Bandpass (Gaussian) Signals

In communication systems, information-bearing signals often have the form

$$X(t) = R_X(t)\cos[2\pi f_0 t + \Theta_X(t)] \qquad (5.141.a)$$

where $X(t)$ is a bandpass signal; $R_X(t)$ and $\Theta_X(t)$ are lowpass signals. $R_X(t)$ is called the envelope of the bandpass signal $X(t)$, $\Theta_X(t)$ is the phase, and f_0 is the carrier or center frequency. $X(t)$ can also be expressed as

$$\begin{aligned}X(t) &= R_X(t)\cos\Theta_X(t)\cos 2\pi f_0 t - R_X(t)\sin\Theta_X(t)\sin 2\pi f_0 t \\ &= X_c(t)\cos 2\pi f_0 t - X_s(t)\sin 2\pi f_0 t \end{aligned} \qquad (5.141.b)$$

$X_c(t) = R_X(t)\cos\Theta_X(t)$ and $X_s(t) = R_X(t)\sin\Theta_X(t)$ are called the quadrature components of $X(t)$.

If the noise in the communication system is additive, then the receiver observes and processes $X(t) + N(t)$ and attempts to extract $X(t)$. In order to analyze the performance of the receiver, it is useful to derive a time domain

318 SPECIAL CLASSES OF RANDOM PROCESSES

representation of $N(t)$ in envelope and phase, or in quadrature form. We now show that any arbitrary bandpass stationary process $N(t)$ with zero mean can be expressed in quadrature (and in envelope and phase) form.

We start with the quadrature representation of a stationary, zero-mean process

$$N(t) = N_c(t)\cos 2\pi f_0 t - N_s(t)\sin 2\pi f_0 t \tag{5.142}$$

where $N_c(t)$ and $N_s(t)$ are two jointly stationary random processes. We now attempt to find the properties of $N_c(t)$ and $N_s(t)$ such that the quadrature representation yields zero mean of $N(t)$ and the correct autocorrelation function $R_{NN}(\tau)$ (and hence the correct psd).

Taking the expected value on both sides of Equation 5.142, we have

$$E\{N(t)\} = E\{N_c(t)\}\cos 2\pi f_0 t - E\{N_s(t)\}\sin 2\pi f_c t$$

Since $E\{N(t)\} = 0$ for all t, the preceding equation requires that

$$E\{N_c(t)\} = E\{N_s(t)\} = 0 \tag{5.143}$$

Also, using Equation 5.142 and starting with

$$N(t)N(t + \tau) = [N_c(t)\cos 2\pi f_0 t - N_s(t)\sin 2\pi f_0 t]$$
$$\cdot [N_c(t + \tau)\cos 2\pi f_0(t + \tau) - N_s(t + \tau)\sin 2\pi f_0(t + \tau)]$$

we can show that

$$2E\{N(t)N(t + \tau)\} = [R_{N_c N_c}(\tau) + R_{N_s N_s}(\tau)]\cos 2\pi f_0 \tau$$
$$+ [R_{N_s N_c}(\tau) - R_{N_c N_s}(\tau)]\sin 2\pi f_0 \tau$$
$$+ [R_{N_c N_c}(\tau) - R_{N_s N_s}(\tau)]\cos 2\pi f_0(2t + \tau)$$
$$- [R_{N_c N_s}(\tau) + R_{N_s N_c}(\tau)]\sin 2\pi f_0(2t + \tau)$$

Since $N(t)$ is stationary, the left-hand side of the preceding equation does not depend on t. For the right-hand side to be independent of t, it is required that

$$R_{N_c N_c}(\tau) = R_{N_s N_s}(\tau) \tag{5.144}$$

and

$$R_{N_c N_s}(\tau) = -R_{N_s N_c}(\tau) \qquad (5.145)$$

Hence

$$R_{NN}(\tau) = R_{N_c N_c}(\tau)\cos 2\pi f_0 \tau + R_{N_s N_c}(\tau)\sin 2\pi f_0 \tau \qquad (5.146)$$

Now, in order to find the relationship between $R_{N_c N_c}(\tau)$, $R_{N_s N_s}(\tau)$, and $R_{NN}(\tau)$, we introduce the *Hilbert transform* $\check{N}(t)$, where

$$\check{N}(t) \triangleq \frac{1}{\pi} \int_{-\infty}^{\infty} \frac{N(\alpha)}{t - \alpha} \, d\alpha = N(t) * \frac{1}{\pi t} \qquad (5.147)$$

and the corresponding analytic signal

$$N_A(t) = N(t) + j\check{N}(t) \qquad (5.148)$$

The reader can show that

1. $\check{N}(t)$ is obtained from $N(t)$ by passing $N(t)$ through an LTIV system (called a quadrature filter) with $H(f) = \begin{cases} -j, & f > 0 \\ +j, & f < 0 \end{cases}$
2. $R_{\check{N}\check{N}}(\tau) = R_{NN}(\tau)$ \hfill (5.149.a)
3. $R_{\check{N}N}(\tau) = -R_{N\check{N}}(\tau)$ \hfill (5.149.b)
4. $R_{N_A N_A}(\tau) = 2[R_{NN}(\tau) + jR_{\check{N}N}(\tau)]$ \hfill (5.149.c)
5. $S_{N\check{N}}(f) = \begin{cases} -jS_{NN}(f), & f > 0 \\ jS_{NN}(f), & f < 0 \end{cases}$ \hfill (5.149.d)

Using $N(t)$ and $\check{N}(t)$, we can form the processes $N_c(t)$ and $N_s(t)$ as

$$\begin{aligned} N_c(t) &= \text{Re}\{N_A(t)\exp(-j2\pi f_0 t)\} \\ &= N(t)\cos 2\pi f_0 t + \check{N}(t)\sin 2\pi f_0 t \end{aligned} \qquad (5.150.a)$$

and

$$\begin{aligned} N_s(t) &= \text{Im}\{N_A(t)\exp(-j2\pi f_0 t)\} \\ &= \check{N}(t)\cos 2\pi f_0 t - N(t)\sin 2\pi f_0 t \end{aligned} \qquad (5.150.b)$$

320 SPECIAL CLASSES OF RANDOM PROCESSES

(Multiplying Equation 5.150.a by $\cos 2\pi f_0 t$ and subtracting Equation 5.150.b multiplied by $\sin 2\pi f_0 t$ will show these processes are the same as N_c and N_s as defined in Equation 5.142.)

Starting with Equation 5.150.a and taking $E\{N_c(t)N_c(t+\tau)\}$, we can show that

$$R_{N_c N_c}(\tau) = R_{NN}(\tau)\cos 2\pi f_0 \tau + R_{\tilde{N}N}(\tau)\sin 2\pi f_0 \tau \qquad (5.151)$$

and inserting Equation 5.146 in Equation 5.151, we obtain

$$R_{N_s N_c}(\tau) = -R_{NN}(\tau)\sin 2\pi f_0 \tau + R_{\tilde{N}N}(\tau)\cos 2\pi f_0 \tau \qquad (5.152)$$

Finally, combining Equations 5.149.c, 5.151, and 5.152, we have

$$R_{N_c N_c}(\tau) = \frac{1}{2} \text{Re}[R_{N_A N_A}(\tau)\exp(-j2\pi f_0 \tau)] \qquad (5.153.\text{a})$$

and

$$R_{N_s N_c}(\tau) = \frac{1}{2} \text{Im}[R_{N_A N_A}(\tau)\exp(-j2\pi f_0 \tau)] \qquad (5.153.\text{b})$$

In the frequency domain, the power spectral density of the analytic signal $N_A(t)$ can be obtained by taking the Fourier transform of Equation 5.149.c as

$$S_{N_A N_A}(f) = 2[S_{NN}(f) + jS_{\tilde{N}N}(f)]$$

Substituting Equation 5.149.d in the previous equation, we can show that

$$S_{N_A N_A}(f) = 4S_{NN}(f)U(f) \qquad (5.154)$$

where $U(f)$ is the unit step function, that is

$$U(f) = \begin{cases} 1, & f > 0 \\ 0, & f \leq 0 \end{cases}$$

Substituting Equation 5.154 in the transform of Equation 5.153 we obtain the following relationship between the power spectral density of $N(t)$, which is

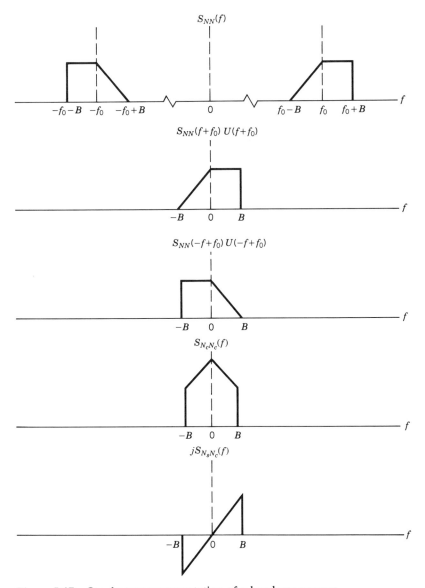

Figure 5.17 Quadrature representation of a bandpass process.

assumed to be known, and the power spectral densities of the quadrature components of $N_c(t)$ and $N_s(t)$.

$$S_{N_cN_c}(f) = S_{N_sN_s}(f)$$
$$= S_{NN}(f + f_0)U(f + f_0) + S_{NN}(-f + f_0)U(-f + f_0) \quad (5.155.\text{a})$$

and

$$jS_{N_sN_c}(f) = S_{NN}(f + f_0)U(f + f_0) - S_{NN}(-f + f_0)U(-f + f_0) \quad (5.155.\text{b})$$

An example of the relationship between the spectra is shown in Figure 5.17. We note that if the positive part of $S_{NN}(f)$ is symmetrical about f_0, then $S_{N_sN_c}(f) = 0$. In this case, $N_c(t)$ and $N_s(t)$ are uncorrelated.

Summarizing, a zero-mean stationary *bandpass* random process can be represented in quadrature form. The quadrature components are themselves zero mean stationary *lowpass* random processes whose spectral densities are related to the spectral density of the parent process as given in Equation 5.155. In general, the quadrature representation is not unique since we have specified only the mean and the spectral density functions of $N_c(t)$ and $N_s(t)$.

Although we have not derived the distribution functions of the quadrature components, it is easy to show that for the Gaussian case, if $N(t)$ is Gaussian then $N_c(t)$ and $N_s(t)$ are also Gaussian.

Given the representation in quadrature form

$$N(t) = N_c(t)\cos 2\pi f_0 t - N_s(t)\sin 2\pi f_0 t$$

we can convert it easily to envelope and phase form as

$$N(t) = R_N(t)\cos[2\pi f_0 t + \Theta_N(t)]$$

where

$$R_N(t) = \sqrt{N_c^2(t) + N_s^2(t)} \quad (5.156)$$

and

$$\Theta_N(t) = \tan^{-1}\left\{\frac{N_s(t)}{N_c(t)}\right\} \quad (5.157)$$

It is left as an exercise for the reader to show that the envelope R_N of a bandpass Gaussian process has a Rayleigh pdf and the phase Θ_N has a uniform pdf.

5.5.5 Effects of Noise in Analog Communication Systems

Successful electrical communication depends on how accurately the receiver in a communication system can determine the transmitted signal. Perfect signal

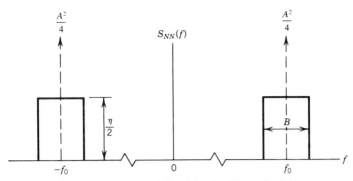

Figure 5.18a Power spectral densities for Example 5.10.

extraction at the receiver might be possible in the absence of "noise" and other contaminations. But thermal noise is always present in electrical systems, and it usually corrupts the desired signal in an additive fashion. While thermal noise is present in all parts of a communication system, its effects are most damaging at the input to the receiver, because it is at this point in the system that the information-bearing signal is the weakest. Any attempt to increase signal power by amplification will increase the noise power also. Thus, the additive noise at the receiver input will have a considerable amount of influence on the quality of the output signal.

EXAMPLE 5.10.

The signal in a communication system is a deterministic "tone" of the form $s(t) = A \cos(2\pi f_0 t)$ where A and f_0 are constants. The signal is corrupted by additive band-limited white Gaussian noise with the spectral density shown in Figure 5.18. Using the quadrature representation for noise, analyze the effect of noise on the amplitude and phase of $s(t)$ assuming $A^2/2 >> E\{N^2(t)\}$

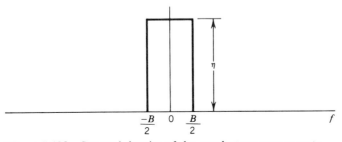

Figure 5.18b Spectral density of the quadrature components.

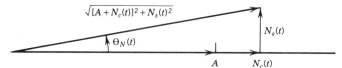

Figure 5.18c Phasor diagram for Example 5.10.

SOLUTION: The signal plus noise can be expressed as

$$s(t) + N(t) = A \cos 2\pi f_0 t + N_c(t) \cos 2\pi f_0 t - N_s(t) \sin 2\pi f_0 t$$

Using the relationship shown in the phasor diagram (Figure 5.18c), we can express $s(t) + N(t)$ as

$$s(t) + N(t) = \sqrt{[A + N_c(t)]^2 + [N_s(t)]^2} \\ \times \cos\left[2\pi f_0 t + \tan^{-1}\left(\frac{N_s(t)}{A + N_c(t)}\right)\right]$$

We are given that $A^2/2 \gg E\{N^2(t)\}$, which implies

$$E\{N_c^2(t)\} = E\{N_s^2(t)\} \ll \frac{A^2}{2}$$

Hence we can use the approximations (see Figure 5.18c) $A + N_c(t) \approx A$, and $\tan^{-1} \Theta \approx \Theta$ when $\Theta \ll 1$ and rewrite $s(t) + N(t)$ as

$$s(t) + N(t) \approx A[1 + A_N(t)]\cos[2\pi f_0 t + \Theta_N(t)]$$

where

$$A_N(t) = \frac{N_c(t)}{A}$$

and

$$\Theta_N(t) = \frac{N_s(t)}{A}$$

are the noise-induced perturbations in the amplitude and phase of the signal $s(t)$. The mean and variance of these perturbations can be computed as

$$E\{A_N(t)\} = \frac{E\{N_c(t)\}}{A} = 0$$

$$\text{var}\{A_N(t)\} = \frac{1}{A^2}\text{var}\{N_c(t)\}$$
$$= \frac{\eta B}{A^2}$$

and

$$E\{\Theta_N(t)\} = 0$$

$$\text{var}\{\Theta_N(t)\} = \frac{\eta B}{A^2}$$

Receiver Design—Analog Communication Systems. A simple model for analog communication systems is shown in Figure 5.19. Here, the information-bearing signal is modeled as a continuous-time random process $S(t)$, which passes through a communication channel with a transfer function $H_c(f)$. Band-limited thermal noise enters the receiver with the received signal $X(t)$, and the receiver, which is modeled as an LTIV system with a transfer function $H_R(f)$, produces an estimate $\hat{S}(t)$ of $S(t)$ as shown in Figure 5.20.

To analyze the effects of noise, let us use a simpler model of an analog communication system in which the channel is assumed to be ideal except for the additive noise (Figure 5.19b). Now, the input to the receiver is

$$Y(t) = S(t) + N(t)$$

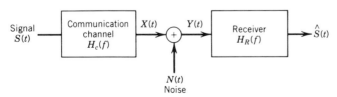

Figure 5.19a Model of an analog communication system.

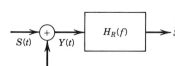

Figure 5.19b Simplified model.

and the receiver output is given by

$$\hat{S}(t) = Y(t) * h_R(t)$$

where $h_R(t)$ is the impulse response of the receiver and * denotes convolution. The receiver transfer function is chosen such that $\hat{S}(t)$ is a "good" estimate of $S(t)$.

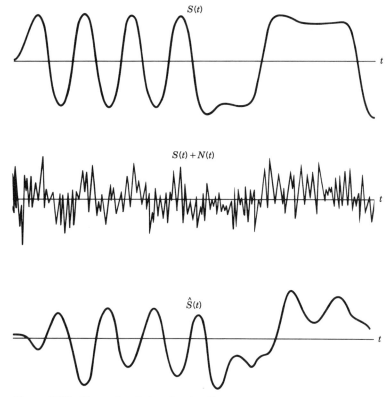

Figure 5.20 Example of signal extraction.

GAUSSIAN PROCESSES

In analog communication systems the mean squared error (MSE)

$$\text{MSE} = E\{[S(t) - \hat{S}(t)]^2\}$$

can be used to judge the effectiveness of the receiver. The receiver output can be written as

$$\hat{S}(t) = S(t) + D(t) + N_0(t)$$

where $D(t)$ is the signal distortion given by

$$D(t) = S(t) * [h_R(t) - 1]$$

and $N_0(t)$ is the output noise

$$N_0(t) = N(t) * h_R(t)$$

Optimum receiver design consists of choosing $h_R(t)$ or $H_R(f)$ so that the MSE is minimized. In general, this minimization is a difficult task, and we present optimal solutions in Chapter 7. Here, we illustrate a simpler suboptimal approach in which a functional form for $H_R(f)$ is assumed and the parameters of the assumed transfer function are chosen to minimize the MSE.

EXAMPLE 5.11.

Given that $S(t)$ and $N(t)$ are two independent stationary random processes with

$$S_{SS}(f) = \frac{2}{(2\pi f)^2 + 1} \quad \text{and} \quad S_{NN}(f) = 1$$

(as shown in Figure 5.21.a), find the best first-order lowpass filter that produces a minimum MSE estimator of $S(t)$ from $S(t) + N(t)$, i.e., find the time constant T of the filter $H_R(f) = 1/(1 + j2\pi fT)$.

SOLUTION: Using Figure 5.21b, we can show that the MSE is given by

$$\text{MSE} = E\{[S(t) - \hat{S}(t)]^2\} = E\{e^2(t)\} = R_{ee}(0)$$

328 SPECIAL CLASSES OF RANDOM PROCESSES

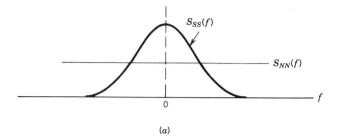

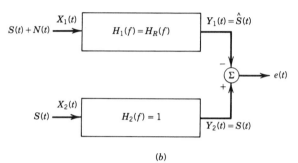

Figure 5.21 Filtering signal and noise: Example 5.11.

where

$$R_{ee}(\tau) = E\{[Y_2(t) - Y_1(t)][Y_2(t + \tau) - Y_1(t + \tau)]\}$$
$$= R_{Y_2Y_2}(\tau) - R_{Y_1Y_2}(\tau) - R_{Y_1Y_2}(-\tau) + R_{Y_1Y_1}(\tau)$$

Using the results given in Section 4.3.5, we find

$$R_{Y_1Y_1}(\tau) = [R_{SS}(\tau) + R_{NN}(\tau)] * h_R(\tau) * h_R(-\tau)$$
$$R_{Y_1Y_2}(\tau) = R_{SS}(\tau) * h_R(-\tau)$$
$$R_{Y_1Y_2}(-\tau) = R_{SS}(\tau) * h_R(\tau)$$
$$R_{Y_2Y_2}(\tau) = R_{SS}(\tau)$$

and transforming to the frequency domain, we have

$$S_{ee}(f) = [1 - H_R(f)][1 - H_R^*(f)]S_{SS}(f) + H_R(f)H_R^*(f)S_{NN}(f)$$

and

$$\text{MSE} = E\{e^2(t)\} = R_{ee}(0)$$
$$= \int_{-\infty}^{\infty} S_{ee}(f)\, df$$

Substituting the form of $H_R(f)$ and using the integration technique given in Section 4.3.4, we obtain

$$\text{MSE} = \frac{T}{1+T} + \frac{1}{2T}$$

The MSE can be minimized with respect to T by setting

$$\frac{d(\text{MSE})}{dT} = 0$$

The value of T that minimizes the MSE is

$$T = \frac{1}{(\sqrt{2}-1)}$$

and minimum value of the MSE is equal to 0.914.

In some communication systems, it might be possible to introduce a filter at the transmitting end as well as at the receiving end (Figure 5.22). With two filters, called the transmitting (or preemphasis) filter and the receiving (or deem-

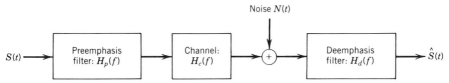

Figure 5.22 Preemphasis/deemphasis filtering. Note that $H_p(f)H_c(f)H_d(f) = 1$, and $\hat{S}(t) = S(t) + N_0(t)$.

phasis) filter, both signal distortion and output noise can be minimized. Indeed we can choose $H_p(f)$ and $H_d(f)$ such that the mean squared error

$$\text{MSE} = E\{[S - \hat{S}(t)]^2\}$$

is minimized subjected to the constraint that

$$H_p(f)H_c(f)H_d(f) = 1 \tag{5.158}$$

Equation 5.158 guarantees that the output can be written as

$$\hat{S}(t) = S(t) + N_0(t)$$

and minimizing the MSE is equivalent to minimizing $E\{N_0^2(t)\}$, where $N_0(t)$ is the output noise or maximizing the signal-to-noise power ratio $(S/N)_0$ at the receiver output, where $(S/N)_0$ is defined as

$$(S/N)_0 = \frac{E\{S^2(t)\}}{E\{N_0^2(t)\}} \tag{5.159}$$

This design is extensively used in audio systems (for example, in commercial FM broadcasting), and audio recording (Dolby type noise-reduction systems). The interested reader is referred to Reference [14] for design details.

5.5.6 Noise in Digital Communication Systems

In binary digital communication systems (Figure 5.23), the transmitted information is a random sequence D_k where the kth transmitted symbol D_k takes on one of two values (for example 0 or 1). The random sequence D_k is first converted to a random process $S(t)$ using a transformation called modulation and $S(t)$ is transmitted over the communication channel. The receiver processes the received signal plus noise and produces an estimate $\hat{D}_k$ of D_k. Because of noise,

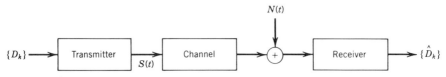

Figure 5.23 Model of a digital communication system.

the receiver occasionally makes an error in determining the transmitted symbol. The end-to-end performance of the digital communication system is measured by the probability of error P_e

$$P_e = P(D_k \neq \hat{D}_k)$$

and the receiver structure and parameters are chosen to minimize P_e. Analysis of the effects of noise on the performance of digital communication systems and the optimal designs for the receivers are treated in detail in the following chapter.

5.5.7 Summary of Noise Models

This section introduced models of noise and signals that are commonly used in communication systems. First, Gaussian processes were described because of their general use and their use to model thermal noise. Next, white and band-limited white noise were defined for continuous processes, and white noise was defined for sequences. The response of LTIV systems to white noise was also given. Narrowband signals were modeled using the quadrature representation by introducing Hilbert transforms. Such a representation was used to analyze the effect of noise on the amplitude and phase of a narrowband signal plus noise.

The design of a receiver, that is, filter, to "separate" the signal from the noise was introduced when the form of the filter is fixed. This idea is expanded in Chapter 7. Similarly, receiver design for digital systems was introduced and will be expanded in Chapter 6.

5.6 SUMMARY

This chapter presented four special classes of random processes. First, autoregressive moving average models were presented. Such models of random processes are very useful when a model of a random process is to be estimated from data. The primary purpose of this section was to familiarize the reader with such models and their characteristics.

Second, Markov models were presented. Such models are often used in analyzing a variety of electrical engineering problems including queueing, filtering, reliability, and communication channels. The emphasis in this chapter was placed on finding multistage transition probabilities, probabilities of each state being occupied at various points in time, and on steady-state probabilities for Markov chains.

Third, point processes were introduced, and the Poisson process was empha-

332 SPECIAL CLASSES OF RANDOM PROCESSES

sized. Applications to queueing and to modeling shot noise were suggested. The time of occurrence, time between occurrences, and number of occurrences were all suggested as possible events of interest. Probabilities of these events were the primary focus of the analysis.

Finally, models of noise and signals that are usually employed in analysis of communication systems were introduced. In particular, Gaussian white noise and bandlimited white noise models were introduced. The quadrature model of narrowband noise was also introduced. Analog communication system design was illustrated using an example of minimizing the mean-square error.

5.7 REFERENCES

References are listed according to the topics discussed in this chapter.

Discrete Linear Models

Reference [2] is the most widely used for ARMA processes. In fact, these processes are often called Box–Jenkins models.

[1] O. D. Anderson, *Time Series Analysis and Forecasting, The Box–Jenkins Approach,* Butterworth, Boston, 1976.

[2] G. E. P. Box and G. M. Jenkins, *Time Series Analysis: Forecasting and Control,* Holden-Day, San Francisco, 1976.

[3] W. A. Fuller, *Introduction to Statistical Time Series,* John Wiley & Sons, New York, 1976.

[4] G. M. Jenkins and D. G. Watts, *Spectral Analysis and Its Applications,* Holden-Day, San Francisco, 1968.

Markov Chains and Processes

[5] R. Billington and R. Allen, *Reliability Evaluation of Engineering Systems: Concepts and Techniques,* Pitman, London, 1983.

[6] D. Isaacson and R. Madsen, *Markov Chains Theory and Applications,* John Wiley & Sons, New York, 1976.

[7] H. Larson and A. Shubert, *Probabilistic Models in Engineering Sciences,* Vol. II, *Random Noise Signals and Dynamic Systems,* John Wiley & Sons, New York, 1979.

[8] R. Markland, *Topics in Management Science,* John Wiley & Sons, New York, 1979.

[9] E. Parzen, *Stochastic Processes,* Holden-Day, San Francisco, 1962.

[10] K. Trivedi, *Probability and Statistics with Reliability, Queueing and Computer Science Applications,* Prentice-Hall, Englewood Cliffs, N.J., 1982.

Point Processes

[11] L. Klienrock, *Queueing Systems,* Vols. I and II, John Wiley & Sons, New York, 1975.

[12] D. Snyder, *Random Point Processes,* John Wiley & Sons, New York, 1975.

Gaussian Processes

[13] A. Papoulis, *Probability, Random Variables and Stochastic Processes*, McGraw-Hill, New York, 1984.

[14] K. S. Shanmugan, *Digital and Analog Communication Systems*, John Wiley & Sons, New York, 1979.

5.8 PROBLEMS

5.1 Convert Equation 5.1 to a state model of the form

$$\mathbf{X}(n) = \mathbf{\Phi}\mathbf{X}(n-1) + \mathbf{E}(n)$$

That is, find $\mathbf{\Phi}$.

5.2 Find the Z transform of Equation 5.1, and the Z transform $H_Z(z)$ of the digital filter shown in Figure 5.1. Assume that $e(n)$ and $X(n)$ are nonzero only for $n \geq 0$, and that their Z transforms exist.

5.3 $X(n) = (1/2)X(n-1) + e(n)$, $n = 0, 1, \ldots$, where $e(n)$ is stationary white Gaussian noise with zero mean and $\sigma_N^2 = 1$. $X(0)$ is a Gaussian random variable, which is independent of $e(n)$, $n \geq 1$. Find the mean and variance of $X(0)$ in order for the process to be stationary.

5.4 Refer to Problem 5.3, and assume stationarity. Find, for the sequence $X(n)$, (a) μ_X, (b) σ_X^2, (c) $R_{XX}(k)$, (d) $r_{XX}(k)$, (e) $S_{XX}(f)$.

5.5 Find the spectral density function of a random sequence that has the autocorrelation function $\phi_{1,1}^m \sigma_X^2$, $m \geq 0$ by using the Fourier transform; that is, show that

$$S_{XX}(f) = \sigma_X^2 + 2 \sum_{k=1}^{\infty} \sigma_X^2 \phi_{1,1}^k \cos 2\pi kf, \quad |f| < \frac{1}{2}$$

5.6 With $\sigma_X^2 = 1$, plot $S_{XX}(f)$ for (a) $\phi_{1,1} = .8$; (b) $\phi_{1,1} = -.9$.

5.7 Show that Equations 5.20.a and 5.20.b are possible solutions to Equation 5.19.

5.8 $R_{XX}(m) = .4R_{XX}(m-1) - .2R_{XX}(m-2)$. Find $R_{XX}(m)$ and $r_{XX}(m)$.

5.9 $R_{XX}(m) = 2R_{XX}(m-1) - R_{XX}(m-2)$. Find $R_{XX}(m)$ and $r_{XX}(m)$ if possible.

5.10 Show that if the inequalities following Equation 5.27 are true, then σ_X^2 as defined by Equation 5.27 will be positive. Is the converse true?

5.11 Find $S_{XX}(f)$ for Problem 5.8. Plot $S_{XX}(f)$.

5.12 For the second-order autoregressive process, show that

$$\phi_{2,1} = r_{XX}(1)\frac{1 - r_{XX}(2)}{1 - r_{XX}^2(1)}; \qquad \phi_{2,2} = \frac{r_{XX}(2) - r_{XX}^2(1)}{1 - r_{XX}^2(1)}$$

5.13 $X(n)$ is a second-order autoregressive process with $r_{XX}(1) = .5$, $r_{XX}(2) = .1$. Find (a) the parameters $\phi_{2,1}$ and $\phi_{2,2}$; (b) $r_{XX}(m)$, $m \geq 2$.

5.14 For a second-order autoregressive model with one repeated root of the characteristic equation, find (a) the relationship between $\phi_{2,1}$ and $\phi_{2,2}$; (b) b_1 and b_2 in the equation $r_{XX}(m) = b_1\lambda^m + b_2 m\lambda^m$

5.15 Given that $r_{XX}(1) = \frac{3}{4}$, $r_{XX}(2) = \frac{1}{2}$, and $r_{XX}(3) = \frac{1}{4}$, find the third-order autoregressive model for $X(n)$.

5.16 If $\phi_{3,1} = \frac{5}{6}$, $\phi_{3,2} = 0$, $\phi_{3,3} = -\frac{1}{6}$ in a third-order autoregressive model, find $S_{XX}(f)$.

5.17 For $\phi_{2,1} = .5$, $\phi_{2,2} = .06$ find $S_{XX}(f)$ by the two methods given for second-order autoregressive models.

5.18 Assume a second-order autoregressive process and show that $\phi_{3,3} = 0$.

5.19 For a third-order autoregressive process, find $\phi_{1,1}$, $\phi_{2,2}$, $\phi_{3,3}$, and $\phi_{4,4}$.

5.20 Show that a first-order autoregressive model is equivalent to an infinite moving average model.

5.21 $X(n) = 0.8e(n - 1) + e(n)$ where $e(n)$ is white noise with $\sigma_N^2 = 1$ and zero mean. Find (a) μ_X, (b) σ_X^2, (c) $r_{XX}(m)$, (d) $S_{XX}(f)$, (e) $\phi_{i,i}$.

5.22 Show that Equation 5.42 is correct for $k = 1, 2, 3, 4, 5$.

5.23 Verify Equation 5.43 by using

$$S_{XX}(f) = |H(f)|^2 S_{ee}(f)$$

and

$$H(f) = \theta_{1,1}\exp(-j2\pi f) + 1$$

5.24 For the qth order moving-average model, find (a) $r_{XX}(m)$ and (b) the Z transform of $R_{XX}(m)$.

5.25 Find the first three partial autocorrelation coefficients of the ARMA (1, 1) model.

5.26 Find $R_{xx}(k)$, $r_{xx}(k)$, and $S_{xx}(f)$ of the following model:

$$X(n) = .7X(n-1) + .2e(n-1) + e(n)$$

5.27 Find the Z transform of an ARMA (2, 3) model.

5.28 Find the digital filter diagram and the state model of an ARMA (3, 1) model.

5.29 Describe the random walk introduced in Section 3.4.2 by a Markov process. Is it a chain? Is it homogeneous? What is $\mathbf{p}(0)$? What is $\mathbf{P}(1)$?

5.30 A binary symmetric communication system consists of five tandem links; in the ith link the error probability is e_i. Let $X(0)$ denote the transmitted bit, $X(i)$ denote the output of the ith link and $X(5)$ denote the received bit. The sequence $X(1), X(2), \ldots, X(5)$ then can be modeled as a discrete parameter Markov chain. Show the ith transition matrix, $i = 1, \ldots, 5$. Is the chain homogeneous? Find $\mathbf{P}(0, 5)$.

5.31 Let $X(n)$ denote the number of messages waiting to be transmitted in a buffer at time n. Assume that the time required for transmission is one second for each message. Thus, if no new messages arrive in one second and if $X(n) > 0$, then

$$X(n+1) = X(n) - 1$$

Now consider that the message arrival rate is specified by $U(n)$. If $P[U(n) = 0] = P[U(n) = 1] = P[U(n) = 2] = \frac{1}{3}$ for all n, describe $\mathbf{P}(1)$. If $X(0) = 1$, find $\mathbf{p}(1)$ and $\mathbf{p}(2)$.

5.32 A communication source can generate one of three possible messages, 1, 2, and 3. Assume that the generation can be described by a homogeneous Markov chain with the following transition probability matrix:

Current Message	Next Message		
	1	2	3
1	.5	.3	.2
2	.4	.2	.4
3	.3	.3	.4

and that $\mathbf{p}^T(0) = [.3 \quad .3 \quad .4]$. Draw the state diagram. Find $\mathbf{p}(4)$.

336 SPECIAL CLASSES OF RANDOM PROCESSES

5.33 A two-state nonhomogeneous Markov process has the following one-step transition probability matrix:

$$\mathbf{P}(n-1, n) = \begin{bmatrix} 1 - \left(\frac{1}{2}\right)^n & \left(\frac{1}{2}\right)^n \\ \left(\frac{1}{3}\right)^n & 1 - \left(\frac{1}{3}\right)^n \end{bmatrix}$$

Find $\mathbf{P}(0, 2)$ and $\mathbf{P}(2, 4)$

5.34 Assume that the weather in a certain location can be modeled as the homogeneous Markov chain whose transition probability matrix is shown:

Today's Weather		Tomorrow's Weather	
	Fair	Cloudy	Rain
Fair	.8	.15	.05
Cloudy	.5	.3	.2
Rain	.6	.3	.1

If $\mathbf{p}^T(0) = [.7 \quad .2 \quad .1]$, find $\mathbf{p}(1), \mathbf{p}(2), \mathbf{p}(4), \lim_{n\to\infty} \mathbf{p}(n)$.

5.35 A certain communication system is used once a day. At the time of use it is either operative G or inoperative B. If it is found to be inoperative, it is repaired. Assume that the operational state of the system can be described by a homogeneous Markov chain that has the following transition probability matrix:

$$\begin{bmatrix} 1 - q & q \\ r & 1 - r \end{bmatrix}$$

Find $\mathbf{P}(2), \mathbf{P}(4), \lim_{n\to\infty} \mathbf{P}(n)$. If $\mathbf{p}^T(0) = [a, 1 - a]$, find $\mathbf{p}(n)$.

5.36 For a continuous-time homogeneous Markov chain, show that Equation 5.85 is true.

5.37 Starting with

$$\lim_{\epsilon \to 0} P_{k,j}(\epsilon) = \lambda_{k,j}\epsilon, \qquad k \neq j$$

$$= 1 + \lambda_{j,j}\epsilon, \qquad k = j$$

Show that as t approaches τ, $t < \tau$

$$P_{i,j}(\tau) = \left[\sum_{\text{all } k} P_{i,k}(t)\lambda_{k,j}(\tau - t)\right] + P_{i,j}(t)$$

or

$$\mathbf{P}'(t) = \mathbf{P}(t)\boldsymbol{\lambda}$$

which has the solution

$$\mathbf{P}(t) = \mathbf{p}(0)\exp(\boldsymbol{\lambda} t)$$

5.38 For the two-state continuous-time Markov chain using the notation of Example 5.7, find the probability of being in each state, given that the system started at time $= 0$ in the inoperative state.

5.39 For Example 5.7, find the average time in each state if the process is in steady state and operates for T seconds.

5.40 For Example 5.7, what should $p_1(0)$ be in order for the process to be stationary?

5.41 The clock in a digital system emits a regular stream of pulses at the rate of one pulse per second. The clock is turned on at $t = 0$, and the first pulse appears after a random delay D whose density is

$$f_D(d) = 2(1 - d), \qquad 0 \le d < 1$$

Describe the point process generated by the times of occurrences of the clock pulses. Verify whether the assumptions stated on page 297 are satisfied.

5.42 Suppose the sequence of clock pulses is subjected to "random jitter" so that each pulse after the first one is advanced or retarded in time. Assume that the displacement of each pulse from its jitter-free location is uniformly distributed in the interval $(-0.1, 0.1)$ independently for each pulse. Describe the ensemble of the resulting point process and find its probability distribution.

5.43 Derive Equations 5.110 and 5.111.

5.44 (a) Show that the sum of two Poisson processes results in a Poisson process.
(b) Extend the proof to n processes by induction.

5.45 Assume that a circuit has an IC whose time to failure is an exponentially distributed random variable with expected lifetime of three months. If there are 10 spare ICs and time from failure to replacement is zero, what is the probability that the circuit can be kept operational for at least one year?

5.46 Assume that an office switchboard has five telephone lines and that starting at 8 A.M. on Monday, the time that a call arrives on each line is an exponential random variable with parameter λ. Also assume that calls arrive independently on the lines and show that the time of arrival of the first call (irrespective of which line it arrives on) is exponential with parameter 5λ.

5.47 Consider a point process on $\Gamma = (-\infty, \infty)$ such that for any $0 < t < \infty$,

$$P[X(t) = k] = \exp(-\alpha t)[1 - \exp(-\alpha t)]^k, \quad k = 0, 1, 2, \ldots$$

where $\alpha > 0$ is a constant. Find $\mu_X(t)$ and the "intensity function" of the process $\lambda(t)$, where $\lambda(t)$ is defined as

$$\lambda(t) = \frac{E\{X(t + \Delta t) - X(t)\}}{\Delta t}, \quad \Delta t > 0$$

5.48 Jobs arrive at a computing facility at an average rate of 50 jobs per hour. The arrival distribution is Poisson. The time it takes to process a job is a random variable with an exponential distribution with a mean processing time of 1 minute per job.

a. Find the mean delay between the time a job arrives and the time it is finished.

b. If the processing capacity is doubled, what is the mean delay in processing a job?

c. If both the arrival rate and the processing capacity increase by a factor of 2, what is the mean delay in processing a job?

5.49 Passengers arrive at a terminal for boarding the next bus. The times of their arrival are Poisson with an average arrival rate of two per minute. The times of departure of each bus are Poisson with an average departure rate of four per hour. Assume that the capacity of the bus is large.

a. Find the average number of passengers in each bus.

b. Find the average number of passengers in the first bus that leaves after 9 A.M.

5.50 A random process $Y(t)$ starts at the value $Y(0) = 1$ at time $t = 0$. At random times $\tau_1, \tau_2, \ldots, \tau_n, \ldots$ thereafter the process changes sign and takes on alternating values of $+1$ or -1. Assuming the switching times constitute a Poisson point process with rate λ, find $E\{Y(t)\}$ and $C_{YY}(t_1, t_2)$. [$Y(t)$ is called the *random telegraph waveform*.]

5.51 The individual pulses in a shot-noise process

$$X(t) = \sum_k h(t - \tau_k)$$

are rectangular

$$h(t) = \begin{cases} 1, & 0 < t < T \\ 0 & \text{elsewhere} \end{cases}$$

Find $P[X(t) = k]$.

5.52 Verify the properties of band-limited white Gaussian noise (Page 315).

5.53 Verify Equations 5.149.a through 5.149.d.

5.54 $N(t)$ is a zero-mean stationary Gaussian random process with the power spectral density

$$S_{NN}(f) = \begin{cases} 10^{-12} & \text{for } |f - 10^7| < 10^6 \\ 0 & \text{elsewhere} \end{cases}$$

Find $P\{|N(t)| > 5(10^{-3})\}$.

5.55 Let $R_N(t)$ and $\Theta_N(t)$ be the envelope and phase of $N(t)$ described in Problem 5.54.

    a. Find the joint pdf of R_N, and Θ_N.

    b. Find the marginal pdfs of R_N and Θ_N.

    c. Show that R_N and Θ_N are independent.

5.56 Let $Z(t) = A \cos(2\pi f_c t) + N(t)$ where A and f_c are constants, and $N(t)$ is a zero-mean stationary Gaussian random process with a bandpass psd $S_{NN}(f)$ centered at f_c. Rewrite $Z(t)$ as

$$Z(t) = R(t)\cos[2\pi f_c t + \Theta(t)]$$

and

    a. Find the joint pdf of R and Θ.

    b. Find the marginal pdf of R.

    c. Are Θ and R independent?

5.57 Let $Y(t) = A \cos(2\pi f_c t + \Theta) + N(t)$ where A is a constant and Θ is a random variable with a uniform distribution in the interval $[-\pi, \pi]$. $N(t)$ is a band-limited Gaussian white noise with a power spectral density

$$S_{NN}(f) = \begin{cases} \eta/2 & \text{for } |f - f_c| < B, \quad f_c \gg B \\ 0 & \text{elsewhere} \end{cases}$$

Find and sketch the power spectral density function of $Y(t)$. [Assume that $N(t)$ and Θ are independent.]

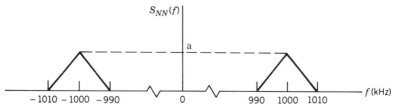

Figure 5.24 Spectral density of $S_{NN}(f)$.

5.58 The spectral density of a narrowband Gaussian process $N(t)$ is shown in Figure 5.24. Find the following spectral densities associated with the quadrature representation of $N(t)$ using $f_0 = 10^6$ Hz: (a) $S_{N_cN_c}(f)$ and (b) $S_{N_cN_s}(f)$.

5.59 $X(t)$ is a zero-mean lowpass process with a bandwidth of B Hertz. Show that

a. $R_{XX}(0) - R_{XX}(\tau) \leq \frac{1}{2}R_{XX}(0)(2\pi B\tau)^2$

b. $E\{[X(t + \tau) - X(t)]^2\} \leq [2\pi B\tau]^2 E\{X^2(t)\}$

5.60 Consider a narrowband process

$$X(t) = X_c(t)\cos 2\pi f_c t + X_s(t)\sin 2\pi f_c t$$

where $X_c(t)$ and $X_s(t)$ are stationary, uncorrelated, lowpass processes with

$$S_{X_cX_c}(f) = S_{X_sX_s}(f) = \begin{cases} g(f), & |f| < B, \quad B \ll f_c \\ 0 & |f| > B \end{cases}$$

a. Show that $S_{XX}(f) = \dfrac{g(f - f_c) + g(f + f_c)}{2}$

b. Show that the envelope $R(t)$ of $X(t)$ satisfies

$$E\{[R(t + \tau) - R(t)]^2\} \leq (2\pi B\tau)^2 E\{R^2(t)\}$$

(which implies that the envelope varies slowly with a time constant of the order of $1/B$).

5.61 Show that if the process $X(t)$ is bandlimited to B Hz, then

$$R_{XX}(\tau) \geq R_{XX}(0) \cos 2\pi B\tau, \quad |\tau| < \frac{1}{4B}$$

5.62 $X(t)$ is zero-mean Gaussian process with

$$R_{XX}(\tau) = \exp(-\alpha|\tau|)$$

Show that $X(t)$ is Markov.

CHAPTER SIX

Signal Detection

6.1 INTRODUCTION

In Chapters 3 and 5 we developed random process models for signals and noise. We are now ready to use these models to derive optimum signal processing algorithms for extracting information contained in signals that are mixed with noise. These algorithms may be used, for example, to determine the sequence of binary digits transmitted over a noisy communication channel, to detect the presence and estimate the location of objects in space using radars, or to filter a noisy audio waveform.

There are two classes of information-extraction algorithms that are considered in this book: (1) signal detection and (2) signal estimation. Examples of signal detection and signal estimation are shown in Figure 6.1a and 6.1b. In signal detection (Figure 6.1a), the "receiver" observes a waveform for T seconds and decides on the basis of this observed waveform the symbol that was transmitted during the time interval. The transmitter *and* the receiver know *a priori* the set of symbols and the waveform associated with each of the symbols. In the case of a binary communication system, the receiver knows that either a "1" or "0" is transmitted every T seconds and that a "1" is represented by a positive pulse and a "0" is represented by a negative pulse. What the receiver does not know in advance is which one of the two symbols is transmitted during a given interval. The receiver makes this "decision" by processing the received waveform. Since the received waveform is usually distorted and masked by noise, the receiver will occasionally make errors in determining which symbol was present in an observation interval. In this chapter we will examine this "decision" (or detection) problem and derive algorithms that can be used to determine

342 SIGNAL DETECTION

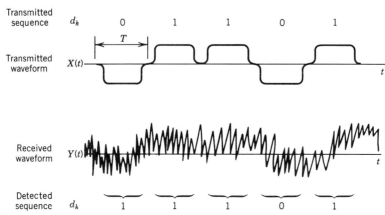

Figure 6.1a Signal detection.

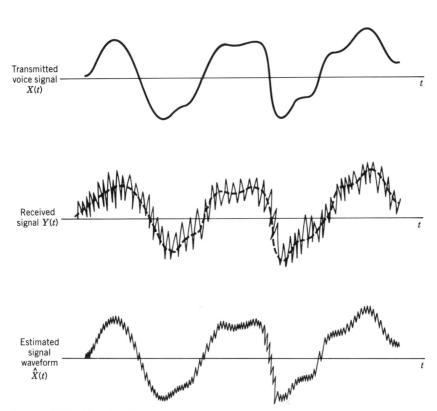

Figure 6.1b Signal estimation.

which one of M "waveforms" (whose "shapes" are known a priori) is present during an observation interval.

Signal estimation usually involves determining the value of an analog signal for all values of time (Figure 6.1b). Examples of signal estimation include the problem of estimating the value of an audio signal that has been transmitted over a noisy channel or tracking the location of an object in space based on noisy radar returns from the object. Unlike the detection problem in which the receiver knows what to expect (within one of M possibilities), in estimation problems the signal to be estimated can have a continuum of values for each value of time. The receiver now has the unenviable task of estimating the values of a waveform that may be viewed as a sample function of a random process. We deal with the estimation problem, which is also referred to as the "filtering" problem, in the following chapter.

In this chapter, we will address both analysis and design of signal-processing algorithms for detection. Analysis involves the evaluation of how well an algorithm "performs," whereas design consists of synthesizing an algorithm that leads to an acceptable level of performance. In detection problems, a typical performance measure is to minimize the probability of an incorrect decision and sometimes the cost of incorrect decisions is included in the performance measure. Mean squared error $E\{[\hat{X}(t) - X(t)]^2\}$ is normally used as the measure of performance of estimation algorithms.

Some applications require both estimation and detection. For example, in some detection problems, the value of one or more parameters might not be known. These parameter values can be estimated from data and the estimated values can be used in the detection algorithm.

6.2 BINARY DETECTION WITH A SINGLE OBSERVATION

There are many applications in which we have to make a choice or decision based on observations. For example, in radar or sonar detection, the return signal is observed and a decision has to be made as to whether a target was present or not. In digital communication systems, the receiver has to make a decision as to which one of M possible signals is actually present based on noisy observations. In each case the solution involves making a decision based on observations or data that are random variables. The theory behind the solutions for these problems has been developed by statisticians and falls under the general area of statistics known as statistical inference, decision theory, or hypothesis testing. We present here a brief review of the principles of hypothesis testing with a simple alternative hypothesis and apply these principles to solve detection problems. Later, in Chapter 8, we discuss hypothesis testing when the alternative hypothesis is *composite*.

We start our discussion with the formulation of the binary detection problem as one of hypothesis testing (i.e., choosing between two possibilities), and derive a decision rule that maximizes a performance measure based on the probabilities of correct and incorrect decisions. The decision rule is then applied to the detection problem when decisions are based on single or multiple observations. Finally, we extend the results to the case of M-ary detection (choosing between one of M possibilities) and continuous observations.

6.2.1 Decision Theory and Hypothesis Testing

In hypothesis testing, a decision has to be made as to which of several hypotheses to accept. The two hypotheses in the binary data transmission example are "0" transmitted, and "1" transmitted. We label these hypotheses as

$$H_0: \text{ "0" was transmitted}$$

$$H_1: \text{ "1" was transmitted}$$

In the language of hypothesis testing, H_0 is called the "null" hypothesis and H_1 is called the "alternate" hypothesis. In the case of target detection, the hypotheses are

$$H_0: \text{ target not present}$$

$$H_1: \text{ target present}$$

Corresponding to each hypothesis, we have one or more observations that are random variables, and the decision is made on the basis of the observed values of these random variables. The rule that is used to form the decision is called the decision rule, and it is derived by maximizing some measure of performance.

Let us first consider the case where we have to make a choice between H_0 and H_1 based on a single observation Y. Note that Y is a random variable and y is a particular value. The probability density functions of Y corresponding to each hypothesis, $f_{Y|H_0}(y|H_0)$ and $f_{Y|H_1}(y|H_1)$, are usually known. If $P(H_i|y)$, $i = 0, 1$, denotes the *a posteriori* probability that H_i was the true hypothesis given the particular value of the observation y, then we can decide in favor of H_0 or H_1 based on whether $P(H_0|y)$ or $P(H_1|y)$ is larger. That is, we choose a decision that maximizes the probability of correct decision, and such a decision rule can be stated as

$$\text{choose } H_0 \text{ if } P(H_0|y) > P(H_1|y) \text{ and}$$
$$\text{choose } H_1 \text{ if } P(H_1|y) > P(H_0|y)$$

or, in a concise form,

$$\frac{P(H_1|y)}{P(H_0|y)} \underset{H_0}{\overset{H_1}{\gtrless}} 1 \qquad (6.1)$$

which means choose H_1 when the ratio is >1, and choose H_0 when the ratio is <1.

6.2.2 MAP Decision Rule and Types of Errors

The conditional probability $P(H_i|y)$ is called an *a posteriori* probability, that is, a probability that is computed after an observation has been made, and the decision criterion stated in Equation 6.1 is called the maximum *a posteriori* probability (MAP) criterion. This decision rule usually leads to a partition of the observation space into two regions R_0 and R_1, and H_0 or H_1 is chosen depending on whether a given observation $y \in R_0$ or R_1.

Using Bayes' rule we can write $P(H_i|y)$ as

$$P(H_i|y) = \frac{f_{Y|H_i}(y|H_i)P(H_i)}{f_Y(y)}$$

and the decision rule given in Equation 6.1 can be rewritten as

$$\frac{P(H_1)f_{Y|H_1}(y|H_1)}{P(H_0)f_{Y|H_0}(y|H_0)} \underset{H_0}{\overset{H_1}{\gtrless}} 1$$

or

$$\frac{f_{Y|H_1}(y|H_1)}{f_{Y|H_0}(y|H_0)} \underset{H_0}{\overset{H_1}{\gtrless}} \frac{P(H_0)}{P(H_1)} \qquad (6.2)$$

(Note that the equal sign may be included in favor of either H_1 or H_0). The ratio

$$L(y) = \frac{f_{Y|H_1}(y|H_1)}{f_{Y|H_0}(y|H_0)} \qquad (6.3)$$

is called a *likelihood ratio* and the MAP decision rule consists of comparing this ratio with the constant $P(H_0)/P(H_1)$, which is called the *decision threshold*. $L(Y)$ is called the likelihood statistic, and $L(Y)$ is a random variable.

In classical hypothesis testing, decisions are based on the likelihood function $L(y)$, whereas in the decision theoretic approach, the *a priori* probabilities and costs associated with various decisions will also be included in the decision rule. In signal detection we usually take into account costs and *a priori* probabilities.

If we have multiple observations, say $Y_1, Y_2, \ldots, Y_n$, on which to base our decision, then the MAP decision rule will be based on the likelihood ratio

$$L(y_1, y_2, \ldots, y_n) = \frac{f_{Y_1, Y_2, \ldots, Y_n | H_1}(y_1, y_2, \ldots, y_n | H_1)}{f_{Y_1, Y_2, \ldots, Y_n | H_0}(y_1, y_2, \ldots, y_n | H_0)} \qquad (6.4)$$

When decisions are based on noisy observations, there is always a chance that some of the decisions may be incorrect. In the binary hypothesis testing problem, we have four possibilities:

1. Decide in favor of H_0 when H_0 is true.
2. Decide in favor of H_1 when H_1 is true.
3. Decide in favor of H_1 when H_0 is true.
4. Decide in favor of H_0 when H_1 is true.

Let D_i represent the decision in favor of H_i. Then, the first two conditional probabilities denoted by $P(D_0|H_0)$ and $P(D_1|H_1)$ correspond to correct choices and the last two, $P(D_1|H_0)$ and $P(D_0|H_1)$, represent probabilities of incorrect decisions. Error 3 is called a *"type-I"* error and 4 is called a *"type-II"* error. In radar terminology, $P(D_1|H_0)$ is called a *"false alarm probability"* (P_F) and $P(D_0|H_1)$ is called the *"probability of a miss"* (P_M). The average *probability of error* P_e is often chosen as the performance measure to be minimized, where

$$P_e = P(H_0)P(D_1|H_0) + P(H_1)P(D_0|H_1)$$

EXAMPLE 6.1.

In a binary communication system, the decision about the transmitted bit is made on the basis of a single noisy observation

$$Y = X + N$$

where $X = 0, 1$ represents the transmitted bit, and N is additive noise, which has a Gaussian distribution with a mean of zero and variance of $\frac{1}{9}$ (X and N are

BINARY DETECTION WITH A SINGLE OBSERVATION 347

independent). Given that $P(X = 0) = \frac{3}{4}$, and $P(X = 1) = \frac{1}{4}$

(a) Derive the MAP decision rule.
(b) Calculate the error probabilities.

SOLUTION

(a) Let

$$H_0: \text{ 0 transmitted}$$
$$H_1: \text{ 1 transmitted}$$

We are given that $P(H_0) = \frac{3}{4}$, $P(H_1) = \frac{1}{4}$

$$f_{Y|H_0}(y|H_0) = f_{Y|X}(y|0)$$
$$= \sqrt{\frac{9}{2\pi}} \exp\left(-\frac{9}{2} y^2\right)$$

and

$$f_{Y|H_1}(y|H_1) = f_{Y|X}(y|1)$$
$$= \sqrt{\frac{9}{2\pi}} \exp\left[-\frac{9}{2}(y-1)^2\right]$$

From Equation 6.3 we obtain the likelihood ratio as

$$L(y) = \frac{\exp\left[-\frac{9}{2}(y-1)^2\right]}{\exp\left(-\frac{9}{2} y^2\right)}$$

and the decision rule becomes

$$\exp\left[\frac{9}{2}(2y - 1)\right] \underset{H_0}{\overset{H_1}{\gtrless}} \left(\frac{3}{4}\right)\Big/\left(\frac{1}{4}\right)$$

Taking logarithms on both sides, the decision rule becomes

$$y \underset{H_0}{\overset{H_1}{\gtrless}} \frac{1}{2} + \frac{1}{9} \ln(3) \approx 0.622$$

That is, the decision rule partitions the observation space, $y \in (-\infty, \infty)$, into two regions $R_0 = (-\infty, 0.622)$, and $R_1 = (.622, \infty)$. H_0 is accepted when $y \in R_0$ and H_1 is chosen when $y \in R_1$.

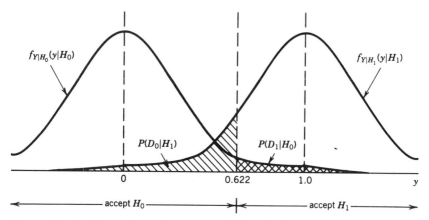

Figure 6.2 Decision rule and error probabilities for Example 6.1.

(b) From Figure 6.2, we obtain the error probabilities as
$$P(D_0|H_1) = P(y \in R_0|H_1) = \int_{R_0} f_{Y|H_1}(y|H_1)\, dy$$
$$= \int_{-\infty}^{.622} \frac{3}{\sqrt{2\pi}} \exp\left[-\frac{9}{2}(y-1)^2\right] dy$$
$$\approx Q(1.134) \approx 0.128$$

and
$$P(D_1|H_0) = P(y \in R_1|H_0) = \int_{R_1} f_{Y|H_0}(y|H_0)\, dy$$
$$\approx Q(1.866) \approx .031$$

The average probability of an incorrect decision is
$$P_e = P(H_0)P(D_1|H_0) + P(H_1)P(D_0|H_1)$$
$$\approx \frac{3}{4} Q(1.866) + \frac{1}{4} Q(1.134)$$
$$\approx .055$$

6.2.3 Bayes' Decision Rule—Costs of Errors

In many engineering applications, costs have to be taken into account in the design process. In such applications, it is possible to derive decision rules that minimize certain average costs. For example, in the context of radar detection, the costs and consequences of a miss are quite different from that of a false alarm and an optimum decision rule has to take into account the relative costs and minimize the average cost. We now derive a decision rule that minimizes the average cost.

If we denote the decisions made in the binary hypothesis problem by D_i, $i = 0, 1$, where D_0 and D_1 denote the decisions in favor of H_0 and H_1, respectively, then we have the following four possibilities: (D_i, H_j), $i = 0, 1$, and $j = 0, 1$. The pair (D_i, H_j) denotes H_j being the true hypothesis and a decision of D_i. Pairs (D_0, H_0) and (D_1, H_1) denote correct decisions, and (D_1, H_0) and (D_0, H_1) denote incorrect decisions. If we associate a cost C_{ij} with each pair (D_i, H_j), then the average cost can be written as

$$\overline{C} = \sum_{\substack{i=0,1 \\ j=0,1}} C_{ij} P(D_i, H_j)$$

$$= \sum_{\substack{i=0,1 \\ j=0,1}} C_{ij} P(H_j) P(D_i | H_j)$$

(6.5)

where $P(D_i | H_j)$ is the probability of deciding in favor of H_i when H_j is the true hypothesis.

A decision rule that minimizes the average cost $\overline{C}$, called the Bayes' decision rule, can be derived as follows. If R_0 and R_1 are the partitions of the observation space, and decisions are made in favor of H_0 if $y \in R_0$ and H_1 if $y \in R_1$, then the average cost can be expressed as

$$\overline{C} = C_{00} P(D_0 | H_0) P(H_0) + C_{10} P(D_1 | H_0) P(H_0)$$

$$+ C_{01} P(D_0 | H_1) P(H_1) + C_{11} P(D_1 | H_1) P(H_1)$$

$$= C_{00} P(H_0) \int_{R_0} f_{Y|H_0}(y|H_0) \, dy$$

$$+ C_{10} P(H_0) \int_{R_1} f_{Y|H_0}(y|H_0) \, dy$$

$$+ C_{01} P(H_1) \int_{R_0} f_{Y|H_1}(y|H_1) \, dy$$

$$+ C_{11} P(H_1) \int_{R_1} f_{Y|H_1}(y|H_1) \, dy$$

350 SIGNAL DETECTION

If we make the reasonable assumption that the cost of making an incorrect decision is higher than the cost of a correct decision

$$C_{10} > C_{00} \quad \text{and} \quad C_{01} > C_{11}$$

and make use of the fact that $R_1 \cup R_0 = (-\infty, \infty)$, and $R_1 \cap R_0 = \emptyset$, then we have

$$\overline{C} = C_{10}P(H_0) + C_{11}P(H_1) + \int_{R_0} \{[P(H_1)(C_{01} - C_{11})f_{Y|H_1}(y|H_1)] \\ - [P(H_0)(C_{10} - C_{00})f_{Y|H_0}(y|H_0)]\} \, dy \tag{6.6}$$

Since the decision rule is specified in terms of R_0 (and R_1), the decision rule that minimizes the average cost is derived by choosing R_0 that minimizes the integral on the right-hand side of Equation 6.6. Note that the first two terms in Equation 6.6 do not depend on R_0.

The smallest value of $\overline{C}$ is achieved when the integrand in Equation 6.6 is negative for all values of $y \in R_0$. Since $C_{01} > C_{11}$ and $C_{10} > C_{00}$, the integrand will be negative and $\overline{C}$ will be minimized if R_0 is chosen such that for every value of $y \in R_0$,

$$P(H_1)(C_{01} - C_{11})f_{Y|H_1}(y|H_1) < P(H_0)(C_{10} - C_{00})f_{Y|H_0}(y|H_0)$$

which leads to the decision rule:

Choose H_0 if

$$P(H_1)(C_{01} - C_{11})f_{Y|H_1}(y|H_1) < P(H_0)(C_{10} - C_{00})f_{Y|H_0}(y|H_0)$$

In terms of the likelihood ratio, we can write the decision rule as

$$L(y) = \frac{f_{Y|H_1}(y|H_1)}{f_{Y|H_0}(y|H_0)} \underset{H_0}{\overset{H_1}{\gtrless}} \frac{P(H_0)(C_{10} - C_{00})}{P(H_1)(C_{01} - C_{11})} \tag{6.7}$$

Note that, with the exception of the decision threshold, the form of the Bayes' decision rule given before is the same as the form of the MAP decision rule given in Equation 6.2! It is left as an exercise for the reader to show that the two decision rules are identical when $C_{10} - C_{00} = C_{01} - C_{11}$ and that the Bayes' decision rule minimizes the probability of making incorrect decisions when $C_{00} = C_{11} = 0$ and $C_{10} = C_{01} = 1$.

6.2.4 Other Decision Rules

In order to use the MAP and Bayes' decision rules, we need to know the *a priori* probabilities $P(H_0)$ and $P(H_1)$ as well as relative costs. In many engineering applications, these quantities may not be available. In such cases, other decision rules are used that do not require $P(H_0)$, $P(H_1)$, and costs. Two rules that are quite commonly used in these situations are the *minmax* rule and the *Neyman–Pearson* rule.

The minmax rule is used when the costs are given, but the *a priori* probabilities $P(H_0)$ and $P(H_1)$ are not known. This decision rule is derived by obtaining the decision rule that minimizes the expected cost corresponding to the value of $P(H_1)$ for which the average cost is maximum.

The Neyman–Pearson (N–P) rule is used when neither *a priori* probabilities nor cost assignments are given. The N–P rule is derived by keeping the probability of false alarm, $P(D_1|H_0)$, below some specified value and minimizing the probability of a miss $P(D_0|H_1)$.

Details of the derivation of the minmax and N–P decision rules are omitted. Both of these decision rules also lead to the form

$$L(y) = \frac{f_{Y|H_1}(y|H_1)}{f_{Y|H_0}(y|H_0)} \underset{H_0}{\overset{H_1}{\gtreqless}} \gamma \qquad (6.8)$$

where γ is the decision threshold. Thus, only the value of the threshold with which the likelihood ratio is compared varies with the criterion that is optimized.

In many applications including radar systems, the performance of decision rules are displayed in terms of a graph of the detection probability

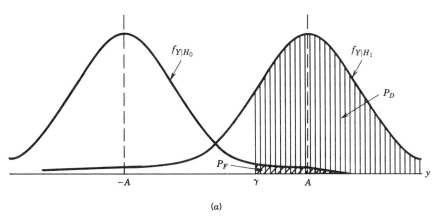

Figure 6.3a Conditional pdfs (assumed to be Gaussian with variance $= \sigma^2$); P_F; and P_D.

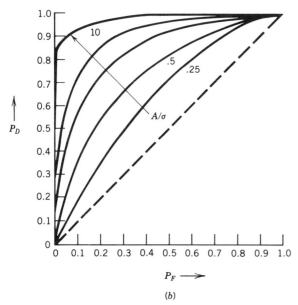

Figure 6.3b Receiver operating characteristic (ROC).

$P(D_1|H_1) = P_D$ versus the probability of false alarm $P(D_1|H_0) = P_F$ for various values of the threshold. These curves are called the receiver operating characteristic (ROC) curves. (See problems 6.7–6.13.) An example is shown in Figure 6.3. Figure 6.3a shows the conditional probabilities with P_D and P_F indicated. Note that 2A is the difference in the means. Figure 6.3b shows example ROC curves as a function of the mean divided by the standard deviation.

6.3 BINARY DETECTION WITH MULTIPLE OBSERVATIONS

In the binary detection problem shown in Figure 6.1a, the receiver makes its decision by observing the received waveform for the duration of T seconds. In the T-second interval, the receiver can extract many samples of $Y(t)$, process the samples, and make a decision based on the information contained in all of the observations taken during the symbol interval. We now develop an algorithm for processing multiple observations and making decisions. We will first address the problem of deciding between one of two *known* waveforms that are corrupted by additive noise. That is, we will assume that the receiver observes

$$Y(t) = X(t) + N(t), \quad 0 \le t \le T$$

BINARY DETECTION WITH MULTIPLE OBSERVATIONS

where

$$X(t) = \begin{cases} s_0(t) & \text{under hypothesis } H_0 \\ s_1(t) & \text{under hypothesis } H_1 \end{cases}$$

and $s_0(t)$ and $s_1(t)$ are the waveforms used by the transmitter to represent 0 and 1, respectively. In a simple case $s_0(t)$ can be a rectangular pulse of duration T and amplitude -1, and $s_1(t)$ can be a rectangular pulse of duration T and amplitude 1. In any case, $s_0(t)$ and $s_1(t)$ are known deterministic waveforms; the only thing the receiver does not know in advance is which one of the two waveforms is transmitted during a given interval.

Suppose the receiver takes m samples of $Y(t)$ denoted by $Y_1, Y_2, \ldots, Y_m$. Note that $Y_k = Y(t_k)$, $t_k \in (0, T)$, is a random variable and y_k is a particular value of Y_k. These samples under the two hypothesis are given by

$$y_k = s_{0,k} + n_k \quad \text{under } H_0$$
$$y_k = s_{1,k} + n_k \quad \text{under } H_1$$

where $s_{0,k} = s_0(t_k)$, and $s_{1,k} = s_1(t_k)$ are known values of the deterministic waveforms $s_0(t)$ and $s_1(t)$.

Let

$$\mathbf{Y} = [Y_1, Y_2, \ldots, Y_m]^T \quad \text{and} \quad \mathbf{y} = [y_1, y_2, \ldots, y_m]^T$$

and assume that the distributions of $\mathbf{Y}$ under H_0 and H_1 are given (the distribution of $Y_1, Y_2, \ldots, Y_m$ will be the same as the joint distribution of $N_1, N_2, \ldots, N_m$ except for a translation of the means by $s_{0,k}$ or $s_{1,k}$, $k = 1, 2, \ldots, m$).

A direct extension of the MAP decision rule discussed in the preceding section leads to a decision algorithm of the form

$$L(\mathbf{y}) = \frac{f_{\mathbf{Y}|H_1}(\mathbf{y}|H_1)}{f_{\mathbf{Y}|H_0}(\mathbf{y}|H_0)} \underset{H_0}{\overset{H_1}{\gtrless}} \frac{P(H_0)}{P(H_1)} \tag{6.9}$$

6.3.1 Independent Noise Samples

In many of the applications involving radio communication channels, the noise that accompanies the signal can be modeled as a zero-mean Gaussian random process with the power spectral density

$$S_{NN}(f) = \begin{cases} \dfrac{N_0}{2}, & |f| < B \\ 0 & \text{elsewhere} \end{cases} \quad (6.10)$$

where $N_0/2$ is the two-sided noise power spectral density and B is the receiver bandwidth, which is usually the same as the signal bandwidth. $N(t)$ with the foregoing properties is called bandlimited white Gaussian noise. It was shown in Section 5.5 that

$$E\{N^2(t)\} = N_0 B \quad (6.11)$$

and

$$R_{NN}(\tau) = N_0 B \frac{\sin 2\pi B \tau}{(2\pi B \tau)} \quad (6.12)$$

If the receiver samples are taken at $t_k = k(1/2B)$, then $N_k = N(t_k)$, $k = 1, 2, \ldots, m$, are uncorrelated (since $R_{NN}(k/2B) = 0$) and independent, and hence the conditional distribution of $\mathbf{Y}$ is n-variate Gaussian with

$$E\{\mathbf{Y}|H_0\} = [s_{0,1}, s_{0,2}, \ldots, s_{0,m}]^T; \quad \text{covar}\{Y_k, Y_j|H_0\} = \delta_{kj}(N_0 B)$$

and

$$E\{\mathbf{Y}|H_1\} = [s_{1,1}, s_{1,2}, \ldots, s_{1,m}]^T; \quad \text{covar}\{Y_k, Y_j|H_1\} = \delta_{kj}(N_0 B)$$

Hence

$$f_{\mathbf{Y}|H_i}(\mathbf{y}|H_i) = \prod_{k=1}^{m} \frac{1}{\sqrt{2\pi}\sigma} \exp\left[-\frac{(y_k - s_{ik})^2}{2\sigma^2}\right] \quad (6.13)$$

where $\sigma^2 = N_0 B$.

Substituting Equation 6.13 into 6.9, taking logarithms, and rearranging the terms yields the following decision rule:

$$\sum_{k=1}^{m} y_k s_{1,k} - \sum_{k=1}^{m} y_k s_{0,k} \underset{H_0}{\overset{H_1}{\gtreqless}} \sigma^2 \ln\left(\frac{P(H_0)}{P(H_1)}\right) + \frac{1}{2} \sum_{k=1}^{m} (s_{1k}^2 - s_{0k}^2) \quad (6.14)$$

or

$$\mathbf{y}^T[\mathbf{s}_1 - \mathbf{s}_0] \underset{H_0}{\overset{H_1}{\gtrless}} \sigma^2 \ln\left(\frac{P(H_0)}{P(H_1)}\right) + \frac{1}{2}(\mathbf{s}_1^T\mathbf{s}_1 - \mathbf{s}_0^T\mathbf{s}_0) \quad (6.15)$$

where

$$\mathbf{s}_i = [s_{i,1}, s_{i,2}, \ldots, s_{i,m}]^T, \quad i = 0, 1$$

The weighted averaging scheme given in Equation 6.14 is also called a *matched filtering* algorithm. The "dot" products, $\mathbf{y}^T\mathbf{s}_1$ and $\mathbf{y}^T\mathbf{s}_0$, indicate how well the received vector $\mathbf{y}$ matches with the two signals $\mathbf{s}_1$ and $\mathbf{s}_0$.

6.3.2 White Noise and Continuous Observations

When the bandwidth of the noise is much larger than $1/T$, then we can treat the noise as white with a power spectral density

$$S_{NN}(f) = \frac{N_0}{2} \quad \text{for all frequencies} \quad (6.16)$$

The autocorrelation of the zero-mean Gaussian white noise becomes

$$R_{NN}(\tau) = \frac{N_0}{2}\delta(\tau) \quad (6.17)$$

and the receiver can obtain and process an uncountably infinite number of independent samples observed in the interval $(0, T)$.

As $m \to \infty$, the samples become dense, and the left-hand side of the decision rule stated in Equation 6.14 becomes

$$\int_0^T y(t)s_1(t)\,dt - \int_0^T y(t)s_0(t)\,dt$$

This is the logarithm of the likelihood ratio, and the receiver has to make its decision by comparing this quantity against the decision threshold γ. So the decision rule becomes

$$\int_0^T y(t)s_1(t)\, dt - \int_0^T y(t)s_0(t)\, dt \underset{H_0}{\overset{H_1}{\gtrless}} \gamma \qquad (6.18)$$

A block diagram of the processing operations that take place in the receiver is shown in Figure 6.4. The receiver consists of two "cross-correlators" (two multiplier and integrator combinations) whose outputs are sampled once every T seconds. When the output of the correlator for signal $s_1(t)$ exceeds the output of the correlator for $s_0(t)$ by γ, a decision in favor of $s_1(t)$ is reached by the receiver. Once a decision is reached, the integration is started again with zero initial condition, and continued for another T seconds when another decision is made. The entire procedure is repeated once every T seconds with the receiver outputing a sequence of 1s and 0s, which are the estimates of the transmitted sequence.

The value of the decision threshold γ can be determined as follows. Let

$$Z = \int_0^T Y(t)[s_1(t) - s_0(t)]\, dt$$

Then, the conditional distribution of the decision variable Z under the two hypotheses can be determined from

$$Z = \int_0^T [s_1(t) + N(t)][s_1(t) - s_0(t)]\, dt \qquad \text{under } H_1$$

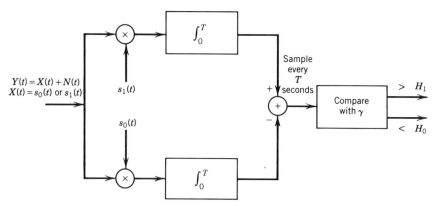

Figure 6.4 Correlation receiver for detecting deterministic signals $s_0(t)$ and $s_1(t)$ corrupted by additive white Gaussian noise.

and

$$Z = \int_0^T [s_0(t) + N(t)][s_1(t) - s_0(t)] \, dt \quad \text{under } H_0$$

Now, Z has a Gaussian distribution and the binary detection problem is equivalent to deciding on H_0 or H_1 based upon the single observation Z. The MAP decision rule for deciding between H_0 and H_1 based on Z reduces to

$$L(z) = \frac{f_{Z|H_1}(z|H_1)}{f_{Z|H_0}(z|H_0)} \overset{H_1}{\underset{H_0}{\gtrless}} \frac{P(H_0)}{P(H_1)}$$

If $f_{Z|H_1}$ is Gaussian with mean μ_1 and variance σ^2 (which can be computed using the results given in Section 3.8), and $f_{Z|H_0}$ is Gaussian with mean μ_0 and variance σ^2, the decision rule reduces to

$$z \overset{H_1}{\underset{H_0}{\gtrless}} \frac{\mu_1 + \mu_0}{2} + \frac{\sigma^2}{\mu_1 - \mu_0} \ln\left(\frac{P(H_0)}{P(H_1)}\right) \qquad (6.19)$$

where

$$z = \int_0^T y(t)s_1(t) \, dt - \int_0^T y(t)s_0(t) \, dt$$

Comparing the right-hand side of Equation 6.19 with Equation 6.18, we see that the decision threshold γ is given by

$$\gamma = \frac{\mu_0 + \mu_1}{2} + \frac{\sigma^2}{\mu_1 - \mu_0} \ln\left(\frac{P(H_0)}{P(H_1)}\right) \qquad (6.20)$$

EXAMPLE 6.2.

A binary communication system uses the signaling waveforms shown in Figure 6.5 for transmitting equiprobable sequences of 1's and 0's. At the receiver input, the signal arrives with additive zero-mean Gaussian noise $N(t)$ with the power spectral density

$$S_{NN}(f) = \begin{cases} 10^{-7}, & |f| < 10^6 \\ 0 & \text{elsewhere} \end{cases}$$

358 SIGNAL DETECTION

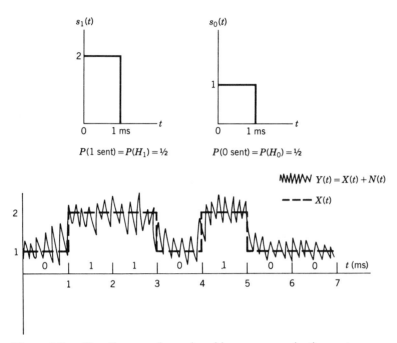

Figure 6.5a Signaling waveforms in a binary communication system.

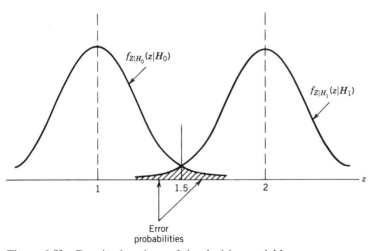

Figure 6.5b Density functions of the decision variable.

(a) Suppose the receiver samples the received waveform once every 0.1 millisecond and makes a decision once every millisecond based on the 10 samples taken during each bit interval. What is the optimum decision rule and the average probability of error?

(b) Develop a receiver structure and a decision rule based on processing the analog signal continuously during each signaling interval. Compare the probability of error with (a).

SOLUTION:

(a) We are given that $Y(t) = X(t) + N(t)$. With the $S_{NN}(f)$ given, we can show that samples of $N(t)$ taken 0.1 milliseconds apart will be independent, and hence, the decision rule becomes (Equation 6.15),

$$\mathbf{y}^T[\mathbf{s}_1 - \mathbf{s}_0] \underset{H_0}{\overset{H_1}{\gtrless}} \frac{1}{2}[\mathbf{s}_1^T\mathbf{s}_1 - \mathbf{s}_0^T\mathbf{s}_0]$$

with

$$\mathbf{y}^T = [y_1, y_2, \ldots, y_m], \quad m = 10$$
$$\mathbf{s}_1^T = [2, 2, \ldots, 2]$$
$$\mathbf{s}_0^T = [1, 1, \ldots, 1]$$

or

$$\sum_{i=1}^{m} y_i \underset{H_0}{\overset{H_1}{\gtrless}} \frac{3m}{2}$$

Dividing both sides by the number of samples, we can rewrite the decision rule as

$$\frac{1}{m}\sum_{i=1}^{m} y_i \underset{H_0}{\overset{H_1}{\gtrless}} 1.5, \quad m = 10$$

which implies that the receiver should average all the samples taken during a bit interval, compare it against a threshold value of 1.5, and make its decision based on whether the average value is greater than or less than 1.5. The decision variable Z,

$$Z = \frac{1}{m}\sum_{i=1}^{m} Y_i$$

has a Gaussian distribution with a mean of 2 under H_1 and 1 under H_0. The variance of Z can be computed as

$$\sigma_Z^2 = \text{variance of } \left\{ \frac{1}{m}\sum_{i=1}^{m} N(t_i) \right\}$$
$$= \sigma_N^2 / m$$

where

$$\sigma_N^2 = \int_{-\infty}^{\infty} S_{NN}(f)\, df = 0.2$$

The average probability of error is given by

$$P_e = P(H_0) \int_{1.5}^{\infty} f_{Z|H_0}(z|H_0)\, dz$$

$$+ P(H_1) \int_{-\infty}^{1.5} f_{Z|H_1}(z|H_1)\, dz$$

$$= Q\left(\frac{0.5}{\sqrt{0.02}}\right) \approx .0002 \qquad \text{when } m = 10$$

$$= Q\left(\frac{0.5}{\sqrt{0.2}}\right) \approx 0.13 \qquad \text{when } m = 1$$

Comparison of the average error probabilities for $m = 1$ and 10 reveals that averaging 10 samples reduces the error probability considerably.

(b) While the correlation receiver structure shown in Figure 6.4 is applicable only when $N(t)$ is white, we can use the receiver here as an approximation since the bandwidth of $N(t)$ is much larger than the bandwidth of the signaling waveform $X(t)$. The optimal decision and processing algorithm for the case of nonwhite or colored noise is rather complicated. A brief discussion of the algorithm will be presented in the next section. However, when the noise bandwidth is much larger than the signal bandwidth, the correlation receiver performance is near optimal.

The decision rule (from Equation 6.18) is

$$\int_0^T y(t)[s_1(t) - s_0(t)]\, dt \underset{H_0}{\overset{H_1}{\gtrless}} \gamma, \qquad T = 1 \text{ ms}$$

or

$$\frac{1}{T}\int_0^T y(t)\, dt \underset{H_0}{\overset{H_1}{\gtrless}} \gamma'$$

where $\gamma' = \gamma/T$ is the decision threshold whose value can be determined from the distribution of the decision variable.

In this example the decision variable Z,

$$Z = \frac{1}{T}\int_0^T Y(t)\, dt$$

has a Gaussian pdf with a mean of 2 under H_1, and a mean of 1 under H_0. The variance of Z can be calculated using the results given in Example 3.18 as

$$\sigma_Z^2 \approx \frac{\sigma_N^2}{2BT}$$

Where the BT product is equal to $(10^6)(10^{-3})$. Hence

$$\sigma_Z^2 = \frac{\sigma_N^2}{2000} = \frac{.2}{2000} = 10^{-4}$$

The reader can show that $\gamma' = 1.5$ and

$$P_e = Q\left(\frac{0.5}{\sqrt{10^{-4}}}\right)$$

$$\approx 0$$

6.3.3 Colored Noise

In the previous sections dealing with binary detection problems, we assumed that the noise samples and hence the observations are independent. While this assumption is valid in most cases, there are many situations in which the observations may be correlated. For example, consider the binary detection problem described in Section 6.3.1 but assume that the noise $N(t)$ has an arbitrary power spectral density function $S_{NN}(f)$, which is nonzero in the interval $[-B, B]$ and zero outside. Now, if the receiver makes its decision based on m observations, then the MAP decision rule is

$$L(\mathbf{y}) = \frac{f_{\mathbf{Y}|H_1}(\mathbf{y}|H_1)}{f_{\mathbf{Y}|H_0}(\mathbf{y}|H_0)} \underset{H_0}{\overset{H_1}{\gtrless}} \frac{P(H_0)}{P(H_1)} \quad (6.21)$$

The conditional pdfs are Gaussian with

$$E\{\mathbf{Y}|H_0\} = \mathbf{s}_0; \quad E\{\mathbf{Y}|H_1\} = \mathbf{s}_1$$

$$E\{[\mathbf{Y} - \mathbf{s}_i][\mathbf{Y} - \mathbf{s}_i]^T|H_i\} = \Sigma_N \quad \text{for } i = 0, 1 \quad (6.22)$$

where Σ_N is the covariance (autocorrelation) matrix of the noise samples $N(t_i)$, $i = 1, 2, \ldots, m$. That is the (i, j)th entry in Σ_N is

$$[\Sigma_N]_{ij} = E\{N(t_i)N(t_j)\}$$

(In the case of uncorrelated samples, Σ_N will be a diagonal matrix with identical entries of σ_N^2).

SIGNAL DETECTION

Substituting the multivariate Gaussian pdfs and taking the logarithm, we obtain

$$\ln[L(\mathbf{y})] = \frac{1}{2}(\mathbf{y} - \mathbf{s}_0)^T \boldsymbol{\Sigma}_N^{-1}(\mathbf{y} - \mathbf{s}_0) - \frac{1}{2}(\mathbf{y} - \mathbf{s}_1)^T \boldsymbol{\Sigma}_N^{-1}(\mathbf{y} - \mathbf{s}_1)$$

$$= \mathbf{y}^T \boldsymbol{\Sigma}_N^{-1} \mathbf{s}_1 - \mathbf{y}^T \boldsymbol{\Sigma}_N^{-1} \mathbf{s}_0 + \frac{1}{2} \mathbf{s}_0^T \boldsymbol{\Sigma}_N^{-1} \mathbf{s}_0 - \frac{1}{2} \mathbf{s}_1^T \boldsymbol{\Sigma}_N^{-1} \mathbf{s}_1 \quad (6.23)$$

Now, suppose that $\lambda_1, \lambda_2, \ldots, \lambda_m$, are distinct eigenvalues of $\boldsymbol{\Sigma}_N$ and $\mathbf{V}_1, \mathbf{V}_2, \ldots, \mathbf{V}_m$ are the corresponding eigenvectors of $\boldsymbol{\Sigma}_N$. Note that the m eigenvalues are scalars, and the normalized eigenvectors are $m \times 1$ column vectors with the following properties:

$$\boldsymbol{\Sigma}_N \mathbf{V}_i = \lambda_i \mathbf{V}_i, \quad i = 1, 2, \ldots, m \quad (6.24.\text{a})$$

$$\mathbf{V}_i^T \mathbf{V}_j = \delta_{ij}, \quad i, j = 1, 2, \ldots, m \quad (6.24.\text{b})$$

$$\mathbf{V}^T \boldsymbol{\Sigma}_N \mathbf{V} = \boldsymbol{\lambda} \quad (6.24.\text{c})$$

$$\mathbf{V}^T \mathbf{V} = \mathbf{I} \quad (6.24.\text{d})$$

where

$$\mathbf{V} = [\mathbf{V}_1, \mathbf{V}_2, \ldots, \mathbf{V}_m]_{m \times m} \quad (6.25)$$

and

$$\boldsymbol{\lambda} = \begin{bmatrix} \lambda_1 & 0 & \cdots & 0 \\ 0 & \lambda_2 & \cdots & 0 \\ \vdots & \vdots & & \vdots \\ 0 & 0 & \cdots & \lambda_m \end{bmatrix}_{m \times m} \quad (6.26)$$

Substituting Equations 6.24.a–d in Equation 6.23 and combining it with Equation 6.21, we obtain the decision rule as

$$\sum_{k=1}^{m} \frac{y_k'(s_{1,k}' - s_{0,k}')}{\lambda_k} \underset{H_0}{\overset{H_1}{\gtrless}} \ln\left(\frac{P(H_0)}{P(H_1)}\right) + \frac{1}{2} \sum_{k=1}^{m} \frac{s_{1,k}'^2 - s_{0,k}'^2}{\lambda_k} \quad (6.27)$$

where

$$\mathbf{s}'_0 = [\mathbf{V}_1^T\mathbf{s}_0, \mathbf{V}_2^T\mathbf{s}_0, \ldots, \mathbf{V}_m^T\mathbf{s}_0]^T, \quad s'_{0,k} = \mathbf{V}_k^T\mathbf{s}_0$$

$$\mathbf{s}'_1 = [\mathbf{V}_1^T\mathbf{s}_1, \mathbf{V}_2^T\mathbf{s}_1, \ldots, \mathbf{V}_m^T\mathbf{s}_1]^T, \quad s'_{1,k} = \mathbf{V}_k^T\mathbf{s}_1$$

$$\mathbf{y}' = [\mathbf{V}_1^T\mathbf{y}, \mathbf{V}_2^T\mathbf{y}, \ldots, \mathbf{V}_m^T\mathbf{y}]^T, \quad y'_k = \mathbf{V}_k^T\mathbf{y} \qquad (6.28)$$

are the transformed signal vectors and the observations.

The decision rule given in Equation 6.27 is similar to the decision rule given in Equations 6.14 and 6.15. The transformations given in Equation 6.28 reduce the problem of binary detection with correlated observations to an equivalent problem of binary detection with uncorrelated observations discussed in the preceding section. With colored noise at the receiver input, the components of the transformed noise vector are uncorrelated and hence the decision rule is similar to a matched-filter scheme described earlier with one exception. In the white noise case, and in the case of uncorrelated noise samples at the receiver input, the noise components were assumed to have equal variances. Now, after transformation, the variance of the transformed noise components are unequal and they appear as normalizing factors in the decision rule.

The extension of the decision rule given in Equation 6.27 to the case of continuous observations is somewhat complicated even though the principle is simple. We now make use of the continuous version of the Karhunen–Loeve expansion. Suppose $g_1(t), g_2(t), \ldots,$ and $\lambda_1, \lambda_2, \cdots$ satisfy

$$\int_0^T R_{NN}(t, u)g_i(u)\, du = \lambda_i g_i(t), \qquad i = 1, 2, \ldots \qquad (6.29)$$

Then we can expand $N(t)$ as

$$N(t) = \lim_{K \to \infty} \sum_{i=1}^{K} N_i g_i(t)$$

$$N_i = \int_0^T N(t)g_i(t)\, dt$$

where N_i, $i = 1, 2, 3, \ldots, K$, are uncorrelated random variables. Because $N(t)$ has a mean of zero it follows that $E\{N_i\} = 0$, and from Section 3.9.3 $E\{N_i N_j\} = \lambda_i \delta_{ij}$. We can expand the signals $s_1(t)$, $s_0(t)$, and the receiver input

$Y(t)$ in terms of $g_1(t), g_2(t), \cdots$ and represent each one by a set of coefficients $s'_{1i}, s'_{0i}, y'_i, i = 1, 2, \ldots,$ where

$$s'_{1i} = \int_0^T s_1(t)g_i(t)\, dt, \quad i = 1, 2, \ldots \quad (6.30)$$

$$s'_{0i} = \int_0^T s_0(t)g_i(t)\, dt, \quad i = 1, 2, \ldots \quad (6.31)$$

$$Y'_i = \int_0^T Y(t)g_i(t)\, dt, \quad i = 1, 2, \ldots \quad (6.32)$$

Equation 6.32 transforms the continuous observation $Y(t), t \in [0, T]$ into a set of uncorrelated random variables, $Y'_1, Y'_2, \ldots$ In terms of the transformed variables, the decision rule is similar to the one given in Equation 6.27 and has the form

$$\sum_{k=1}^{\infty} \frac{y'_k(s'_{1,k} - s'_{0,k})}{\lambda_k} \underset{H_0}{\overset{H_1}{\gtrless}} \ln\left(\frac{P(H_0)}{P(H_1)}\right) + \frac{1}{2}\sum_{k=1}^{\infty} \frac{s'^2_{1,k} - s'^2_{0,k}}{\lambda_k} \quad (6.33)$$

The only difference between Equation 6.27 and Equation 6.33 is the number of transformed observations that are used in arriving at a decision. While Equation 6.33 implies an infinite number of observations, in practice only a finite number of transformed samples are used. The actual number N is determined on the basis the value of N that yields

$$\lambda_1 + \lambda_2 + \cdots + \lambda_N \gg \lambda_{N+1} + \lambda_{N+2} + \cdots$$

That is, we truncate the sum when the eigenvalues become very small.

6.4 DETECTION OF SIGNALS WITH UNKNOWN PARAMETERS

In the preceding sections of this chapter we discussed the problem of detecting known signals. The only uncertainty was caused by additive noise. In many applications this assumption about known signals may not be valid. For example, in radar and sonar applications, the return pulse we attempt to detect might have a random amplitude and phase. In such cases the signals would be functions of one or more unknown parameters.

If the unknown parameter is assumed to be a constant, then its value can be estimated from the observations and the estimated value can be used to formulate a decision rule. Techniques for estimating parameter values from data are discussed in Chapter 8.

Another approach involves modeling the parameter as a random variable with a known probability density function. In this case the solution is rather straightforward. We compute the (conditional) likelihood ratio as a function of the unknown parameter and then average it with respect to the known distribution of the unknown parameter. For example, if there is one unknown parameter Θ, we compute the likelihood ratio as

$$L(z) = \int_\Theta L(z|\theta) f_\Theta(\theta) \, d\theta \tag{6.34}$$

and then derive the decision rule. We illustrate this approach with an example.

Suppose we have a binary detection problem in which

$$Y(t) = X(t) + N(t)$$

$$X(t) = \begin{cases} s_1(t) = A & \text{under } H_1, \\ s_0(t) = 0 & \text{under } H_0 \end{cases} \quad 0 \leq t \leq T$$

$N(t)$ is additive zero-mean white Gaussian noise with a power spectral density $S_{NN}(f) = N_0/2$, and A is the unknown signal amplitude with a pdf

$$f_A(a) = \frac{2a}{R} \exp\left(\frac{-a^2}{R}\right), \quad 0 < a; \quad R > 0$$

$$= 0 \quad a \leq 0$$

For a given value of $A = a$, the results of Section 6.3.2 leads to a decision rule based on Z,

$$Z = \int_0^T [X(t) + N(t)] \, dt$$

$$= \begin{cases} aT + W & \text{under } H_1 \\ 0 + W & \text{under } H_0 \end{cases}$$

where W is Gaussian with

$$\mu_W = 0$$

and

$$\sigma_W^2 = \frac{N_0 T}{2}$$

Hence the conditional likelihood function is given by

$$L(z|A = a) = \frac{f_{Z|H_1,a}}{f_{Z|H_0,a}}$$

$$= \exp\left\{-\frac{(a^2 T^2 - 2aTz)}{2\sigma_W^2}\right\}$$

and

$$L(z) = \int_a L(z|A = a) f_A(a) \, da$$

Completing the integration, taking the logarithm, and rearranging the terms, it can be shown that the decision rule reduces to

$$z^2 \underset{H_0}{\overset{H_1}{\gtrless}} \gamma \qquad (6.35)$$

This decision rule, called a *square-law detector*, is implemented as shown in Figure 6.6. Variations of this decision rule are used in many communication systems for detecting signals of the form $A \cos(\omega_0 t + \Theta)$ where A and Θ are unknown (random) amplitude and phase, respectively.

6.5 M-ARY DETECTION

Thus far we have considered the problem of deciding between one of two alternatives. In many digital communication systems the receiver has to make a

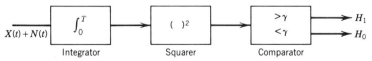

Figure 6.6 Square-law detector.

choice between one of many, say M, possibilities. For example, in a digital communication system, each transmitted pulse may have one of four possible amplitude values and the receiver has to decide which one of the four amplitudes or symbols was transmitted during an observation interval. Deciding between M alternate hypotheses, $M > 2$, is labeled as an M-ary detection problem.

If the decision is made on the basis of a single observation y, the MAP decision rule will involve choosing the hypothesis H_i that maximizes the *a posteriori* probability $P(H_i|y)$. Using Bayes' rule, we can write $P(H_i|y)$ as

$$P(H_i|y) = \frac{P(H_i) f_{Y|H_i}(y|H_i)}{f_Y(y)}$$

and the decision rule becomes

$$\text{Accept} \quad H_i \quad \text{if} \max_{H_j} \{P(H_j) f_{Y|H_j}(y|H_j)\} = P(H_i) f_{Y|H_i}(y|H_i) \quad (6.36)$$

The decision rule given in Equation 6.36 can be easily extended to the case of multiple and continuous observations. We illustrate the M-ary detection problem with a simple example.

EXAMPLE 6.3.

A digital communication system uses the signaling waveform shown in Figure 6.7a. Each transmitted pulse can have one of four equally likely amplitude levels,

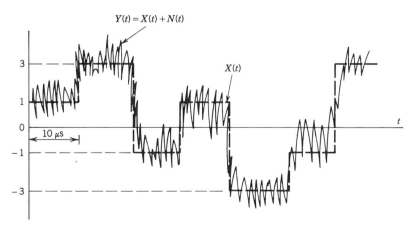

Figure 6.7a Signaling waveform.

368 SIGNAL DETECTION

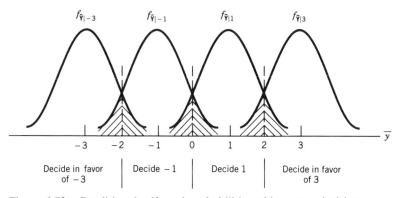

Figure 6.7b Conditional pdfs and probabilities of incorrect decisions.

−3, −1, 1, or 3 volts. The received signal has the form

$$Y(t) = X(t) + N(t)$$

where $N(t)$ is a zero-mean Gaussian random process with the power spectral density function

$$S_{NN}(f) = \begin{cases} (2)(10^{-7}), & |f| \leq 1 \text{ MHz} \\ 0 & |f| > 1 \text{ MHz} \end{cases}$$

The receiver takes five equally spaced samples in each signaling interval and makes a decision based on these five samples. Derive the decision rule and calculate the probability of an incorrect decision.

SOLUTION: During the first "symbol" interval [0, 10 μs], $X(t)$ is constant, and the samples of $Y(t)$ taken 2 μs apart are independent (why?). The MAP decision rule will be based on

$$f_{Y_1,Y_2,Y_3,Y_4,Y_5|H_j}(y_1, y_2, y_3, y_4, y_5|H_j) = \prod_{k=1}^{5} \frac{1}{\sqrt{2\pi}\sigma} \exp\left[\frac{-(y_k - A_j)^2}{2\sigma^2}\right]$$

where A_j is the transmitted amplitude corresponding to the jth hypothesis and σ^2 is the variance of $N(t)$. Taking the logarithm of the conditional pdf and collecting only those terms that involve A_j, we have the decision rule,

$$\text{choose } H_i \text{ if } \sum_{k=1}^{5} (y_k - A_j)^2 \text{ is minimum when } A_j = A_i$$

or

choose H_i if $(\bar{y} - A_j)^2$ is minimum when $A_j = A_i$

where $\bar{y} = \dfrac{1}{5}\sum_{k=1}^{5} y_k$

Thus, the decision rule reduces to averaging the five observations and choosing the level "closest" to $\bar{y}$ as the output. The conditional pdfs of the decision variable $\bar{Y}$ and the decision boundaries are shown in Figure 6.7b. Each shaded area in Figure 6.7b represents the probability of an incorrect decision and the average probability of incorrect decision is

$$P_e = \sum_{i=1}^{5} P(\text{incorrect decision}|H_i)P(H_i)$$

$$= \dfrac{1}{5}[\text{sum of the shaded areas shown in Figure 6.7b}]$$

$$= \dfrac{1}{5}(6)\,Q\left(\dfrac{1}{\sqrt{\sigma^2/5}}\right)$$

The factor 6 in the preceding expression represents the six shaded areas and $\sigma^2/5$ is the variance of $\bar{Y}$ in the conditional pdfs,

$$\sigma^2 = \int_{-\infty}^{\infty} S_{NN}(f)\,df = 0.4$$

Hence, the average probability of error is

$$P_e = \dfrac{6}{5}Q(\sqrt{12.5}) \approx .00025$$

6.5 SUMMARY

In this chapter, we developed techniques for detecting the presence of known signals that are corrupted by noise. The detection problem was formulated as a hypothesis testing problem, and decision rules that optimize different performance criteria were developed. The MAP decision rule that maximizes the *a posteriori* probability of a given hypothesis was discussed in detail. The decision rules were shown to have the general form

$$L(\mathbf{y}) = \frac{f_{\mathbf{Y}|H_1}(\mathbf{y}|H_1)}{f_{\mathbf{Y}|H_0}(\mathbf{y}|H_0)} \underset{H_0}{\overset{H_1}{\gtreqless}} \gamma$$

where **y** is a function of the observations, and γ is the decision threshold.

In the case of detecting known signals in the presence of additive white Gaussian noise the decision rule reduces to

$$\mathbf{y}^T[\mathbf{s}_1 - \mathbf{s}_0] \underset{H_0}{\overset{H_1}{\gtreqless}} \gamma'$$

This form of the decision rule is called a matched filter in which the received waveform is correlated with the known signals. The correlator output is compared with a threshold, and a decision is made on the basis of the comparison. Extension to *M*-ary detection and detecting signals mixed with colored noise were also discussed briefly.

6.7 REFERENCES

Several good references that deal with detection theory in detail are available to the interested reader. Texts on mathematical statistics (Reference [1] for example) provide good coverage of the theory of hypothesis testing and statistical inference, which is the basis of the material discussed in this chapter. Applications to signal detection are covered in great detail in References [2]–[4]. Reference [2] provides coverage at the intermediate level and References [3] and [4] provide a more theoretical coverage at the graduate level.

[1] R. V. Hogg and A. T. Craig, *Introduction to Mathematical Statistics,* Macmillan, New York, 1978.

[2] M. D. Srinath and P. K. Rajasekaran, *An Introduction to Statistical Signal Processing with Applications,* Wiley Interscience, New York, 1979.

[3] J. M. Wozencraft and I. M. Jacobs, *Principles of Communication Engineering,* John Wiley & Sons, New York, 1965.

[4] H. T. Van Trees, *Detection and Estimation,* John Wiley & Sons, New York, 1968.

6.8 PROBLEMS

6.1 Given the following conditional probability density functions of an observation Y

$$f_{Y|H_0}(y|H_0) = 1, \quad 0 \le y \le 1$$

and
$$f_{Y|H_1}(y|H_1) = 2y, \quad 0 \le y \le 1$$

a. Derive the MAP decision rule assuming $P(H_0) = \frac{1}{2}$.

b. Find the average probability of error.

6.2 Suppose that we want to decide whether or not a coin is fair by tossing it eight times and observing the number of heads showing up. Assume that we have to decide in favor of one of the following two hypotheses:

$$H_0: \text{ Fair coin, } P(\text{head}) = \frac{1}{2}$$

$$H_1: \text{ Unfair coin, } P(\text{head}) = 0.4$$

(The case where H_1: $P(\text{head}) \ne \frac{1}{2}$ is called a composite alternative hypothesis and is discussed in Chapter 8.)

a. Derive the MAP decision rule assuming $P(H_0) = \frac{1}{2}$.

b. Calculate the average probability of error.

6.3 In a radar system, decision about the presence (H_1) or absence (H_0) of a target is made on the basis of an observation Y that has the following conditional probability density functions:

$$f_{Y|H_0}(y|H_0) = \frac{y}{N_0} \exp\left(-\frac{y^2}{2N_0}\right), \quad y > 0$$

and

$$f_{Y|H_1}(y|H_1) \simeq \sqrt{\frac{y}{2\pi A N_0}} \exp\left[-\frac{(y-A)^2}{2N_0}\right], \quad y > 0$$

where A is the "signal amplitude" when a target is present and N_0 is the variance of the noise.

Derive the MAP decision rule assuming $P(H_0) = 0.999$.

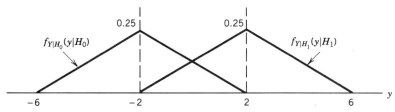

Figure 6.8 Conditional pdfs for Problem 6.4.

372 SIGNAL DETECTION

6.4 Figure 6.8 shows the conditional pdfs of an observation based on which binary decisions are made. The cost function for this decision problem is

$$C_{00} = 1, \quad C_{01} = C_{10} = 3, \quad C_{11} = 1$$

Assume that $P(H_0) = \frac{1}{3}$ and $P(H_1) = \frac{2}{3}$.

    a. Derive the decision rule that minimizes the average cost, and find the minimum cost.

    b. Find the average cost associated with the decision rule

$$y \underset{H_0}{\overset{H_1}{\gtreqless}} 0$$

and compare it with the average cost of the decision rule derived in part a.

6.5 Assume that under hypothesis H_1 we observe a signal of amplitude 2 volts corrupted by additive noise N. Under hypothesis H_0 we observe only the noise N. The noise is zero-mean Gaussian with a variance of $\frac{1}{9}$. Thus, we observe

$$Y = 2 + N \quad \text{under} \quad H_1 \quad \text{and} \quad Y = N \quad \text{under} \quad H_0$$

with

$$f_N(n) = \frac{3}{\sqrt{2\pi}} \exp\left(-\frac{9}{2} n^2\right)$$

Assuming $P(H_0) = P(H_1) = \frac{1}{2}$, derive the MAP decision rule and calculate the average probability of error.

6.6 With reference to Problem 6.5, find the decision rule that minimizes the average cost with respect to the cost function

$$C_{00} = 0, \quad C_{11} = 1, \quad C_{10} = C_{01} = 2$$

6.7   a. Show that the average cost associated with a decision rule can be expressed as

$$\overline{C} = C_{00}(1 - P_F) + C_{10}P_F + P(H_1)[(C_{11} - C_{00}) + (C_{01} - C_{11})P_M \\ - (C_{10} - C_{00})P_F]$$

where P_F is the false alarm probability and P_M is the miss probability.

    b. Show that the value of $P(H_1)$ that maximizes $\overline{C}$ (worst case) requires

$$(C_{11} - C_{00}) + (C_{01} - C_{11})P_M - (C_{10} - C_{00})P_F = 0$$

(The preceding equation is called the *minmax equation*).

PROBLEMS 373

6.8 In a detection problem, $f_{Y|H_1} \sim N(-1, 1)$ and $f_{Y|H_0} \sim N(1, 1)$. (The notation $N(\mu, \sigma^2)$ denotes a Gaussian distribution with mean μ and variance σ^2.)

  a. Assuming $P(H_0) = \frac{1}{3}$ and $P(H_1) = \frac{2}{3}$, find the MAP decision rule.

  b. With $C_{00} = C_{11} = 0$, $C_{01} = 6$, and $C_{10} = 1$, $P(H_0) = \frac{1}{3}$, $P(H_1) = \frac{2}{3}$, find the decision rule that minimizes $\overline{C}$ and the value of $\overline{C}_{min}$.

6.9 With reference to Problem 6.8, assume that the a priori probabilities are not known. Derive the minmax decision rule and calculate $\overline{C}$. (Refer to Problem 6.7 for the derivation of the minmax decision rule.)

6.10 In the Neyman–Pearson (N-P) decision rule, P_F is fixed at some value α and the decision rule is chosen to minimize P_M. That is, choose the decision rule that minimizes the objective function

$$J = P_M + \lambda(P_F - \alpha)$$

where $\lambda \geq 0$ is the Lagrange multiplier.

  a. Show that J can be reduced to

$$J = \lambda(1 - \alpha) + \int_{R_0} [f_{Y|H_1}(y|H_1) - \lambda f_{Y|H_0}(y|H_0)]\, dy$$

  b. Show that the N-P decision rule reduces to

$$\frac{f_{Y|H_1}(y|H_1)}{f_{Y|H_0}(y|H_0)} \overset{H_1}{\underset{H_0}{\gtrless}} \lambda$$

and explain how λ should be chosen.

6.11 Suppose we want to construct a N-P decision rule with $P_F = 0.001$ for a detection application in which

$$f_{Y|H_1} \sim N(-2, 4) \quad \text{and} \quad f_{Y|H_0} \sim N(2, 4)$$

  a. Find the decision threshold.

  b. Find $P_D = (1 - P_M)$ and comment on the value of P_D.

6.12 In a receiver for binary data

$$f_{Y|H_1}(y|H_1) \sim N(A, \sigma^2), \quad f_{Y|H_0}(y|H_0) \sim N(-A, \sigma^2)$$

$$P(H_0) = P(H_1) = \frac{1}{2}$$

Plot the receiver operating characteristic curves for $A^2/\sigma^2 = 1$ and $A^2/\sigma^2 = 16$. (A^2/σ^2 is a measure of the signal quality at the input to the detector. This ratio is referred to as the signal-to-noise power ratio.)

374 SIGNAL DETECTION

6.13 The conditional pdfs corresponding to two hypothesis are given:

$$f_{Y|H_0}(y|H_0) = \frac{1}{2} \exp\left(-\frac{y}{2}\right), \quad 0 < y$$

$$f_{Y|H_1}(y|H_1) = \frac{1}{4} \exp\left(-\frac{y}{4}\right), \quad 0 < y$$

Suppose we want to test these hypothesis based on two independent samples Y_1 and Y_2. Assume equally likely priors.

   a. Derive the MAP decision rule for the test.

   b. Calculate P_M and P_F.

6.14 The signaling waveforms used in a binary communication system are

$$s_1(t) = 4 \sin(2\pi f_0 t), \quad 0 \le t \le T, \quad T = 1 \text{ ms}$$

$$s_0(t) = -4 \sin(2\pi f_0 t), \quad 0 \le t \le T$$

$$P[s_1(t)] = P[s_0(t)] = \frac{1}{2}$$

where T is the signal duration and $f_0 = 10/T$.

Assume that the signal is accompanied by zero-mean additive white Gaussian noise with power spectral density of

$$S_{NN}(f) = 10^{-3} \text{ W/Hz}$$

   a. Find the decision rule that minimizes the average probability of error P_e.

   b. Find the value of P_e.

6.15 Repeat Problem 6.14 with

$$s_1(t) = \sqrt{8}, \quad 0 \le t \le T, \quad T = 1 \text{ ms}$$

$$s_0(t) = -\sqrt{8}, \quad 0 \le t \le T$$

and compare the results.

6.16 Show that the eigenvector transformation (or projection given in Equation 6.24) transforms the correlated components of the observation vector **Y** to a set of uncorrelated random variables.

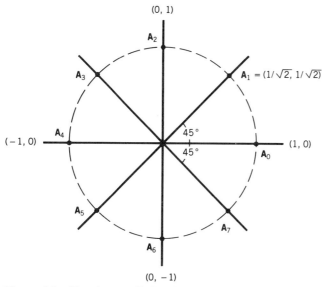

Figure 6.9 Signal constellation for Problem 6.19.

6.17 Given that $f_{\mathbf{Y}|H_0}$ and $f_{\mathbf{Y}|H_1}$, $\mathbf{Y} = [Y_1, Y_2]^T$, have bivariate Gaussian distributions with

$$E\{\mathbf{Y}|H_0\} = [1, 1]^T, \quad E\{\mathbf{Y}|H_1\} = [4, 4]^T$$

$$\text{covar}\{\mathbf{Y}|H_0\} = \text{covar}\{\mathbf{Y}|H_1\} = \begin{bmatrix} 2 & 1 \\ 1 & 2 \end{bmatrix}$$

$$P(H_0) = P(H_1) = \frac{1}{2}$$

a. Find the MAP decision rule.

b. Find P_e.

6.18 Consider the M-ary detection problem with cost functions

$$C_{ij} = 1 \quad \text{for} \quad i \neq j \quad \text{and} \quad C_{ii} = 0 \quad \text{for} \quad i, j = 0, 1, 2, \ldots, M - 1.$$

Show that the optimum decision criterion (that minimizes the average cost) is to choose the hypothesis for which the conditional cost $C(H_i|y)$ is smallest, where

$$C(H_i|y) = \sum_{\substack{j=0 \\ j \neq i}}^{M-1} C_{ij} P(H_j|y)$$

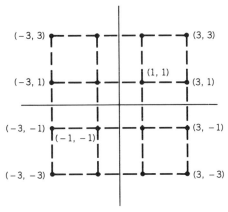

Figure 6.10 Signal constellation for Problem 6.20.

6.19 In an M-ary detection problem,
$$\mathbf{Y} = \mathbf{A}_j + \mathbf{N} \quad \text{under} \quad H_j, \quad j = 0, 1, 2, \ldots, 7$$
where $\mathbf{A}_j$ is the signal value associated with the jth hypothesis and $\mathbf{N}$ is a zero-mean bivariate Gaussian noise vector. The signal values corresponding to $H_0, H_1, \ldots, H_7$ are shown in Figure 6.9. Given that $P(H_i) = 1/M$, and
$$\boldsymbol{\Sigma}_N = \begin{bmatrix} 0.1 & 0 \\ 0 & 0.1 \end{bmatrix}$$

a. Find the decision boundaries that lead to a minimum P_e.

b. Bound the value of P_e using the Union Bound.

6.20 Repeat Problem 6.19 for the "signal-constellation" shown in Figure 6.10.

CHAPTER SEVEN

Linear Minimum Squared Error Filtering

7.1 INTRODUCTION

The basic idea of this chapter is that a noisy version of a signal is observed and the "true" value of a signal is to be estimated. This is often viewed as the problem of separating the signal from the noise. We will follow the standard approach whereby the theory is presented in terms of developing an estimator that minimizes the expected squared error between the estimator and the true but unknown signal. Historically, this signal estimation problem was viewed as *filtering* narrow-band signals from wide-band noise; hence the name "filtering" for signal estimation. Figure 7.1 shows an example of filtering $X(t) = S(t) + N(t)$ to produce an estimator $\hat{S}(t)$ of $S(t)$.

In the first section, we consider estimating the signal $S(t)$, based on a finite number of observations of a related random process $X(t)$. This is important in its own right, and it also forms the basis for both Kalman and Wiener filters. We then introduce innovations as estimation residuals or the unpredictable part of the latest sample. In the third section we consider digital Wiener filters, and then in the fourth section, the filter is extended to an infinite sequence of observations where we emphasize a discrete recursive filter called the Kalman filter. Finally, we consider continuous observations and discuss the Wiener filter. Both Kalman and Wiener filters are developed by using orthogonality and innovations.

In all cases in this chapter, we seek to minimize the expected (or mean) squared error (MSE) between the estimator and the value of the random variable

378 LINEAR MINIMUM MEAN SQUARED ERROR FILTERING

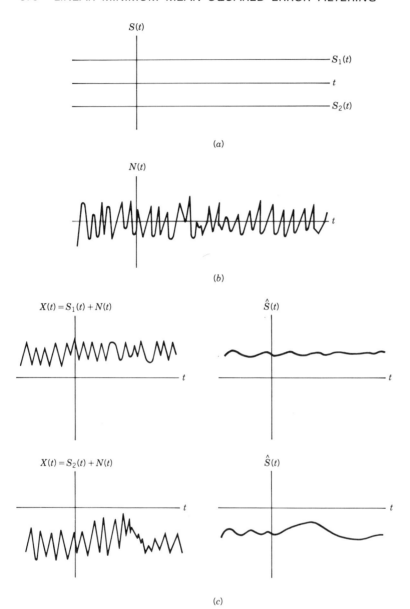

Figure 7.1 Filtering. (*a*) The two possible values of $S(t)$. (*b*) One sample function of $N(t)$. (*c*) Two examples of processing $X(t)$ to produce $\hat{S}(t)$.

(or signal) being estimated. In addition, with the exception of some theoretical development, we restrict our attention to linear estimators or linear filters.

One other important assumption needs to be mentioned. Throughout this chapter, we assume that the necessary means and correlation functions are known. The case where these moments must be estimated from data is discussed in Chapters 8 and 9.

7.2 LINEAR MINIMUM MEAN SQUARED ERROR ESTIMATORS

First we consider estimating a random variable by a constant, and then we base the estimator on a number of observations of a related random variable in order to reduce the mean squared error (MSE) of the estimator.

7.2.1 Estimating a Random Variable with a Constant

We show that the mean μ_S of random variable S is the estimator that minimizes the MSE. This is done by showing that the expected squared value of $(S - a)$ is always greater than or equal to the expected squared value of $(S - \mu_S)$ where

$$\mu_S = E\{S\} \tag{7.1}$$

and

$$\sigma_{SS} = E\{(S - \mu_S)^2\} \tag{7.2}$$

Indeed

$$\begin{aligned} E\{(S-a)^2\} &= E\{[(S-\mu_S) + (\mu_S - a)]^2\} \\ &= E\{(S-\mu_S)^2\} + E\{(\mu_S - a)^2\} + E\{2(S-\mu_S)(\mu_S - a)\} \\ &= \sigma_{SS} + (\mu_S - a)^2 + 2(\mu_S - a)E\{S - \mu_S\} \end{aligned}$$

The last term is zero, thus

$$E\{(S-a)^2\} = \sigma_{SS} + (\mu_S - a)^2 \tag{7.3}$$

showing that the minimum MSE is the variance, and if the constant a is not the mean, then the squared error is the variance plus the squared difference between the estimator a and the "best" estimator μ_S.

7.2.2 Estimating S with One Observation X

Suppose that X and S are related random variables. Having observed X we seek a linear* estimator $\hat{S} = a + hX$ of S such that $E\{(S - \hat{S})^2\}$ is minimized by

---

*This is actually an affine transformation of X because of the constant a, but we follow standard practice and call the affine transformation "linear." As we shall see it is linear with respect to a and h.

380   LINEAR MINIMUM MEAN SQUARED ERROR FILTERING

selection of the constants a and h. Note that our estimator $\hat{S}$ is a linear (affine) function of X and that the criterion is to minimize the MSE; hence the name, *linear minimum mean squared estimation*. We assume that all necessary moments of X and S are known.

We now show that the best estimator is

$$h = \frac{\sigma_{XS}}{\sigma_{XX}} \tag{7.4}$$

$$a = \mu_S - h\mu_X = \mu_S - \frac{\sigma_{XS}}{\sigma_{XX}}\mu_X \tag{7.5}$$

where

$$\sigma_{XS} = E\{(X - \mu_X)(S - \mu_S)\} \tag{7.6}$$

Taking derivatives of the MSE and setting them equal to zero we have, because the linear operations of expectation and differentiation can be interchanged

$$\frac{\partial E\{(S - a - hX)^2\}}{\partial a} = -2E\{S - a - hX\} = 0 \tag{7.7.a}$$

$$\frac{\partial E\{(S - a - hX)^2\}}{\partial h} = -2E\{(S - a - hX)X\} = 0 \tag{7.7.b}$$

The coefficients a and h are now the values that satisfy Equation 7.7 and should be distinguished by, say a_0 and h_0, but we stay with the simpler notation. Equation 7.7.a can be written

$$\mu_S - a - h\mu_X = 0 \tag{7.8}$$

which is algebraically equivalent to Equation 7.5. Equation 7.7.b becomes

$$E\{XS\} - a\mu_X - hE\{X^2\} = 0 \tag{7.9}$$

Recognizing that

$$E\{X^2\} = \sigma_{XX} + \mu_X^2$$

and

$$\sigma_{XS} = E\{(X - \mu_X)(S - \mu_S)\} = E\{XS\} - \mu_X\mu_S$$

Equation 7.9 becomes

$$\sigma_{XS} + \mu_X\mu_S = a\mu_X + h\sigma_{XX} + h\mu_X^2$$

Using Equation 7.5 in the foregoing equation produces

$$\sigma_{XS} + \mu_X\mu_S = \mu_X\mu_S - h\mu_X^2 + h\sigma_{XX} + h\mu_X^2$$

or

$$h = \frac{\sigma_{XS}}{\sigma_{XX}}$$

Thus, the solutions given do provide at least a stationary point. We now show that it is indeed a minimum.

Assume $\hat{S} = c + dX$, where c and d are arbitrary constants then

$$\begin{aligned}
E\{(S - c - dX)^2\} &= E\{[S - a - hX + (a - c) + (h - d)X]^2\} \\
&= E\{(S - a - hX)^2\} + E\{[(a - c) \\
&\quad + (h - d)X]^2\} + 2(a - c)E\{S - a - hX\} \\
&\quad + 2(h - d)E\{(S - a - hX)X\}
\end{aligned}$$

The last two terms are zero if a and h are the constants chosen as shown in Equation 7.7. Thus

$$E\{(S - c - dX)^2\} = E\{(S - a - hX)^2\} + E\{[(a - c) + (h - d)X]^2\}$$

Now $E\{[(a - c) + (h - d)X]^2\}$ is the expected value of a nonnegative random variable (there are no real values of a, h, c, d, or X that make $\{[(a - c) + (h - d)X]^2\}$ negative) and thus

$$E\{(S - c - dX)^2\} \geq E\{(S - a - hX)^2\}$$

which proves that a and h from Equations 7.4 and 7.5 minimize the MSE.

LINEAR MINIMUM MEAN SQUARED ERROR FILTERING

We now obtain a simpler expression for the minimum MSE. Since (see Equation 7.7)

$$E\{(S - a - hX)a\} = 0 = E\{(S - a - hX)hX\}$$

the minimum MSE is given by

$$\begin{aligned}
E\{(S - a - hX)^2\} &= E\{(S - a - hX)S\} \\
&= E\{S^2\} - aE\{S\} - hE\{XS\} \\
&= \sigma_{SS} + \mu_S^2 - \mu_S^2 + \frac{\sigma_{XS}}{\sigma_{XX}}\mu_X\mu_S - \frac{\sigma_{XS}}{\sigma_{XX}}(\sigma_{XS} + \mu_X\mu_S) \\
&= \sigma_{SS} - \frac{\sigma_{XS}^2}{\sigma_{XX}}
\end{aligned} \quad (7.10)$$

or

$$E\{(S - a - hX)^2\} = \sigma_{SS} - \frac{\rho_{XS}^2 \sigma_{XX} \sigma_{SS}}{\sigma_{XX}} = \sigma_{SS}(1 - \rho_{XS}^2) \quad (7.11)$$

where

$$\rho_{XS} = \frac{\sigma_{XS}}{\sigma_X \sigma_S} \quad \text{and} \quad \sigma_X = \sqrt{\sigma_{XX}}; \quad \sigma_S = \sqrt{\sigma_{SS}}$$

From Equation 7.11 it follows that if the correlation coefficient is near ± 1, then the expected squared error is nearly zero. If ρ_{XS} is zero, then the variance of S is not reduced by using the observation; thus observing X and using it in a linear estimator is of no value in estimating S.

EXAMPLE 7.1.

The observation X is made up of the true signal S and noise. It is known that $\mu_X = 10$, $\mu_S = 11$, $\sigma_{XX} = 1$, $\sigma_{SS} = 4$, $\rho_{XS} = .9$. Find the minimum linear MSE estimator of S in terms of X. What is the expected squared error with this estimator? Find the estimate of the true signal if we observe $X = 12$.

SOLUTION:

Using Equations 7.4 and 7.5 we obtain

$$h = \frac{(1)(2)(.9)}{1} = 1.8$$

$$a = 11 - (1.8)(10) = -7$$

Thus $\hat{S} = -7 + 1.8X$ is the best linear MSE estimator.
Using Equation 7.11, the minimum MSE is

$$E\{(S - \hat{S})^2\} = 4[1 - (.9^2)] = .76$$

If $X = 12$, then the estimate is

$$\hat{S} = -7 + (1.8)(12) = 14.6$$

7.2.3 Vector Space Representation

If the *inner product* of two n-dimensional random vectors is defined by

$$d(\mathbf{X}, \mathbf{Y}) \triangleq E\{\mathbf{X}^T\mathbf{Y}\} = E\left\{\sum_{i=1}^{n} X_i Y_i\right\}$$

where

$$\mathbf{X}^T = (X_1, \ldots, X_n) \quad \text{and} \quad \mathbf{Y}^T = (Y_1, \ldots, Y_n)$$

and T represents transpose. The length of a random vector $\mathbf{X}$ or *norm* is

$$d(\mathbf{X}, \mathbf{X}) = E\{\mathbf{X}^T\mathbf{X}\} = E\left\{\sum X_i^2\right\}$$

Then with these definitions $d(\mathbf{X}, \mathbf{Y}) = 0$ implies the vectors $\mathbf{X}$ and $\mathbf{Y}$ are orthogonal. The results of Section 7.2.2 can be visualized in a way that will be

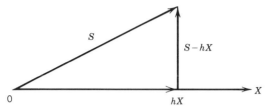

Figure 7.2 Orthogonality Principle.

useful in future developments by referring to Figure 7.2. In this figure, all means are assumed to equal 0.

Note that:

1. The error $S - hX$ is orthogonal to X. (Equation 7.7.b with the means $= 0$).
2. The best estimator of S is the projection of S on the X axis.

In more complicated linear filtering problems that follow, we shall see that these two conditions continue to be fundamental descriptors of the optimum linear filters.

7.2.4 Multivariable Linear Minimum Mean Squared Error Estimation

We now assume that values of the random sequence $X(1), \ldots, X(n)$ are observed and that the value of S is to be estimated with a linear estimator of the form

$$\hat{S} = h_0 + \sum_{i=1}^{n} h_i X(i)$$

such that $E\{(S - \hat{S})^2\}$ is minimized by the choice of the h_i, $i = 1, \ldots, n$.

Differentiating $E\{(S - \hat{S})^2\}$ with respect to each h_j, $j = 0, 1, \ldots, n$, results in

$$E\left\{S - h_0 - \sum_{i=1}^{n} h_i X(i)\right\} = 0 \qquad (7.12.\text{a})$$

$$E\left\{\left[S - h_0 - \sum_{i=1}^{n} h_i X(i)\right] X(j)\right\} = 0, \qquad j = 1, \ldots, n \qquad (7.12.\text{b})$$

Note that Equation 7.12.b can be visualized as in the previous section as stating that the error is orthogonal to each observation. In addition, Equation 7.12.a can be visualized as stating that the error is orthogonal to a constant. These two equations are called the *orthogonality conditions*.

Equation 7.12.a can be converted to

$$h_0 = \mu_S - \sum_{i=1}^{n} h_i \mu_{X(i)} \tag{7.13}$$

and using this in the n equations represented by Equation 7.12.b produces

$$\sum_{i=1}^{n} h_i C_{XX}(i, j) = \sigma_{SX(j)}, \quad j = 1, 2, \ldots, n \tag{7.14}$$

where

$$C_{XX}(i, j) = E\{[X(i) - \mu_{X(i)}][X(j) - \mu_{X(j)}]\}$$

and

$$\sigma_{SX(j)} = E\{(S - \mu_S)[X(j) - \mu_{X(j)}]\}$$

Equations 7.13 and 7.14 can be shown to result in a minimum (not just a saddle point) using the technique used in Section 7.2.2. (See Problem 7.8.)

The n equations of Equation 7.14 will now be written in matrix form. Defining

$$\mathbf{X}^T = [X(1), X(2), \ldots, X(n)]$$
$$\mathbf{h}^T = [h_1, h_2, \ldots, h_n]$$
$$\mathbf{\Sigma}_{XX} = [C_{XX}(i, j)]$$
$$\mathbf{\Sigma}_{SX}^T = [\sigma_{SX(1)}, \sigma_{SX(2)}, \ldots, \sigma_{SX(n)}]$$

Then Equation 7.14 can be written as

$$\mathbf{\Sigma}_{XX} \mathbf{h} = \mathbf{\Sigma}_{SX} \tag{7.15}$$

LINEAR MINIMUM MEAN SQUARED ERROR FILTERING

The solution to Equation 7.15 (or Equation 7.14) is

$$\mathbf{h} = \Sigma_{XX}^{-1}\Sigma_{SX} \qquad (7.16)$$

provided the covariance matrix Σ_{XX} has an inverse.

EXAMPLE 7.2.

The signal S is related to the observed sequence $X(1)$, $X(2)$. Find the linear MSE estimator, $\hat{S} = h_0 + h_1 X(1) + h_2 X(2)$. The needed moments are:

$$\mu_S = 1$$
$$\sigma_S = .01$$
$$\mu_{X(1)} = .02$$
$$C_{XX}(1, 1) = (.005)^2$$
$$\mu_{X(2)} = .006$$
$$C_{XX}(2, 2) = (.0009)^2$$
$$C_{XX}(1, 2) = 0$$
$$\sigma_{SX(1)} = .00003$$
$$\sigma_{SX(2)} = .000004$$

SOLUTION:

Using Equation 7.16

$$\begin{bmatrix} h_1 \\ h_2 \end{bmatrix} = \begin{bmatrix} (.005)^2 & 0 \\ 0 & (.0009)^2 \end{bmatrix}^{-1} \begin{bmatrix} .00003 \\ .000004 \end{bmatrix}$$

or

$$h_1 = \frac{.00003}{.000025} = \frac{6}{5}$$

$$h_2 = \frac{.000004}{.00000081} = \frac{400}{81}$$

Using Equation 7.13, we obtain

$$h_0 = 1 - \frac{6}{5}(.02) - \frac{400}{81}(.006) \approx .946$$

Thus the best estimator is

$$\hat{S} = .946 + \frac{6}{5}X_1 + \frac{400}{81}X_2$$

Now consider the case where $S(n)$ is the variable to be estimated by $\hat{S}(n)$, and the N observations are $S(n-1), S(n-2), S(n-3), \ldots, S(n-N)$, that is, $X(1) = S(n-1), \ldots, X(N) = S(n-N)$. Also assume $S(n)$ is a stationary random sequence with 0 mean. Then $h_0 = 0$, and Equation 7.14 becomes

$$\sum_{i=1}^{N} h_i R_{SS}(i-j) = R_{SS}(j), \quad j = 1, 2, \ldots, N \qquad (7.17)$$

where h_i is the weight of $S(n-i)$, $i = 1, \ldots, N$. Thus

$$\begin{bmatrix} R_{SS}(0) & R_{SS}(1) & R_{SS}(2) & \cdots & R_{SS}(N-1) \\ R_{SS}(1) & R_{SS}(0) & R_{SS}(1) & \cdots & R_{SS}(N-2) \\ R_{SS}(2) & R_{SS}(1) & R_{SS}(0) & \cdots & R_{SS}(N-3) \\ \vdots & \vdots & \vdots & & \vdots \\ R_{SS}(N-1) & & & \cdots & R_{SS}(0) \end{bmatrix} \begin{bmatrix} h_1 \\ h_2 \\ \cdot \\ \cdot \\ \cdot \\ h_N \end{bmatrix} = \begin{bmatrix} R_{SS}(1) \\ R_{SS}(2) \\ \cdot \\ \cdot \\ \cdot \\ R_{SS}(N) \end{bmatrix} \qquad (7.18)$$

Equation 7.17 or 7.18 is called the Yule–Walker equation.

EXAMPLE 7.3.

If $R_{SS}(n) = a^n$, find the best linear estimator of $S(n)$ in terms of the previous N observations, that is, find

$$\hat{S}(n) = h_1 S(n-1) + h_2 S(n-2) + \cdots + h_N S(n-N)$$

388 LINEAR MINIMUM MEAN SQUARED ERROR FILTERING

SOLUTION:

Equation 7.18 becomes

$$\begin{bmatrix} 1 & a & a^2 & \cdots & a^{N-1} \\ a & 1 & a & \cdots & a^{N-2} \\ \vdots & & & & \vdots \\ a^{N-1} & & \cdots & & 1 \end{bmatrix} \begin{bmatrix} h_1 \\ h_2 \\ \vdots \\ h_N \end{bmatrix} = \begin{bmatrix} a \\ a^2 \\ a^3 \\ \vdots \\ a^N \end{bmatrix}$$

It is obvious that

$$\mathbf{h}^T = [a, 0, \ldots, 0]$$

satisfies this set of equations. Thus

$$\hat{S}(n) = aS(n-1)$$

is the minimum MSE estimator. Because $\hat{S}(n)$ is the "best" estimator, and because it ignores all but the most recent value of S, the geometric autocorrelation function results in a Markov type of estimator.

Returning to the problem of estimating S based on $X(i), \ldots X(n)$, we find the residual error, given that the best estimator has been selected. This is very important because not only is the best estimate of a signal needed, but one needs to know whether the best estimate is good enough.

We now examine the MSE when the h_is are chosen in accordance with Equations 7.13 and 7.14. The MSE is given by

$$E\left\{\left[S - h_0 - \sum_{i=1}^{n} h_i X(i)\right]\left[S - h_0 - \sum_{j=1}^{n} h_j X(j)\right]\right\}$$

$$= E\left\{\left[S - h_0 - \sum_{i=1}^{n} h_i X(i)\right] S\right\} - h_0 E\left\{S - h_0 - \sum_{i=1}^{n} h_i X(i)\right\}$$

$$- \sum_{j=1}^{n} h_j E\left\{\left[S - h_0 - \sum_{i=1}^{n} h_i X(i)\right] X(j)\right\}$$

The last two terms are zero by the orthogonality condition, Equation 7.12. Thus

$$E\{(S - \hat{S})^2\} = E\{S^2\} - h_0\mu_S - \sum_{i=1}^{n} h_i E\{X(i)S\}$$

Using Equation 7.13

$$E\{(S - \hat{S})^2\} = \sigma_S^2 + \mu_S^2 - \mu_S^2 - \sum_{i=1}^{n} h_i[E\{X(i)S\} - \mu_{X(i)}\mu_S]$$

$$P(n) \triangleq E\{(S - \hat{S})^2\} = \sigma_S^2 - \sum_{i=1}^{n} h_i \sigma_{X(i)S} \tag{7.19}$$

The notation $P(n)$ is introduced to correspond to the symbol that will be used in Kalman filtering. The minimum MSE $P(n)$ when the best linear estimator is used is also called the residual MSE.

It is important to observe that

$$E\{(S - \hat{S})X(i)\} = 0$$

and

$$E\{(S - \hat{S})\} = 0$$

by the orthogonality conditions. Also

$$\hat{S} = h_0 + \sum_{i=1}^{n} h_i X(i)$$

and hence

$$E\{(S - \hat{S})\hat{S}\} = 0 \tag{7.20}$$

Thus with the optimal linear estimator

$$E\{(S - \hat{S})^2\} = E\{(S - \hat{S})S\} \tag{7.21}$$

LINEAR MINIMUM MEAN SQUARED ERROR FILTERING

One other equivalent form of the residual error is important. From Equation 7.20

$$E\{S\hat{S}\} = E\{\hat{S}^2\} \tag{7.22}$$

Using this result in Equation 7.21 produces

$$E\{(S - \hat{S})^2\} = E\{S^2\} - E\{\hat{S}^2\} \tag{7.23}$$

This equation can be visualized using Figure 7.2. That is, if $\hat{S} = hX$ then the square of the error, $S - \hat{S}$, is equal to $S^2 - \hat{S}^2$. If $\hat{S}$ is the projection of S on a hyperplane, then the same conceptual picture remains valid.

We again consider $\hat{S}(n)$ and note that Equations 7.18 define the weights in a finite linear digital filter when $S(n)$ is a zero-mean wide-sense stationary process and $X(i) = S(n - i)$, $i = 1, 2, \ldots, N$. In this case Equation 7.19 becomes

$$P(N) = E\{[S(n) - \hat{S}(n)]^2\} = R_{SS}(0) - \sum_{i=1}^{N} h_i R_{SS}(i) \tag{7.24}$$

EXAMPLE 7.4 (CONTINUATION OF EXAMPLE 7.2).

Find the residual MSE for Example 7.2 with the constants chosen in Example 7.2.

SOLUTION:

Using Equation 7.19

$$P(2) = E\{(S - \hat{S})^2\} = (.01)^2 - \left(\frac{6}{5}\right)(.00003) - \left(\frac{400}{81}\right)(.000004)$$
$$\approx .0001 - .000036 - .000020 = .000044.$$

EXAMPLE 7.5 (CONTINUATION OF EXAMPLE 7.3).

Find the residual MSE for Example 7.3.

LINEAR MINIMUM MEAN SQUARED ERROR ESTIMATORS

SOLUTION:

Using Equation 7.24

$$P(N) = E\{[S(n) - \hat{S}(n)]^2\} = R_{SS}(0) - aR_{SS}(1)$$
$$= 1 - a^2$$

Note that this MSE is independent of N.

7.2.5 Limitations of Linear Estimators

We illustrate the limitations of a linear estimator with an example.

EXAMPLE 7.6.

Suppose we tried to find a linear estimator $\hat{S} = h_0 + h_1 X$ when, *unknown to us*, S is related to X by the equation

$$S = X^2$$

and

$$f_X(x) = \frac{1}{2}, \quad -1 \leq x \leq 1,$$
$$= 0 \quad \text{elsewhere}$$

SOLUTION:

In this case from the assumed density of X and the relation $S = X^2$, the moments are

$$E\{X\} = 0$$

$$E\{X^2\} = \sigma_X^2 = \int_{-1}^{1} \frac{1}{2} x^2 \, dx = \frac{1}{3}$$

$$E\{S\} = E\{X^2\} = \frac{1}{3}$$

$$E\{XS\} = E\{X^3\} = \int_{-1}^{1} \frac{1}{2} x^3 \, dx = 0$$

392 LINEAR MINIMUM MEAN SQUARED ERROR FILTERING

Thus, using Equations 7.14 and 7.13

$$h_1 = 0$$

$$h_0 = \frac{1}{3}$$

or

$$\hat{S} = \frac{1}{3}$$

and using Equation 7.19

$$P(1) = E\{(S - \hat{S})^2\} = \sigma_S^2$$

Thus the best fit is a constant, and the MSE is simply the variance. The best linear estimator is obviously a very poor estimator in this case.

Suppose we tried the estimator

$$\hat{S} = h_0 + h_1 X + h_2 X^2$$

This estimator would result in a perfect fit for the Example 7.6. (See Problem 7.10.) However, is this a linear estimator? It appears to be nonlinear, but suppose we define $X_0 = 1$, $X_1 = X$ and $X_2 = X^2$; then

$$\hat{S} = h_0 X_0 + h_1 X_1 + h_2 X_2$$

is a linear estimator of the type discussed.

Thus, if one is willing to consider multivariate equations, then any polynomial in X can be used, and in addition estimators such as

$$\hat{S} = h_0 + h_1 X + h_2 X^2 + h_3 \cos X + h_4[X^3 - 3X + 19 \tan^{-1}(X)]$$

can be used. This estimator is linear in the h_i's and thus is within the theory developed.

7.2.6 Nonlinear Minimum Mean Squared Error Estimators

In the last section limitations of linear estimators were illustrated. We now show that the minimum mean squared error estimator S^* of S, given $X(1), \ldots, X(n)$, is

$$S^* = E\{S|X(1), \ldots, X(n)\} \quad (7.25)$$

Indeed, letting $\hat{S}$ represent any estimator based upon the observations $X_1, \ldots, X_n$

$$E\{(S - \hat{S})^2\} = E_\mathbf{X}\{E_{S|\mathbf{X}}\{(S - \hat{S})^2 | X(1), \ldots, X(n)\}\}$$
$$= \int_{-\infty}^{\infty} \cdots \int_{-\infty}^{\infty} E_{S|\mathbf{X}}\{(S - \hat{S})^2 | X(1) = x_1, \ldots, X(n) = x_n\}$$
$$\times f_{X(1),\ldots,X(n)}(x_1, \ldots, x_n) \, dx_1, \ldots, dx_n$$

Thus in order to minimize the multiple integral, because $f_{X(1),\ldots,X(n)} \geq 0$, it is sufficient to minimize the conditional expected value of $(S - \hat{S})^2$. We show that S^* as given by Equation 7.25 does minimize the conditional expected value. Let $\hat{S}$ be another estimator (perhaps a linear estimator).

$$E\{(S - \hat{S})^2 | X(1) = x_1, \ldots, X(n) = x_n\}$$
$$= E\{(S - S^* + S^* - \hat{S})^2 | X(1) = x_1, \ldots, X(n) = x_n\}$$
$$= E\{(S - S^*)^2 | X(1) = x_1, \ldots, X(n) = x_n\}$$
$$+ E\{(S^* - \hat{S})^2 | X(1) = x_1, \ldots, X(n) = x_n\}$$
$$+ 2E\{(S - S^*)(S^* - \hat{S}) | X(1) = x_1, \ldots, X(n) = x_n\}$$

Note that given $X(1) = x_1, \ldots, X(n) = x_n$, the second term in the last equation is a nonnegative constant and the third term is

$$2(S^* - \hat{S})[E\{S|X(1) = x_1, \ldots, X(n) = x_n\} - S^*] = 0$$

Thus

$$E\{(S - \hat{S})^2 | X(1), \ldots, X(n)\} \geq E\{(S - S^*)^2 | X(1), \ldots, X(n)\} \quad (7.26)$$

Equation 7.26 shows that the conditional expectation has an expected mean squared error that is less than or equal to any estimator including the best linear estimator.

EXAMPLE 7.7.

Find (a) the minimum MSE estimator (b) the linear minimum MSE estimator of S based on observing X and the MSEs if

$$f_{S|X}(s|x) = \frac{1}{\sqrt{2\pi}} \exp\left\{-\frac{1}{2}\left(s - x - \frac{x^2}{2}\right)^2\right\}$$

$$f_X(x) = e^{-x}, \quad x \geq 0$$
$$= 0 \quad x < 0$$

SOLUTION:

(a) By identification of the mean of a Gaussian random variable

$$S^* = E\{S|X\} = X + \frac{X^2}{2}$$

and by identification the MSE is the conditional variance, thus $E\{(S - S^*)^2\} = 1$

(b) X is exponential. Thus

$$E\{X\} = 1; \quad E\{X^2\} = 2; \quad E\{X^3\} = 6$$

and

$$E\{S\} = E_X\left\{X + \frac{X^2}{2}\right\} = 2$$

$$E\{SX\} = E_X\left\{X\left(X + \frac{X^2}{2}\right)\right\} = 5$$

Using Equations 7.4 and 7.5

$$h = \frac{5 - 2}{2 - 1} = 3$$

$$a = 2 - 3 = -1$$

Thus

$$\hat{S} = -1 + 3X$$

and the reader can show that $E\{(S - \hat{S})^2\} = 2$

The reader might wonder why linear mean squared estimators are stressed, when we have just shown that the conditional expectation is a better estimator. The answer is that the conditional expectation is not known unless the joint distribution functions are known, a situation that is very rare in practice. On the other hand, a linear minimum MSE estimator is a function of only first and second moments, which may be known or can be estimated.

However, in one important special case, the minimum MSE estimator is a linear estimator. This case is discussed next.

7.2.7 Jointly Gaussian Random Variables

If X_1 and X_2 are jointly Gaussian, then a special case of Equation 2.66.a is

$$E\{X(1)|X(2)\} = \mu_1 + \rho \frac{\sigma_1}{\sigma_2}[X(2) - \mu_2]$$

or

$$E\{X(1)|X(2)\} = \left(\mu_1 - \mu_2\rho\frac{\sigma_1}{\sigma_2}\right) + \rho\frac{\sigma_1}{\sigma_2}X(2) \qquad (7.27)$$

where ρ is the correlation coefficient between $X(1)$ and $X(2)$.

Note that the conditional expected value of jointly Gaussian random variables is a linear estimator, that is, $E\{X(1)|X(2)\} = h_0 + h_1 X(2)$. Moreover, comparison of Equation 7.27 with Equation 7.4 and Equation 7.5 shows that the conditional expectation is exactly the same as the linear estimator.

Also we see from Equation 2.66.b that the conditional variance of $X(1)$ given $X(2)$ is $\sigma_1^2(1 - \rho^2)$, which is in agreement with Equation 7.11.

Note that the conditional variance of $X(1)$ given $X(2)$ does not depend upon $X(2)$. Indeed, since

$$E\{[X(1) - E\{X(1)|X(2)\}]^2|X(2)\} = g(X(2)) = \sigma_1^2(1 - \rho)^2$$

is a constant and thus independent of $X(2)$ and because

$$E\{[X(1) - E\{X(1)|X(2)\}]^2\} = E\{g[X(2)]\} = E\{\sigma_1^2(1 - \rho^2)\} = \sigma_1^2(1 - \rho^2)$$

the conditional variance and the MSE's are identical.

If a MSE estimator is based on N Gaussian random variables, then the estimator (conditional mean) and MSE (conditional variance) are given in Equations 2.66.a and 2.66.b, respectively.

EXAMPLE 7.8.

The joint density function of two random variables X and S is given by

$$f_{X,S}(x, s) = \frac{1}{\pi\sqrt{3}} \exp\left\{-\frac{2}{3}(x^2 - xs + s^2)\right\}$$

Find:

(a) The marginal density functions of X and of S.
(b) $f_{S|X}$.
(c) Simplify the expression found in (b) and identify the conditional mean of S given X.
(d) Identify the conditional variance, that is

$$E_{S|X}\{[S - E_{S|X}(S|X)]^2 | X\}$$

(e) Find $E_X\{E_{S|X}[S - E_{S|X}(S|X)]^2 | X\}$

SOLUTION:

(a)

$$f_X(x) = \int_{-\infty}^{\infty} \frac{1}{\pi\sqrt{3}} \exp\left\{-\frac{2}{3}(x^2 - xs + s^2)\right\} ds = \frac{1}{\sqrt{2\pi}} \exp\left(-\frac{x^2}{2}\right)$$

Similarly

$$f_S(s) = \frac{1}{\sqrt{2\pi}} \exp\left(-\frac{s^2}{2}\right)$$

(b)

$$f_{S|X}(s|x) = \frac{\frac{1}{\pi\sqrt{3}} \exp\left\{-\frac{2}{3}(x^2 - xs + s^2)\right\}}{\frac{1}{\sqrt{2\pi}} \exp\left\{-\frac{1}{2}x^2\right\}}$$

$$= \frac{1}{\sqrt{\frac{2\pi 3}{4}}} \exp\left\{-\frac{2}{3}\left(x^2 - \frac{3}{4}x^2 - xs + s^2\right)\right\}$$

$$= \frac{1}{\sqrt{\frac{2\pi 3}{4}}} \exp\left\{-\frac{1}{3/2}\left(s - \frac{x}{2}\right)^2\right\}$$

(c) The density is Gaussian with conditional mean $x/2$. Thus

$$E\{S|X = x\} = \frac{x}{2}$$

and

$$E\{S|X\} = \frac{X}{2}, \quad \text{a random variable}$$

(d) The conditional variance is 3/4 and in this case is a constant (i.e. independent of x).
(e) $E_X\{3/4\} = 3/4$

7.3 INNOVATIONS

We now reconsider estimating S by observations, $X(1), X(2), \ldots, X(n)$ using the concept of innovations. We do this in order to resolve this relatively simple problem by the same method that will be used to develop both the Kalman and the Wiener filters.

For simplification we assume that

$$E\{S\} = E\{X(i)\} = 0, \quad i = 1, \ldots, n$$

and that all variables are jointly Gaussian. If the variables are not jointly Gaussian but the estimators are restricted to linear estimators, then the same results would be obtained as we develop here.

Now we can find the optimal estimator $\hat{S}_1$ in terms of $X(1)$ by the usual method (Note that subscripts are used to differentiate the linear estimators when innovations are used.)

$$\hat{S}_1 = h_1 X(1) \tag{7.28}$$

The orthogonality condition requires that

$$E\{[S - h_1 X(1)]X(1)\} = 0 \tag{7.29}$$

or

$$h_1 = \frac{E\{SX(1)\}}{E\{X^2(1)\}} \tag{7.30}$$

Now we want to improve our estimate of S by also considering $X(2)$. This was done in Section 7.2.4, but we now consider the same problem using innovations.

398 LINEAR MINIMUM MEAN SQUARED ERROR FILTERING

We define the innovation $V_X(2)$ of X_2 to be the unpredictable part of $X(2)$, that is

$$V_X(2) \triangleq X(2) - E\{X(2)|X(1)\} \tag{7.31}$$

Because $X(2)$ and $X(1)$ both have zero means and because they are jointly Gaussian we know that

$$E\{X(2)|X(1)\} = a_1 X(1) \tag{7.32}$$

where (see Equations 7.4 and 7.27)

$$a_1 = \rho \frac{\sigma_2}{\sigma_1} = \frac{\sigma_{X(1)X(2)}}{\sigma_{X(1)X(1)}} \tag{7.33}$$

and ρ is the correlation coefficient between $X(1)$ and $X(2)$ and $\sigma_i = \sqrt{\sigma_{X(i)X(i)}}$. Thus

$$V_X(2) = X(2) - \frac{\rho\sigma_2}{\sigma_1} X(1) \tag{7.34}$$

Note that

1. $V_X(2)$ is a linear combination of $X(1)$ and $X(2)$.
2. $E\{V_X(2)\} = E\left\{X(2) - \frac{\rho\sigma_2}{\sigma_1} X(1)\right\} = 0$ (7.35)
3. $E\{V_X(2) X(1)\} = E\left\{\left[X(2) - \frac{\rho\sigma_2}{\sigma_1} X(1)\right] X(1)\right\}$

$$= \rho\sigma_1\sigma_2 - \frac{\rho\sigma_2}{\sigma_1} \sigma_1^2 = 0 \tag{7.36}$$

4. $X(2) = V_X(2) + \frac{\rho\sigma_2}{\sigma_1} X(1)$ (7.37)

We now seek the estimator, $\hat{S}_2$, of S based on $X(1)$ and the innovation of $X(2)$, that is

$$\hat{S}_2 = h_1 X(1) + h_2 V_X(2)$$

where $X(1)$ is the "old" observation and $V_X(2)$ is the innovation or unpredictable [from $X(1)$] part of the new observation $X(2)$.

We know that $V_X(2)$ and $X(1)$ are orthogonal (Equation 7.36). Thus Equation 7.15 becomes

$$\begin{bmatrix} \sigma_{X(1)X(1)} & 0 \\ 0 & \sigma_{V_X(2)V_X(2)} \end{bmatrix} \begin{bmatrix} h_1 \\ h_2 \end{bmatrix} = \begin{bmatrix} \sigma_{SX(1)} \\ \sigma_{SV_X(2)} \end{bmatrix} \qquad (7.38)$$

and h_1 is as given in Equation 7.30 whereas

$$h_2 = \frac{E\{SV_X(2)\}}{E\{V_X(2)^2\}} = \frac{\sigma_{SV_X(2)}}{\sigma_{V_X(2)V_X(2)}} \qquad (7.39)$$

Note that these expected values can be found using Equation 7.34. Furthermore, $E\{SV_X(2)\}$ is called the *partial covariance* of S and $X(2)$, and the partial covariance properly normalized is called the partial correlation coefficient. The partial correlation coefficient was defined in Chapter 5.

We now proceed to include the effect of $X(3)$ by including its innovation, $V_X(3)$. By definition

$$V_X(3) = X(3) - E\{X(3)|X(1), X(2)\} \qquad (7.40)$$

Now we note from Equation 7.37 that $X(2)$ is a linear combination of $X(1)$ and $V_X(2)$. Indeed

$$X(2) = V_X(2) + \frac{\rho\sigma_2}{\sigma_1} X(1)$$

Thus, knowing $X(1)$ and $X(2)$ is equivalent to knowing $X(1)$ and $V_X(2)$. So

$$E\{X(3)|X(1), X(2)\} = E\{X(3)|X(1), V_X(2)\} \qquad (7.41)$$

and using this in Equation 7.40 produces

$$V_X(3) = X(3) - E\{X(3)|X(1), V_X(2)\} \qquad (7.42)$$

Now, because of the zero-mean Gaussian assumption

$$E\{X(3)|X(1), V_X(2)\} = b_1 X(1) + b_2 V_X(2) \qquad (7.43)$$

where

$$b_1 = \frac{E\{X(3)X(1)\}}{E\{X(1)^2\}} \quad (7.44)$$

$$b_2 = \frac{E\{X(3)V_X(2)\}}{E\{V_X(2)^2\}} \quad (7.45)$$

and

$$V_X(3) = X(3) - b_1 X(1) - b_2 V_X(2) \quad (7.46)$$

Note that

1. $V_X(3)$ is a linear combination of $X(1)$, $X(2)$, and $X(3)$.
2. $E\{V_X(3)\} = 0$
3. $E\{V_X(3)X(1)\} = 0. \quad E\{V_X(3)V_X(2)\} = 0 \quad (7.47)$

Now in the estimator

$$\hat{S}_3 = h_1 X(1) + h_2 V_X(2) + h_3 V_X(3) \quad (7.48)$$

h_1 and h_2 were found in Equations 7.30 and 7.39 and

$$h_3 = \frac{E\{SV_X(3)\}}{E\{V_X(3)^2\}} \quad (7.49)$$

The general case is discussed next.

7.3.1 Multivariate Estimator Using Innovations

We now describe the general linear estimator based on innovations.

$$\hat{S}_n = \sum_{j=1}^{n} h_j V_X(j) \quad (7.50)$$

where

$$V_X(1) = X(1) \quad (7.51)$$

$$V_X(j) = X(j) - E\{X(j)|X(1), \ldots, X(j-1)\}, \quad j = 2, 3, \ldots, n \quad (7.52)$$

$$h_j = \frac{E\{SV_X(j)\}}{E\{V_X^2(j)\}} \quad (7.53)$$

Furthermore, the residual MSE, $P(n)$ is, using Equation 7.23

$$P(n) = E\{[S - \hat{S}_n]^2\} = E\{S^2\} - E\{\hat{S}_n^2\}$$

$$= \sigma_S^2 - E\left\{\left[\sum_{j=1}^{n} h_j V_X(j)\right]^2\right\} \quad (7.54)$$

$$= \sigma_S^2 - \sum_{j=1}^{n} h_j^2 \sigma_{V_X(j)V_X(j)} \quad (7.55)$$

where the last equation follows from orthogonality of $V_X(j)$ and $V_X(i)$, $i \neq j$.

At this stage, we have a general form for both the estimator and the error in terms of innovations. It can be shown that this estimator is equivalent to the linear estimator of Section 7.2. (See Problems 7.11 and 7.12.)

7.3.2 Matrix Definition of Innovations

The remainder of this section describes the general relation between the set of observations $X(1), X(2), \ldots, X(n)$ and the innovations $V_X(1), V_X(2), \ldots, V_X(n)$.

We define the innovation, $V_X(i)$ of $X(i)$ to be

$$V_X(1) = X(1) \quad (7.56)$$

$$V_X(i) = X(i) - E\{X(i)|X(1), \ldots, X(i-1)\}, \quad i = 1, 2, \ldots, n \quad (7.57)$$

and we further restrict our attention to the zero-mean Gaussian case such that $E\{X(j)|X(1), \ldots, X(j-1)\}$ is a linear function, that is

$$E\{X(j)|X(1), \ldots, X(j-1)\} = \sum_{i=1}^{j-1} b_i X(i) \quad (7.58)$$

In this case, $V_X(i)$ is a linear function of $X(i), X(i-1), \ldots, X(1)$. Equation 7.57 can be rewritten as

$$X(i) = E\{X(i)|X(1), \ldots, X(i-1)\} + V_X(i) \quad (7.59)$$

Note that $V_X(i)$ is the unpredictable part of $X(i)$. Equation 7.59 thus serves as a decomposition of $X(i)$ into a predictable part, which is the conditional expectation, and an unpredictable part, $V_X(i)$.

Thus, by Equations 7.57 and 7.58, the innovation $V_X(i)$ is a linear function of $X(n)$. We argue later that $X(n)$ is also a linear function of $V_X(n)$. Indeed:

$$V_X(1) = X(1)$$
$$V_X(2) = X(2) + \gamma(2,1)X(1) \qquad (7.60)$$
$$V_X(3) = X(3) + \gamma(3,1)X(1) + \gamma(3,2)X(2) \qquad (7.61)$$
$$\vdots \qquad \vdots$$

where the $\gamma(i,j)$ are constants, which will be specified later.

Thus

$$X(1) = V_X(1) \qquad (7.62)$$
$$X(2) = V_X(2) + l(2,1)V_X(1) \qquad (7.63)$$
$$X(3) = V_X(3) + l(3,1)V_X(1) + l(3,2)V_X(2) + l(3,1)V_X(1) \qquad (7.64)$$
$$\vdots \qquad \vdots$$

and the $l(i,j)$ are constants which can be derived from set of $\gamma(i,j)$.

We now summarize the properties of $V_X(n)$ if $X(n)$ is a zero-mean Gaussian random sequence.

1. $E\{V_X(n) | V_X(1), \ldots, V_X(n-1)\} = 0 \qquad (7.65)$
2. $E\{V_X(n) V_X(i)\} = 0, \quad i \neq n \qquad (7.66)$
3. $\mathbf{V}_X(n) = \boldsymbol{\Gamma} \mathbf{X}(n)$ where $\boldsymbol{\Gamma}$ is a linear transformation or filter
4. $\mathbf{X}(n) = \mathbf{L} \mathbf{V}_X(n)$ where $\mathbf{L}$ is another linear transformation or filter

We can find $\boldsymbol{\Gamma}$ and $\mathbf{L}$ as follows. We first seek a linear transformation $\boldsymbol{\Gamma}$ such that

$$\boldsymbol{\Gamma} \mathbf{X} = \mathbf{V}_X \qquad (7.67)$$

where

$\mathbf{X}^T = [X(1), \ldots, X(n)]$
$\boldsymbol{\Gamma}$ is an $n \times n$ matrix
$\mathbf{V}_X^T = [V_X(1), \ldots, V_X(n)]$

where the covariance matrix of $\mathbf{V}_X$ is diagonal.

Note that in order to be a realizable (i.e., causal—see Section 4.1.1) filter, Γ must be a lower triangular matrix because $V_X(1)$ must rely only on $X(1)$, $V_X(2)$ must rely only on $X(1)$ and $X(2)$, and so on. That is

$$\begin{bmatrix} \gamma(1,1) & 0 & 0 & \cdots & 0 \\ \gamma(2,1) & \gamma(2,2) & 0 & \cdots & 0 \\ \gamma(3,1) & \gamma(3,2) & \gamma(3,3) & 0 & \cdots & 0 \\ \cdot & & & & & \cdot \\ & \cdot & \cdot & \cdot & & \cdot \\ \cdot & & & & & \cdot \\ & & & & & 0 \\ \gamma(n,1) & \gamma(n,2) & \gamma(n,3) & \cdots & \gamma(n,n) \end{bmatrix} \begin{bmatrix} X(1) \\ X(2) \\ X(3) \\ \cdot \\ \cdot \\ \cdot \\ X(n) \end{bmatrix} = \begin{bmatrix} V_X(1) \\ V_X(2) \\ \cdot \\ \cdot \\ \cdot \\ \cdot \\ V_X(n) \end{bmatrix} \qquad (7.68)$$

Note that Equations 7.67 and 7.68 are identical equations that generalize Equations 7.59–7.61.

We now recall that

$$E\{X(i)\} = 0, \quad i = 1, \ldots, n$$

and we will call $R_{XX}(i, j)$ the covariance between $X(1)$ and $X(2)$, that is

$$E\{X(i)X(j)\} = R_{XX}(i, j)$$

In order for the covariance matrix of $\mathbf{V}_X$ to be a diagonal matrix and to satisfy Equation 7.56 and Equation 7.57, Γ must be a lower triangular matrix with 1's on the diagonal. Thus, $\gamma(i, i) = 1$, and in particular

$$\gamma(1, 1) = 1 \qquad (7.69)$$

We can find $\gamma(2, 1)$ as follows

$$E\{V_X(1)V_X(2)\} = 0 = E\{X(1)[\gamma(2,1)X(1) + X(2)]\}$$
$$= \gamma(2,1)R_{XX}(1,1) + R_{XX}(1,2) = 0$$

$$\gamma(2, 1) = \frac{-R_{XX}(1, 2)}{R_{XX}(1, 1)} \qquad (7.70)$$

404 LINEAR MINIMUM MEAN SQUARED ERROR FILTERING

To solve for $\gamma(3, 1)$ and $\gamma(3, 2)$ we use the equations

$$E\{V_X(1)V_X(3)\} = 0$$
$$E\{V_X(2)V_X(3)\} = 0$$

In a similar fashion, the entire Γ matrix can be found.

EXAMPLE 7.9.

Find the transformation Γ when $E\{X(n)\} = 0$ and the covariance matrix of $X(1)$, $X(2)$, $X(3)$ is

$$\Sigma_X = \begin{bmatrix} 1 & \frac{1}{2} & \frac{1}{4} \\ \frac{1}{2} & 1 & \frac{1}{2} \\ \frac{1}{4} & \frac{1}{2} & 1 \end{bmatrix}$$

SOLUTION:

We know

$$\Gamma = \begin{bmatrix} 1 & 0 & 0 \\ \gamma(2, 1) & 1 & 0 \\ \gamma(3, 1) & \gamma(3, 2) & 1 \end{bmatrix}$$

We find $\gamma(2, 1)$ from Equation 7.70 using Σ_X.

$$\gamma(2, 1) = -\tfrac{1}{2}$$

We find $\gamma(3, 1)$ and $\gamma(3, 2)$ from

$$E\{V_X(1)V_X(3)\} = 0$$
$$E\{X(1)[\gamma(3, 1)X(1) + \gamma(3, 2)X(2) + X(3)]\} = 0$$
$$\gamma(3, 1) + \tfrac{1}{2}\gamma(3, 2) + \tfrac{1}{4} = 0 \tag{7.71}$$

and

$$E\{V_X(2)V_X(3)\} = 0$$
$$E\{[-\tfrac{1}{2}X(1) + X(2)][\gamma(3, 1)X(1) + \gamma(3, 2)X(2) + X(3)]\} = 0$$
$$-\tfrac{1}{2}\gamma(3, 1) - \tfrac{1}{4}\gamma(3, 2) - \tfrac{1}{8} + \tfrac{1}{2}\gamma(3, 1) + \gamma(3, 2) + \tfrac{1}{2} = 0$$

or

$$\gamma(3, 2) = -\tfrac{1}{2}$$

Using this result in Equation 7.71 produces

$$\gamma(3, 1) = 0$$

Thus

$$\Gamma = \begin{bmatrix} 1 & 0 & 0 \\ -\tfrac{1}{2} & 1 & 0 \\ 0 & -\tfrac{1}{2} & 1 \end{bmatrix}$$

We have demonstrated a procedure whereby a transformation (or filter) may be found that produces white noise (the V_X are independent), which is the innovation of X if X has zero-mean (what would occur if X had a nonzero mean?). If we were considering other than Gaussian processes, then this transformation would produce uncorrelated and orthogonal but not necessarily independent random variables.

It can be shown that the lower triangular matrix Γ has an inverse that will also be lower triangular. Call

$$\Gamma^{-1} = L \qquad (7.72)$$

Then

$$X = L V_X$$

Thus, Equation 7.72 defines L and we can recover X from V_X by passing it through the filter L.

EXAMPLE 7.10.

Find L and X in terms of V_X using the results of Example 7.9.

SOLUTION:

$$\mathbf{L} = \mathbf{\Gamma}^{-1} = \begin{bmatrix} 1 & 0 & 0 \\ \frac{1}{2} & 1 & 0 \\ \frac{1}{4} & \frac{1}{2} & 1 \end{bmatrix}$$

Thus

$$X(1) = V_X(1)$$
$$X(2) = \tfrac{1}{2}V_X(1) + V_X(2)$$
$$X(3) = \tfrac{1}{4}V_X(1) + \tfrac{1}{2}V_X(2) + V_X(3)$$

7.4 REVIEW

In the preceding sections we have shown how to find the "best" linear estimator of a random variable S or a random sequence $S(n)$ at time index n in terms of another correlated sequence $X(n)$. The "best" estimator is one that minimizes the MSE between $S(n)$ and the estimator $\hat{S}(n)$. However, the "best" estimator is actually "best" only under one of two limitations; either it is "best" among all linear estimators, or it is "best" among all estimators when the joint densities are normal or Gaussian.

The best linear estimators are based on projecting the quantity to be estimated, S, onto the space defined by the observations $X(1), X(2), \ldots, X(n)$ in such a fashion that the error $S - \hat{S}$ is orthogonal to the observations. The equations for the weights are given in Equation 7.16 and the constant, if needed, is given in Equation 7.13. It is also important to find how good the best estimator is, and Equation 7.19 describes the residual MSE.

Finally, the concept of innovations was introduced in order to solve this relatively simple problem in the same fashion that will be used to derive both the Kalman and the Wiener filters. The innovation $V_X(n)$ of a Gaussian sequence $X(1), X(2), \ldots, X(n)$ is the unpredictable part of $X(n)$. That is, the best estimator of $X(n)$ given $X(1), X(2), \ldots, X(n-1)$ is $E\{X(n)|X(1), \ldots, X(n-1)\}$ and the innovation of $X(n)$ is the difference between $X(n)$ and $E\{X(n)|X(1), \ldots, X(n-1)\}$. The innovations of a sequence are mutually orthogonal. Thus, from a vector space picture, $V_X(n)$ is the part of $X(n)$ that is orthogonal to the space defined by $X(1), \ldots, X(n-1)$.

The weights of the best linear filter can be described in terms of innovations and Equation 7.53 gives these weights. Equation 7.55 gives the error in terms of innovations.

Because of the orthogonality of the innovations, the weights of each innovation can be computed independently of the other weights. This can be done in a recursive form as will be demonstrated in the section on Kalman Filtering.

In the next two sections we discuss the form of these best estimators when predicting, smoothing, and digitally filtering random sequences. Particular emphasis is placed on recursive (Kalman) filters when the signal can be modeled as an autoregressive sequence.

7.5 DIGITAL WIENER FILTERS

In this section we consider the problem of estimating a random sequence, $S(n)$, while observing another random sequence, $X(m)$. This problem is usually addressed using the following terminology:

1. If we can observe a sample sequence $X(m)$, where $m < n$ and we want to estimate the value of $S(n)$, then this problem is called *prediction*. Further, if $X(j) = S(j)$, then the problem is called *pure prediction*.
2. If we can observe a sample sequence $X(m)$, where $m \geq n$, and we want to estimate the value of $S(n)$, the problem is called *filtering*. Only in the case where $n = m$ can the process be accomplished in real time or on-line, and in this case the filter is called a *causal* or realizable filter. When $m > n$, the data must be stored and in this case the filter is often called a *smoothing filter* or a noncausal (unrealizable) filter.

In this section, all processes are assumed to be wide-sense stationary and to have zero mean unless stated otherwise.

7.5.1 Digital Wiener Filters with Stored Data

We assume a simple model of the relationship between the signal S and the measurement X. This is called the additive noise model and is represented by

$$X(n) = S(n) + v(n) \tag{7.73}$$

where $S(n)$ is the zero-mean signal random sequence, $v(n)$ is the zero-mean white noise sequence, and $S(n)$ and $v(n)$ are uncorrelated. Thus, since they have zero means, they are also orthogonal.

We now assume that the data are stored and $X(n)$ is available for all n, that is, the filter is no longer finite. We seek the best, that is, minimum linear mean squared error estimator $\hat{S}$, where

$$\hat{S}(n) = \sum_{k=-\infty}^{\infty} h(k) X(n - k) \tag{7.74}$$

408 LINEAR MINIMUM MEAN SQUARED ERROR FILTERING

Now for practical cases, because memories are finite, the sum in Equation 7.74 will be finite, and the method of Section 7.2 can be used as illustrated in Example 7.10 except that in some cases the required matrix inversion of Equation 7.16 would be either impossible or too slow. In this section, we explore other possibilities.

The MSE given by

$$E\{[S(n) - \hat{S}(n)]^2\}$$

can be minimized by applying the orthogonality conditions:

$$E\left\{\left[S(n) - \sum_{k=-\infty}^{\infty} h(k)X(n-k)\right]X(n-i)\right\} = 0 \quad \text{for all } i \quad (7.75)$$

or

$$R_{XS}(i) = \sum_{k=-\infty}^{\infty} h(k)R_{XX}(i-k) \quad \text{for all } i \quad (7.76)$$

Because of the infinite summation, matrix inversion is impossible and thus will not result in a solution for $h(k)$. However, if $R_{XX}(m) = 0$, $|m| > M$, then Equation 7.76 reduces to a $2M + 1$ matrix, which can be inverted resulting in the weights $h(-M), h(-M+1), \ldots, h(0), \ldots, h(M)$.

In the case of an infinite sum, we take the Fourier transform of both sides of Equation 7.76 resulting in

$$S_{XS}(f) = H(f)S_{XX}(f), \quad |f| < \tfrac{1}{2} \quad (7.77)$$

or

$$H(f) = \frac{S_{XS}(f)}{S_{XX}(f)}, \quad |f| < \tfrac{1}{2} \quad (7.78)$$

The resulting MSE is

$$\begin{aligned}
P &= E\{[S(n) - \hat{S}(n)]^2\} \\
&= E\left\{\left[S(n) - \sum_{k=-\infty}^{\infty} h(k)X(n-k)\right]S(n)\right\} \\
&= R_{SS}(0) - \sum_{k=-\infty}^{\infty} h(k)R_{SX}(-k)
\end{aligned} \quad (7.79)$$

Defining

$$\overline{P}(m) \stackrel{\Delta}{=} R_{SS}(m) - \sum_{k=-\infty}^{\infty} h(k) R_{SX}(m-k) \qquad (7.80)$$

Then $\overline{P}(0) = P$. Taking the Fourier transforms of both sides of Equation 7.80 results in

$$\overline{P}_F(f) = S_{SS}(f) - H(f) S_{SX}(f), \qquad |f| < \tfrac{1}{2} \qquad (7.81)$$

where $\overline{P}_F(f)$ is the Fourier transform of $\overline{P}(m)$.
Thus

$$\overline{P}(m) = \int_{-1/2}^{+1/2} \overline{P}_F(f) \exp(+j2\pi f m) \, df$$

and

$$P = \overline{P}(0) = \int_{-1/2}^{+1/2} [S_{SS}(f) - H(f) S_{SX}(f)] \, df \qquad (7.82)$$

EXAMPLE 7.11.

If $X(n) = S(n)$, i.e., no noise, find the optimum unrealizable $H(f)$ and P.

SOLUTION:

From Equation 7.78 we obtain

$$H(f) = \frac{S_{SS}(f)}{S_{SS}(f)} = 1$$

and using Equation 7.82

$$P = \int_{-1/2}^{+1/2} [S_{SS}(f) - S_{SS}(f)] \, df = 0$$

410 LINEAR MINIMUM MEAN SQUARED ERROR FILTERING

Obviously, if $S(n)$ can be observed with no noise, then $\hat{S}(n) = S(n)$, that is, $H = 1$, $h(n) = \delta(n)$, and the residual error is zero.

EXAMPLE 7.12.

Assume that $X(n) = S(n) + v(n)$, where $v(n)$ is white with $\sigma_v^2 = 1$. $S(n)$ is a zero-mean Markov or first-order autoregressive sequence, that is

$$R_{SS}(n) = a^{|n|}\sigma_S^2$$

(a) Find R_{SX}, S_{SX}, R_{XX}, S_{XX}, and $H(f)$, the optimum unrealizable filter.
(b) If $a = 1/2$, $\sigma_S^2 = 1$, find $h(n)$, the optimum unrealizable or noncausal weights in the time domain.

SOLUTION:

(a)
$$R_{SX}(m) = E\{S(n)X(n + m)\}$$
$$= E\{S(n)[S(n + m) + v(n + m)]\}$$
$$= R_{SS}(m) = a^{|m|}\sigma_S^2$$

$$S_{SX}(f) = S_{SS}(f) = \sigma_S^2 \sum_{i=-\infty}^{\infty} a^{|i|} \exp(-j2\pi i f)$$

$$= \frac{\sigma_S^2(1 - a^2)}{1 - 2a \cos 2\pi f + a^2}, \qquad |f| < \tfrac{1}{2}$$

(See Equation 5.12.)

$$R_{XX}(n) = R_{SS}(n) + R_{vv}(n)$$
$$= a^{|n|}\sigma_S^2 + \delta(n)$$

$$S_{XX}(f) = \frac{\sigma_S^2(1 - a^2)}{1 - 2a \cos 2\pi f + a^2} + 1, \qquad |f| < \tfrac{1}{2}$$

$$= \frac{1 + a^2 + \sigma_S^2(1 - a^2) - 2a \cos 2\pi f}{1 - 2a \cos 2\pi f + a^2}, \qquad |f| < \tfrac{1}{2}$$

Finally using Equation 7.78

$$H(f) = \frac{\sigma_S^2(1 - a^2)}{1 + a^2 + \sigma_S^2(1 - a^2) - 2a \cos 2\pi f}, \qquad |f| < \tfrac{1}{2}$$

(b) If $\sigma_S^2 = 1$, $a = 1/2$

$$H(f) = \frac{\tfrac{3}{4}}{2 - \cos 2\pi f}, \qquad |f| < \tfrac{1}{2}$$

This can be seen to be the Fourier transform of

$$h(n) \approx .433 \, (.268)^{|n|}$$

by using the result that

$$h(n) = cb^{|n|}$$

has the Fourier transform

$$\frac{c(1-b^2)}{1+b^2-2b\cos 2\pi f} = \frac{\dfrac{c(1-b^2)}{2b}}{\dfrac{1+b^2}{2b}-\cos 2\pi f}$$

Note that $h(n) \neq 0$ when $n < 0$. This is unrealizable in real time or noncausal.

7.5.2 Real-time Digital Wiener Filters

We consider the same problem of Section 7.5.1 except that now we seek an estimator of the form

$$\hat{S}(n) = \sum_{k=0}^{\infty} h_k X(n-k) \qquad (7.83.\text{a})$$

The filter defined by Equation 7.83.a is a realizable filter. That is, only present and past values of X are used in estimating $S(n)$. In this case, the filter can be conceptually realized as shown in Figure 7.3. If, in fact, the correlation between X and S is zero after a delay of more than M, that is

$$R_{SX}(m) = 0, \qquad |m| > M$$

then the filter can be made up of M delays and $M+1$ amplifiers resulting in a *realizable* or *causal finite* response filter.

We seek a *finite* linear filter when $X(0), \ldots, X(n)$ can be observed

$$\hat{S}(n) = \sum_{k=0}^{n} h_k X(n-k)$$

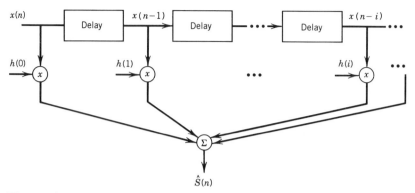

Figure 7.3 Realization of a digital filter.

and the solution is of the same form given in Equation 7.16, that is

$$\mathbf{h} = \mathbf{R}_{XX}^{-1}\mathbf{R}_{XS} \tag{7.83.b}$$

where

$$\mathbf{h}^T = [h_0, h_1, \ldots, h_n]$$

$$\mathbf{R}_{XX} = \begin{bmatrix} R_{XX}(0) & R_{XX}(1) & \cdots & R_{XX}(n) \\ R_{XX}(1) & R_{XX}(0) & \cdots & R_{XX}(n-1) \\ \vdots & \vdots & & \vdots \\ R_{XX}(n) & R_{XX}(n-1) & \cdots & R_{XX}(0) \end{bmatrix}$$

$$\mathbf{R}_{XS}^T = E\{S(n)[X(n), X(n-1), \ldots, X(0)]\}$$
$$= [R_{SS}(0), R_{SS}(1), \ldots, R_{SS}(n)]$$

In this case, Equation 7.16 is said to define a finite discrete Wiener filter. The MSE is given in Equation 7.24.

EXAMPLE 7.13.

Let $S(n)$ be an ensemble of constant functions of time with $E\{S(n)\} = 0$, and $E\{S^2(n)\} = \sigma_S^2$. $X(n) = S(n) + v(n)$ where $v(n)$ is a zero-mean white noise sequence with $E\{v^2(n)\} = \sigma_v^2$. Define $\gamma = \sigma_v^2/\sigma_S^2$ as the noise-to-signal ratio (1/SNR). Find $\mathbf{h}$ and $P(n)$ as a function of n, if $X(n-1)$ is the last observation, i.e. $h_0 = 0$.

SOLUTION:

$$R_{xs}(k) = R_{ss}(k) = \sigma_S^2$$
$$R_{xx}(k) = \sigma_S^2 + \delta(k)\sigma_v^2$$

where

$$\delta(k) = \begin{cases} 1, & k = 0 \\ 0 & k \neq 0 \end{cases}$$

From Equation 7.83.b

$$\sigma_S^2 \sum_{j=1}^{n} h_j + h_i \sigma_v^2 = \sigma_S^2$$

Summing this equation for $i = 1$ to n results in

$$n\sigma_S^2 \sum_{j=1}^{n} h_j + \sigma_v^2 \sum_{i=1}^{n} h_i = n\sigma_S^2$$

Thus

$$\sum_{j=1}^{n} h_j = \frac{n\sigma_S^2}{n\sigma_S^2 + \sigma_v^2}$$

and using this result in the first equation, we have

$$\frac{n\sigma_S^4}{n\sigma_S^2 + \sigma_v^2} + h_i \sigma_v^2 = \sigma_S^2$$

or

$$h_i = \frac{\sigma_S^2}{n\sigma_S^2 + \sigma_v^2} = \frac{1}{n + \gamma}, \quad i = 1, \ldots, n$$

Note that all h_i are equal, as should be expected because S is a constant and the noise is stationary. Also note that the weights h_i change (decrease) as n increases.

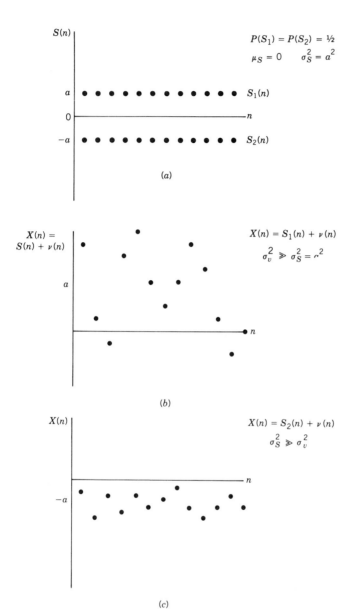

Figure 7.4 Different cases for Example 7.13. (a) The two possible values of $S(n)$. (b) $X(n)$ given $S(n) = S_1(n)$ and $\sigma_\nu^2 \gg a^2$. (c) $X(n)$ given $S(n) = S_2(n)$ and $\sigma_\nu^2 \ll a^2$.

Also note that this finite discrete Wiener filter is

$$\hat{S}(n) = \sum_{i=1}^{n} \frac{1}{n+\gamma} X(i) = \left[\frac{1}{n}\sum_{i=1}^{n} X_i\right] \frac{n}{n+\gamma} + 0 \frac{\gamma}{n+\gamma}$$

The last equation shows $\hat{S}(n)$ to be the weighted average of the mean of the observations and the *a priori* estimate of 0. If the noise to signal ratio γ is small compared with n, then $\hat{S}$ is the simple average of the observations. Figure 7.4 shows different cases for this example.

The MSE $P(n)$ is given by Equation 7.24 as

$$P(n) = \sigma_S^2 - \sum_{j=1}^{n} \sigma_S^2 \frac{1}{n+\gamma}$$

$$= \sigma_S^2 \left[1 - \frac{n}{n+\gamma}\right] = \frac{\sigma_v^2}{n+\gamma}$$

which decreases with n.

EXAMPLE 7.14.

Given the observations $X(0), \ldots, X(n)$, $n \geq 2$, with

$$R_{ss}(0) = \tfrac{3}{4}$$
$$R_{ss}(1) = \tfrac{1}{2}$$
$$R_{ss}(2) = \tfrac{1}{4}$$
$$R_{ss}(n) = 0, \quad n > 2$$
$$X(n) = S(n) + v(n), \quad \text{S and v are uncorrelated}$$
$$R_{vv}(n) = \tfrac{1}{4}, \quad n = 0$$
$$\phantom{R_{vv}(n)} = 0, \quad \text{elsewhere}$$

Find the optimum realizable filter for $S(n)$.

LINEAR MINIMUM MEAN SQUARED ERROR FILTERING

SOLUTION:

$$R_{XS}(n) = R_{SS}(n)$$
$$R_{XX}(0) = 1$$
$$R_{XX}(1) = \tfrac{1}{2}$$
$$R_{XX}(2) = \tfrac{1}{4}$$
$$R_{XX}(n) = 0, \quad n > 2$$

There are only two nonzero weights and these weights (h's) are the solution to (see Equation 7.83.b)

$$\begin{bmatrix} 1 & \tfrac{1}{2} & \tfrac{1}{4} \\ \tfrac{1}{2} & 1 & \tfrac{1}{2} \\ \tfrac{1}{4} & \tfrac{1}{2} & 1 \end{bmatrix} \begin{bmatrix} h_0 \\ h_1 \\ h_2 \end{bmatrix} = \begin{bmatrix} \tfrac{3}{4} \\ \tfrac{1}{2} \\ \tfrac{1}{4} \end{bmatrix}$$

or

$$\begin{bmatrix} h_0 \\ h_1 \\ h_2 \end{bmatrix} = \begin{bmatrix} \tfrac{4}{3} & -\tfrac{2}{3} & 0 \\ -\tfrac{2}{3} & \tfrac{5}{3} & -\tfrac{2}{3} \\ 0 & -\tfrac{2}{3} & \tfrac{4}{3} \end{bmatrix} \begin{bmatrix} \tfrac{3}{4} \\ \tfrac{1}{2} \\ \tfrac{1}{4} \end{bmatrix} = \begin{bmatrix} \tfrac{2}{3} \\ \tfrac{1}{6} \\ 0 \end{bmatrix}$$

This is represented by the filter shown in Figure 7.5, and is usually called a finite discrete Wiener filter.

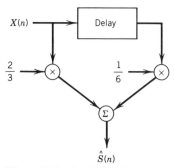

Figure 7.5 Optimum filter for Example 7.14.

We now return to the general case represented by Equation 7.83. Unfortunately this cannot be solved by inverting a matrix.

The orthogonality condition

$$E\{[S(n) - \sum_{k=0}^{\infty} h_k X(n - k)] X(n - i)\} = 0, \quad i = 1, 2, \ldots \quad (7.84)$$

results in

$$R_{XS}(i) = \sum_{k=0}^{\infty} h_k R_{XX}(i - k), \quad i = 1, 2, \ldots \quad (7.85)$$

Equation 7.85 is the discrete version of an equation called the Wiener–Hopf equation, and it cannot be solved by transform methods using power spectral densities because Equation 7.85 is defined only for $i > 0$. Instead, it will be solved by a method based on innovations that is similar to a solution for the continuous Wiener–Hopf equation due to Bode and Shannon [4]. This solution is outlined next. However, practical digital filters are not usually based on this solution. Recursive or Kalman filters are more common primarily because they require less computational capability.

Outline of the Solution to the Optimum Infinite Realizable Filtering Problem. The optimum unrealizable filter described by

$$\hat{S}(n) = \sum_{k=-\infty}^{\infty} h(k) X(n - k)$$

can be written as the sum of two parts

$$\hat{S}(n) = \sum_{k=-\infty}^{-1} h(k) X(n - k) + \sum_{k=0}^{\infty} h(k) X(n - k) \quad (7.86)$$

where the first sum requires future values of the observations, $X(i)$, which are not available in real time. The second sum is the realizable part.

Temporarily assume that $X(n)$ is the zero-mean white noise, then the conditional expected value of $X(n - k)$ for negative k (i.e., values of X that occur after time n) would be zero, that is

$$E\{X(n + 1)|X(n), X(n - 1), \ldots\} = 0$$

418 LINEAR MINIMUM MEAN SQUARED ERROR FILTERING

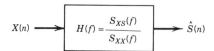

Figure 7.6a Optimum unrealizable filter.

because by the definition of zero-mean white noise, each value is independent of all previous values and has an expected value of zero. Thus, because the conditional expected value of $X(n - k)$, $k \leq -1$, given $\ldots X(-2), X(-1), X(0)$ is zero, the best guess (conditional expected value) of the first sum of Equation 7.86 is zero. Thus, those parts of the filter involving delays should be kept while those involving advances should be eliminated (i.e., set equal to zero). The problem would be solved if the input were white noise.

When the input is not white noise, X, the input, is first put into a digital filter (an innovations filter), which effectively converts it to white noise, called the innovations of X. Then the realizable, or real-time, part of the remaining optimum filter is kept. The steps are shown in Figure 7.6.

Note that in Figure 7.6b

$$S_{V_X V_X}(f) = S_{XX}(f)|H_1(f)|^2$$

and if $H_1(f)$ can be chosen such that

1. It is realizable, and
2. $|H_1(f)|^2 = [S_{XX}(f)]^{-1}$

then

$$S_{V_X V_X}(f) = S_{XX}(f) \frac{1}{S_{XX}(f)} = 1$$

or V_X is white noise. V_X is called the innovations of X, which were described previously.

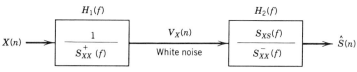

Figure 7.6b Optimum unrealizable filter.

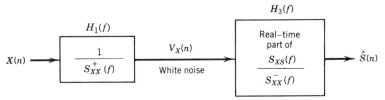

Figure 7.6c Optimum realizable filter.

Here we assume that such a filter can be found and we call it $S_{XX}^+(f)$. Finding it in this discrete case is not discussed in this book. However, a similar continuous filter is described in Section 7.7.

Also note in Figures 7.6a and 7.6b that

$$H_1(f) \cdot H_2(f) = H(f); \text{ or } S_{XX}^{(f)} = S_{XX}^+(f) \, S_{XX}^-(f)$$

Thus, the filters of Figure 7.6a and Figure 7.6b are the same and each is that given by Equation 7.78.

In the final step, shown in Figure 7.6c, the optimum realizable (causal or real-time) filter is H_3, which is the real-time part of H_2 because the input to H_2 is white noise, that is the innovations of X, and Equation 7.86 was used to justify taking only the real-time part of $\hat{S}(n)$ in the case of white-noise input. This H_3 is multiplied by H_1 assuming H_1 meets the conditions specified previously.

As stated, we do not emphasize the solution described for digital filters. A solution is outlined in Problems 7.17–7.20. We emphasize recursive digital filters, which are described in the next section.

7.6 KALMAN FILTERS

Kalman filters have been effective in a number of practical problems. They allow nonstationarity, and their recursive form is easy to implement. We describe the general model, discuss the advantages of recursive filters, and then find the optimum filter in the scalar case. Finally, we find the optimum Kalman filter in the more general vector case.

We adopt the following notation for Kalman filtering: $S(n)$, the signal, is a zero-mean Gaussian sequence described by an autoregressive or Markov model of the form

$$S(n) = a(n)S(n-1) + W(n) \tag{7.87}$$

where $a(n)$ is a series of known constants and $W(n)$ is white noise. Furthermore, if $S(n)$ is to be a stationary random sequence, then $a(n)$ will be constant with

n. However, $a(n)$ can vary with n resulting in a nonstationary sequence. (Wiener filters are not applicable to nonstationary sequences.)

The model given in Equation 7.87 can be extended to a higher order autoregressive moving-average model of the form

$$S(n) = \sum_{i=1}^{m} [a(n)_i S(n-i) + W(n+1-i)]$$

In the later case, the higher order difference equation will be converted to a state model of the form

$$\mathbf{S}(n) = \mathbf{A}(n)\mathbf{S}(n-1) + \mathbf{W}(n) \qquad (7.88)$$

where the input or driving function, $\mathbf{W}(n)$, is white noise.

Equations such as Equations 7.87 and 7.88 that describe $S(n)$ will be called *signal models* or *state models*. Further, the observation $X(n)$ is assumed to be of the form

$$X(n) = S(n) + v(n) \qquad (7.89)$$

where $v(n)$ is white noise that is independent of both $W(n)$ and $S(n)$, and $v(n)$ is called *observation noise*. Equation 7.89 will be called the *observation model*.

We know that the optimal (minimum MSE) estimator of $S(n+1)$ using observations $X(1), \ldots, X(n)$ is

$$\hat{S}(n+1) = E\{S(n+1)|X(1), \ldots, X(n)\}$$

If we assume that S and W are Gaussian, then this estimator will be linear. If we delete the Gaussian assumption and seek the minimum MSE *linear* predictor, the result will be unchanged. We want to find this optimum filter in a recursive form so that it can be easily implemented on a computer using a minimum amount of storage.

7.6.1 Recursive Estimators

Suppose $S(n)$ is an unknown constant. Then we could estimate $S(n)$ having observed $X(1), \ldots, X(n-1)$ by

$$\hat{S}(n) = \frac{X(1) + X(2) + \cdots + X(n-1)}{n-1}$$

if $X(n) = S(n) + v(n)$ and $v(n)$ is stationary zero-mean white noise and we agree that a linear estimator is appropriate, for example if X is Gaussian. That is, $\hat{S}(n)$ is the average of the observations. Then, the estimate of $S(n + 1)$ would be

$$\hat{S}(n + 1) = \frac{X(1) + \cdots + X(n)}{n}$$

The problem with this form of estimator is that the memory required to store the observations grows with n, leading to a very large (eventually too large a) memory requirement. However, if we simply express $\hat{S}(n + 1)$ in recursive form, then

$$\hat{S}(n + 1) = \frac{X(1) + \cdots + X(n - 1)}{n} + \frac{X(n)}{n}$$
$$= \frac{n - 1}{n} \hat{S}(n) + \frac{1}{n} X(n)$$

This implementation of the filter requires that only the previous estimate and the new observation be stored. Further, note that the estimator is a convex linear combination of the new observation and the old estimate. We will find this is also the form of the Kalman filter.

7.6.2 Scalar Kalman Filter

We reiterate the assumptions

1. $S(n) = a(n)S(n - 1) + W(n)$ \hfill (7.87)
2. $W(n)$ is a zero-mean white Gaussian noise sequence, called the system noise driving function. Its variance is assumed to be known and will be denoted $\sigma_W^2(n)$.
3. $S(1)$ is a zero-mean Gaussian random variable independent of the sequence $W(n)$. Its variance is also assumed to be known.
4. $a(n)$ is a known series of constants.

Thus, $S(n)$ is a Gauss–Markov sequence, or a first-order autoregressive sequence. It also has the form of a first-order state equation.

5. We observe

$$X(n) = S(n) + v(n) \hspace{1cm} (7.89)$$

where $v(n)$ is also zero-mean white Gaussian noise with variance, $\sigma_v^2(n)$, which is independent of both the sequence $W(n)$ and $S(1)$.

422 LINEAR MINIMUM MEAN SQUARED ERROR FILTERING

Our goal is to find the minimum MSE linear predictor of $S(n+1)$ given observations $X(1), \ldots, X(n)$. Because of the Gaussian assumption, we know that

$$\hat{S}(n+1) = E\{S(n+1) | X(1), \ldots, X(n)\} \tag{7.90}$$

will be linear and will be the optimum estimator. We want to find this conditional expected value in a recursive form.

We know that the innovations process V_X and the observations X are linearly equivalent. Thus,

$$\hat{S}(n+1) = E\{S(n+1) | V_X(1), \ldots, V_X(n)\} \tag{7.91}$$

will be equivalent to Equation 7.90 and will be a linear combination of the innovations, that is

$$\hat{S}(n+1) = \sum_{j=1}^{n} c(n, j) V_X(j) \tag{7.92}$$

Note that this is not yet in recursive form.

We now seek the values of $c(n, j)$. Note that

$$E\{V_X(j) V_X(k)\} = 0, \quad j \neq k$$
$$= \sigma_V^2(j), \quad j = k \tag{7.93}$$

We now seek the values of $c(n, j)$ in this scalar case. A similar treatment of the vector case is presented in the next section. In both cases we will:

1. Show that the optimum Kalman filter is a linear combination of (a) an updating (projection) of the previous best estimator; and (b) an updating (projection) of the innovation due to the most recent observation.
2. Find a recursive expression for the error.
3. Find a recursive expression for the weights to be used in the linear combination.

Linear Combination of Previous Estimator and Innovation. From Equations 7.92 and 7.93 it follows that for $j = 1, \ldots, n$

$$E\{\hat{S}(n+1) V_X(j)\} = c(n, j) \sigma_V^2(j) \tag{7.94}$$

We now show that for all $j = 1, \ldots, n$

$$E\{S(n + 1)V_X(j)\} = E\{\hat{S}(n + 1)V_X(j)\}$$

This follows because the error, $S(n + 1) - \hat{S}(n + 1)$, is orthogonal to the observations $X(j)$, $j \leq n$, due to the orthogonality condition. Because the innovation $V_X(j)$ is a linear combination of $X(i)$, $i \leq j$, the error is orthogonal to $V_X(j)$, $j \leq n$, or

$$E\{[S(n + 1) - \hat{S}(n + 1)]V_X(j)\} = 0$$

This produces the stated result and when this result is used in Equation 7.94 we have

$$c(n, j) = \frac{E\{S(n + 1)V_X(j)\}}{\sigma_V^2(j)} \tag{7.95}$$

Equations 7.95 and 7.92 result in

$$\hat{S}(n + 1) = \sum_{j=1}^{n} \frac{E\{S(n + 1)V_X(j)\}}{\sigma_V^2(j)} V_X(j) \tag{7.96}$$

Now we use the state model, Equation 7.87, in order to produce a recursive form

$$E\{S(n + 1)V_X(j)\} = E\{[a(n + 1)S(n) + W(n + 1)]V_X(j)\}$$
$$= a(n + 1)E\{S(n)V_X(j)\} + E\{W(n + 1)V_X(j)\} \tag{7.97}$$

But the last term is zero. To see this, note that $W(n + 1)$ is independent of $S(k)$ for $k \leq n$ and $E[W(n + 1)] = 0$. That is

$$E\{W(n + 1)S(k)\} = 0, \quad k \leq n$$

Now the observation model, Equation 7.89, produces

$$E\{W(n + 1)[X(k) - v(k)]\} = 0, \quad k \leq n$$

424 LINEAR MINIMUM MEAN SQUARED ERROR FILTERING

But $v(k)$ is independent of $W(n + 1)$ for all k and all n, thus

$$E\{W(n + 1)X(k)\} = 0, \quad k \le n$$

Now because $V_X(j)$ is a linear combination of $X(k)$, $k \le j$, then

$$E\{W(n + 1)V_X(j)\} = 0, \quad j \le n$$

Thus returning to Equation 7.97

$$E\{S(n + 1)V_X(j)\} = a(n + 1)E\{S(n)V_X(j)\}, \quad j \le n \quad (7.98)$$

Using Equation 7.98 in Equation 7.95 produces

$$c(n, j) = \frac{a(n + 1)E\{S(n)V_X(j)\}}{\sigma_V^2(j)}$$

and this result used in Equation 7.92 results in

$$\hat{S}(n + 1) = a(n + 1) \sum_{j=1}^{n} \frac{E\{S(n)V_X(j)\}}{\sigma_V^2(j)} V_X(j) \quad (7.99)$$

Equation 7.96 is valid for all n and in particular, let $n = n + 1$; then

$$\hat{S}(n) = \sum_{j=1}^{n-1} \frac{E\{S(n)V_X(j)\}}{\sigma_V^2(j)} V_X(j)$$

Using this result in Equation 7.99 produces the recursive form

$$\hat{S}(n + 1) = a(n + 1)\left[\hat{S}(n) + \frac{E\{S(n)V_X(n)\}}{\sigma_V^2(n)} V_X(n)\right] \quad (7.100)$$

$$= \hat{S}_1(n + 1) + b(n)V_X(n) \quad (7.101)$$

where

$$\hat{S}_1(n + 1) \stackrel{\Delta}{=} a(n + 1)\hat{S}(n) \quad (7.101.a)$$

$$b(n) = a(n+1)\frac{E\{S(n)V_X(n)\}}{\sigma_V^2(n)} \quad \text{is yet to be determined} \quad (7.101.\text{b})$$

$$V_X(n) = X(n) - \hat{S}(n) \quad (7.101.\text{c})$$

(See Problem 7.22.)

We now have accomplished the first objective because Equation 7.101 shows that the best estimator is a combination of $\hat{S}_1(n+1)$, which is $[a(n+1)$ multiplied by $\hat{S}(n)]$, that is, an updating of $\hat{S}(n)$, and $b(n)$ times the innovation, $V_X(n)$, of $X(n)$. However, some elaboration will be instructive.

It will be helpful in order to prepare for the vector case to define

$$\hat{S}_2(n+1) \triangleq b(n)V_X(n) \quad (7.102)$$

Then we can write Equation 7.101 as

$$\hat{S}(n+1) = \hat{S}_1(n+1) + \hat{S}_2(n+1)$$

Finally we define

$$k(n) = \frac{b(n)}{a(n+1)} \quad (7.103)$$

then Equation 7.101 becomes, using Equation 7.101.a

$$\hat{S}(n+1) = a(n+1)\hat{S}(n) + a(n+1)k(n)V_X(n) \quad (7.104)$$

This important equation can be rewritten as (see Problem 7.24)

$$\hat{S}(n+1) = a(n+1)\{\hat{S}(n) + k(n)[X(n) - \hat{S}(n)]\} \quad (7.105)$$

This form shows that the initial estimate $\hat{S}(n)$ at stage n is improved by the innovation of the observation $X(n)$ at stage n. The term within the brackets is the best (after observation) estimator at stage n. Then $a(n+1)$ is used to project the best estimator at stage n to the estimator at stage $n+1$. This estimator would be used if there were no observation at stage $n+1$ and can be called the prediction form of the Kalman filter.

This equation can also be written as

$$\hat{S}(n+1) = a(n+1)\{[1 - k(n)]\hat{S}(n) + k(n)X(n)\} \quad (7.106)$$

426 LINEAR MINIMUM MEAN SQUARED ERROR FILTERING

showing that the best estimator at stage n, that is the term in brackets, is a convex linear combination of the estimator $\hat{S}(n)$ and the observation $X(n)$.

At this stage we have shown that the optimum filter is a linear combination of the previous best estimator $\hat{S}(n)$ and the innovation $V_X(n) = X(n) - \hat{S}(n)$ of $X(n)$. However, we do not yet know $k(n)$. We will find $k(n)$ by minimizing the MSE. First we find a recursive expression for the error.

Recursive Expression for Error. Defining the error $\tilde{S}(n)$ as

$$\tilde{S}(n) \stackrel{\Delta}{=} S(n) - \hat{S}(n) \quad (7.107)$$

$$P(n) \stackrel{\Delta}{=} E\{\tilde{S}^2(n)] \quad (7.108)$$

Now from the definition of Equation 7.101.c and using the observation model, Equation 7.89

$$V_X(n) = S(n) + v(n) - \hat{S}(n)$$

Using the definition from Equation 7.107 results in

$$V_X(n) = \tilde{S}(n) + v(n) \quad (7.109)$$

Thus

$$E\{S(n)V_X(n)\} = E\{S(n)\tilde{S}(n)\} + E\{S(n)v(n)\}$$
$$= E\{S(n)\tilde{S}(n)\}$$
$$= E\{[\tilde{S}(n) + \hat{S}(n)]\tilde{S}(n)\}$$

The term $E\{\hat{S}(n)\tilde{S}(n)\}$ is zero by orthogonality; thus,

$$E\{S(n)V_X(n)\} = P(n) \quad (7.110)$$

Now

$$P(n+1) = E\{\tilde{S}^2(n+1)\} = E\{[\hat{S}(n+1) - S(n+1)]^2\} \quad (7.111)$$

We use Equation 7.104 for $\hat{S}(n+1)$ and the state model, Equation 7.87, for

$S(n + 1)$ resulting in

$$P(n + 1) = E\{[a(n + 1)\hat{S}(n) + a(n + 1)k(n)V_X(n) \\ - a(n + 1)S(n) - W(n + 1)]^2\} \\ = E\{[a(n + 1)[\hat{S}(n) - S(n)] \\ + a(n + 1)k(n)V_X(n) - W(n + 1)]^2\} \\ = a^2(n + 1)P(n) + a^2(n + 1)k^2(n)E\{V_X^2(n)\} \\ + \sigma_W^2(n + 1) + 2a^2(n + 1)k(n)E\{\hat{S}(n)V_X(n)\} \\ - 2a^2(n + 1)k(n)E\{S(n)V_X(n)\} \\ - 2a(n + 1)E\{W(n + 1)[\hat{S}(n) - S(n)]\} \\ - 2a(n + 1)k(n)E\{V_X(n)W(n + 1)\} \quad (7.112)$$

Now, the fourth term of Equation 7.112 is zero because of the definition of innovations and the fact that $\hat{S}(n)$ is a linear combination of $V_X(1), \ldots, V_X(n - 1)$. Similarly, the last two terms are zero due to the fact that $W(n + 1)$ is orthogonal to $S(n)$, $\hat{S}(n)$, and $V_X(n)$. Using Equation 7.110 in Equation 7.112 results in

$$P(n + 1) = a^2(n + 1)P(n) + a^2(n + 1)k^2(n)E\{V_X^2(n)\} + \sigma_W^2(n + 1) \\ - 2a^2(n + 1)k(n)P(n)$$

Now

$$E\{V_X^2(n)\} = E\{[S(n) + v(n) - \hat{S}(n)]^2\} \\ = P(n) + \sigma_v^2(n)$$

Thus

$$P(n + 1) = a^2(n + 1)k^2(n)[P(n) + \sigma_v^2(n)] - 2a^2(n + 1)k(n)P(n) \\ + a^2(n + 1)P(n) + \sigma_W^2(n + 1) \\ = a^2(n + 1)\{[1 - k(n)]^2 P(n) + k^2(n)\sigma_v^2(n)\} \\ + \sigma_W^2(n + 1) \quad (7.113)$$

We have now accomplished objective two; Equation 7.113 is a recursive expression for the mean-square error. $P(n + 1)$ is the error at stage $n + 1$ using all observations through stage n.

428 LINEAR MINIMUM MEAN SQUARED ERROR FILTERING

Minimum MSE Weight. Finally, we could differentiate $P(n + 1)$ with respect to $k(n)$ and set the result equal to zero in order to find a minimum. (See Problem 7.26.) However, in order to prepare for the vector case, we take a different approach in order to find the value of $k(n)$ that minimizes $P(n + 1)$.

Equation 7.113 can be rewritten as

$$P(n + 1) = a^2(n + 1)[k^2(n)D^2 - 2P(n)k(n) + C] \qquad (7.114)$$

where

$$D = \sqrt{P(n) + \sigma_v^2(n)} \qquad (7.115)$$

$$C = P(n) + \frac{\sigma_w^2(n + 1)}{a^2(n + 1)} \qquad (7.116)$$

Completing the square in $k(n)$

$$P(n + 1) = a^2(n + 1)[k(n)D - B]^2 + C_1 \qquad (7.117)$$

where C_1 is some term not involving $k(n)$. Comparing Equations 7.114 and 7.117 we find that

$$k(n)BD = P(n)k(n)$$

or

$$B = \frac{P(n)}{D} \qquad (7.118)$$

In order to minimize $P(n + 1)$ by choice of $k(n)$, the best one can do is to make the term in the brackets of Equation 7.117 equal to zero. Thus choose

$$k(n) = \frac{B}{D} \qquad (7.119)$$

Using Equations 7.118 and 7.115 in Equation 7.119 we obtain

$$k(n) = \frac{P(n)}{P(n) + \sigma_v^2(n)} \qquad (7.120)$$

This third and last of the defining equations of the Kalman filter can be viewed, using Equation 7.106, as stating that the optimal estimator weights $X(n)$ directly proportional to the variance of $\hat{S}(n)$, while $\hat{S}(n)$ has a weight

$$1 - k(n) = \frac{\sigma_v^2(n)}{P(n) + \sigma_v^2(n)}$$

which is directly proportional to the conditional variance of $X(n)$.

If the expression for $k(n)$ is now used in Equation 7.113, we find

$$P(n+1) = a^2(n+1) \frac{[\sigma_v^2(n) P(n)]}{P(n) + \sigma_v^2(n)} + \sigma_W^2(n+1) \tag{7.121}$$

This can also be written

$$P(n+1) = a^2(n+1) \left[\frac{\sigma_v^4(n)}{[P(n) + \sigma_v^2(n)]^2} P(n) + \frac{P^2(n)}{[P(n) + \sigma_v^2(n)]^2} \sigma_v^2(n) \right] + \sigma_W^2(n+1) \tag{7.122}$$

Initialize
$n = 1$
$P(1) = \sigma^2$ (Assumed value, usually larger than σ_v^2 and σ_W^2)
$\hat{S}(1) = S$ (Assumed value, usually zero)

1. Start Loop
   Get data: $\sigma_v^2(n)$, $\sigma_W^2(n)$, $a(n+1)$; $X(n)$
   $k(n) = P(n)/[P(n) + \sigma_v^2(n)]$
   $\hat{S}(n+1) = a(n+1) \{\hat{S}(n) + k(n) [X(n) - \hat{S}(n)]\}$
   $P(n+1) = a^2(n+1)[1 - k(n)]P(n) + \sigma_W^2(n+1)$
   $n = n + 1$
   Go to 1.

7.7(a) Algorithm

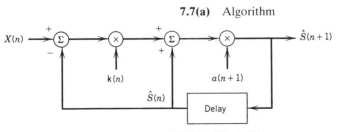

7.7(b) Block Diagram

Figure 7.7 Kalman filtering (scalar).

430 LINEAR MINIMUM MEAN SQUARED ERROR FILTERING

The term in the large brackets of Equation 7.122 shows that the revision of $\hat{S}(n)$ using $X(n)$ produces a linear combination of their variances, which is less than either variance. This reduced variance is then projected to the next stage as would be expected from the state model, Equation 7.87, where the variance is multiplied by the square of $a(n + 1)$ and the variance of $W(n + 1)$ is added, as would be expected. Finally, the term in brackets of Equation 7.121 can be expressed in terms of $k(n)$ to find

$$P(n + 1) = a^2(n + 1)[1 - k(n)]P(n) + \sigma_W^2(n + 1) \qquad (7.123)$$

Summary of Scalar Kalman Filtering. We now summarize in Figure 7.7a the Kalman filtering algorithm by using Equation 7.105, Equation 7.120, and Equation 7.123. Figure 7.7b shows it as a block diagram.

In the first step of the algorithm, $k(n)$ is calculated using Equation 7.120. In the next step, $\hat{S}(n)$ is revised by including the new information, that is, the innovation due to the measurement $X(n)$, and the revision is projected to the next stage using $a(n + 1)$. The next step multiplies $P(n)$ by $[1 - k(n)]$, projects it to the next stage using $a^2(n + 1)$, and adds the variance contribution of the projection noise.

Note that in order to start the algorithm when $n = 1$, corresponding to the first observation, $X(1)$; $P(1)$ and $\hat{S}(1)$ are needed. These are usually assumed in practice. Usual assumptions are $\hat{S}(1) = 0$ and $P(1) \geq \sigma_W^2(1)$, $P(1) \geq \sigma_v^2(1)$.

EXAMPLE 7.15.

$$S(n) = .6S(n - 1) + W(n)$$
$$X(n) = S(n) + v(n)$$

$W(n)$ and $v(n)$ are independent stationary white-noise sequences with

$$\sigma_W^2(n) = \frac{1}{4}, \qquad \sigma_v^2(n) = \frac{1}{2}$$

The first observation is $X(1)$ and we start with the assumed values, $\hat{S}(1) = 0$, $P(1) = 1$, and find the scalar Kalman filter.

SOLUTION: Following the steps in Figure 7.7

$$k(1) = \frac{P(1)}{P(1) + \sigma_v^2} = \frac{1}{1 + \dfrac{1}{2}} = \frac{2}{3}$$

$$\hat{S}(2) = (.6)\left\{0 + \frac{2}{3}[X(1) - 0]\right\} = .4X(1)$$

$$P(2) = (.6)^2\left[1 - \frac{2}{3}\right]1 + \frac{1}{4} = .37$$

We now perform one more cycle

$$k(2) = \frac{.37}{.37 + \frac{1}{2}} = \frac{37}{87} \approx .425$$

$$\hat{S}(3) = .6\left\{\hat{S}(2) + \left(\frac{37}{87}\right)[X(2) - \hat{S}(2)]\right\}$$
$$\approx .6[.4X(1) + .425X(2) - .17X(1)]$$
$$\approx .6[.23X(1) + .425X(2)]$$
$$\approx .138X(1) + .255X(2)$$

$$P(3) \approx (.36)\frac{50}{87}(.37) + \frac{1}{4} \approx .326$$

Problem 7.28 extends this example.

If $a(n)$ does not vary with n, and $W(n)$ and $v(n)$ are both stationary, that is, $\sigma_W^2(n)$ and $\sigma_v^2(n)$ are both constants, then both $k(n)$ and $P(n)$ will approach limits as n approaches infinity. These limits can be found by assuming that $P(n + 1) = P(n) = P$ in Equation 7.121 and using Equation 7.120.

$$P = a^2(1 - k)P + \sigma_W^2$$

$$k = \frac{P}{P + \sigma_v^2}$$

Thus

$$P = a^2\left(\frac{\sigma_v^2}{P + \sigma_v^2}\right)P + \sigma_W^2$$

Solving the resulting quadratic equation for P and taking the positive value results in

$$\lim_{n\to\infty} P(n) = P = \frac{\sigma_W^2 + \sigma_v^2(a^2 - 1) + \sqrt{[\sigma_W^2 + \sigma_v^2(a^2 - 1)]^2 + 4\sigma_v^2\sigma_W^2}}{2}$$

432 LINEAR MINIMUM MEAN SQUARED ERROR FILTERING

Using this value in Equation 7.120 produces the *steady-state Kalman gain*, k.

EXAMPLE 7.16.

Using the data from Example 7.15, find the limits P and k.

SOLUTION:

$$P \approx \frac{.25 + .5(.36 - 1) + \sqrt{[.25 + .5(.36 - 1)]^2 + .5}}{2} \approx .320$$

$$k \approx \frac{.32}{.32 + .5} \approx .390$$

7.6.3 Vector Kalman Filter

We now assume a state model of the form

$$\mathbf{S}(n) = \mathbf{A}(n)\mathbf{S}(n - 1) + \mathbf{W}(n) \tag{7.124}$$

and an observation model of the form

$$\mathbf{X}(n) = \mathbf{H}(n)\mathbf{S}(n) + \mathbf{\nu}(n) \tag{7.125}$$

where

$\mathbf{S}(n)$ = ($m \times 1$) signal or state vector; a zero-mean Gaussian vector random sequence.

$\mathbf{A}(n)$ = ($m \times m$) state matrix of constants, which is a description of the mth order difference equation model of the signal.

$\mathbf{W}(n)$ = ($m \times 1$) vector sequence of zero-mean Gaussian white noise uncorrelated with both $\mathbf{S}(1)$ and the sequence $\mathbf{\nu}(n)$.

$\mathbf{X}(n)$ = ($j \times 1$) vector measurement or observation; a Gaussian random sequence.

$\mathbf{H}(n)$ = ($j \times m$) matrix of constants describing the relationship between the signal vector and the observation vector.

$\mathbf{\nu}(n)$ = ($j \times 1$) vector of measurement error; a zero-mean Gaussian white-noise sequence uncorrelated with both $\mathbf{S}(1)$ and the sequence $\mathbf{W}(n)$.

$\mathbf{S}(1)$ = ($m \times 1$) initial state or signal vector; a zero-mean Gaussian random variable with covariance matrix $\mathbf{P}(1)$.

KALMAN FILTERS

We assume that the following covariance matrices are known.

$$E[\mathbf{W}(k)\mathbf{W}^T(i)] = \mathbf{Q}(k), \quad k = i$$
$$= \mathbf{0} \quad k \neq i \quad (7.126)$$

$$E[\mathbf{v}(k)\mathbf{v}^T(i)] = \mathbf{R}(k), \quad k = i$$
$$= \mathbf{0} \quad k \neq i \quad (7.127)$$

We now proceed as in the scalar case and the interested reader can compare the steps in this vector case with those of the earlier scalar case.

The optimal vector estimator will be

$$\hat{\mathbf{S}}(n+1) = E\{\mathbf{S}(n+1)|\mathbf{X}(1), \ldots, \mathbf{X}(n)\} \quad (7.128)$$

Because of the Gaussian assumptions it will be a linear combination of $\mathbf{X}(1)$, $\ldots, \mathbf{X}(n)$. Defining the vector innovations

$$\mathbf{V}_X(n) = \mathbf{X}(n) - E\{\mathbf{X}(n)|\mathbf{X}(1), \ldots, \mathbf{X}(n-1)\} \quad (7.129.\text{a})$$
$$= \mathbf{X}(n) - E\{\mathbf{X}(n)|\mathbf{V}_X(1), \ldots, \mathbf{V}_X(n-1)\} \quad (7.129.\text{b})$$

we have

$$\hat{\mathbf{S}}(n+1) = \sum_{k=1}^{n} \mathbf{C}(n, k)\mathbf{V}_X(k) \quad (7.130)$$

where $\mathbf{C}(n, k)$ is an $(m \times j)$ matrix of constants to be determined.

We first assume $\hat{\mathbf{S}}(n)$ is known where $\hat{\mathbf{S}}(n)$ is given by Equation 7.128 or an equivalent form with n set equal to $n - 1$, that is,

$$\hat{\mathbf{S}}(n) = E\{\mathbf{S}(n)|\mathbf{X}(1), \ldots, \mathbf{X}(n-1)\}$$
$$= E\{\mathbf{S}(n)|\mathbf{V}_X(1), \ldots, \mathbf{V}_X(n-1)\} \quad (7.131)$$

First we find

$$\hat{\mathbf{S}}_1(n+1) = E\{\mathbf{S}(n+1)|\mathbf{V}_X(1), \ldots, \mathbf{V}_X(n-1)\} \quad (7.132)$$

where the subscript 1 indicates the optimal estimator that does not include $\mathbf{V}_X(n)$. Using the state Equation 7.124 in Equation 7.132

$$\hat{\mathbf{S}}_1(n+1) = E\{\mathbf{A}(n+1)\mathbf{S}(n) + \mathbf{W}(n+1)|\mathbf{V}_X(1), \ldots \mathbf{V}_X(n-1)\}$$
$$\hat{\mathbf{S}}_1(n+1) = \mathbf{A}(n+1)\hat{\mathbf{S}}(n) \quad (7.133)$$

434 LINEAR MINIMUM MEAN SQUARED ERROR FILTERING

where the last term is zero because $\mathbf{W}(n + 1)$ is independent of both $\boldsymbol{\nu}(n)$ and $\mathbf{S}(1), \ldots, \mathbf{S}(n)$ and hence $\mathbf{W}(n)$ is independent of $\mathbf{X}(1), \ldots, \mathbf{X}(n)$, and finally because the innovations $\mathbf{V}_X$ are a linear combination of $\mathbf{X}$

$$E\{\mathbf{W}(n + 1)|\mathbf{V}_X(1), \ldots, \mathbf{V}_X(n)\} = E\{\mathbf{W}(n + 1)\} = \mathbf{0}$$

We now find

$$\hat{\mathbf{S}}_2(n + 1) = E\{\mathbf{S}(n + 1)|\mathbf{V}_X(n)\} \qquad (7.134)$$

where the subscript 2 indicates an estimator based only on $\mathbf{V}_X(n)$.
The observation model, Equation 7.125, results in

$$\begin{aligned}
E\{\mathbf{X}(n)|\mathbf{V}_X(1), \ldots, \mathbf{V}_X(n - 1)\} \\
= E\{\mathbf{H}(n)\mathbf{S}(n) + \boldsymbol{\nu}(n)|\mathbf{V}_X(1), \ldots, \mathbf{V}_X(n - 1)\} \\
= \mathbf{H}(n)E\{\mathbf{S}(n)|\mathbf{V}_X(1), \ldots, \mathbf{V}_X(n - 1)\} \\
= \mathbf{H}(n)\hat{\mathbf{S}}(n)
\end{aligned} \qquad (7.135)$$

Using this result in Equation 7.129.b

$$\mathbf{V}_X(n) = \mathbf{X}(n) - \mathbf{H}(n)\hat{\mathbf{S}}(n) \qquad (7.136)$$

Now we rely on the assumption that $\hat{\mathbf{S}}_2$ is a linear estimator and that all means are zero. Thus $\hat{\mathbf{S}}_2(n + 1)$ is

$$\hat{\mathbf{S}}_2(n + 1) = \mathbf{B}(n)[\mathbf{X}(n) - \mathbf{H}(n)\hat{\mathbf{S}}(n)] \qquad (7.137)$$

where $\mathbf{B}(n)$ is an $(m \times j)$ matrix of constants yet to be determined. In addition, we can define $\mathbf{K}(n)$ by

$$\mathbf{B}(n) = \mathbf{A}(n + 1)\mathbf{K}(n)$$

or

$$\mathbf{K}(n) = \mathbf{A}^{-1}(n + 1)\mathbf{B}(n) \qquad (7.138)$$

where we assume that $\mathbf{A}$ has an inverse, and $\mathbf{K}(n)$ is an $(m \times j)$ matrix.

Because $\hat{\mathbf{S}}(n + 1)$ must be a linear combination of $\mathbf{V}_X(1), \ldots, \mathbf{V}_X(n)$, and because $\hat{\mathbf{S}}_1$ and $\hat{\mathbf{S}}_2$ are independent, and both are the best estimators using part of the sequence $\mathbf{V}_X(1), \ldots, \mathbf{V}_X(n)$, we can state (as was shown in the scalar case) that

$$\hat{\mathbf{S}}(n + 1) = \hat{\mathbf{S}}_1(n + 1) + \hat{\mathbf{S}}_2(n + 1)$$
$$= \mathbf{A}(n + 1)\hat{\mathbf{S}}(n) + \mathbf{A}(n + 1)\mathbf{K}(n)[\mathbf{X}(n) - \mathbf{H}(n)\hat{\mathbf{S}}(n)] \quad (7.139)$$

This is the first of the Kalman vector equations, which corresponds with the scalar Equation 7.104. It can be rewritten as

$$\hat{\mathbf{S}}(n + 1) = \mathbf{A}(n + 1)\{\hat{\mathbf{S}}(n) + \mathbf{K}(n)[\mathbf{X}(n) - \mathbf{H}(n)\hat{\mathbf{S}}(n)]\} \quad (7.140)$$

or as

$$\hat{\mathbf{S}}(n + 1) = \mathbf{A}(n + 1)\{\mathbf{K}(n)\mathbf{X}(n) + [\mathbf{I} - \mathbf{K}(n)\mathbf{H}(n)]\hat{\mathbf{S}}(n)\} \quad (7.141)$$

where $\mathbf{I}$ is the $(m \times m)$ identity matrix. These equations, just as in the scalar case, will be viewed as a revision of $\hat{\mathbf{S}}(n)$ based on $\mathbf{X}(n)$ (the expression inside the brackets) and updating.

Now as in the scalar case, we must find the equation for updating the error covariance matrix and find $\mathbf{K}(n)$. We define

$$\tilde{\mathbf{S}}(n) \stackrel{\Delta}{=} \mathbf{S}(n) - \hat{\mathbf{S}}(n) \quad (7.142)$$
$$\mathbf{P}(n) \stackrel{\Delta}{=} E[\tilde{\mathbf{S}}(n)\tilde{\mathbf{S}}^T(n)] \quad (7.143)$$

where $\mathbf{P}(n)$ is an $(m \times m)$ covariance matrix. It follows that

$$\mathbf{P}(n + 1) = E\{\tilde{\mathbf{S}}(n + 1)\tilde{\mathbf{S}}^T(n + 1)\}$$
$$= E\{[\hat{\mathbf{S}}(n + 1) - \mathbf{S}(n + 1)][\hat{\mathbf{S}}(n + 1) - \mathbf{S}(n + 1)]^T\} \quad (7.144)$$

Using Equation 7.139 and the state model Equation 7.124 in Equation 7.144 results in

$$\mathbf{P}(n + 1) = E\{[\mathbf{A}(n + 1)\hat{\mathbf{S}}(n) + \mathbf{A}(n + 1)\mathbf{K}(n)[\mathbf{X}(n) - \mathbf{H}(n)\hat{\mathbf{S}}(n)]$$
$$- \mathbf{A}(n + 1)\mathbf{S}(n) - \mathbf{W}(n + 1)][\mathbf{A}(n + 1)\hat{\mathbf{S}}(n)$$
$$+ \mathbf{A}(n + 1)\mathbf{K}(n)[\mathbf{X}(n) - \mathbf{H}(n)\hat{\mathbf{S}}(n)]$$
$$- \mathbf{A}(n + 1)\mathbf{S}(n) - \mathbf{W}(n + 1)]^T\}$$

436   LINEAR MINIMUM MEAN SQUARED ERROR FILTERING

Now, using the definition of $\mathbf{V}_X(n)$ from Equation 7.136 and Equation 7.142

$$\mathbf{P}(n+1) = E\{[-\mathbf{A}(n+1)\tilde{\mathbf{S}}(n) + \mathbf{A}(n+1)\mathbf{K}(n)\mathbf{V}_X(n) - \mathbf{W}(n+1)] \\ \times [-\mathbf{A}(n+1)\tilde{\mathbf{S}}(n) + \mathbf{A}(n+1)\mathbf{K}(n)\mathbf{V}_X(n) - \mathbf{W}(n+1)]^T\}$$

This quadric form is now expanded using the definition in Equations 7.143 and 7.126:

$$\begin{aligned}\mathbf{P}(n+1) = {} & \mathbf{A}(n+1)\mathbf{P}(n)\mathbf{A}^T(n+1) + \mathbf{A}(n+1)\mathbf{K}(n)E\{\mathbf{V}_X(n)\mathbf{V}_X^T(n)\} \\ & \times \mathbf{K}^T(n)\mathbf{A}^T(n+1) + \mathbf{Q}(n+1) + \mathbf{A}(n+1)\,E\{\hat{\mathbf{S}}(n)\mathbf{V}_X^T(n)\}\mathbf{K}^T(n)\mathbf{A}^T(n+1) \\ & - \mathbf{A}(n+1)E\{\mathbf{S}(n)\mathbf{V}_X^T(n)\}\mathbf{K}^T(n)\mathbf{A}^T(n+1) \\ & + \mathbf{A}(n+1)\mathbf{K}(n)E\{\mathbf{V}_X(n)\hat{\mathbf{S}}^T(n)\}\mathbf{A}^T(n+1) \\ & - \mathbf{A}(n+1)\mathbf{K}(n)E\{\mathbf{V}_X(n)\mathbf{S}^T(n)\}\mathbf{A}^T(n+1) \\ & + \mathbf{A}(n+1)E\{\tilde{\mathbf{S}}(n)\mathbf{W}^T(n+1)\} + E\{\mathbf{W}(n+1)\tilde{\mathbf{S}}^T(n)\}\mathbf{A}^T(n+1) \\ & - \mathbf{A}(n+1)\mathbf{K}(n)E\{\mathbf{V}_X(n)\mathbf{W}^T(n+1)\} \\ & - E\{\mathbf{W}(n+1)\mathbf{V}_X^T(n)\}\mathbf{K}^T(n)\mathbf{A}^T(n+1) \end{aligned} \quad (7.145)$$

Now the fourth and sixth terms are zero because $\hat{\mathbf{S}}(n)$ is a linear combination of $\mathbf{V}_X(1), \ldots \mathbf{V}_X(n-1)$, which are orthogonal to $\mathbf{V}_X(n)$. The eighth and ninth terms are zero because $\mathbf{W}(n+1)$ is orthogonal to $\mathbf{S}(n)$ and $\hat{\mathbf{S}}(n)$. The tenth and eleventh terms are zero because $\mathbf{W}(n+1)$ is orthogonal to $\mathbf{V}_X(n)$. We now evaluate the fifth and seventh terms. Using Equation 7.136 and the observation Equation 7.125

$$\begin{aligned}\mathbf{V}_X(n) & = \mathbf{H}(n)\mathbf{S}(n) + \boldsymbol{v}(n) - \mathbf{H}(n)\hat{\mathbf{S}}(n) \\ & = \mathbf{H}(n)[\mathbf{S}(n) - \hat{\mathbf{S}}(n)] + \boldsymbol{v}(n)\end{aligned}$$

Now, the definition in Equation 7.142 produces

$$\mathbf{V}_X(n) = \mathbf{H}(n)\tilde{\mathbf{S}}(n) + \boldsymbol{v}(n) \quad (7.146)$$

Thus

$$\begin{aligned}E[\mathbf{S}(n)\mathbf{V}_X^T(n)] & = E\{\mathbf{S}(n)\tilde{\mathbf{S}}^T(n)\}\mathbf{H}^T(n) + E\{\mathbf{S}(n)\boldsymbol{v}^T(n)\} \\ & = E\{\mathbf{S}(n)\tilde{\mathbf{S}}^T(n)\}\mathbf{H}^T(n) \\ & = E\{[\tilde{\mathbf{S}}(n) + \hat{\mathbf{S}}(n)]\tilde{\mathbf{S}}^T(n)\}\mathbf{H}^T(n)\end{aligned}$$

KALMAN FILTERS 437

and the last product is zero by the orthogonality condition. Thus

$$E\{\mathbf{S}(n)\mathbf{V}_X^T(n)\} = \mathbf{P}(n)\mathbf{H}^T(n) \quad (7.147)$$

Similarly

$$E\{\mathbf{V}_X(n)\mathbf{S}^T(n)\} = \mathbf{H}(n)\mathbf{P}(n) \quad (7.148)$$

Thus Equation 7.145 becomes

$$\begin{aligned}\mathbf{P}(n+1) = {} & \mathbf{A}(n+1)\mathbf{P}(n)\mathbf{A}^T(n+1) \\ & + \mathbf{A}(n+1)\mathbf{K}(n)E\{\mathbf{V}_X(n)\mathbf{V}_X^T(n)\}\mathbf{K}^T(n)\mathbf{A}^T(n+1) \\ & + \mathbf{Q}(n+1) - \mathbf{A}(n+1)\mathbf{P}(n)\mathbf{H}^T(n)\mathbf{K}^T(n)\mathbf{A}^T(n+1) \\ & - \mathbf{A}(n+1)\mathbf{K}(n)\mathbf{H}(n)\mathbf{P}(n)\mathbf{A}^T(n+1) \end{aligned} \quad (7.149)$$

Now, Equations 7.146 and 7.127 result in

$$\begin{aligned} E\{\mathbf{V}_X(n)\mathbf{V}_X^T(n)\} &= E\{[\mathbf{H}(n)\tilde{\mathbf{S}}(n) + \boldsymbol{\nu}(n)][\mathbf{H}(n)\tilde{\mathbf{S}}(n) + \boldsymbol{\nu}(n)]^T\} \\ &= \mathbf{H}(n)\mathbf{P}(n)\mathbf{H}^T(n) + \mathbf{R}(n) \end{aligned}$$

This result in Equation 7.149 produces

$$\begin{aligned}\mathbf{P}(n+1) = {} & \mathbf{A}(n+1)\mathbf{P}(n)\mathbf{A}^T(n+1) \\ & + \mathbf{A}(n+1)\mathbf{K}(n)\mathbf{H}(n)\mathbf{P}(n)\mathbf{H}^T(n)\mathbf{K}^T(n)\mathbf{A}^T(n+1) \\ & + \mathbf{A}(n+1)\mathbf{K}(n)\mathbf{R}(n)\mathbf{K}^T(n)\mathbf{A}^T(n+1) \\ & + \mathbf{Q}(n+1) - \mathbf{A}(n+1)\mathbf{P}(n)\mathbf{H}^T(n)\mathbf{K}^T(n)\mathbf{A}^T(n+1) \\ & - \mathbf{A}(n+1)\mathbf{K}(n)\mathbf{H}(n)\mathbf{P}(n)\mathbf{A}^T(n+1) \end{aligned} \quad (7.150)$$

Equation 7.150 is the second of the updating equations; it updates the covariance matrix. We now wish to minimize by the choice of $\mathbf{K}(n)$ the sum of the expected squared errors, which is the trace of $\mathbf{P}(n+1)$. We proceed as follows. For simplification, Equation 7.150 will be rewritten dropping the argument n and rearranging

$$\begin{aligned}\mathbf{P}(n+1) = {} & \mathbf{APA}^T + \mathbf{Q} + \mathbf{AK}(\mathbf{HPH}^T + \mathbf{R})\mathbf{K}^T\mathbf{A}^T \\ & - \mathbf{APH}^T\mathbf{K}^T\mathbf{A}^T - \mathbf{AKHPA}^T \end{aligned} \quad (7.151)$$

438 LINEAR MINIMUM MEAN SQUARED ERROR FILTERING

Now we use the fact that $\mathbf{HPH}^T + \mathbf{R}$ is a covariance matrix and thus a matrix $\mathbf{D}$ can be found such that

$$\mathbf{HPH}^T + \mathbf{R} = \mathbf{DD}^T \tag{7.152}$$

Then

$$\begin{aligned}\mathbf{P}(n+1) = {}&\mathbf{APA}^T + \mathbf{Q} - \mathbf{APH}^T\mathbf{K}^T\mathbf{A}^T \\ &- \mathbf{AKHPA}^T + \mathbf{AKDD}^T\mathbf{K}^T\mathbf{A}^T\end{aligned} \tag{7.153}$$

Completing the square in $\mathbf{K}$ and introducing the matrix $\mathbf{B}$

$$\begin{aligned}\mathbf{P}(n+1) = {}&\mathbf{APA}^T + \mathbf{Q} + \mathbf{A}(\mathbf{KD} - \mathbf{B})(\mathbf{KD} - \mathbf{B})^T\mathbf{A}^T \\ &- \mathbf{ABB}^T\mathbf{A}^T\end{aligned} \tag{7.154}$$

Now, for Equation 7.153 to equal Equation 7.154, it is required that

$$\mathbf{BD}^T\mathbf{K}^T + \mathbf{KDB}^T = \mathbf{PH}^T\mathbf{K}^T + \mathbf{KHP}$$

or

$$\mathbf{B} = \mathbf{PH}^T(\mathbf{D}^T)^{-1} \tag{7.155}$$

We return to Equation 7.154 and seek to minimize the trace of $\mathbf{P}(n+1)$ by a choice of $\mathbf{K}$. The first two terms and the last term are independent of $\mathbf{K}$. The third term is the product of a matrix and its transpose insuring that the terms on the diagonal are nonnegative. Thus, the minimum is achieved when

$$\mathbf{KD} - \mathbf{B} = 0$$

or

$$\mathbf{K} = \mathbf{BD}^{-1}$$

Using Equation 7.155

$$\mathbf{K} = \mathbf{PH}^T(\mathbf{D}^T)^{-1}\mathbf{D}^{-1}$$

Initialize
 $n = 1$
 $\mathbf{P}(1) = \sigma^2 \mathbf{I}$ (Assumed value)
 $\hat{\mathbf{S}}(1) = \mathbf{S}$ (Assumed value)

1. Start Loop
 Get data: $\mathbf{R}(n)$, $\mathbf{H}(n)\mathbf{Q}(n)$; $\mathbf{A}(n + 1)$; $\mathbf{X}(n)$
 $\mathbf{K}(n) = \mathbf{P}(n)\mathbf{H}^T(n)[\mathbf{H}(n)\mathbf{P}(n)\mathbf{H}^T(n) + \mathbf{R}(n)]^{-1}$
 $\hat{\mathbf{S}}(n + 1) = \mathbf{A}(n + 1)\{\hat{\mathbf{S}}(n) + \mathbf{K}(n)[\mathbf{X}(n) - \mathbf{H}(n)\hat{\mathbf{S}}(n)]\}$
 $\mathbf{P}(n + 1) = \mathbf{A}(n + 1)\{[\mathbf{I} - \mathbf{K}(n)\mathbf{H}(n)]\mathbf{P}(n)\}\mathbf{A}^T(n + 1) + \mathbf{Q}(n + 1)$
 $n = n + 1$
 Go to 1.

Figure 7.8 Kalman filtering algorithm (vector)

or

$$\mathbf{K} = \mathbf{PH}^T(\mathbf{DD}^T)^{-1}$$

Using Equation 7.152

$$\mathbf{K} = \mathbf{PH}^T(\mathbf{HPH}^T + \mathbf{R})^{-1} \tag{7.156}$$

Equation 7.156 defines the Kalman gain and corresponds with the scalar Equation 7.120. It, along with Equations 7.150 and 7.140, defines the Kalman filter.

If Equation 7.156 is used in Equation 7.150, after some matrix algebra we arrive at

$$\mathbf{P}(n + 1) = \mathbf{A}(n + 1)[\mathbf{P}(n) - \mathbf{K}(n)\mathbf{H}(n)\mathbf{P}(n)] \\ \times \mathbf{A}^T(n + 1) + \mathbf{Q}(n + 1) \tag{7.157}$$

or

$$\mathbf{P}(n + 1) = \mathbf{A}(n + 1)\{[\mathbf{I} - \mathbf{K}(n)\mathbf{H}(n)]\mathbf{P}(n)\}\mathbf{A}^T(n + 1) + \mathbf{Q}(n + 1)$$

As in the scalar case, the term inside the brackets will be interpreted as the revised MSE, and the remainder of the equation will be viewed as updating. The vector Kalman filter is summarized in Figure 7.8.

Figure 7.8 is completely analogous with Figure 7.7 with matrix operations replacing the scalar operations. In addition, in the vector case, the observation $\mathbf{X}$ is a linear transformation $\mathbf{H}$ of the signal, while h was assumed to be 1 in the scalar case. (See Problem 7.23.)

EXAMPLE 7.17.

Assume that $S_1(t)$ represents the position of a particle, and $S_2(t)$ represents its velocity. Then we can write

$$S_1(t) = \int_0^t S_2(\tau)\, d\tau + S_1(0)$$

or with a sampling time of one second, this can be converted to the approximate difference equation

$$S_1(n) = S_2(n-1) + S_1(n-1)$$

We assume that the velocity can be described as a constant plus stationary zero-mean white noise with unity variance

$$S_2(n) = S_2(n-1) + W(n)$$

These equations form the state model

$$\mathbf{S}(n) = \begin{bmatrix} 1 & 1 \\ 0 & 1 \end{bmatrix} \mathbf{S}(n-1) + \mathbf{W}(n)$$

where

$$\mathbf{Q}(n) = \begin{bmatrix} 0 & 0 \\ 0 & 1 \end{bmatrix}$$

We observe the position S_1 with an error, which is also stationary zero-mean white noise with unity variance. Thus, the observation model is

$$X(n) = \mathbf{H}(n)\mathbf{S}(n) + v(n)$$

where $X(n)$ is scalar, $\mathbf{H}(n) = [1 \ \ 0]$ and $\mathbf{R}(k) = 1$ is also scalar.
Assume

$$\mathbf{P}(1) = \begin{bmatrix} 1 & 0 \\ 0 & 1 \end{bmatrix}$$

and $\hat{\mathbf{S}}(1) = \begin{bmatrix} 0 \\ 0 \end{bmatrix}$ find $\mathbf{K}(n)$ and $\mathbf{P}(n)$ for $n = 1, 2, 3, 4$. Also find $\hat{\mathbf{S}}(2)$, $\hat{\mathbf{S}}(3)$, and $\hat{\mathbf{S}}(4)$.

SOLUTION:

$$\mathbf{K}(1) = \begin{bmatrix} 1 & 0 \\ 0 & 1 \end{bmatrix} \begin{bmatrix} 1 \\ 0 \end{bmatrix} \left\{ \begin{bmatrix} 1 & 0 \end{bmatrix} \begin{bmatrix} 1 & 0 \\ 0 & 1 \end{bmatrix} \begin{bmatrix} 1 \\ 0 \end{bmatrix} + 1 \right\}^{-1}$$

$$= \begin{bmatrix} \frac{1}{2} \\ 0 \end{bmatrix}$$

$$\hat{\mathbf{S}}(2) = \begin{bmatrix} 1 & 1 \\ 0 & 1 \end{bmatrix} \left\{ \begin{bmatrix} \frac{1}{2} \\ 0 \end{bmatrix} [X(1)] \right\}$$

$$= \begin{bmatrix} \frac{X(1)}{2} \\ 0 \end{bmatrix}$$

$$\mathbf{P}(2) = \begin{bmatrix} 1 & 1 \\ 0 & 1 \end{bmatrix} \left\{ \left(\begin{bmatrix} 1 & 0 \\ 0 & 1 \end{bmatrix} - \begin{bmatrix} \frac{1}{2} \\ 0 \end{bmatrix} \begin{bmatrix} 1 & 0 \end{bmatrix} \right) \begin{bmatrix} 1 & 0 \\ 0 & 1 \end{bmatrix} \right\} \begin{bmatrix} 1 & 0 \\ 1 & 1 \end{bmatrix} + \begin{bmatrix} 0 & 0 \\ 0 & 1 \end{bmatrix}$$

$$= \begin{bmatrix} 1.5 & 1 \\ 1 & 2 \end{bmatrix}$$

The Kalman gains for $n \geq 2$ can be computed as

$$\mathbf{K}(2) = \begin{bmatrix} 1.5 & 1 \\ 1 & 2 \end{bmatrix} \begin{bmatrix} 1 \\ 0 \end{bmatrix} \left\{ \begin{bmatrix} 1 & 0 \end{bmatrix} \begin{bmatrix} 1.5 & 1 \\ 1 & 2 \end{bmatrix} \begin{bmatrix} 1 \\ 0 \end{bmatrix} + 1 \right\}^{-1}$$

$$= \begin{bmatrix} .6 \\ .4 \end{bmatrix}$$

$$\mathbf{P}(3) = \begin{bmatrix} 1 & 1 \\ 0 & 1 \end{bmatrix} \left\{ \left(\begin{bmatrix} 1 & 0 \\ 0 & 1 \end{bmatrix} - \begin{bmatrix} .6 \\ .4 \end{bmatrix} \begin{bmatrix} 1 & 0 \end{bmatrix} \right) \begin{bmatrix} 1.5 & 1 \\ 1 & 2 \end{bmatrix} \right\} \begin{bmatrix} 1 & 0 \\ 1 & 1 \end{bmatrix} + \begin{bmatrix} 0 & 0 \\ 0 & 1 \end{bmatrix}$$

$$= \begin{bmatrix} 3 & 2 \\ 2 & 2.6 \end{bmatrix}$$

$$\mathbf{K}(3) = \begin{bmatrix} 3 & 2 \\ 2 & 2.6 \end{bmatrix} \begin{bmatrix} 1 \\ 0 \end{bmatrix} \{3 + 1\}^{-1}$$

$$= \begin{bmatrix} .75 \\ .5 \end{bmatrix}$$

$$P(4) = \begin{bmatrix} 1 & 1 \\ 0 & 1 \end{bmatrix} \left\{ \left(\begin{bmatrix} 1 & 0 \\ 0 & 1 \end{bmatrix} - \begin{bmatrix} .75 \\ .5 \end{bmatrix} [1 \ 0] \right) \begin{bmatrix} 3 & 2 \\ 2 & 2.6 \end{bmatrix} \right\}$$
$$\times \begin{bmatrix} 1 & 0 \\ 1 & 1 \end{bmatrix} + \begin{bmatrix} 0 & 0 \\ 0 & 1 \end{bmatrix}$$
$$= \begin{bmatrix} 3.35 & 2.1 \\ 2.1 & 2.6 \end{bmatrix}$$

$$K(4) = \begin{bmatrix} 3.35 & 2.1 \\ 2.1 & 2.6 \end{bmatrix} \begin{bmatrix} 1 \\ 0 \end{bmatrix} \{3.35 + 1\}^{-1}$$
$$= \begin{bmatrix} .7701 \\ .4828 \end{bmatrix}$$

The reader can show that $\hat{S}(n + 1)$ has the form

$$\hat{S}(n + 1) = \begin{bmatrix} [k_1(n) + k_2(n)]X(n) + [1 - k_1(n) - k_2(n)]\hat{S}_1(n) + \hat{S}_2(n) \\ k_2(n)X(n) - k_2(n)\hat{S}_1(n) + \hat{S}_2(n) \end{bmatrix}$$

and hence

$$\hat{S}(3) = \begin{bmatrix} X(2) \\ .6X(2) - .2X(1) \end{bmatrix}$$

and

$$\hat{S}(4) = \begin{bmatrix} 1.25X(3) + .35X(2) - .2X(1) \\ 0.5X(3) + .1X(2) - .2X(1) \end{bmatrix}$$

7.7 WIENER FILTERS

In this section, we consider estimating the value of a continuous-time stationary random processes. The ideas of the previous sections are again used. That is, we seek to minimize the MSE and use the orthogonality condition and innovations.

We want to find the estimator of

$$G(t) = S(t + \alpha) \tag{7.158}$$

The estimator $\hat{G}(t)$ will be restricted to be linear, that is

$$\hat{G}(t) = \hat{S}(t + \alpha) = \int_a^b h(t, \lambda)X(\lambda)\, d\lambda \qquad (7.159)$$

Note that this is the continuous analogue of the linear estimator of Section 7.5.

If the lower limit, a, in the integral of Equation 7.159 is $-\infty$ and $b = t$, that is, we can observe the infinite past up to the present, then the estimator is usually given the following names:

$$\alpha > 0 \quad \text{prediction}$$
$$\alpha = 0 \quad \text{filtering}$$
$$\alpha < 0 \quad \text{smoothing}$$

The most common measurement model and the only one considered in this book is

$$X(t) = S(t) + N(t) \qquad (7.160)$$

where $S(t)$ and $N(t)$ are independent.

We will also restrict our attention to stationary random processes, in which case h will be time invariant and Equation 7.159 can be written as the convolution integral.

$$\hat{G}(t) = \hat{S}(t + \alpha) = \int_a^b h(t - \lambda)X(\lambda)\, d\lambda \qquad (7.161)$$

The first step in finding the optimum solution is obtained quite simply from the orthogonality condition

$$E\left\{\left[G(t) - \int_a^b h(t - \lambda)X(\lambda)\, d\lambda\right] X(\xi)\right\} = 0, \qquad a \leq \xi \leq b$$

where h now indicates the optimum filter, that is, the h that satisfies the orthogonality condition. Now, using the standard notation for the expected values

$$R_{XG}(t - \xi) = \int_a^b R_{XX}(\xi - \lambda)h(t - \lambda)\, d\lambda, \qquad a \leq \xi \leq b \qquad (7.162)$$

444 LINEAR MINIMUM MEAN SQUARED ERROR FILTERING

This basic equation can be rewritten if $G(t) = S(t + \alpha)$ to

$$R_{XS}(t + \alpha - \xi) = \int_a^b R_{XX}(\xi - \lambda)h(t - \lambda)\, d\lambda, \qquad a \leq \xi \leq b \quad (7.163)$$

The MSE is

$$P(t) = P = E\{[G(t) - \hat{G}(t)]^2\}$$

Because both $G(t)$ and $\hat{G}(t)$ are stationary, $P(t)$ will not vary with time. Recalling that the observations and the error are orthogonal

$$P = E\left\{\left[G(t) - \int_a^b X(\lambda)h(t - \lambda)\, d\lambda\right] G(t)\right\}$$

$$= R_{GG}(0) - \int_a^b R_{XG}(t - \lambda)h(t - \lambda)\, d\lambda \quad (7.164)$$

7.7.1 Stored Data (Unrealizable Filters)

We consider the case where the data are stored, and the data processing need not be done in real time. In this case, the filter is called unrealizable or noncausal, and the description of the optimum filter is easy. We let $a = -\infty$ and $b = +\infty$ in Equation 7.163

$$R_{XS}(t + \alpha - \xi) = \int_{-\infty}^{\infty} R_{XX}(\xi - \lambda)h(t - \lambda)\, d\lambda \quad (7.165)$$

For simplification, we will let $t - \xi = \tau$. Then

$$R_{XS}(\tau + \alpha) = \int_{-\infty}^{\infty} R_{XX}(t - \tau - \lambda)h(t - \lambda)\, d\lambda$$

Next, let $\beta = t - \lambda$ and

$$R_{XS}(\tau + \alpha) = \int_{-\infty}^{\infty} R_{XX}(\tau - \beta)h(\beta)\, d\beta \quad (7.166)$$

This integral equation is easily solved for the optimum filter via Fourier transforms. Indeed

$$S_{XS}(f)\exp(j2\pi f\alpha) = S_{XX}(f) \cdot H(f) \tag{7.167}$$

or

$$H(f) = \frac{S_{XS}(f)\exp(j2\pi f\alpha)}{S_{XX}(f)}, \quad S_{XX}(f) \neq 0 \tag{7.168}$$

With H chosen according to Equation 7.168 the residual mean-square error is (using the orthogonality condition)

$$P = E\left\{\left[S(t+\alpha) - \int_{-\infty}^{\infty} X(t-\lambda)h(\lambda)\,d\lambda\right]S(t+\alpha)\right\}$$

$$= R_{SS}(0) - \int_{-\infty}^{\infty} R_{SX}(-\alpha - \lambda)h(\lambda)\,d\lambda \tag{7.169}$$

If $\overline{P}(\tau) \triangleq R_{SS}(\tau) - \int_{-\infty}^{\infty} R_{SX}(\tau - \alpha - \lambda)h(\lambda)\,d\lambda \tag{7.170}$

then $\overline{P}(0) = P$.

Taking transforms of Equation 7.170 produces

$$P_F(f) = S_{SS}(f) - S_{SX}(f)\exp(-j2\pi f\alpha)H(f)$$

where $P_F(f)$ is the Fourier transform of $\overline{P}(\tau)$ and using Equation 7.167

$$P_F(f) = S_{SS}(f) - S_{SX}(f)\frac{S_{XS}(f)}{S_{XX}(f)}$$

$$= S_{SS}(f) - \frac{|S_{XS}(f)|^2}{S_{XX}(f)} \tag{7.171}$$

Finally

$$P = \overline{P}(0) = \int_{-\infty}^{\infty}\left[S_{SS}(f) - \frac{|S_{XS}(f)|^2}{S_{XX}(f)}\right]df \tag{7.172}$$

We now consider some examples.

EXAMPLE 7.18.

We want to estimate $S(t)$ (i.e., $\alpha = 0$) with stored data, when the observation is

$$X(t) = S(t) + N(t)$$

and

$$R_{SS}(\tau) = \exp{-|\tau|}$$
$$R_{NN}(\tau) = \delta(\tau) + 3\exp(-|\tau|)$$
$$R_{SN}(\tau) = 0$$

Find the optimum unrealizable filter and the residual error.

SOLUTION:

$$S_{SS}(f) = \frac{2}{(2\pi f)^2 + 1}$$

$$S_{XS}(f) = S_{SS}(f)$$

$$S_{XX}(f) = \frac{2}{(2\pi f)^2 + 1} + 1 + \frac{6}{(2\pi f)^2 + 1} = \frac{(2\pi f)^2 + 9}{(2\pi f)^2 + 1}$$

Using Equation 7.168

$$H(f) = \frac{S_{XS}(f)}{S_{XX}(f)} = \frac{2}{(2\pi f)^2 + 9}$$

Although network synthesis is not considered in this text, it is important that this $H(f)$ is not a positive real function and therefore cannot be synthesized.

The impulse response of the optimum filter can be found by inverting $H(f)$ as follows:

$$H(f) = \frac{2}{(2\pi f)^2 + 9} = \frac{\frac{1}{3}}{j2\pi f + 3} + \frac{\frac{1}{3}}{(-j2\pi f) + 3}$$

Thus

$$h(t) = \frac{1}{3}\exp(-3t), \quad t \geq 0$$
$$= \frac{1}{3}\exp(+3t), \quad t < 0$$

Finally the error P is (using Equation 7.172)

$$P = \int_{-\infty}^{\infty} \left[S_{SS}(f) - \frac{|S_{SS}(f)|^2}{S_{SS}(f) + S_{NN}(f)} \right] df$$
$$= \int_{-\infty}^{\infty} \frac{S_{SS}(f) S_{NN}(f)}{S_{SS}(f) + S_{NN}(f)} df$$
$$= \int_{-\infty}^{\infty} \frac{2[(2\pi f)^2 + 7]}{[(2\pi f)^2 + 1][(2\pi f)^2 + 9]} df \approx .833$$

where the integral is evaluated using I_2 from Table 4.1 (page 235).

EXAMPLE 7.19.

We want to estimate $S(t)$ using $X(t) = S(t) + N(t)$; $S(t)$ and $N(t)$ are uncorrelated. $S_{SS}(f)$ and $S_{NN}(f)$ do not overlap, that is, $S_{SS}(f) \cdot S_{NN}(f) = 0$. Figure 7.9 shows a typical case.

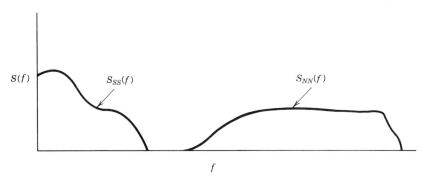

Figure 7.9 Spectral densities of signal and noise for Example 7.19.

SOLUTION:

In this case using Equation 7.168

$$H(f) = \frac{S_{ss}(f)}{S_{ss}(f) + S_{NN}(f)}$$

$$H(f) = 1 \qquad S_{ss}(f) > 0, \quad S_{NN}(f) = 0$$
$$= 0 \qquad S_{NN}(f) > 0, \quad S_{ss}(f) = 0$$
$$= \text{undetermined} \qquad S_{NN}(f) = 0 = S_{ss}(f)$$

and

$$P = \int_{-\infty}^{\infty} \frac{S_{ss}(f) S_{NN}(f)}{S_{ss}(f) + S_{NN}(f)} \, df = 0$$

7.7.2 Real-time or Realizable Filters

We now turn our attention toward the same problem considered in the last section with a seemingly slight but important change. Only the past and present values of data are now assumed to be available, that is

$$\hat{G}(t) = \hat{S}(t + \alpha) = \int_{-\infty}^{t} X(\lambda) h_R(t - \lambda) \, d\lambda \qquad (7.173)$$

where the subscript R denotes a realizable (causal) filter. Then the orthogonality principle implies (see Equation 7.163)

$$R_{XS}(t + \alpha - \xi) = \int_{-\infty}^{t} R_{XX}(\xi - \lambda) h_R(t - \lambda) \, d\lambda, \qquad \xi \le t \quad (7.174)$$

Letting $t - \xi = \tau$

$$R_{XS}(\tau + \alpha) = \int_{-\infty}^{t} R_{XX}(t - \tau - \lambda) h_R(t - \lambda) \, d\lambda, \qquad 0 \le \tau \quad (7.175)$$

Now letting $\beta = t - \lambda$

$$R_{XS}(\tau + \alpha) = \int_0^\infty R_{XX}(\tau - \beta)h_R(\beta)\, d\beta, \qquad 0 \leq \tau \qquad (7.176)$$

Equation 7.176 is called the Wiener–Hopf equation. This integral equation appears similar to Equation 7.166. However it is more difficult to solve because of the range of integration and the restriction $\tau \geq 0$ precludes the use of Fourier transform to directly find the solution. We will solve it using innovations, and then we will prove that the solution does satisfy Equation 7.176. The steps that follow in order to solve Equation 7.176 are

1. Assume the input is white noise and find the optimum realizable filter.
2. Discuss spectrum factorization in order to find a realizable innovations filter.
3. Find the optimum filter with a more general input.

Optimum Realizable (Real-time) Filters with White Noise Input. If $X(t)$, the input to the filter, were white noise $\hat{S}(t + \alpha)$ could, if all data were available, be written as

$$\begin{aligned}\hat{S}(t + \alpha) &= \int_{-\infty}^{\infty} h(t - \lambda)X(\lambda)\, d\lambda \\ &= \int_{-\infty}^{t} h(t - \lambda)X(\lambda)\, d\lambda + \int_{t}^{\infty} h(t - \lambda)X(\lambda)\, d\lambda \qquad (7.177)\end{aligned}$$

where the first integral represents the past and the second integral represents the future.

In this case, the optimum filter is defined by Equation 7.167. We will call this filter the "optimum unrealizable filter."

Now assume that $X(\lambda)$ cannot be observed for $\lambda > t$, that is, future $X(\lambda)$ are not available now. Then it seems reasonable that if $X(\lambda)$ cannot be observed, we should substitute for $X(\lambda)$ the best estimate of $X(\lambda)$, that is

$$E\{X(\lambda)|X(\xi); \xi \leq t\}, \lambda > t$$

And if $X(\lambda)$ is zero-mean white noise, then

$$E\{X(\lambda)|X(\xi); \xi \leq t\} = 0, \qquad \lambda > t \qquad (7.178)$$

Thus, referring to Equation 7.177, the last integral that represents the future or unrealizable part of the filter should be set equal to zero.

We have argued in this case where the input is white noise that the optimum realizable filter is the positive time portion of the optimum unrealizable filter, $h(t)$, and the negative time portion of $h(t)$ set equal to zero. In equation form the optimum realizable filter h_{RW} with white noise input is

$$h_{RW}(t) = h(t), \quad t \geq 0$$
$$= 0 \quad t < 0 \qquad (7.179)$$

where

$$h(t) = \int_{-\infty}^{\infty} \frac{S_{XS}(f)\exp(j2\pi f\alpha)}{S_{XX}(f)} \exp(j2\pi ft) \, df \qquad (7.180)$$

(see Equation 7.168)

We next discuss spectrum factorization in order to characterize in the frequency domain the positive time portion of $h(t)$.

Note finally that if $h(t) \neq 0$ for $t < 0$, then (referring to Equation 7.177) this would require the filter to respond to an input that has not occurred. Such filters, which correspond to a person laughing before being tickled, are theoretically as well as practically impossible to build and hence are called unrealizable or noncausal filters.

Spectrum Factorization. In this introductory book we consider only rational functions, that is, we consider

$$Q(f) \triangleq \frac{P_1(f)}{P_2(f)}$$

where P_1 and P_2 are polynomials in f.

If $Q(f)$ is a power density spectrum, then by the fact that it is even, we know that P_1 and P_2 can contain only even powers of f.

Although the following argument could be made by letting f be a complex variable, we will instead discuss spectrum factorization by finding the two-sided Laplace transform that corresponds with $Q(f)$. That is, let $s = j2\pi f$, and

$$L(s) = Q\left(\frac{s}{j2\pi}\right)$$

The two-sided Laplace transform, if it exists, is defined by

$$L(s) \triangleq \int_{-\infty}^{\infty} l(t)\exp(-st)\, dt$$

EXAMPLE 7.20.

Find the Fourier transform $Q(f)$ and the two-sided Laplace transform $L(s)$ of

$$l(t) = \exp(-|t|)$$

SOLUTION:

$$Q(f) = \int_{-\infty}^{\infty} \exp(-|t|)\exp(-j2\pi ft)\, dt$$

$$= \int_{-\infty}^{0} \exp(t)\exp(-j2\pi ft)\, dt + \int_{0}^{\infty} \exp(-t)\exp(-j2\pi ft)\, dt$$

$$= \frac{1}{1 - j2\pi f}\exp[(1 - j2\pi f)t]\Big|_{-\infty}^{0} + \frac{-1}{1 + j2\pi f}\exp[-(1 + j2\pi f)t]\Big|_{0}^{\infty}$$

$$= \frac{1}{1 - j2\pi f} + \frac{1}{1 + j2\pi f}$$

$$= \frac{2}{1 + (2\pi f)^2}$$

Note that the $1/(1 - j2\pi f)$ came from the negative time part of $l(t)$, which might be called $l^-(t)$, and $1/(1 + j2\pi f)$ came from the positive time part, $l^+(t)$, of $l(t)$. Similarly

$$L(s) = \frac{2}{1 - s^2} = \frac{2}{(1 + s)(1 - s)}$$

Now, because $Q(f)$ is even, $L(s)$ will also be even. That is, both the numerator and denominator of $L(s)$ will be polynomials in $-s^2$. Thus, if s_i is a singularity

452 LINEAR MINIMUM MEAN SQUARED ERROR FILTERING

(pole or zero) of $L(s)$, then $-s_i$ is also a singularity. Then $L(s)$ can be factored into

$$L(s) = L^+(s)L^-(s)$$

where $L^+(s)$ contains all of the poles and zeros in the left-half s-plane that correspond with the positive time portion of $l(t)$, and $L^-(s)$ contains all of the poles and zeros in the right-half s-plane that correspond with the negative time portion of $l(t)$. Also

$$|L^+(j\omega)|^2 = |L^-(j\omega)|^2 = L(j\omega) \tag{7.181}$$

EXAMPLE 7.21.

Refer to Example 7.20. Find $L^+(s)$ and $L^-(s)$ and $l^+(t)$ and $l^-(t)$.

SOLUTION:

$$L(s) = \frac{2}{(1+s)(1-s)} = \left(\frac{\sqrt{2}}{1+s}\right)\left(\frac{\sqrt{2}}{1-s}\right)$$

$$L^+(s) = \frac{\sqrt{2}}{1+s}, \quad L^-(s) = \frac{\sqrt{2}}{1-s}$$

Further note that if

$$l^+(t) = \sqrt{2}\exp(-t), \quad t \geq 0$$
$$= 0 \quad t < 0$$

then

$$\int_0^\infty \sqrt{2}\exp(-t)\exp(-st)\,dt = \frac{\sqrt{2}}{1+s}$$

and that this integral will converge whenever $\text{Re}(s) > -1$. Also

$$\int_{-\infty}^0 \sqrt{2}\exp(+t)\exp(-st)\,dt = \frac{\sqrt{2}}{1-s}$$

and that this integral will converge whenever $\mathrm{Re}(s) < +1$. Thus

$$l^+(t) = \sqrt{2}\exp(-t), \quad t \geq 0$$
$$= 0 \quad t > 0$$
$$l^-(t) = \sqrt{2}\exp(+t), \quad t \leq 0$$
$$= 0 \quad t > 0$$

EXAMPLE 7.22.

Find $L^+(s)$ and $L^-(s)$ for

$$Q(f) = \frac{49 + 25(2\pi f)^2}{(2\pi f)^2[(2\pi f)^2 + 169]}$$

SOLUTION:

$$L(s) = \frac{49 - 25s^2}{-s^2(-s^2 + 169)}$$

$$L^+(s) = \frac{(7 + 5s)}{(s)(s + 13)}$$

$$L^-(s) = \frac{(7 - 5s)}{(-s)(-s + 13)}$$

We will now simplify the notation and say that

$$Q(f) = Q^+(f)Q^-(f)$$

where $Q^+(f) \triangleq L^+(j2\pi f)$, and $Q^-(f) \triangleq L^-(j2\pi f)$.

Optimum Realizable Filters. Now suppose that $X(t)$ is a signal that has a rational power spectral density function. If $X(t)$ is the input to an innovations filter, then the white-noise innovation of $X(t)$, $V_X(t)$ is the output of this filter. With the white noise $V_X(t)$ as input, the positive time portion of the remaining optimum filter should be the optimum realizable filter, just as was discussed in the first subsection of Section 7.5.2. That is, compare the two systems shown in Figure 7.10.

454 LINEAR MINIMUM MEAN SQUARED ERROR FILTERING

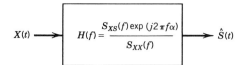

Figure 7.10a Optimum unrealizable filter.

Note that $H_1 \cdot H_2$ in Figure 7.10b equal H in Figure 7.10a using the factorization discussed in the preceding section, that is

$$S_{xx}(f) = S_{xx}^+(f) S_{xx}^-(f) \qquad (7.182)$$

Where $S_{xx}^+(s/j2\pi)$ contains the poles and zeros in the left-half of the s-plane and its inverse transform is nonzero for positive time. Similarly $S_{xx}^-(s/j2\pi)$ corresponds with the Laplace transform that has roots in the right-half plane and whose inverse transform is nonzero for negative time. We can easily show that V_X of Figure 7.10b is white noise. Indeed

$$S_{V_X V_X}(f) = \left| \frac{1}{S_{xx}^+(f)} \right|^2 S_{xx}(f), \qquad S_{xx}(f) > 0$$

$$= \frac{1}{S_{xx}^-(f)} S_{xx}(f) = 1$$

We now have the optimum unrealizable filter $H(f)$, as developed in Equation 7.167, described by

$$H(f) = H_1(f) H_2(f) \qquad (7.183)$$

where

$$H_1(f) = \frac{1}{S_{xx}^+(f)} \qquad (7.184)$$

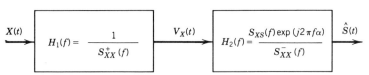

Figure 7.10b Another version of optimum unrealizable filter.

WIENER FILTERS   455

$$H_2(f) = \frac{S_{XS}(f)\exp(j2\pi f\alpha)}{S_{XX}^-(f)} \qquad (7.185)$$

$$h_2(t) = \int_{-\infty}^{\infty} H_2(f)\exp(j2\pi ft) \, df \qquad (7.186)$$

We now turn our attention toward the optimum realizable filter. $V_X(t)$ is white noise and the best estimate (conditional expected value) of future values of $V_X(t)$ is zero. This justifies taking the positive time portion of $H_2(f)$ and implies that the equations that describe H_R, the optimum realizable filter, are

$$H_R(f) = H_1(f)H_3(f) \qquad (7.187)$$

where

$$\begin{aligned} h_3(t) &= h_2(t), & t \geq 0 \\ &= 0 & t < 0 \end{aligned} \qquad (7.188)$$

$$H_3(f) = \int_{-\infty}^{\infty} h_3(t)\exp(-j2\pi ft) \, dt = \int_0^{\infty} h_2(t)\exp(-j2\pi ft) \, dt \qquad (7.189)$$

When working problems, it may be easier to find H_1 by Equation 7.184; find H_2 by Equation 7.185; find H_3 via spectrum factorization; and finally use Equation 7.187 to find the optimum realizable Wiener filter. Figure 7.11 shows the optimum realizable filter. We formally combine these steps here.

$$\begin{aligned} H_R(f) &= H_1(f)H_3(f) \\ &= \frac{1}{S_{XX}^+(f)} \int_0^{\infty} h_2(t)\exp(-j2\pi ft) \, dt \\ &= \frac{1}{S_{XX}^+(f)} \int_0^{\infty} \int_{-\infty}^{\infty} \frac{S_{XS}(\lambda)\exp(j2\pi\lambda\alpha)}{S_{XX}^-(\lambda)} \\ &\quad \times \exp(j2\pi\lambda t) \, d\lambda \, \exp(-j2\pi ft) \, dt \end{aligned} \qquad (7.190)$$

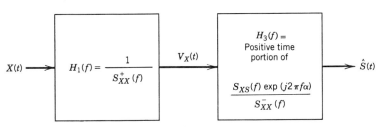

Figure 7.11   Optimum realizable filter.

456 LINEAR MINIMUM MEAN SQUARED ERROR FILTERING

We now find the mean squared error with the optimum realizable filter.

$$P = E\left\{\left[S(t + \alpha) - \int_{-\infty}^{\infty} h_R(\xi)X(t - \xi)\,d\xi\right]S(t + \alpha)\right\}$$

$$= R_{SS}(0) - \int_{-\infty}^{\infty} h_R(\xi) R_{SX}(-\alpha - \xi)\,d\xi$$

$$= R_{SS}(0) - \int_{-\infty}^{\infty}\left[\int_{-\infty}^{\infty} \frac{1}{S_{XX}^{+}(f)} \int_{0}^{\infty} h_2(t)\exp(-j2\pi ft)\,dt\right.$$
$$\left.\times \exp(j2\pi f\xi)\,df\right] R_{SX}(-\alpha - \xi)\,d\xi$$

where

$$h_2(t) = \int_{-\infty}^{\infty} \frac{S_{XS}(\lambda)}{S_{XX}^{-}(\lambda)} \exp(j2\pi\lambda\alpha)\exp(j2\pi\lambda t)\,d\lambda$$

Thus

$$P = R_{SS}(0) - \int_{0}^{\infty}\int_{-\infty}^{\infty}\left[\int_{-\infty}^{\infty} \exp(j2\pi f\xi)\, R_{SX}(-\alpha - \xi)\,d\xi\right]$$
$$\times \frac{\exp(-j2\pi ft)}{S_{XX}^{+}(f)}\,df\, h_2(t)\,dt$$

$$= R_{SS}(0) - \int_{0}^{\infty}\int_{-\infty}^{\infty} S_{SX}(f)\exp(-j2\pi f\alpha)\frac{\exp(-j2\pi ft)}{S_{XX}^{+}(f)}\,df\, h_2(t)\,dt \quad (7.191)$$

Let $f' = -f$, then

$$P = R_{SS}(0) - \int_{0}^{\infty}\int_{-\infty}^{\infty}\left[\frac{S_{XS}(f')}{S_{XX}^{-}(f')}\exp(j2\pi f'\alpha)\right]\exp(j2\pi f't)\,df'\, h_2(t)\,dt$$

The term inside the brackets is $H_2(f')$ by Equation 7.185; thus

$$P = R_{SS}(0) - \int_{0}^{\infty} h_2^2(t)\,dt \qquad (7.192.\text{a})$$

WIENER FILTERS

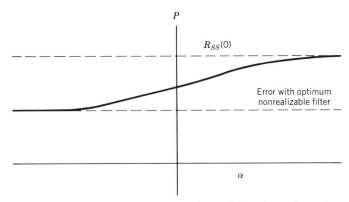

Figure 7.12 MSE versus α when $S(t + \alpha)$ is to be estimated.

An alternate expression for P is (from pages 328, 329)

$$P = \int_{-\infty}^{\infty} [1 - H_R(f)][1 - H_R^*(f)]S_{SS}(f) + H_R(f)H_R^*(f)S_{NN}(f)\, df \quad (7.192.b)$$

A plot of the typical mean-square error in estimating $S(t + \alpha)$ versus α is shown in Figure 7.12. We now present some examples of optimum filters, and then conclude this section with a proof that Equation 7.190 is indeed the optimum realizable filter.

EXAMPLE 7.23 PURE PREDICTION.

Estimate $G(t) = S(t + \alpha)$, given $X(t) = S(t)$ (no noise), and

$$R_{SS}(\tau) = k \exp(-c|\tau|), \quad c > 0, \quad k > 0$$

SOLUTION:

$$S_{SS}(f) = k \int_{-\infty}^{0} \exp(c\tau)\exp(-j2\pi f\tau)\, d\tau$$
$$+ k \int_{0}^{\infty} \exp(-c\tau)\exp(-j2\pi f\tau)\, d\tau$$
$$= \frac{k}{c - j2\pi f} + \frac{k}{c + j2\pi f} = \frac{2ck}{c^2 + (2\pi f)^2}$$

$$S_{SS}\left(\frac{s}{j2\pi}\right) = \frac{2ck}{c^2 - s^2} = \frac{\sqrt{2ck}}{c - s} \cdot \frac{\sqrt{2ck}}{c + s}$$

458 LINEAR MINIMUM MEAN SQUARED ERROR FILTERING

Using Equation 7.184

$$H_1(f) = \frac{c + j2\pi f}{\sqrt{2ck}}$$

Now, Equation 7.185 results in

$$H_2(f) = \frac{\left[\dfrac{2ck}{c^2 - (j2\pi f)^2}\right]}{\left[\dfrac{\sqrt{2ck}}{c - j2\pi f}\right]} \exp(j2\pi f\alpha)$$

$$= \frac{\sqrt{2ck}\,\exp(j2\pi f\alpha)}{c + (j2\pi f)}$$

and

$$h_2(t) = \sqrt{2ck}\,\exp[-c(t + \alpha)], \qquad t \geq -\alpha$$
$$= 0 \qquad\qquad\qquad\qquad\qquad t < -\alpha$$

(See Figure 7.13.)
If $\alpha > 0$, then

$$h_3(t) = \sqrt{2ck}\,\exp(-c\alpha)\exp(-ct), \qquad t \geq 0$$
$$= 0 \qquad\qquad\qquad\qquad\qquad\qquad t < 0$$

$$H_3(f) = \sqrt{2ck}\,\exp(-c\alpha)\left[\frac{1}{c + j2\pi f}\right]$$

$$H_R(f) = H_1(f)H_3(f) = \exp(-c\alpha)$$

Thus, the optimum filter is a constant and can be implemented as a voltage transfer function as shown in Figure 7.14, where

$$\frac{R_1}{R_1 + R_2} = \exp(-c\alpha)$$

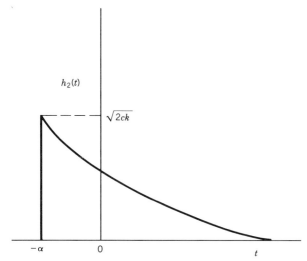

Figure 7.13 $h_2(t)$ for Example 7.23.

The mean-square error, using Equation 7.192 for $\alpha > 0$, is

$$P = k - \int_0^\infty 2ck \exp[-2c(t + \alpha)]\, dt$$

$$P = k[1 - \exp(-2c\alpha)]$$

If $\alpha = 0$, there is no error, and as $\alpha \to \infty$, $P \to k$, where k is the variance of $S(t)$. As $\alpha \to \infty$, $S(t)$ and $S(t + \alpha)$ are uncorrelated; hence the minimum MSE estimator of $S(t + \alpha)$ is its mean, and the MSE is its variance.

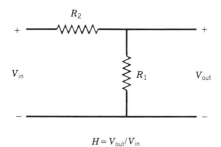

Figure 7.14 Voltage transfer function for Example 7.23.

EXAMPLE 7.24.

Given

$$X(t) = S(t) + N(t), \quad S \text{ and } N \text{ are independent}$$

$$S_{SS}(f) = \frac{3600}{(2\pi f)^2[(2\pi f)^2 + 169]}$$

$$S_{NN}(f) = 1$$

Find the optimum unrealizable and the optimum realizable filters for estimating

$$G(t) = S(t), \quad \text{i.e.,} \quad \alpha = 0$$

SOLUTION:

$$S_{XS}(f) = S_{SS}(f)$$
$$S_{XX}(f) = S_{SS}(f) + S_{NN}(f)$$
$$= \frac{(2\pi f)^4 + 169(2\pi f)^2 + 3600}{(2\pi f)^2[(2\pi f)^2 + 169]}$$

$$S_{XX}^+(f) = \frac{(j2\pi f)^2 + 17(j2\pi f) + 60}{(j2\pi f)(j2\pi f + 13)}$$

Using Equation 7.168, the optimum unrealizable filter is

$$H(f) = \frac{3600}{(2\pi f)^4 + (169)(2\pi f)^2 + 3600}$$

Now, we seek the optimum realizable filter beginning with Equation 7.184:

$$H_1(f) = \frac{(j2\pi f)(j2\pi f + 13)}{(j2\pi f)^2 + 17(j2\pi f) + 60}$$

Using Equation 7.185

$$H_2(f) = \frac{3600}{(2\pi f)^2[(2\pi f)^2 + 169]} \frac{(-j2\pi f)(-j2\pi f + 13)}{[(j2\pi f)^2 - 17(j2\pi f) + 60]}$$

$$= \frac{3600}{(j2\pi f)(j2\pi f + 13)[(j2\pi f)^2 - 17j2\pi f + 60]}$$

The partial fraction expansion is

$$H_2(f) = \frac{A_1}{j2\pi f} + \frac{A_2}{j2\pi f + 13} + \frac{Y(f)}{(j2\pi f)^2 - 17j2\pi f + 60}$$

where

$$A_1 = \frac{60}{13} \quad A_2 = -\frac{8}{13}$$

and $Y(f)$ is not evaluated because these poles correspond with right-half plane poles of the corresponding Laplace transform.

$h_3(t)$ comes from the poles of the corresponding Laplace transform that are in the left-half plane. Thus

$$H_3(f) = \frac{\frac{60}{13}}{j2\pi f} - \frac{\frac{8}{13}}{j2\pi f + 13} = \frac{4(j2\pi f) + 60}{(j2\pi f)(j2\pi f + 13)}$$

and

$$H_R(f) = H_1(f)H_3(f) = \frac{4[j2\pi f + 15]}{(j2\pi f)^2 + 17(j2\pi f) + 60}$$

It can be shown using Equation 7.192.b that the MSE is approximately 4.0. Although synthesis is not directly a part of this book, it is interesting to note that $H_R(f)$ can be implemented as a driving point impedance as shown in Figure

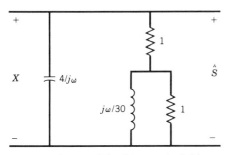

Figure 7.15a $H(\omega)$ of Example 7.24 as a driving point impedence.

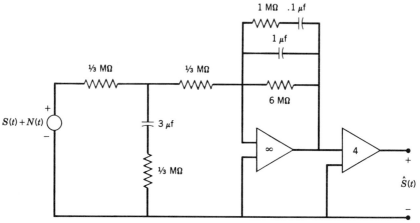

Figure 7.15b H_R of Example 7.24 as voltage transfer function.

7.15a. It may also be realized as a voltage transfer function shown in Figure 7.15b.

EXAMPLE 7.25.

This is the same problem as Example 7.18, except now a realizable (real-time) filter is required. The error with this optimum realizable filter is to be compared with the error obtained in Example 7.18 with the unrealizable filter.

SOLUTION:

As in Example 7.18

$$S_{XX}(f) = \frac{(2\pi f)^2 + 9}{(2\pi f)^2 + 1}$$

Using Equation 7.184

$$H_1(f) = \frac{1}{S_{XX}^+(f)} = \frac{j2\pi f + 1}{j2\pi f + 3}$$

Equation 7.185 produces

$$H_2(f) = \frac{2}{[(2\pi f)^2 + 1]} \frac{(-j2\pi f + 1)}{[-j2\pi f + 3]}$$

$$= \frac{2}{(j2\pi f + 1)(-j2\pi f + 3)} = \frac{A}{j2\pi f + 1} + \frac{B}{-j2\pi f + 3}$$

$$A = \frac{2}{(1 + 3)} = \frac{1}{2}$$

Using Equations 7.188 and 7.189

$$H_3(f) = \frac{1/2}{j2\pi f + 1}$$

Finally Equation 7.187 produces

$$H_R(f) = \frac{1/2}{j2\pi f + 3}$$

Now the MSE, P, with this filter, using Equation 7.192.a, is

$$P = R_{SS}(0) - \int_0^\infty \left[\frac{1}{2} \exp(-t)\right]^2 dt$$

$$= 1 - \frac{1}{4} \int_0^\infty \exp(-2t) \, dt$$

$$= 1 - \frac{1}{4} \cdot \frac{1}{2} = 1 - \frac{1}{8} = .875$$

This error is somewhat greater than the .833 found for the unrealizable filter that used stored data.

Proof that the Solution Satisfies the Wiener–Hopf Equation. In the previous section we argued that the solution given in Equation 7.190 solved the Wiener–Hopf Equation 7.176. Here we formally show that it does.

464 LINEAR MINIMUM MEAN SQUARED ERROR FILTERING

Theorem. If

$$H_R(f) = \frac{1}{S_{XX}^+(f)} \int_0^\infty \exp(-j2\pi ft) \left[\int_{-\infty}^\infty \frac{S_{XS}(\lambda)\exp(j2\pi\lambda\alpha)}{S_{XX}^-(\lambda)} \exp(j2\pi\lambda t)\, d\lambda \right] dt$$

then

$$R_{XS}(\tau + \alpha) = \int_0^\infty h_R(\beta) R_{XX}(\tau - \beta)\, d\beta, \qquad 0 \le \tau \le \infty$$

Proof. We show RHS → LHS where $h_R(\beta) = \int_{-\infty}^\infty H_R(f)\exp(j2\pi f\beta)\, df$.

$$\text{RHS} = \int_0^\infty R_{XX}(\tau - \beta) \left\{ \int_{-\infty}^\infty \exp(j2\pi f\beta)\, df \frac{1}{S_{XX}^+(f)} \int_0^\infty \exp(-j2\pi ft)\, dt \right.$$
$$\left. \times \int_{-\infty}^\infty \frac{S_{XS}(\lambda)\exp(j2\pi\lambda\alpha)}{S_{XX}^-(\lambda)} \exp(j2\pi\lambda t)\, d\lambda \right\} d\beta$$

$$= \int_{-\infty}^\infty \frac{S_{XS}(\lambda)\exp(j2\pi\lambda\alpha)}{S_{XX}^-(\lambda)}\, d\lambda \int_0^\infty \exp(j2\pi\lambda t)\, dt \int_{-\infty}^\infty \frac{1}{S_{XX}^+(f)}$$
$$\times \exp(-j2\pi ft)\, df \int_0^\infty R_{XX}(\tau - \beta)\exp(j2\pi f\beta)\, d\beta$$

$$= \int_{-\infty}^\infty \text{''}\, d\lambda \int_0^\infty \text{''}\, dt \int_{-\infty}^\infty \frac{1}{S_{XX}^+(f)} \exp[-j2\pi f(t-\tau)]\, df$$
$$\times \int_{-\infty}^\infty R_{XX}(v)\exp(-j2\pi fv)\, dv, \qquad v = \tau - \beta$$

$$= \int_{-\infty}^\infty \text{''}\, d\lambda \int_0^\infty \text{''}\, dt \int_{-\infty}^\infty \frac{1}{S_{XX}^+(f)} S_{XX}(f)\exp[-j2\pi f(t-\tau)]\, df$$

$$= \int_{-\infty}^\infty \text{''}\, d\lambda \int_0^\infty \text{''}\, dt \int_{-\infty}^\infty S_{XX}^-(f)\exp[-j2\pi f(t-\tau)]\, df$$

$$= \int_{-\infty}^\infty \text{''}\, d\lambda \int_0^\infty R_{XX}^-(\tau - t)\exp(+j2\pi\lambda t)\, dt$$

where $R_{XX}^-(\tau - t) = 0$ for $\tau - t \ge 0$.
Now let $\tau - t = y$ in the last integral:

$$\text{RHS} = \int_0^\infty h_R(\beta) R_{XX}(\tau - \beta)\, d\beta = \int_{-\infty}^\infty \frac{S_{XS}(\lambda)\exp(j2\pi\lambda\alpha)}{S_{XX}^-(\lambda)} \exp(j2\pi\lambda\tau)\, d\lambda$$
$$\times \int_{-\infty}^\tau R_{XX}^-(y)\exp(-j2\pi\lambda y)\, dy$$

But by hypothesis $\tau > 0$ and $R_{\bar{X}X}(\tau - t) = R_{\bar{X}X}(y) = 0$, $y > 0$, thus

$$\int_{-\infty}^{0} R_{X\bar{X}}(y)\exp(-j\pi\lambda y)\, dy = S_{X\bar{X}}(\lambda)$$

Using this result produces

$$\int_{0}^{\infty} h_R(\beta) R_{XX}(\tau - \beta)\, d\beta = \int_{-\infty}^{\infty} \frac{S_{XS}(\lambda)\exp(j2\pi\lambda\alpha)}{S_{X\bar{X}}(\lambda)} \exp(j2\pi\lambda\tau)\, S_{\bar{X}X}(\lambda)\, d\lambda$$

$$= \int_{-\infty}^{\infty} S_{XS}(\lambda)\exp[j2\pi\lambda(\alpha + \tau)]\, d\lambda$$

$$= R_{XS}(\alpha + \tau) \qquad \text{Q.E.D.}$$

7.7.3 Relation Between Kalman and Wiener Filters

If the state model of the signal results in a stationary signal and if the observation noise is stationary, then the Kalman filter approaches a steady-state form. This special case of the discrete Kalman filter is a recursive form of the discrete Wiener filter. Note that the state model together with a description of the input noise can be converted to an autocorrelation function of the signal.

In addition to its increased generality (if a linear state model for the signal may be found), the discrete Kalman filter has found many practical applications because its recursive nature is a natural for computer implementation. Also note that the calculation for the weights **K**(n) and the covariances **P**(n), can be done off-line before the data **X**(i) are available. Thus real-time estimation or filtering is feasible.

A continuous version of the Kalman filter exists, but it was not discussed in this introductory text. It is described, for instance, in Chapter 7 of Reference [4].

Thus, both Kalman and Wiener filters exist in discrete and continuous forms. Under stationary conditions the Kalman filters reduce to equivalent Wiener filters. The primary advantages of the discrete Kalman filter are the recursive form of the filter and its ability to handle nonstationary data.

7.8 SUMMARY

This chapter introduced the problem of how to estimate the value of a random signal $S(t)$ when observations of a related random process $X(t)$ are available. The usual case is that the observation is the sum of signal and noise. We assumed that the objective of the estimation was to minimize the mean squared error (MSE) between $S(t)$ and the estimator $\hat{S}(t)$.

We begin by estimating S with a finite linear (affine) combination of observations. We showed that the optimal estimator was one that resulted in the error being orthogonal to all observations. This orthogonality condition was used throughout the chapter as a basis for linear minimum MSE estimation.

Next the concept of innovations, or the unpredictable part, of observations was introduced. The innovations of a sequence of observations are uncorrelated. In addition, the innovations of observations can be used to estimate S, and the optimal linear estimator based on the innovations is algebraically equivalent to the optimal linear estimator based directly on the observations. Because the innovations are uncorrelated, estimators using the innovations have some computational advantages.

Optimal Wiener digital filters are exactly the same as those already discussed when the number of observations are finite. When there are more than a finite number of observations, the optimal digital filter was suggested, but practical solutions were not developed.

Instead, recursive digital filters in the form of Kalman filters were suggested. The scalar case was considered first, and the minimal MSE recursive filter was shown to be a weighted average of (1) the old estimate projected forward via the state equation and (2) the innovation of the new observation. The weights are inversely proportional to the mean squared errors of these two estimators.

Finally, continuous observations were considered, and continuous Wiener filters were derived from the orthogonality condition based on continuous innovations of the observations.

If the state model or equivalently the autocorrelation function is unknown, then the model structure and parameters must be estimated and the estimated model is used to design the filter. This case is considered in the last chapter of the book.

The common theme of this chapter was the estimation of an unknown signal in the presence of noise such that the MSE is minimized. The estimators were limited to linear estimators, which are optimum in the Gaussian case. Innovations and orthogonality were used in developing both Kalman and Wiener estimators.

7.9 REFERENCES

Optimum filtering or estimation began with Reference [11]. Reference [7] contains a very readable exposition of this early work, and Reference [2] explains this work on "Wiener" filtering in a manner that forms a basis for the presentation in this text. Reference [5] introduced "Kalman" filtering. Reference [8] emphasizes the orthogonality principle, and Reference [6] emphasizes innovations. Reference [1] is a concise summary of "Kalman" theory. References [3] and [9] and [10] present alternate treatments of filtering, and Reference [4] is a very readable account of the material contained in this chapter, and it also contains some interesting practical examples and problems.

[1] A. V. Balakrishnan, *Kalman Filtering Theory*, Optimization Software, Inc., 1984.

[2] H. W. Bode and C. E. Shannon, "A Simplified Derivation of Linear Least Squares Smoothing and Prediction Theory," *Proceedings of the IRE*, Vol. 38, April 1950, pp. 417–424.

[3] S. M. Bozic, *Digital and Kalman Filtering*, John Wiley & Sons, New York, 1979.

[4] R. G. Brown, *Introduction to Random Signal Analysis and Kalman Filtering*, John Wiley & Sons, New York, 1983.

[5] R. E. Kalman, "A New Approach to Linear Filtering and Prediction Problems," *Transactions of the ASME—Journal of Basic Engineering*, Ser. D, Vol. 82, March 1960, pp. 35–45.

[6] H. J. Larson and B. O. Shubert, *Probabilistic Models in Engineering and Science*, Vol. II, John Wiley & Sons, New York, 1979.

[7] Y. W. Lee, *Statistical Theory of Communication*, John Wiley & Sons, New York, 1960.

[8] A. Papoulis, *Probability, Random Variables and Stochastic Processes*, McGraw-Hill, New York, 1965 and 1984.

[9] M. D. Srinath and P. K. Rajasekaran, *An Introduction to Statistical Signal Processing with Applications*, John Wiley & Sons, New York, 1979.

[10] M. Schwartz and L. Shaw, *Signal Processing; Discrete Spectral Analysis, Detection, and Estimation*, McGraw-Hill, New York, 1975.

[11] N. Wiener, *Extrapolation, Interpolation, and Smoothing of Stationary Time Series*, John Wiley & Sons, New York, 1949.

7.10 PROBLEMS

7.1 The median m of X is defined by $F_X(m) = .5$. Show that m minimizes $E\{|X - m|\}$ by showing for $a < m$

$$E\{|X - a|\} = E\{|X - m|\} + 2 \int_a^m (x - a) f_X(x) \, dx$$

and for $a > m$

$$E\{|X - a|\} = E\{|X - m|\} + 2 \int_m^a (a - x) f_X(x) \, dx$$

7.2 Assume that X is an observation of a constant signal plus noise, and that

$$f_S(s) = \frac{1}{\sqrt{2\pi}} \exp\left[\frac{-(s - 1)^2}{2}\right].$$

468   LINEAR MINIMUM MEAN SQUARED ERROR FILTERING

We observe $X = 2$. Use Bayes' rule to find the new estimate of S. Assume S and N are statistically independent and that

$$f_N(n) = \frac{1}{\sqrt{2\pi}} e^{-n^2/2}.$$

Hint: First find $f_{X|S}$ and use Bayes' rule to find $f_{S|X}$. Then use the mean as the estimate.

7.3 We observe a signal $S(1)$ at time 1 and we want the best linear estimator of the signal $S(2)$ at a later time. Assume $E\{S(1)\} = E\{S(2)\} = 0$. Find the best linear estimator in terms of moments.

7.4 Assume the same situation as in Problem 7.3, but now the derivative, $S'(1)$, of the signal at time 1 can also be observed. $E\{S'(1)\} = 0$. Find the best estimator of the form

$$\hat{S}(2) = a_1 S(1) + a_2 S'(1)$$

7.5 We wish to observe a state variable X of a control system. But the observation Y is actually

$$Y = mX + N$$

where m is some known constant and N is a normal random variable, which is independent of X. Find the best linear estimator of X

$$\hat{X} = a + bY$$

in terms of m and the moments of X and N.

7.6 Show that if Y and X are jointly normal, then

$$E\{Y|X = x\} = a + bx$$

and

$$E\{[Y - (a + bX)]^2\} = \sigma_Y^2(1 - \rho_{XY}^2)$$

Note that a and b are as prescribed by Equations 7.4 and 7.5 for the best linear estimator.

7.7 Find the estimator of $S(n)$, a stationary sequence, given $S(n - 1)$, $S(n - 2)$, and $S(n - 3)$ if $\mu_S = 0$, $R_{SS}(0) = 1$, $R_{SS}(1) = .9$, $R_{SS}(2) = .7$, $R_{SS}(3) = .6$; that is, find h_1, h_2, and h_3 in the equation $\hat{S}(n) = h_1 S(1) + h_2 S(2) + h_3 S(3)$.

7.8 Show that Equations 7.13 and 7.14 do minimize the mean-square error by showing that

$$E\left\{\left[S - b_0 - \sum_{i=1}^{n} b_i X(i)\right]^2\right\} \geq E\left\{\left[S - h_0 - \sum_{i=1}^{n} h_i X(i)\right]^2\right\}$$

where h_0 and h_i are as defined by Equations 7.13 and 7.14.

7.9 Find the residual MSE for data of Problem 7.7.

7.10 $S = X^2$, and

$$f_X(x) = \frac{1}{2}, \quad -1 \leq x \leq 1$$
$$= 0 \quad \text{elsewhere}$$

Find h_0, h_1, and h_2 in the model

$$\hat{S} = h_0 + h_1 X + h_2 X^2$$

such that the mean-square error is minimized. Find the resulting minimum mean-square error.

7.11 Show that $\hat{S}_2 = h_1 X(1) + h_2 V_X(2)$ where $V_X(2)$ as defined by Equation 7.34 is identical to $\hat{S}_2$ as defined by Equation 7.16 when the means are zero.

7.12 Show that $\hat{S}_3 = h_1 X(1) + h_2 V_{X(2)} + h_3 V_{X(3)}$ is identical to $\hat{S}_3$ as defined by Equation 7.16 when the means are zero.

7.13 Show that Equations 7.44 and 7.45 are correct.

7.14 For $\mathbf{X}$, a Gaussian random vector with zero means and

$$\Sigma_{XX} = \begin{bmatrix} 2 & 1 & \frac{1}{2} \\ 1 & 3 & 1 \\ \frac{1}{2} & 1 & 4 \end{bmatrix}$$

Find $\mathbf{\Gamma}$ and $\mathbf{L}$.

7.15 If Equation 7.74 is restricted to a symmetrical finite sum, that is,

$$\hat{S}(n) = \sum_{k=-M}^{M} h(k) X(n - k)$$

Find the optimum filter $h(k)$, $k = -M, -M + 1, \ldots, M$.

470 LINEAR MINIMUM MEAN SQUARED ERROR FILTERING

7.16 Find the optimum realizable filter for the signal given in Example 7.12 with $\sigma_s^2 = 1$ and $a = \frac{1}{2}$.

7.17 Show that if $S(f)$ is the Fourier transform of $R(i)$ and

$$S(f) = \frac{5 - 4\cos 2\pi f}{10 - 6\cos 2\pi f}$$

then

$$S_Z(z) = \frac{5 - 2(z + z^{-1})}{10 - 3(z + z^{-1})}$$

where

$$S_Z(z) = \sum_{i=-\infty}^{\infty} R(i) z^{-i}$$

and

$$S_Z[\exp(j2\pi f)] = S(f)$$

7.18 Show that the function

$$S_Z(z) = \frac{5 - 2(z + z^{-1})}{10 - 3(z + z^{-1})}$$

has two zeros $\frac{1}{2}$ and 2, and two poles $\frac{1}{3}$ and 3.

7.19 Draw the digital filter corresponding to the poles and zeros inside the unit circle of $S(z)$ of Problem 7.18. That is, draw the digital filter corresponding to

$$H(z) = \frac{2z - 1}{3z - 1}$$

7.20 Show that if the signal corresponding to the psd $S(f)$ of Problem 7.17 is put into a digital filter

$$\frac{1}{H(z)} = \frac{3z - 1}{2z - 1}$$

then the output has the spectrum of white noise, that is, the output is the innovation of the signal corresponding to the psd $S(f)$.

7.21 Assume that $S(0)$ is a Gaussian random variable with mean 0 and variance σ^2. $W(n), n = 0, 1, \ldots,$ is a stationary uncorrelated zero-mean Gaussian random sequence that is also uncorrelated with $S(0)$.

$$S(n + 1) = a(n)S(n) + W(n)$$

where $a(n)$ is a sequence of known constants

a. Show that $S(n)$ is a Gaussian random sequence and find its mean and variance.

b. What are the necessary conditions on $a(n)$, and the variances of $S(0)$ and $W(n)$ in order for $S(n)$ to be stationary?

c. Find the autocorrelation function of $S(n)$.

7.22 Show that

$$E\{X(n)|V_X(1), \ldots, V_X(n-1)\} = \hat{S}(n)$$

where $\hat{S}(n)$ is the linear minimum MSE estimator, and thus Equation 7.101.c is an alternate definition of the innovation.

7.23 If $X(n) = h(n)S(n) + v(n)$, where the $h(n)$ are known constants, is used in place of Equation 7.89 and other assumptions remain unchanged, show that Equation 7.102 becomes

$$\hat{S}_2(n+1) = b'(n)\left[\frac{X(n)}{d(n)} - \hat{S}(n)\right]$$

where $b'(n)$ and $d(n)$ are constants

7.24 Equation 7.105 gives the best estimator of $S(n+1)$ using all observations through $X(n)$. Show that the best recursive estimator $\hat{\hat{S}}(n)$ using all observations through $X(n)$ is

$$\hat{\hat{S}}(n) = \hat{S}(n) + k(n)[X(n) - \hat{S}(n)]$$
$$= k(n)X(n) + [1 - k(n)]\hat{S}(n)$$

where $\hat{S}(n)$ and $k(n)$ are as defined in the text.

Furthermore, show that

$$\hat{S}(n+1) = a(n+1)\hat{\hat{S}}(n)$$

This is an alternate form of the Kalman filter, that is, $\hat{\hat{S}}(n)$, is often used.

7.25 Show that the expected squared error $P'(n)$ of $\hat{\hat{S}}(n)$ is given by

$$P'(n) = [1 - k(n)]P(n)$$

Also, show that

$$P(n+1) = a^2(n+1)P'(n) + \sigma_W^2(n+1)$$

7.26 Minimize with respect to $k(n)$, $P(n+1)$ as given in Equation 7.113 by differentiating and setting the derivative equal to zero. Show that the resulting stationary point is a minimum.

472 LINEAR MINIMUM MEAN SQUARED ERROR FILTERING

7.27 Let $Z = k_1 X + k_2 Y$, where X and Y are independent with means μ_X and μ_Y and variance σ_X^2 and σ_Y^2. k_1 and k_2 are constants.

   a. Find μ_Z and σ_Z^2.

   b. Assume $\mu_X = \mu_Y$. Find a condition on k_1 and k_2 that makes $\mu_X = \mu_Y = \mu_Z$.

   c. Using the result of (b), find k_1 and k_2 such that σ_Z^2 is minimized.

   d. Using the results of (b) and (c) find σ_Z^2. Compare this with Equations 7.121 and 7.122.

7.28 Refer to Example 7.15. Find $\hat{S}(4)$, $P(4)$, $\hat{S}(5)$, and $P(5)$.

7.29 Refer to Example 7.15

   a. Find $\hat{S}(1)$, $\hat{S}(2)$, $\hat{S}(3)$, and $\hat{S}(4)$ as defined in Problem 7.24 in terms of $X(1)$, $X(2)$, $X(3)$, and $X(4)$. Show that the weight on $\hat{S}(1)$ decreases to zero. Thus, a bad guess of $\hat{S}(1)$ has less effect.

   b. How would you adjust $P(1)$ in order to decrease the weight attached to $\hat{S}(1)$?

7.30 Assume the model $S(n) = .9S(n-1) + W(n)$

$\sigma_W^2(n) = \sigma_v^2(n) = 1$

$X(n) = S(n) + v(n)$

$\hat{S}(1) = 0, \quad P(1) = 10$

$X(1) = 1, \quad X(2) = 1.1, \quad X(3) = 1.2, \quad X(4) = .9, \quad X(5) = 1.2$

Find $\hat{S}(i)$, $i = 2, 3, 4, 5, 6$.

7.31 What is the steady-state gain, i.e. $\lim_{n \to \infty} k(n)$ for Problem 7.30?

7.32 Show in the vector case that Equation 7.139 is true following the steps used from Equations 7.94–7.100 in the scalar case.

7.33 Explain why the cross-product terms are zero in the equation following Equation 7.149.

7.34 Use Equation 7.150 and Equation 7.156 to produce Equation 7.157.

7.35 Assume $X(t) = S(t) + N(t)$, with

$$S_{SS}(f) = \frac{1}{(2\pi f)^2 + 1}$$

$$S_{NN}(f) = \frac{4}{(2\pi f)^2 + 4}$$

Signal and noise are uncorrelated. Find the optimum unrealizable Wiener filter and the MSE, when $\alpha = 0$.

7.36 Assume that

$$S_{SS}(f) = \frac{2}{(2\pi f)^2 + 1}$$

$$S_{NN}(f) = 1$$

Signal and noise are uncorrelated. Find the minimum MSE unrealizable filter, $H(f)$ for $\alpha = 0$ and show that $P = .577$.

7.37 Assume that the signal and noise are as specified in Problem 7.36. However, the filter form is specified as

$$H(f) = \frac{1}{1 + j2\pi f RC} = \frac{1}{1 + j2\pi f T}$$

Find T for minimum mean squared error and show that the MSE is .914 if the T is correctly chosen.

7.38 Find the optimum realizable filter for the situation described in Problems 7.36 and 7.37, and show that $P = .732$. Discuss the three filters and their errors.

7.39 Find $L^+(s)$, $L^-(s)$, $l^+(t)$, and $l^-(t)$, where

$$L(s) = \frac{-4s^2 + (a+b)^2}{(a^2 - s^2)(b^2 - s^2)}$$

7.40 $S(t)$ and $N(t)$ are uncorrelated with

$$S_{SS}(f) = \frac{(2\pi f)^2}{[1 + (2\pi f)^2]^2}$$

$$S_{NN}(f) = \frac{1}{16}$$

Find the optimum realizable Wiener filter with $\alpha = 0$.

7.41 Contrast the assumptions of discrete Wiener and discrete Kalman filters. Contrast their implementation. Derive the Kalman filter for Problem 7.36 and compare this with the results of Problem 7.38.

7.42 An alternate form of the vector Kalman filtering algorithm uses the observations at stage n in addition to the previous observations. Show that it can be described by the following algorithm.

$$\mathbf{K}(n) = \mathbf{P}(n)\mathbf{H}^T(n)[\mathbf{H}(n)\mathbf{P}(n)\mathbf{H}^T(n) + \mathbf{R}(n)]^{-1}$$

$$\hat{\hat{\mathbf{S}}}(n) = \hat{\mathbf{S}}(n) + \mathbf{K}(n)[\mathbf{X}(n) - \mathbf{H}(n)\hat{\mathbf{S}}(n)]$$
$$\mathbf{P}'(n) = [\mathbf{I} - \mathbf{K}(n)\mathbf{H}(n)]\mathbf{P}(n)$$
$$\hat{\mathbf{S}}(n+1) = \mathbf{A}(n+1)\hat{\hat{\mathbf{S}}}(n)$$
$$\mathbf{P}(n+1) = \mathbf{A}^T(n+1)\mathbf{P}'(n)\mathbf{A}^T(n+1) + \mathbf{Q}(n+1)$$

This form is often used (see page 200 of Reference [4]). In this form $\mathbf{P}'(n)$ is the covariance matrix reduced from $\mathbf{P}(n)$ by the use of the observation at time n, and $\hat{\hat{\mathbf{S}}}$ is the revised estimator based on this observation.

CHAPTER EIGHT

Statistics

8.1 INTRODUCTION

Statistics deals with methods for making decisions based on measurements (or observations) collected from the results of experiments. The two types of decisions emphasized in this chapter are the decision as to which value to use as the estimate of an unknown parameter and the decision as to whether or not to accept a certain hypothesis.

Typical estimation decisions involve estimating the mean and variance of a specified random variable or estimating the autocorrelation function and power spectral density function of a random process. In such problems of estimating unknown parameters, there are two important questions. What is a "good" method of using the data to estimate the unknown parameter, and how "good" is the resulting estimate?

Typical hypothesis acceptance or rejection decisions are as follows: Is a signal present? Is the noise white? Is the random variable normal? That is, decide whether or not to accept the hypothesis that the random variable is normal. In such problems we need a "good" method of testing a hypothesis, that is, a method that makes a "true" hypothesis likely to be accepted and a "false" hypothesis likely to be rejected. In addition, we would like to know the probability of making a mistake of either type. The methods of hypothesis testing are similar to the methods of decision making introduced in Chapter 6. However, the methods introduced in this chapter are classical in the sense that they do not use *a priori* probabilities and they also do not explicitly use loss functions. Also, composite alternative hypotheses (e.g., $\mu \neq 0$) will be considered in addition to simple alternative hypotheses (e.g., $\mu = 1$).

In this chapter, we discuss the estimators of those parameters that often are needed in electrical engineering applications and particularly those used in Chapter 9 to estimate parameters of random processes. We also emphasize those statistical (i.e., hypothesis) tests that are used in Chapter 9.

After a characterization of a collection of observations or measurements from a probabilistic or statistical viewpoint, some example estimators are introduced and then measures for evaluating the quality of estimators are defined. The method of maximum likelihood estimation is introduced as a general method of determining estimators. The distribution of three estimators is studied in order to portray how estimators may vary from one try or sample set to the next. These distributions are also useful in certain hypothesis tests, which are described next. This chapter concludes with a discussion of linear regression, which is the most widely used statistical technique of curve fitting.

Note that in the first seven chapters of the book, we had assumed that the probability distributions associated with the problem at hand were known. Probabilities, autocorrelation functions, and power spectral densities were either derived from a set of assumptions about the underlying random processes or assumed to be given. In many practical applications, this may not be the case and the properties of the random variables (and random processes) have to be obtained by collecting and analyzing data. In this and the following chapter, we focus our attention on data analysis or statistics.

8.2 MEASUREMENTS

If we want to "determine" an unknown parameter or parameters or to test a hypothesis by using "measurements," in most cases of interest, attempts at repeated measurements will result in different values. Our central problem is to use these measurements in some optimum fashion to estimate the unknown parameter or to test a hypothesis.

In order to use the measurements in an organized fashion to estimate a parameter, it is standard practice to assume an underlying model involving random variables and the unknown parameter being estimated. For example, let m be the unknown parameter, N be the error in the measurement, and X be the measurement. Then, we can assume a model of the form

$$X = m + N \tag{8.1}$$

Note that in this model, m is an unknown constant whereas X and N are random variables.

A very important special case of this model occurs when the expected value of N is zero. In this case, the mean of X is m, and we say that we have an unbiased measurement.

As most of us remember from physics laboratory and from our common experiences, it is often better to repeat the experiment, that is, to make repeated measurements. This idea of repeated measurements is very important and will be quantified later in this chapter. For now, we want to describe these repeated measurements. The assumption is that the first measurement is $X_1 = m + N_1$, the second measurement is $X_2 = m + N_2$; and the nth measurement is $X_n = m + N_n$. The important assumption in models involving repeated measurements is that the random variables, $N_1, N_2, \ldots, N_n$ are independent and identically distributed (i.i.d.), and thus $X_1, \ldots, X_n$ are also i.i.d.

With Equation 8.1 as the underlying model, and with the i.i.d. assumption, the measurements can be combined in various ways. These combinations of $X_1, X_2, \ldots, X_n$ are loosely called *statistics*. A formal definition of a statistic appears in Section 8.2.1.

A very common statistic is

$$\overline{X} = \frac{1}{n} \sum_{i=1}^{n} X_i \tag{8.2}$$

The bar indicates an average of measurements and $\overline{X}$ is used to estimate m; thus, it is also often called

$$\hat{m} = \overline{X} = \frac{1}{n} \sum_{i=1}^{n} X_i$$

where the hat indicates "an estimator of."

Thus, measurements are used to form statistics that are used to estimate unknown parameters and also to test hypotheses. It is important for the statistics to be based on a representative sample or representative measurements. Famous failures to obtain a representative sample have occurred in presidential elections. For example in the 1936 presidential election, the *Literary Digest* forecast that Alf Landon, who carried the electoral votes only of Vermont and Maine, would defeat President Roosevelt by a 3 to 2 margin. Actually, Roosevelt received 62.5% of the vote. This error is attributed to samples chosen in part from the telephone directory and from automobile registration files at a time when only the more wealthy people had telephones and automobiles. This resulted in a sample that accurately represented (we assume) telephone and automobile owners' presidential preferences but was not a representative sample from the general voting public. Further analysis of sampling techniques have resulted in far more accurate methods of choosing the sample to be included in the statistics used to estimate the results of presidential elections. It is always important that the measurements be representative of the ensemble from which parameters are to be estimated. When considering random processes, both stationarity and ergodicity are critical assumptions in choosing a sample or measurements that will result in good estimators.

The basic premise of estimation is to determine the value of an unknown quantity using a statistic, that is, using a function of measurements. The *estimator* $g(X_1, X_2, \ldots, X_n)$ is a random variable. A specific set of measurements will result in $X_i = x_i$ and the resulting value $g(x_i, x_2, \ldots, x_n)$ will be called an *estimate* or an estimated value.

8.2.1 Definition of a Statistic

Let $X_1, X_2, \ldots, X_n$ be n i.i.d. random variables from a given distribution function F_X. Then $Y = g(X_1, X_2, \ldots, X_n)$ is called a statistic if the function g does not depend on any unknown parameter.

For example,

$$\overline{X} = \frac{1}{n} \sum_{i=1}^{n} X_i$$

is a statistic but

$$\widehat{\sigma^2} = \sum_{i=1}^{n} \frac{(X_i - \mu)^2}{n}$$

is not a statistic because it depends upon the unknown parameter, μ.

8.2.2 Parametric and Nonparametric Estimators

We use two classes of estimation techniques: *parametric* and *nonparametric*. To illustrate the basic difference between these two techniques, let us consider the problem of estimating a pdf $f_X(x)$. If we use a parametric method, we might assume, for example, that $f_X(x)$ is Gaussian with parameters μ_X and σ_X^2 whose values are not known. We estimate the values of these parameters from data, and substitute these estimated values in $f_X(x)$ to obtain an estimate of $f_X(x)$ for all values of x. In the nonparametric approach, we do not assume a functional form for $f_X(x)$, but we attempt to estimate $f_X(x)$ directly from data for all values of x.

In the following sections we present parametric and nonparametric estimators of probability density and distribution functions of random variables. Parametric and nonparametric estimators of autocorrelation functions and spectral density functions of random processes are discussed in Chapter 9.

8.3 NONPARAMETRIC ESTIMATORS OF PROBABILITY DISTRIBUTION AND DENSITY FUNCTIONS

In this section we describe three simple and widely used procedures for nonparametric estimation of probability distribution and density functions: (1) the empirical distribution function (or the cumulative polygon); (2) the histogram (or the bar chart, also called a frequency table); and (3) Parzen's estimator for a pdf. The first two are used extensively for graphically displaying measurements, whereas Parzen's estimator is used to obtain a smooth, closed-form expression for the estimated value of the pdf of a random variable.

8.3.1 Definition of the Empirical Distribution Function

Assume that the random variable X has a distribution function F_X and that a sample of i.i.d. measurements $X_1, X_2, \ldots, X_n$, of X is available. From this sample we construct an empirical distribution function, $\hat{F}_{X|X_1,\ldots,X_n}$, which is an estimator of the common unknown distribution function F_X.

Before we define the empirical distribution function, the reader is asked to make his own definition via the following example. Suppose the resistances of 20 resistors from a lot of resistors have been measured to be (the readings have been listed in ascending order) 10.3, 10.4, 10.5, 10.5, 10.6, 10.6, 10.6, 10.7, 10.8, 10.8, 10.9, 10.9, 10.9, 10.9, 11.0, 11.0, 11.0, 11.1, 11.1, 11.2. What is the probability that a resistor selected from the lot has a resistance less than or equal to 10.75?

If your answer is 8/20 because eight out of the 20 measurements were less than 10.75, then you have essentially defined the empirical distribution function. More precisely

$$\hat{F}_{X|X_1,\ldots,X_n}(x|x_1, \ldots, x_n)$$
$$= \frac{\text{number of measurements } x_1, \ldots, x_n \text{ which are no greater than } x}{n}$$

(8.3)

Note that $\hat{F}_{X|X_1,\ldots,X_n}$ is a distribution function; that is

$$\hat{F}_{X|X_1,\ldots,X_n}(-\infty|\cdots) = 0$$
$$\hat{F}_{X|X_1,\ldots,X_n}(\infty|\cdots) = 1$$
$$\hat{F}_{X|X_1,\ldots,X_n}(x|x_1, \ldots, x_n) \geq \hat{F}_{X|X_1,\ldots,X_n}(y|x_1, \ldots, x_n), \quad x > y$$

and

$$\lim_{\substack{\Delta x > 0 \\ \Delta x \to 0}} \hat{F}_{X|X_1,\ldots,X_n}(x + \Delta x | x_1, \ldots, x_n) = \hat{F}_{X|X_1,\ldots,X_n}(x | x_1, \ldots, x_n)$$

A probability mass function can be easily derived either from the empirical distribution function or directly from the data. The empirical distribution function and the empirical probability mass function for the resistance data are shown in Figures 8.1 and 8.2. Problems at the end of this chapter call for construction of empirical distribution functions based on the definition in Equation 8.3.

8.3.2 Joint Empirical Distribution Functions

We consider a sample of size n from two random variables X and Y; (X_1, Y_1), (X_2, Y_2), ..., (X_n, Y_n). That is, for each of the n outcomes we observe the sample values of both random variables. Then the joint empirical distribution

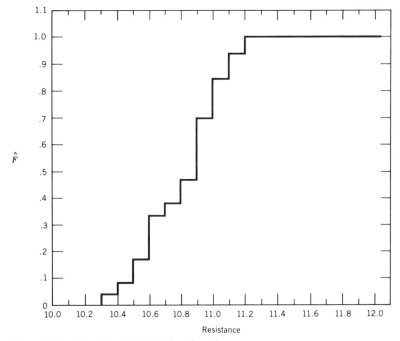

Figure 8.1 Empirical distribution function.

NONPARAMETRIC ESTIMATORS OF PDF'S.

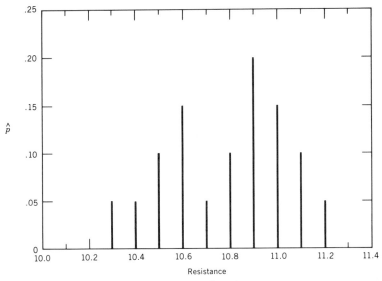

Figure 8.2 Empirical probability mass function.

function is

$$\hat{F}_{X,Y|X_1,Y_1,\ldots,X_n,Y_n}[x_0, y_0|(x_1, y_1), \ldots, (x_n, y_n)]$$
$$= \frac{\text{number of measurements where both } x_i \leq x_0 \text{ and } y_i \leq y_0}{n}$$

Higher dimensional empirical distribution functions can be defined in a similar fashion.

8.3.3 Histograms

When many measurements are available, in order to simplify both data handling and visual presentation, the data are often grouped into cells. That is, the range of data is divided into a number of cells of equal size and the number of data points within each cell is tabulated. This approximation or grouping of the data results in some loss of information. However, this loss is usually more than compensated for by the ease in data handling and interpretation when the goal is visual display.

When the grouped data are plotted as an approximate distribution function, the plot is usually called a cumulative frequency polygon. A graph of the grouped data plotted as an approximate probability mass function in the form of a bar graph is called a *histogram*.

EXAMPLE 8.1.

The values of a certain random sequence are recorded below. Plot a histogram of the values of the random sequence.

3.42	3.51	3.61	3.47	3.36	3.39	3.56
3.48	3.40	3.52	3.59	3.46	3.57	3.45
3.54	3.42	3.50	3.54	3.48	3.47	3.65
3.51	3.46	3.59	3.47	3.40	3.38	3.57
3.48	3.50	3.42	3.54	3.55	3.45	3.43
3.57	3.49	3.50	3.62	3.52	3.61	3.50
3.59	3.48	3.54	3.50	3.41	3.52	3.51
3.63	3.53	3.38	3.49	3.50	3.50	3.58
3.50	3.51	3.47	3.52	3.43	3.49	3.42
3.45	3.44	3.48	3.57	3.49	3.53	3.49
3.51	3.59	3.35	3.60	3.48	3.59	3.61
3.55	3.57	3.40	3.51	3.61	3.49	3.40
3.59	3.55	3.56	3.45	3.56	3.47	3.58
3.50	3.46	3.49	3.41	3.52	3.50	3.47
3.61	3.52					

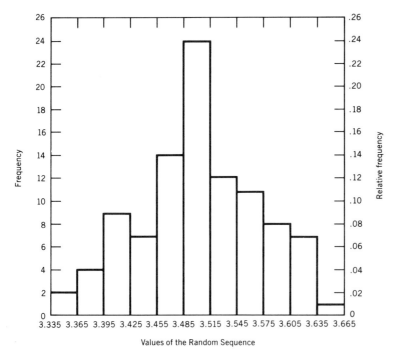

Figure 8.3 Histogram for Example 8.1.

SOLUTION:

The idea is to split the data into groups. The range of the data is 3.35 to 3.65. If each interval is chosen to be .030 units wide, there would be 10 cells and the resulting figure would be reasonable. The center cell of the histogram is chosen to be 3.485 to 3.515 and this, with the interval size chosen, determines all cells. For ease of tabulation it is common practice to choose the ends of a cell to one more significant figure than is recorded in the data so that there is no ambiguity concerning into which cell a reading should be placed. As with most pictorial representations, some of the choices as to how to display the data (e.g., cell size) arc arbitrary; however, the choices do influence the expected errors, as explained in Section 8.5. The histogram is shown in Figure 8.3.

EXAMPLE 8.2.

A random variable has a normal distribution with mean 2 and variance 9. A quantized empirical distribution function is shown along with the true distribution function in Figures 8.4 and 8.5 for samples of size 200 and 2000, respectively.

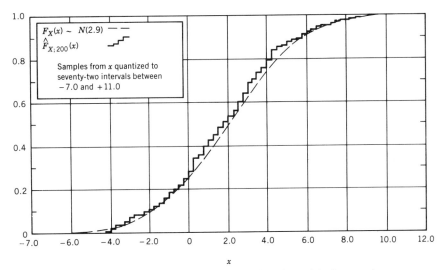

Figure 8.4 Comparison between the theoretical and empirical normal distribution functions, $n = 200$.

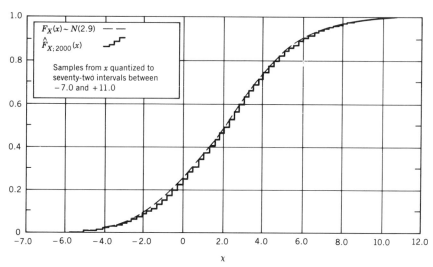

Figure 8.5 Comparison between the theoretical and empirical normal distribution function, $n = 2000$.

8.3.4 Parzen's Estimator for a pdf

The histogram is extensively used to graphically display the estimator of a pdf. Parzen's estimator [9], on the other hand, provides a smoothed estimator of a pdf in analytical form as

$$\hat{f}_X(x) = f_{X|X_1,\ldots,X_n}(x|x_1, \ldots, x_n) \triangleq \frac{1}{nh(n)} \sum_{i=1}^{n} g\left(\frac{x - x_i}{h(n)}\right) \qquad (8.4)$$

In Equation 8.4, n is the sample size, $g(\ )$ is a weighting function, and $h(n)$ is a smoothing factor. In order for $\hat{f}_X(x)$ to be a valid pdf, $g(y)$ and $h(n)$ must satisfy

$$h(n) > 0 \qquad (8.5.\text{a})$$
$$g(y) \geq 0 \qquad (8.5.\text{b})$$

and

$$\int_{-\infty}^{\infty} g(y)\, dy = 1 \qquad (8.5.\text{c})$$

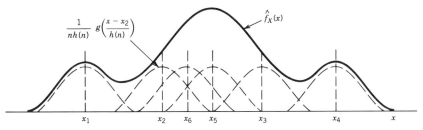

Figure 8.6 Parzen's estimator of a pdf.

While the choice of $h(n)$ and $g(y)$ are somewhat arbitrary, they do influence the accuracy of the estimator as explained in Section 8.5. Recommended choices for $g(y)$ and $h(n)$ are

$$g(y) = \frac{1}{\sqrt{2\pi}} \exp\left(\frac{-y^2}{2}\right) \qquad (8.6.\text{a})$$

$$h(n) = \frac{1}{\sqrt{n}} \qquad (8.6.\text{b})$$

The Parzen estimator is illustrated in Figure 8.6.

8.4 POINT ESTIMATORS OF PARAMETERS

The basic purpose of point estimation is to estimate an unknown parameter with a statistic, that is, with a function of the i.i.d. measurements. Assume that the unknown parameter is θ and that there is a sample that consists of n i.i.d. measurements $X_1, \ldots, X_n$. We then form a statistic $g(X_1, \ldots, X_n)$, which we hope will be close to θ. We will call

$$\hat{\theta} = g(X_1, \ldots, X_n)$$

the *point estimator* of θ. A specific sample will result in $X_i = x_i$, $i = 1, \ldots, n$ and the resulting value of $\hat{\theta}$ will be called the *point estimate* of θ. Thus, the estimator is a random variable that will take on different values depending on the values of the measurements, whereas the estimate is a number.

Before continuing this general discussion of estimation, some specific examples will be given.

8.4.1 Estimators of the Mean

The mean μ_X of a random variable X is usually estimated by the average $\overline{X}$ of the sample, that is

$$\hat{\mu}_X = \overline{X} = \frac{1}{n}\sum_{i=1}^{n} X_i \qquad (8.7)$$

where the X_i's are n i.i.d. measurements or observations from the population with the distribution F_X.

$\overline{X}$ is the most familiar estimator of μ_X. However, the following two estimators of μ_x are sometimes (not often) used:

1. $\dfrac{1}{2}(X_{\max} + X_{\min})$

2. x such that $\hat{F}_{X|X_1,\ldots,X_n}(x|x_1, \ldots, x_n) = \dfrac{1}{2}$

   (called the empirical median)

Methods of comparing the quality of different estimators are introduced in Section 8.5.

8.4.2 Estimators of the Variance

The variance σ_X^2 of a random variable X is commonly estimated by

$$\widehat{\sigma_X^2} = \frac{1}{n}\sum_{i=1}^{n}(X_i - \overline{X})^2 \qquad (8.8.\text{a})$$

or by the estimator S^2 where

$$S^2 = \frac{1}{n-1}\sum_{i=1}^{n}(X_i - \overline{X})^2 \qquad (8.8.\text{b})$$

8.4.3 An Estimator of Probability

Let p be the probability of an event A. Then a statistic that can be used to estimate p is

$$\hat{p} = \frac{N_A}{n}$$

where N_A is the random variable that represents the number of times that event A occurs in n independent trials.

Note that $\hat{p}$ is similar to the relative frequency definition of probability. However, there is an important difference. The relative frequency definition takes the limit as n goes to infinity. This causes N_A/n to become a number whereas for finite n, N_A/n is a random variable with a nonzero variance and a well-defined distribution function.

8.4.4 Estimators of the Covariance

The covariance σ_{XY} is usually estimated by

$$\hat{\sigma}_{XY} = \frac{1}{n} \sum_{i=1}^{n} (X_i - \overline{X})(Y_i - \overline{Y}) \tag{8.9.a}$$

or by

$$\hat{\sigma}_{XY} = \frac{1}{n-1} \sum_{i=1}^{n} (X_i - \overline{X})(Y_i - \overline{Y}) \tag{8.9.b}$$

8.4.5 Notation for Estimators

Our interest in estimating the unknown parameter θ will be reflected by writing the distribution function of the random variable X by

$$F_X(x; \theta)$$

or simply by $F(x; \theta)$ if X is understood to be the random variable. As an example, consider a normal random variable with unit variance and unknown mean μ.

Then

$$f(x; \mu) = f_X(x; \mu) = \frac{1}{\sqrt{2\pi}} \exp\left[\frac{-(x-\mu)^2}{2}\right]$$

The only change is that our present emphasis on estimating the unknown parameter θ causes us to change the notation in order to reflect the fact that we now enlarge the model to include a family of distributions. Each value of θ corresponds with one member of the family. The purpose of the experiment, the concomitant measurements, and the resulting estimator is to select one member of the family as being the "best." One general way of determining estimators is described in the next subsection. Following that is a discussion of criteria for evaluating estimators.

8.4.6 Maximum Likelihood Estimators

We now discuss the most common method of deriving estimators. As before, we use the notation $F(x; \theta)$ to indicate the distribution of the random variable X given the parameter θ. Thus, if $X_1, X_2, \ldots, X_n$ are i.i.d. measurements of X, then $f(x_1, x_2, \ldots, x_n; \theta)$ is the joint pdf of $X_1, \ldots, X_n$ given θ, and because of independence it can be written

$$f(x_1, \ldots, x_n; \theta) = \prod_{i=1}^{n} f(x_i; \theta)$$

Now, if the values of $X_1, \ldots, X_n$ are considered fixed and θ is an unknown parameter, then $f(x_1, \ldots, x_n; \theta)$ is called a *likelihood function* and is usually denoted by

$$L(\theta) = f(x_1, \ldots, x_n; \theta) = \prod_{i=1}^{n} f(x_i; \theta) \qquad (8.10)$$

EXAMPLE 8.3.

X is known to be exponentially distributed, that is

$$f(x; \theta) = \frac{1}{\theta} \exp(-x/\theta), \qquad x \geq 0, \quad \theta > 0$$
$$= 0 \qquad\qquad\qquad\qquad x < 0$$

If a sample consists of five i.i.d. measurements of X that are 10, 11, 8, 12, and 9, find the likelihood function of θ.

SOLUTION:

$$L(\theta) = \prod_{i=1}^{5} f(x_i; \theta) = \prod_{i=1}^{5} \frac{1}{\theta} \exp\left(\frac{-x_i}{\theta}\right) = \theta^{-5} \exp\left(\frac{-50}{\theta}\right), \quad \theta > 0$$

The value $\hat{\theta}$ that maximizes the likelihood function is called a *maximum likelihood estimator* of θ. That is, a value $\hat{\theta}$ such that for all values θ

$$L(\hat{\theta}) = f(x_1, \ldots, x_n; \hat{\theta}) \geq f(x_1, : \ldots, x_n; \theta) = L(\theta) \qquad (8.11)$$

is called a maximum likelihood estimate of θ. Such an estimate is justified on the basis that $\hat{\theta}$ is the value that maximizes the joint probability density (likelihood), given the sample of observations or measurements that was obtained.

EXAMPLE 8.4.

Find the maximum likelihood estimate of θ from Example 8.3.

SOLUTION:

$$L(\theta) = \theta^{-5} \exp\left(\frac{-50}{\theta}\right), \quad \theta > 0$$

$$\frac{dL(\theta)}{d\theta} = -5\theta^{-6} \exp\left(\frac{-50}{\theta}\right) + \theta^{-5}\left(\frac{50}{\theta^2}\right) \exp\left(\frac{-50}{\theta}\right), \quad \theta > 0$$

Setting the derivative equal to zero and solving for $\hat{\theta}$, the value of θ that causes the derivative to equal zero yields the estimate

$$\hat{\theta} = \frac{50}{5} = 10$$

Because a number of the common probability density functions occur in exponential form, it is often more convenient to work with the natural logarithm of the likelihood function rather than the likelihood function. Of course, because the natural logarithm is a strictly increasing function, the value of θ that maximizes the likelihood function is the same value that maximizes the logarithm of the likelihood function.

EXAMPLE 8.5.

There is a sample of n i.i.d. measurements from a normal distribution with known variance σ^2. Find the maximum likelihood estimate of the mean.

SOLUTION:

$$L(\mu) = f(x_1, \ldots, x_n; \mu) = \prod_{i=1}^{n} \frac{1}{\sqrt{2\pi}\sigma} \exp\left[-\frac{(x_i - \mu)^2}{2\sigma^2}\right]$$

Finding the value that maximizes $\ln[L(\mu)]$ is equivalent to finding the value of μ that maximizes $L(\mu)$. Thus

$$g(\mu) = \ln[L(\mu)] = n \ln\left(\frac{1}{\sqrt{2\pi}\sigma}\right) - \frac{1}{2\sigma^2} \sum_{i=1}^{n} (x_i - \mu)^2$$

$$\frac{dg}{d\mu} = -\frac{1}{2\sigma^2} \sum_{i=1}^{n} -2(x_i - \mu)\big|_{\mu=\hat{\mu}} = 0$$

or

$$n\hat{\mu} = \sum_{i=1}^{n} x_i$$

Thus, the maximum likelihood estimator is

$$\hat{\mu} = \frac{1}{n} \sum_{i=1}^{n} X_i$$

Note that in this case the maximum likelihood estimator is simply $\overline{X}$.

EXAMPLE 8.6.

X is uniformly distributed between 0 and θ. Find the maximum likelihood estimator of θ based on a sample of n observations.

SOLUTION:

With

$$f(x; \theta) = \frac{1}{\theta}, \quad 0 \le x \le \theta, \quad 0 < \theta$$
$$= 0 \quad \text{elsewhere}$$

we have

$$L(\theta) = \frac{1}{\theta^n} \quad 0 \le x_i \le \theta, \quad i = 1, 2, \ldots, n$$

where x_i, $i = 1, 2, \ldots, n$ represent the values of the n i.i.d. measurements.

The maximum of $1/\theta^n$ cannot be found by differentiation because the smallest possible value of $\theta > 0$ maximizes $L(\theta)$. The smallest possible value of θ is $\max(x_i)$ because of the constraint $x_i \le \theta$. Thus, the maximum likelihood estimator of θ is

$$\hat{\theta} = \max(X_i)$$

8.4.7 Bayesian Estimators

If the unknown parameter Θ is considered to be a random variable with an appropriate probability space defined and a distribution function F_Θ assigned, then Bayes' rule can be used to find $F_{\Theta|X_1,\ldots,X_n}$. For instance if Θ is continuous and the X_i's are i.i.d. given Θ then

$$f_{\Theta|X_1,\ldots,X_n} = \frac{f_{X_1,\ldots,X_n|\Theta} f_\Theta}{f_{X_1,\ldots,X_n}}$$

$$= \frac{\left[\prod_{i=1}^{n} f_{X_i}(x_i; \theta)\right] f_\Theta(\theta)}{\int_{-\infty}^{\infty} \left[\prod_{i=1}^{n} f_{X_i}(x_i; \lambda)\right] f_\Theta(\lambda) \, d\lambda}$$

492 STATISTICS

The *a posteriori* distribution function, $F_{\Theta|X_1,\ldots,X_n}$, or equivalently the a posteriori density shown above, displays all of the "information" about Θ. However, this distribution combines the information contained in the assigned *a priori* distribution (e.g., f_Θ), as well as the information in the sample. This combination has led to an extensive debate as to whether Bayes' rule is a philosophically sound method for characterizing estimators. In this introductory chapter we do not discuss Bayesian estimation at any length. We conclude this section with an example and we remark that if the a priori distribution of Θ is uniform then the conditional density function of Θ (given $X_1, \ldots, X_n$) is formally the same as the likelihood function.

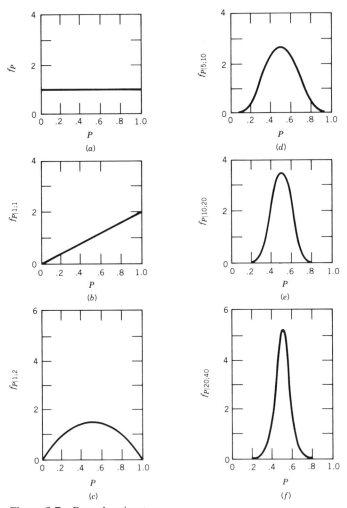

Figure 8.7 Bayes' estimator.

EXAMPLE 8.7.

A thumbtack when tossed has a probability P of landing with the head on the surface and the point up; it has a probability $1 - P$ of landing with both the head and the point touching the surface. Figure 8.7 shows the results of applying Bayes' rule with f_P assigned to be uniform (Figure 8.7a). This figure shows $f_{P|k;n}$, that is, the conditional density function of P when k times out of n tosses the thumbtack has landed with the point up.

A study of Figure 8.7b will reveal that, after one experiment in which the point was up, the a posteriori probability density function of P is zero at $P = 0$ and increases to its maximum value at $P = 1$. This is certainly a reasonable result. Figure 8.7c illustrates, with the ratio of $k/n = \frac{1}{2}$, that the a posteriori density function of P has a peak at $\frac{1}{2}$, and, moreover, that the peak becomes more pronounced as more data are used to obtain the a posteriori density function (Figures 8.7d, 8.7e, and 8.7f).

Point estimators can be derived from the a posteriori distribution. Usually $\hat{\theta} = E\{\Theta | X_1, \ldots, X_n\}$, that is, the mean of the a posteriori distribution is used as the point estimator. Occasionally the mode of the a posteriori distribution is used as the point estimator.

8.5 MEASURES OF THE QUALITY OF ESTIMATORS

As demonstrated in Section 8.4 there can be more than one (point) estimator for an unknown parameter. In this section, we define measures for determining "good" estimators.

What properties do we wish $\hat{\theta}$ to possess? It seems natural to wish that $\hat{\theta} = \theta$, but $\hat{\theta}$ is a random variable (because it is a function of random variables). Thus, we must adopt some probabilistic criteria for measuring how close $\hat{\theta}$ is to θ.

8.5.1 Bias

An estimator $\hat{\theta}$ of θ is called *unbiased* if

$$E\{\hat{\theta}\} = \theta$$

EXAMPLE 8.8.

If $X_1, \ldots, X_n$ are i.i.d. with mean μ then

$$\overline{X} = \hat{\mu} = g(X_1, \ldots, X_n) = \frac{X_1 + X_2 + \cdots + X_n}{n}$$

is an unbiased estimator of the mean μ because

$$E\{\overline{X}\} = E\{\hat{\mu}\} = \frac{E\{X_1 + X_2 \cdots + X_n\}}{n} = \frac{n\mu}{n} = \mu$$

If $E\{\hat{\theta}\} = a \neq \theta$, then $\hat{\theta}$ is said to be *biased* and the *bias* b or bias error, is

$$b[\hat{\theta}] = E\{\hat{\theta}\} - \theta \qquad (8.12)$$

EXAMPLE 8.9.

Compute the expected value of $\widehat{\sigma^2} = \frac{1}{n}\sum_{i=1}^{n}(X_i - \overline{X})^2$ and determine whether it is an unbiased estimator of σ^2. (Note that the subscripts are omitted on both μ and σ^2).

SOLUTION:

First we note, because X_i and X_j are independent, that

$$E\left\{(X_i - \mu)\left(\sum_{j=1}^{n} X_j - \mu\right)\right\}$$

$$= E\{(X_i - \mu)^2\} + \sum_{\substack{j=1 \\ j \neq i}}^{n} E\{X_i - \mu\}E\{X_j - \mu\} = \sigma^2$$

Now, looking at one term of the sum

$$E\{(X_i - \overline{X})^2\} = E\left\{\left[X_i - \frac{1}{n}\sum_{j=1}^{n} X_j\right]^2\right\}$$

$$= E\left\{\left[X_i - \mu + \mu - \frac{1}{n}\sum_{j=1}^{n} X_j\right]^2\right\}$$

$$= E\left\{\left[X_i - \mu - \frac{1}{n}\sum_{j=1}^{n}(X_j - \mu)\right]^2\right\}$$

$$= E\{(X_i - \mu)^2\} - \frac{2}{n} E\left\{(X_i - \mu)\sum_{j=1}^{n}(X_j - \mu)\right\} + \frac{1}{n^2}\sum_{j=1}^{n}\sigma^2$$

Combining this with the first part of the solution, we obtain

$$E\{(X_i - \overline{X})^2\} = \sigma^2 - \frac{2}{n}\sigma^2 + \frac{1}{n}\sigma^2$$

But

$$E\{\widehat{\sigma^2}\} = \frac{1}{n}\sum_i E\{(X_i - \overline{X})^2\} = E\{(X_i - \overline{X})^2\}$$

or

$$E\{\widehat{\sigma^2}\} = \sigma^2 - \frac{2}{n}\sigma^2 + \frac{1}{n}\sigma^2 = \sigma^2\left(1 - \frac{1}{n}\right) = \sigma^2\frac{(n-1)}{n} \quad (8.13)$$

Thus, $\widehat{\sigma^2}$ is a biased estimator of σ^2. However, it is easy to see that

$$S^2 = \frac{1}{n-1}\sum_{i=1}^{n}(X_i - \overline{X})^2$$

is an unbiased estimator of σ^2.

Notice that for large n, $\widehat{\sigma^2}$ and S^2 are nearly the same. The intuitive idea is that one *degree of freedom* is used in determining $\overline{X}$. That is, $\overline{X}, X_1, \cdots, X_{n-1}$ determine X_n. Thus, the sum should be divided by $n - 1$ rather than by n. (See Problem 8.27.)

We have seen that $\overline{X}$ is an unbiased estimator of μ but there are many other unbiased estimators of μ. For instance, the first measurement X_1 is also unbiased. Thus, some other measure(s) of estimators is needed to separate good estimators from those not so good. We now define another measure.

8.5.2 Minimum Variance, Mean Squared Error, RMS Error, and Normalized Errors

If the estimator $\hat{\theta}$ has a mean of θ, then we also desire $\hat{\theta}$ to have a small variation from one sample to the next. This variation can be measured in various ways. For instance, measures of variation are $E\{|\hat{\theta} - \theta|\}$, [maximum $\hat{\theta}$ − minimum $\hat{\theta}$], or $E\{(\hat{\theta} - \theta)^2\}$. Although any of these or other measures might be used, the most common measure is the mean squared error (MSE), or

$$\text{MSE} = E\{(\hat{\theta} - \theta)^2\}$$

If $\hat{\theta}$ is unbiased then the MSE is simply the variance of $\hat{\theta}$. If $E\{\hat{\theta}\} = m$, then

$$E\{(\hat{\theta} - \theta)^2\} = (\theta - m)^2 + \sigma_{\hat{\theta}}^2 \qquad (8.14)$$

This important result may be stated as

$$\text{MSE}(\hat{\theta}) = [\text{Bias}(\hat{\theta})]^2 + \text{Variance}(\hat{\theta})$$

or

$$\text{MSE} = b^2 + \text{Var}[\hat{\theta}]$$

Equation 8.14 may be shown as follows:

$$\begin{aligned}
E\{(\hat{\theta} - \theta)^2\} &= E\{(\hat{\theta} - m + m - \theta)^2\} \\
&= E\{(\hat{\theta} - m)^2\} + 2(m - \theta)E\{\hat{\theta} - m\} + E\{(m - \theta)^2\} \\
&= \sigma_{\hat{\theta}}^2 + 2(m - \theta)(m - m) + (m - \theta)^2 \\
&= \sigma_{\hat{\theta}}^2 + (m - \theta)^2
\end{aligned}$$

We return to the example of estimating the mean of a random variable X when a sample of n i.i.d. measurements $X_1, \cdots, X_n$, are available. Although

$$E\left\{\sum_{i=1}^{n} \frac{X_i}{n}\right\} = E\{X_1\} = \mu$$

the variances of the estimators $\overline{X}$ and X_1 are different. Indeed

$$\mathrm{Var}[X_1] = \sigma^2$$
$$\mathrm{Var}[\overline{X}] = \mathrm{Var}\left[\sum_{i=1}^{n} \frac{X_i}{n}\right] = \frac{\sigma^2}{n} \tag{8.15}$$

The average $\overline{X}$ has a lower variance and by the criterion of minimum variance or of minimum MSE, $\overline{X}$ is a better estimator than a single measurement.

The positive square root of the variance of $\hat{\theta}$ is often called the *standard error* of $\hat{\theta}$ and sometimes also called the random error. The positive square root of the MSE is called the *root mean square error* or the RMS error.

For a given sample size n, $\hat{\theta}_n = g(X_1, \ldots, X_n)$ will be called an *unbiased minimum variance* estimator of the parameter θ if $\hat{\theta}_n$ is unbiased and if the variance of $\hat{\theta}_n$ is less than or equal to the variance of every other unbiased estimator of θ that uses a sample of size n.

If $\theta \neq 0$, the normalized bias, normalized standard deviation (also called the coefficient of variation), and the normalized RMS error are defined as follows:

$$\text{Normalized bias} = \epsilon_b = \frac{b(\hat{\theta})}{\theta} \tag{8.16}$$

$$\text{Normalized standard error} = \epsilon_r = \frac{\sigma_{\hat{\theta}}}{\theta} \tag{8.17}$$

$$\text{Normalized RMS error} = \epsilon = \frac{\sqrt{E\{(\hat{\theta} - \theta)^2\}}}{\theta} \tag{8.18.a}$$

$$\text{Normalized MSE} = \epsilon^2 = \frac{E\{(\hat{\theta} - \theta)^2\}}{\theta^2} \tag{8.18.b}$$

8.5.3 The Bias, Variance, and Normalized RMS Errors of Histograms

In order to illustrate the measures of the quality of estimators introduced in the last section, we use the histogram as an estimator of the true but unknown probability density function and evaluate the bias, variance, and normalized RMS error. The true but unknown probability density function of X is $f(x)$, and there are N measurements to be placed in n cells each of width W.

Now, in order that the area under the histogram equal one, the height of each cell will be normalized by dividing the number N_i of samples in cell i by NW; thus, the area of cell i is

$$(\text{Area})_i = \frac{N_i}{NW} \cdot W = \frac{N_i}{N}$$

Because

$$\sum_{i=1}^{n} N_i = N \quad \text{then} \quad \sum_{i=1}^{n} (\text{Area})_i = 1$$

Thus, the estimate $\hat{f}_i(x)$, of $f(x)$ for x within the ith cell is

$$\hat{f}_i(x) = \frac{N_i}{NW} \quad \text{for} \quad x_{i,L} \le x \le x_{i,U} \qquad (8.19)$$

where $x_{i,L}$ and $x_{i,U}$ are respectively the lower and upper limits of the ith cell. Note that N and W are constants, whereas N_i is a random variable.

Bias of $\hat{f}_i(x)$. The expected value of $\hat{f}_i(x)$ is

$$E\{\hat{f}_i(x)\} = \frac{1}{NW} E\{N_i\}, \quad x_{i,L} \le x \le x_{i,U} \qquad (8.20)$$

N_i is a binomial random variable because each independent sample has a constant probability $P(i)$ of falling in the ith cell. Thus

$$E\{N_i\} = NP(i) \qquad (8.21)$$

where

$$P(i) = \int_{x_{i,L}}^{x_{i,U}} f(x)\, dx \qquad (8.22)$$

Using Equations 8.21 and 8.22 in Equation 8.20

$$E[\hat{f}_i(x)] = \frac{1}{W} \int_{x_{i,L}}^{x_{i,U}} f(x)\, dx, \quad x_{i,L} \le x \le x_{i,U} \qquad (8.23)$$

In general, Equation 8.23 will not equal $f(x)$ for all values of x within the ith cell. Although there will be an x within the ith cell for which this is an unbiased estimator, for most $f(x)$ and most x, Equation 8.19 describes a biased estimator. In order to illustrate the bias, we will choose x_{c_i} to be in the center of the ith cell and expand $f(x)$ in a second-order Tayor series about this point,

MEASURES OF THE QUALITY OF ESTIMATORS

that is

$$f(x) \approx f(x_{c_i}) + f'(x_{c_i})(x - x_{c_i}) + f''(x_{c_i})\frac{(x - x_{c_i})^2}{2}, \quad x_{i,L} \le x \le x_{i,U}$$

In this case, using Equation 8.23

$$E[\hat{f}_i(x)] \approx \frac{1}{W} \int_{x_{c_i}-W/2}^{x_{c_i}+W/2} \left[f(x_{c_i}) + f'(x_{c_i})(x - x_{c_i}) + f''(x_{c_i})\frac{(x - x_{c_i})^2}{2} \right] dx$$

or

$$E[\hat{f}_i(x)] \approx f(x_{c_i}) + f''(x_{c_i})\frac{W^2}{24}, \quad x_{i,L} \le x \le x_{i,U} \quad (8.24)$$

Thus the bias, or bias error, of $\hat{f}_i(x)$ when it is used as an estimator of $f(x_{c_i})$ is approximately

$$b[\hat{f}_i(x)] \approx f''(x_{c_i})\frac{W^2}{24}, \quad x_{i,L} \le x \le x_{i,U} \quad (8.25)$$

The normalized bias is

$$\epsilon_b \approx \frac{f''(x_{c_i})}{f(x_{c_i})}\frac{W^2}{24} \quad (8.26)$$

Note that the bias and the normalized bias increase with cell width.

Variance of $\hat{f}_i(x)$. Using Equation 8.19 and the fact that N_i is a binomial random variable, the variance of $\hat{f}_i(x)$ is

$$\text{Var}[\hat{f}_i(x)] = \left(\frac{1}{NW}\right)^2 N[P(i)][1 - P(i)], \quad x_{i,L} \le x \le x_{i,U} \quad (8.27)$$

where $P(i)$ is given by Equation 8.22 and approximately by

$$P(i) = \int_{x_{i,L}}^{x_{i,U}} f(x)\, dx \approx W f(x_{c_i}) \quad (8.28)$$

Thus, using Equation 8.28 in Equation 8.27

$$\text{Var}[\hat{f}_i(x)] \approx \frac{1}{NW} f(x_{c_i})[1 - Wf(x_{c_i})], \qquad x_{i,L} \le x \le x_{i,U} \qquad (8.29)$$

The normalized standard error ϵ_r assuming $f(x_{c_i})$ is being estimated is

$$\epsilon_r = \frac{\sqrt{\text{Var}\,\hat{f}_i(x)}}{f(x_{c_i})} \approx \frac{1}{\sqrt{NW}\sqrt{f(x_{c_i})}} \sqrt{1 - Wf(x_{c_i})} \qquad (8.30)$$

The variance and normalized standard error decrease with W for N fixed. Thus, the cell width has opposite effects on bias and variance. However, increasing the number N of samples reduces the variance and can also reduce the bias if the cell size is correspondingly reduced.

Normalized MSE of $\hat{f}_i(x)$. The MSE, is

$$\begin{aligned}
\text{MSE} &= b^2 + \text{Var}[\hat{f}_i(x)] \\
&\approx [f''(x_{c_i})]^2 \frac{W^4}{576} + \frac{1}{NW} f(x_{c_i})[1 - Wf(x_{c_i})]
\end{aligned} \qquad (8.31)$$

and the normalized MSE is

$$\epsilon^2 \approx \left[\frac{f''(x_{c_i})}{f(x_{c_i})}\right]^2 \frac{W^4}{576} + \frac{[1 - Wf(x_{c_i})]}{NWf(x_{c_i})} \qquad (8.32)$$

The normalized RMS error, ϵ, is the positive square root of this expression.

Note that $Wf(x_{c_i})$ is less than one; thus increasing W clearly decreases the variance or random error. However, increasing W tends to increase the bias. If $N \to \infty$ and $W \to 0$ in such a way that $NW \to \infty$ (e.g., $W = 1/\sqrt{N}$) then the MSE will approach zero as $N \to \infty$. Note that if $f''(x_{c_i})$ and higher derivatives are small, that is, there are no abrupt peaks in the density function, then the bias will also approach zero.

Usually, when we attempt to estimate the pdf using this nonparametric method, the estimators will be biased. Any attempt to reduce the bias will result in an increase in the variance and attempts to reduce the variance by smoothing (i.e., increasing W) will increase the bias. Thus, there is a trade-off to be made between bias and variance. By a careful choice of the smoothing factor we can make the bias and variance approach zero as the sample size increases (i.e., the estimator can be made to be asymptotically unbiased).

8.5.4 Bias and Variance of Parzen's Estimator

We now consider the bias and variance of the estimator given in Equation 8.4. Parzen has shown (see Reference [9]) that if $g(y)$ satisfies the constraints of Equation 8.5 and the additional constraints

$$|y\, g(y)| \to 0 \quad \text{as} \quad |y| \to \infty$$

and

$$\int_{-\infty}^{\infty} y^2 g^2(y)\, dy < \infty$$

then the bias and variance of the estimator

$$\hat{f}_X(x) = \frac{1}{nh(n)} \sum_{i=1}^{n} g\left(\frac{x - x_i}{h(n)}\right)$$

are

$$\text{Bias}[\hat{f}_X(x)] = -f''_X(x) \frac{h^2(n)}{2} \int_{-\infty}^{\infty} y^2 g(y)\, dy \tag{8.33}$$

$$\text{Variance}[\hat{f}_X(x)] = \frac{f_X(x)}{nh(n)} \int_{-\infty}^{\infty} g^2(y)\, dy \tag{8.34}$$

for large values of n.

The foregoing expressions are similar to expressions for the bias and variance of the histogram type estimator (Equations 8.25 and 8.29). Once again if $h(n)$ is chosen such that

$$h(n) \to 0 \quad \text{as} \quad n \to \infty$$

and

$$nh(n) \to \infty \quad \text{as} \quad n \to \infty$$

then the estimator is asymptotically unbiased and the variance of the estimator approaches zero. $h(n) = 1/\sqrt{n}$ is a reasonable choice.

8.5.5 Consistent Estimators

Any statistic or estimator that converges in probability (see Chapter 2 for definition) to the parameter being estimated is called a consistent estimator of that parameter.

For example

$$\overline{X}_n = \frac{1}{n} \sum_{i=1}^{n} X_i$$

has mean μ and variance σ^2/n. Thus, as $n \to \infty$, $\overline{X}_n$ has mean μ and a variance that approaches 0. Thus, $\overline{X}_n$ converges in probability to μ (by Tchebycheff's inequality) and $\overline{X}_n$ is a consistent estimator of μ. (See Problem 8.15.)

Note also that both

$$\frac{1}{n} \sum (X_i - \overline{X}_n)^2$$

and

$$\frac{1}{n-1} \sum (X_i - \overline{X}_n)^2$$

are consistent estimators of σ^2.

8.5.6 Efficient Estimators

Let $\hat{\theta}_1$ and $\hat{\theta}_2$ be unbiased estimators of θ. Then we define the (relative) *efficiency* of $\hat{\theta}_1$ with respect to $\hat{\theta}_2$ as

$$\frac{\text{Var}(\hat{\theta}_2)}{\text{Var}(\hat{\theta}_1)}$$

In some cases it is possible to find among the unbiased estimators one that has the minimum variance, V. In such a case, the absolute efficiency of an unbiased estimator $\hat{\theta}_1$ is

$$\frac{V}{\text{Var}(\hat{\theta}_1)}$$

For example, $\overline{X}$ is an unbiased estimator of μ and it has a variance of σ^2/n. The mean of a normal distribution could also be estimated by $\hat{m}$, the median of the empirical distribution function. It can be shown that $\hat{m}$ is unbiased and that, for large n

$$\text{Var}[\hat{m}] \approx \frac{\pi\sigma^2}{2n}$$

Thus, the efficiency of $\hat{m}$ with respect to $\overline{X}$ is approximately

$$\left(\frac{\sigma^2}{n}\right) \bigg/ \left(\frac{\pi\sigma^2}{2n}\right) = \frac{2}{\pi} \approx 64\%$$

8.6 BRIEF INTRODUCTION TO INTERVAL ESTIMATES

In most engineering problems the best point estimate of an unknown parameter is needed for design equations. In addition, it is always important to know how much the point estimate can be expected to vary from the true but unknown parameter. We have emphasized the MSE of the estimator as a prime measure of the expected variation. However, in some problems, one is interested in whether the unknown parameter is within a certain interval (with a high probability). In these cases an *interval estimate* is called for. Such interval estimates are often called *confidence limits*. We do not emphasize interval estimates in this book. We will conclude the very brief introduction with one example where the distribution of the statistic is particularly simple. This example defines and illustrates interval estimates. Distributions of other common statistics are found in the next section, and they could be used to find interval estimates or confidence limits for the unknown parameter.

EXAMPLE 8.10.

X is a normal random variable with $\sigma_X^2 = 1$. The mean μ of X is estimated using a sample size of 10. Find the random interval $I = [(\overline{X} - a), (\overline{X} + a)]$ such that $P[\mu \in I] = P[\overline{X} - a \le \mu \le \overline{X} + a] = 0.95$ ("I" is called the 95% confidence interval for μ_X).

SOLUTION:

It is easy to show that

$$\overline{X} = \frac{1}{10}\sum_{i=1}^{10} X_i$$

is normal with mean μ and variance $\sigma^2/n = 1/10$ (see Problem 8.22). Thus, $(\overline{X} - \mu)/\sqrt{.1}$ is a standard normal random variable, and from Appendix D we can conclude that

$$P\left(-1.96 \leq \frac{\overline{X} - \mu}{\sqrt{.1}} \leq 1.96\right) = .95$$

These inequalities can be rearranged to

$$\overline{X} - 1.96\sqrt{.1} \leq \mu \leq \overline{X} + 1.96\sqrt{.1} \approx \overline{X} + .62$$

The interval $[\overline{X} - 0.62, \overline{X} + 0.62]$ is the 95% confidence interval for μ.

8.7 DISTRIBUTION OF ESTIMATORS

In Section 8.5, we discussed the mean and variance of estimators. In some cases, for example in interval estimation, we may be interested in a more complete description of the distribution of the estimators. This description is particularly needed in tests of hypotheses that will be considered in the next section. In this section we find the distribution of four of the most used estimators.

8.7.1 Distribution of $\overline{X}$ with Known Variance

If $X_1, \ldots, X_n$ are i.i.d. observations from a normal distribution X with mean μ and variance σ^2, then it is easy to show that

$$\overline{X} = \frac{1}{n}\sum_{i=1}^{n} X_i$$

is also normal with mean μ and variance σ^2/n (see Problem 8.22). Note also that the central limit theorem implies that, as $n \to \infty$, $\overline{X}$ will be normal, almost regardless of the distribution of X.

Note that $(\overline{X} - \mu)/(\sigma/\sqrt{n})$ is a standard ($\mu = 0$, $\sigma^2 = 1$) normal random variable.

EXAMPLE 8.11.

X_i is normal with mean 1000 and variance 100. Find the distribution of $\overline{X}_n$ when $n = 10, 50, 100$.

SOLUTION:

From the previous results, $\overline{X}_n$ is normal with mean 1000, and the variances of $\overline{X}_n$ are

$$\text{Var}[\overline{X}_{10}] = 10$$
$$\text{Var}[\overline{X}_{50}] = 2$$
$$\text{Var}[\overline{X}_{100}] = 1$$

Note that with a sample of size 100, $\overline{X}_{100}$ has a probability of .997 of being within 1000 ± 3.

8.7.2 Chi-square Distribution

We assume that $X_1, \ldots, X_m$ are i.i.d. normal random variables with mean 0 and variance σ^2. Their joint density is

$$f(x_1, x_2, \ldots, x_m) = \frac{1}{(2\pi)^{m/2}\sigma^m} \exp\left(-\frac{1}{2\sigma^2} \sum_{i=1}^{m} x_i^2\right)$$

We now define $Z_i = X_i/\sigma$, $i = 1, \ldots, m$. Thus, Z_i is standard normal. We first find the probability density function of $Y_i = Z_i^2$. Since

$$P(Y_i \leq y) = P(-\sqrt{y} \leq Z_i \leq \sqrt{y})$$

we have

$$F_{Y_i}(y) = \int_{-\sqrt{y}}^{+\sqrt{y}} \frac{1}{\sqrt{2\pi}} \exp(-z^2/2) \, dz$$

and

$$f_{Y_i}(y) = \frac{dF_Y(y)}{dy} = \frac{1}{\sqrt{2\pi}} y^{-1/2} \exp\left(\frac{-y}{2}\right), \quad y \geq 0$$
$$= 0 \quad y < 0 \quad (8.35)$$

The probability density function given by Equation 8.35 is called chi-square with *one degree of freedom*. (Equation 8.35 may also be found by using the change-of-variable technique introduced in Chapter 2.)

From the characteristic function of a standard normal density function, we have (see Example 2.13)

$$E\{Z_i^2\} = 1$$

and

$$E\{Z_i^4\} = 3$$

Thus

$$E\{Y_i\} = E\{Z_i^2\} = 1 \quad (8.36)$$

and

$$\text{Var}[Y_i] = E\{Y_i^2\} - \mu_{Y_i}^2 = E\{Z_i^4\} - 1 = 2 \quad (8.37)$$

The characteristic function Ψ of Y_i is

$$\Psi_{Y_i}(\omega) = \int_0^\infty \frac{1}{\sqrt{2\pi}} \exp(+j\omega y) y^{-1/2} \exp\left(\frac{-y}{2}\right) dy$$
$$= (1 - j2\omega)^{-1/2} \quad (8.38)$$

Now we define the *chi-square random variable*

$$\chi_m^2 \triangleq \sum_{i=1}^m Y_i = \sum_{i=1}^m Z_i^2 = \sum_{i=1}^m \frac{X_i^2}{\sigma^2}$$

The characteristic function of χ_m^2 is Equation 8.38 raised to the mth power, or

$$\Psi_{\chi_m^2}(\omega) = (1 - j2\omega)^{-m/2} \tag{8.39}$$

and it can be shown (see Problem 8.23) that

$$f_{\chi_m^2}(y) = \frac{1}{2^{m/2}\Gamma(m/2)} y^{(m/2-1)} \exp\left(\frac{-y}{2}\right), \quad y \geq 0$$
$$= 0 \quad\quad y < 0 \tag{8.40}$$

where $\Gamma(n)$ is the *gamma function* defined by

$$\Gamma(n) \triangleq \int_0^\infty y^{n-1} \exp(-y)\, dy, \quad n > 0$$

The gamma function has the following properties:

$$\Gamma(1/2) = \sqrt{\pi}$$
$$\Gamma(1) = 1$$
$$\Gamma(n) = (n-1)\Gamma(n-1), \quad n > 1$$

and

$$\Gamma(n) = (n-1)!, \quad n \text{ a positive integer}$$

Equation 8.40 defines the chi-square pdf; the parameter m is called the *degrees of freedom*.

Note that if $R = \sqrt{\chi_2^2}$ then R is called a Rayleigh random variable. Furthermore, $M = \sqrt{\chi_3^2}$ is called a Maxwell random variable. The density functions for various degrees of freedom are shown in Figure 8.8. Appendix E is a table of the chi-square distribution.

Using Equations 8.36 and 8.37, it is apparent that the mean and variance of a chi-square random variable are

$$E\{\chi_m^2\} = mE\{Z_i^2\} = m \tag{8.41}$$
$$E\{(\chi_m^2 - m)^2\} = m\text{Var}[Y_i] = 2m \tag{8.42}$$

508   STATISTICS

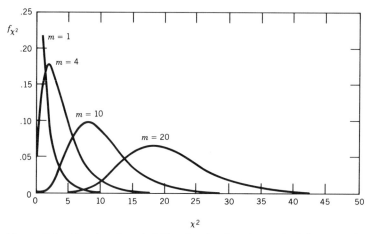

Figure 8.8 Graphs of probability density functions of χ_m^2 for $m = 1$, 4, 10, and 20.

8.7.3 (Student's) t Distribution

Let Z be a standard normal random variable and Y_m be an independent chi-square variable with the density as given in Equation 8.40. Then the joint density of Z and Y_m is

$$f_{Z,Y_m}(z, y) = \frac{1}{\sqrt{2\pi}} \exp\left(\frac{-z^2}{2}\right) \frac{1}{2^{m/2}\Gamma(m/2)} y^{(m/2-1)} \exp\left(\frac{-y}{2}\right), \quad y \geq 0$$
$$= 0 \quad\quad\quad\quad\quad\quad\quad\quad\quad\quad\quad\quad\quad\quad\quad\quad\quad\quad\quad\quad\quad\quad\quad\quad y < 0$$
(8.43)

We define a new random variable T_m by

$$T_m = \frac{Z}{\sqrt{\dfrac{Y_m}{m}}} \tag{8.44}$$

We find the density of T_m by finding the joint density of

$$T_m = \frac{Z}{\sqrt{\dfrac{Y_m}{m}}} \quad \text{and} \quad U = Y_m$$

DISTRIBUTION OF ESTIMATORS

and then finding the marginal density function of T_m. The joint density T_m and U is given by

$$f_{T_m,U}(t, u) = f_{Z,Y_m}\left(t\frac{\sqrt{u}}{\sqrt{m}}, u\right)|J|$$

$$= \frac{1}{\sqrt{2\pi}\, 2^{m/2}\Gamma(m/2)} u^{(m/2-1)}$$

$$\times \exp\left\{-\frac{u}{2}\left(1 + \frac{t^2}{m}\right)\right\}\left|\frac{\sqrt{u}}{\sqrt{m}}\right|, \quad u \geq 0$$

The marginal density of T_m is

$$f_{T_m}(t) = \int_0^\infty \frac{1}{\sqrt{2\pi}\, 2^{m/2}\Gamma(m/2)\sqrt{m}} u^{(m-1)/2} \exp\left\{-\frac{u}{2}\left(1 + \frac{t^2}{m}\right)\right\} du$$

Let

$$w = \frac{u}{2}\left(1 + \frac{t^2}{m}\right)$$

Then

$$f_{T_m}(t) = \frac{1}{\sqrt{2\pi}\, 2^{m/2}\Gamma(m/2)\sqrt{m}} \int_0^\infty \exp(-w)\left[\frac{2w}{\left(1 + \frac{t^2}{m}\right)}\right]^{(m-1)/2} \frac{2}{1 + \frac{t^2}{m}} dw$$

$$= \frac{2^{(m+1)/2}}{\sqrt{2\pi}\, 2^{m/2}\Gamma(m/2)\sqrt{m}\left(1 + \frac{t^2}{m}\right)^{(m+1)/2}} \Gamma\left(\frac{m+1}{2}\right)$$

or

$$f_{T_m}(t) = \frac{\Gamma\left(\frac{m+1}{2}\right)}{\sqrt{\pi m}\,\Gamma(m/2)\left(1 + \frac{t^2}{m}\right)^{(m+1)/2}} \quad (8.45)$$

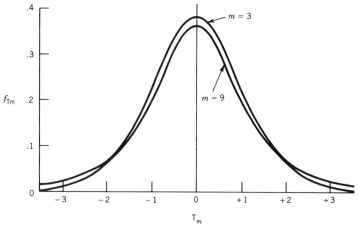

Figure 8.9 Two examples of the T_m density.

The probability density function given in Equation 8.45 is usually called the (*Student's*) t_m *density*. Note that the only parameter in Equation 8.45 is m, the number of independent squared normal random variables in the chi-square. A sketch of f_{T_m} is shown in Figure 8.9 for two values of m and Appendix F is a table of the t_m distribution.

8.7.4 Distribution of S^2 and $\overline{X}$ with Unknown Variance

In the appendix to this chapter, we show that for n i.i.d. normal random variables, $n \geq 2$, $\overline{X}$ and S^2 are independent random variables, and that $(n-1)S^2/\sigma^2$ has the standard chi-square distribution with $m = n-1$ degrees of freedom. Also $\sqrt{n}\,(\overline{X} - \mu)/S$ has the (Student's) t_m distribution with the $m = n-1$.

EXAMPLE 8.12.

If there are 31 measurements from a normal population with $\sigma^2 = 10$, what is the probability that S^2 exceeds 6.17?

SOLUTION:

$$P[S^2 > 6.17] = P\left[\frac{(n-1)S^2}{\sigma^2} > \frac{6.17(n-1)}{\sigma^2}\right]$$
$$= P(\chi^2_{30} > 18.5) \approx .95$$

where .95 is obtained from Appendix E with 30 degrees of freedom.

EXAMPLE 8.13.

Nine i.i.d. measurements from a normal random variable produce a sample mean $\overline{X}$ and a sample standard deviation S. Find the probability that $\overline{X}$ will be more than .77S above its true mean.

SOLUTION:

$$P\{\overline{X} - \mu > 0.77S\} = P\left\{\frac{\overline{X} - \mu}{S} > 0.77\right\}$$

$$= P\left\{\sqrt{9}\,\frac{\overline{X} - \mu}{S} > (0.77)(\sqrt{9})\right\}$$

since $\sqrt{n}\,(\overline{X} - \mu)/S$ has a t_m distribution with $m = 8$, that is, t_8, we find

$$P\{(\overline{X} - \mu) > .77S\} = P\{t_8 > 2.31\} \approx .025$$

where .025 is obtained from Appendix F.

8.7.5 F Distribution

We seek the probability density function of the ratio of two independent chi-square random variables. To this end consider two independent variables, U and V, each of which has a density given by Equation 8.40 where U has the parameter m_1 degrees of freedom, and V has m_2 degrees of freedom. The joint density of these two independent RV's is

$$f_{U,V}(u, v) = \frac{1}{2^{(m_1+m_2)/2}\Gamma(m_1/2)\,\Gamma(m_2/2)}\, u^{(m_1/2-1)}\, v^{(m_2/2-1)}$$

$$\times \exp\left[-\frac{(u+v)}{2}\right], \quad u \geq 0,\ v \geq 0 \quad (8.46)$$

We define new random variables F and Z by

$$F = \frac{\left(\dfrac{U}{m_1}\right)}{\left(\dfrac{V}{m_2}\right)} \quad \text{and} \quad Z = V$$

512 STATISTICS

We now find the joint density of F and Z and then integrate with respect to z to find the marginal density of F. From Equation 8.46 and using the defined transformation

$$f_{F,Z}(\lambda, z) = \frac{1}{\Gamma\left(\frac{m_1}{2}\right)\Gamma\left(\frac{m_2}{2}\right)2^{(m_1+m_2)/2}}\left(\frac{m_1}{m_2}z\lambda\right)^{(m_1/2-1)}$$

$$\times z^{(m_2/2-1)}\exp\left[-\frac{z}{2}\left(\frac{m_1\lambda}{m_2}+1\right)\right]\frac{m_1 z}{m_2}, \quad \lambda, z > 0$$

$$f_F(\lambda) = \int_0^\infty \frac{\left(\frac{m_1}{m_2}\right)^{m_1/2}\lambda^{(m_1/2-1)}}{\Gamma\left(\frac{m_1}{2}\right)\Gamma\left(\frac{m_2}{2}\right)2^{(m_1+m_2)/2}} z^{(m_1+m_2)/2-1}$$

$$\times \exp\left[-\frac{z}{2}\left(\frac{m_1\lambda}{m_2}+1\right)\right] dz$$

Let

$$y = \frac{z}{2}\left(\frac{m_1\lambda}{m_2}+1\right)$$

then

$$f_F(\lambda) = \frac{\left(\frac{m_1}{m_2}\right)^{m_1/2}\lambda^{(m_1/2-1)}}{\Gamma\left(\frac{m_1}{2}\right)\Gamma\left(\frac{m_2}{2}\right)2^{(m_1+m_2)/2}} \int_0^\infty \left[2y\bigg/\left(\frac{m_1\lambda}{m_2}+1\right)\right]^{(m_1+m_2)/2-1}$$

$$\times \exp(-y)\left(\frac{2}{\frac{m_1\lambda}{m_2}+1}\right) dy$$

$$= \frac{\left(\frac{m_1}{m_2}\right)^{m_1/2}\lambda^{(m_1/2-1)}}{\Gamma\left(\frac{m_1}{2}\right)\Gamma\left(\frac{m_2}{2}\right)\left(\frac{m_1\lambda}{m_2}+1\right)^{(m_1+m_2)/2}}\Gamma\left(\frac{m_1+m_2}{2}\right), \quad \lambda \geq 0$$

$$= 0 \qquad\qquad\qquad\qquad\qquad\qquad\qquad\qquad \text{elsewhere} \quad (8.47)$$

The distribution described in Equation 8.47 is called the F distribution. It is determined by the two parameters m_1 and m_2 and is often called F_{m_1,m_2}. Appendix G is a table of the F_{m_1,m_2} distribution.

EXAMPLE 8.14.

Let $F_{10,20}$ be the ratio of the two quantities

$$\sum_{i=1}^{11} \frac{(X_i - \overline{X}_1)^2}{10}$$

and

$$\sum_{j=12}^{32} \frac{(X_j - \overline{X}_2)^2}{20}$$

where $\overline{X}_1$ is the sample mean of first 11 samples and $\overline{X}_2$ is the mean of the next 21 samples. If these 32 i.i.d. samples were from the same normal random variable, find the probability that $F_{10,20}$ exceeds 1.94.

SOLUTION:

From Appendix G

$$P(F_{10,20} > 1.94) \approx .10.$$

8.8 TESTS OF HYPOTHESES

Aside from parameter estimation, the most used area of statistics is statistical tests or tests of hypotheses. Examples of hypotheses tested are (1) The addition of fertilizer causes no change in the yield of corn. (2) There is no signal present in a specified time interval. (3) A certain medical treatment does not affect cancer. (4) The random process is stationary. (5) A random variable has a normal distribution.

We will introduce the subject of hypothesis testing by considering an example introduced in Chapter 6 and as we discuss this example, we will define the general characteristics of hypothesis testing. Then we will introduce the concept of a composite alternative hypothesis. This will be followed by specific examples of hypothesis tests.

8.8.1 Binary Detection

A binary communication system transmits one of two signals, say 0 or 1. The communication channel adds noise to the transmitted signal. The received signal (transmitted signal plus noise) is then operated on by the receiver (e.g., matched filter) and a voltage is the output. If there were no noise, we assume the output would be 0 if 0 were transmitted, and the output would be 1 if 1 were transmitted. However, the noise can cause the voltage to be some other value.

We will test the hypothesis that 0 was the transmitted signal and, as in Chapter 6, we will call this hypothesis H_0. The alternative, that is, that a 1 was transmitted, will be called H_1. The receiver's output voltage Y is the output of the transmitted signal plus the noise assumed to be Gaussian with mean 0 and variance 1. Hence,

$$f_{Y|H_0}(\lambda|H_0) = \frac{1}{\sqrt{2\pi}} \exp\left(-\frac{\lambda^2}{2}\right) \tag{8.48}$$

and

$$f_{Y|H_1}(\lambda|H_1) = \frac{1}{\sqrt{2\pi}} \exp\left[-\frac{(\lambda-1)^2}{2}\right] \tag{8.49}$$

For reasons that will be more apparent when composite alternatives are considered, the first hypothesis stated above, that is, that a 0 was transmitted, will be called the *null hypothesis*, H_0. The alternative that a 1 was transmitted will be called the *alternative hypothesis*, H_1. Obviously, these definitions could be reversed, but in many examples one hypothesis is more attractive for designation as the null hypothesis.

We must decide which alternative to accept and which to reject. We now assume that we will accept H_0, if the received signal Y is below the value γ and we will accept H_1 if the received signal is equal to or above γ. Note that the exact value of γ has not yet been determined. However, we hope that it is intuitively clear that small values of Y suggest H_0, whereas large values of Y suggest H_1 (see Chapter 6 and Figure 8.10).

The set of values for which we reject H_0, that is, $Y > \gamma$, is often called the *critical region* of the test. We can make two kinds of errors:

Type I error: Reject H_0 (accept H_1) when H_0 is true.
Type II error: Reject H_1 (accept H_0) when H_1 is true.

The probability of a type I error, often called the *significance level* of the test, is (in the example)

$$P(\text{type I error}) = P(Y \geq \gamma | H_0)$$
$$= \int_\gamma^\infty \frac{1}{\sqrt{2\pi}} \exp\left[-\frac{\lambda^2}{2}\right] d\lambda \tag{8.50}$$

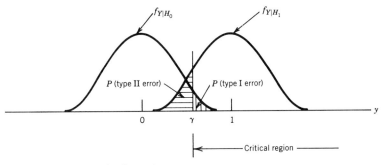

Figure 8.10 Hypothesis testing.

The probability of a type II error is

$$P(\text{type II error}) = P(Y < \gamma | H_1)$$
$$= \int_{-\infty}^{\gamma} \frac{1}{\sqrt{2\pi}} \exp\left[-\frac{(\lambda - 1)^2}{2}\right] d\lambda \qquad (8.51)$$

By choosing γ such that $P(Y \geq \gamma | H_0) = \alpha$ we can control the probability of a type I error. It is usual practice to set α at say .01 or .001 or .05, depending on our desire to control the type I error, or the significance level of the test. In this example, we will choose α to be .05 and say that the significance level is 5%. Thus, using Equation 8.50

$$.05 = \int_{\gamma}^{\infty} \frac{1}{\sqrt{2\pi}} \exp\left[-\frac{\lambda^2}{2}\right] d\lambda$$
$$= Q(\gamma) \qquad (8.52)$$

where $Q(\gamma)$ is one minus the distribution function of a standard Gaussian random variable. Thus, using the table of $Q(\gamma)$ in Appendix D

$$\gamma \approx 1.65$$

With γ set at 1.65, then the probability of a type II error is, using Equation 8.51

$$P(\text{Type II error}) = \int_{-\infty}^{1.65} \frac{1}{\sqrt{2\pi}} \exp\left[-\frac{(\lambda - 1)^2}{2}\right] d\lambda$$
$$= 1 - Q(.65) \approx .7422$$

Thus, in this example, the probability of a type II error seems excessively high. This means that when a 1 is sent, we are quite likely to call it a zero.

Can we reduce this probability of a type II error? Assuming that more than one sample of the received signal can be taken and that the samples are independent, then we could test the same hypothesis with

$$\overline{Y} = \sum_{i=1}^{n} \frac{Y_i}{n}$$

and expect that with the same probability of a type I error, the probability of a type II error would be reduced to an acceptable level. If, for example, $n = 10$, then the variance of $\overline{Y}$ would be $1/10$ and Equation 8.52 would become

$$.05 = \int_{\gamma}^{\infty} \frac{1}{\sqrt{.2\pi}} \exp\left[-\frac{\lambda^2}{.2}\right] d\lambda \qquad (8.53)$$

and with $z = \sqrt{10}\lambda$

$$.05 = \int_{\sqrt{10}\gamma}^{\infty} \frac{1}{\sqrt{2\pi}} \exp\left[-\frac{z^2}{2}\right] dz$$

Thus

$$Q(\sqrt{10}\gamma) = .05 \Rightarrow \gamma \approx \frac{1.65}{\sqrt{10}} \approx .52$$

and the type II error is reduced to

$$P(\text{type II error}) \approx \int_{-\infty}^{.52} \frac{1}{\sqrt{.2\pi}} \exp\left[-\frac{(\lambda-1)^2}{.2}\right] d\lambda$$

$$= \int_{-\infty}^{-.48\sqrt{10}} \frac{1}{\sqrt{2\pi}} \exp\left[-\frac{z^2}{2}\right] dz$$

$$\approx .065$$

Note that not only is the type II error reduced to a more acceptable level, but also the dividing line γ for decision is now approximately halfway between 0 and 1, the intuitively correct setting of the threshold.

The previous example could have been formulated from a decision theory point of view (see Chapter 6). The significant difference is that here we follow the classical hypothesis-testing formulation. *A priori* probabilities and loss functions are not used and the protection against a type I error is given more emphasis than the protection against a type II error. Both approaches, that is, hypothesis testing and decision theory, have useful applications. The next subsection introduces a case where the hypothesis-testing formulation provides some definite advantage.

These ideas are used in the following sections and in problems at the end of this chapter to illustrate the use of t, chi-square, and F distributions in hypothesis tests. Then nonparametric (or distribution-free) hypothesis tests are illustrated by the chi-square goodness-of-fit test.

8.8.2 Composite Alternative Hypothesis

In the example discussed in the previous section the alternative to the null hypothesis was of the same simple form as the null hypothesis, that is, $\mu = 0$ versus $\mu = 1$. Often we are interested in testing a certain null hypothesis versus a less definite or a composite alternative hypothesis, for example, the mean is not zero In this composite alternative case, the probability of a type II error is now a function rather than simply a number.

In order to illustrate the effect of a composite alternative hypothesis, we return to the example of the previous section with this important change. The null hypothesis remains the same

$$H_0: \mu_Y = 0$$

However under the composite alternative hypothesis the exact level or mean of the received signal is now unknown, except we know that the transmitted signal or mean of Y is now greater then zero, that is

$$H_1: \mu_Y = \delta > 0 \tag{8.54}$$

In this case the probability of a type I error and the critical region remain unchanged. However, the probability of a type II error becomes

$$\begin{aligned} P[\text{type II error}] &= P[Y < \gamma | H_1] \\ &= \int_{-\infty}^{\gamma} \frac{1}{\sqrt{2\pi}} \exp\left[-\frac{(\lambda - \delta)^2}{2}\right] d\lambda \\ &= 1 - Q(\gamma - \delta) \end{aligned} \tag{8.55}$$

If the significance level of the test were set at .05 as in the previous section,

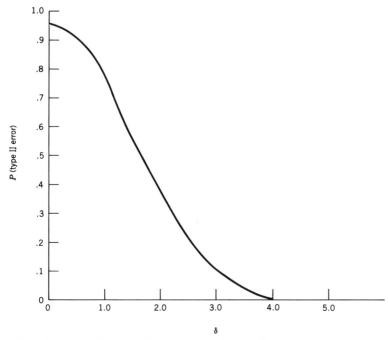

Figure 8.11 Probability of type II error (power function).

then $\gamma \approx 1.65$ as before. In this case, the probability of a type II error is plotted versus δ in Figure 8.11.

A plot like Figure 8.11, which shows the probability of a type II error versus the value of a parameter in a composite alternative hypothesis, is called the *power of the test* or the *power function*.

8.8.3 Tests of the Mean of a Normal Random Variable

Known Variance. Assume that the variance of a normal random variable is 10, and that the mean is unknown. We wish to test the hypothesis that the mean is zero versus the alternative that it is not zero. This is called a *two-sided composite alternative hypothesis*. We might be interested in such a hypothesis if the random variable represented the bias of an instrument. We formulate the problem as follows:

$$H_0: \mu = 0$$
$$H_1: \mu = \delta \neq 0$$

Assume 10 i.i.d. measurements are to be taken and that we accept H_0 if $|\overline{X}| < \gamma$. The variance of $\overline{X}$ is

$$\sigma_{\overline{X}}^2 = \frac{\sigma_X^2}{n} = \frac{10}{10} = 1$$

Now the probability, α, of a type I error, that is, rejecting H_0 when it is true, is the probability that $\overline{X}$ is too far removed from zero.

This does not completely define the critical region because $\overline{X}$ could be too large or too small. However, it is reasonable to assign half of the error probability to $\overline{X}$ being too large and half of the error probability to $\overline{X}$ being too small. Thus

$$P(\overline{X} > \gamma | \mu = 0) = \frac{\alpha}{2} \tag{8.56}$$

$$P(\overline{X} < -\gamma | \mu = 0) = \frac{\alpha}{2} \tag{8.57}$$

$$P(|\overline{X}| \leq \gamma) = 1 - \alpha \tag{8.58}$$

If $\alpha = .01$, then from Appendix D with $\sigma^2 = 1$, $\gamma \approx 2.58$, and the critical region R is

$$R \approx \{\overline{X} < -2.58\} \cup \{\overline{X} > 2.58\} \tag{8.59}$$

Unknown Variance. Next we consider the same null hypothesis, that is $\mu = 0$ versus the same alternative hypothesis $\mu \neq 0$ that was just considered except we now assume that the variance of X is unknown. In this case we use the statistic

$$T = \frac{\overline{X}}{S/\sqrt{n}}$$

to set up the hypothesis test. In Section 8.7.3 it was shown that this random variable has the t_m distribution, and at the α level of significance,

$$P\left[\frac{\overline{X}}{S/\sqrt{n}} > t_{n-1,\alpha/2}\right] = \frac{\alpha}{2}$$

For instance if $n = 10$ and $\alpha = .01$, then $t_{9,.005} \approx 3.25$ (from Appendix F) and the critical region R is

$$R = \left\{\frac{\overline{X}}{S/\sqrt{10}} < -3.25\right\} \cup \left\{3.25 < \frac{\overline{X}}{S/\sqrt{10}}\right\}$$

Thus we would compute $\overline{X}$ and S from the sample data, and then if the ratio

$$\frac{\overline{X}}{S/\sqrt{10}}$$

falls within the critical region, the null hypothesis that $\mu = 0$ would be rejected.

8.8.4 Tests of the Equality of Two Means

As an example of the use of a test of hypothesis, we consider testing whether two means are the "same" when the variances are unknown but assumed to be equal. A typical application would be the following problem. We want to know whether a random sequence is stationary in the mean. We calculate $\overline{X}_1$ and S_1 from one section of the record, and $\overline{X}_2$ and S_2 from a different (independent) section of the record. We hypothesize that the sequence is stationary in the mean and wish to test this null hypothesis.

Our problem is to test the hypothesis

$$H_0: \mu_1 - \mu_2 = 0 \tag{8.60}$$

versus

$$H_1: \mu_1 - \mu_2 = \delta \neq 0 \tag{8.61}$$

The random variable $W = \overline{X}_1 - \overline{X}_2$ will have mean $\mu_1 - \mu_2$, and if $\overline{X}_1$ and $\overline{X}_2$ are independent, it will have variance $\sigma_1^2/n_1 + \sigma_2^2/n_2$. Further if $\sigma_1 = \sigma_2$, then the variance is $[(1/n_1) + (1/n_2)]\sigma_1^2$, and if X_1 and X_2 are normal and we assume that they have the same variance σ_1^2, then the difference in means will be estimated by $\overline{X}_1 - \overline{X}_2$ and σ_1^2 will be estimated by

$$\frac{\left[\sum_{i=1}^{n_1}(X_i - \overline{X}_1)^2 + \sum_{j=1}^{n_2}(X_{n_1+j} - \overline{X}_2)^2\right]}{(n_1 + n_2 - 2)}$$

TESTS OF HYPOTHESES

Thus if $\overline{X}_1$ and $\overline{X}_2$ have the same mean, then

$$T = \frac{\overline{X}_1 - \overline{X}_2}{\left[\left(\frac{1}{n_1} + \frac{1}{n_2}\right)\left(\frac{\sum_{i=1}^{n_1}(X_i - \overline{X}_1)^2 + \sum_{j=1}^{n_2}(X_{n_1+j} - \overline{X}_2)^2}{n_1 + n_2 - 2}\right)\right]^{1/2}} \quad (8.62)$$

will have a t distribution with $n_1 + n_2 - 2$ degrees of freedom. Thus, Equation 8.62 can be used to test the hypothesis that two means are equal. The critical region in this case consists of the set of values for which

$$|T| > t_{n_1+n_2-2;\,\alpha/2}$$

EXAMPLE 8.15.

We wish to test whether the two sets of samples, each of size four, are consistent with the hypothesis that $\mu_1 = \mu_2$ at the 5% significance level. We shall assume that the underlying random variables are normal, that $\sigma_1 = \sigma_2 = \sigma$, $\overline{X}_1$ and $\overline{X}_2$ are independent, and the sample statistics are

$$\overline{X}_1 = 3.0 \qquad \overline{X}_2 = 3.1$$
$$S_1^2 = .0032 \qquad S_2^2 = .0028$$

SOLUTION:

From Appendix F we find that $t_{6;.025} = 2.447$. Thus, the computed value of T must be outside ± 2.447 in order to reject the null hypothesis. In this example, using Equation 8.62

$$T = \frac{3.0 - 3.1}{\left[\left(\frac{1}{4} + \frac{1}{4}\right)\frac{[(.0032)3 + (.0028)3]}{6}\right]^{1/2}}$$
$$\approx -2.58$$

Thus, the hypothesis of stationarity of the mean is narrowly rejected at the 5% level of significance. It would not be rejected at the 1% significance level because in this case the computed value of T must be outside ± 3.707 for rejection.

8.8.5 Tests of Variances

In a number of instances we are interested in comparing two variances. A primary case of interest occurs in regression, which is discussed later in this chapter.

The central idea is to compare the sample variances S_1^2 and S_2^2 by taking their ratio. In Section 8.7 it was shown that such a ratio has an F distribution. This result depends on the assumption that each of the underlying Gaussian random variables has a unity variance. However, if X does not have a unity variance, then $Y = X/\sigma$ will have a unity variance, and it follows that the ratio

$$\frac{(S_1^2/\sigma_1^2)}{(S_2^2/\sigma_2^2)}$$

will have an F distribution. Furthermore, if both samples are from the same distribution, then $\sigma_1^2 = \sigma_2^2$. If this is the case, then the ratio of the sample variances will follow the F distribution. Usually, it is possible to also hypothesize that if the variances are not equal, then S_1 will be larger than S_2. Thus, we usually test the null hypothesis

$$H_0: \sigma_2^2 = \sigma_1^2$$

versus

$$H_1: \sigma_1^2 > \sigma_2^2$$

by using the statistic, S_1^2/S_2^2. If the sample value of S_1^2/S_2^2 is too large, then the null hypothesis is rejected. That is, the critical region is the set of S_1^2/S_2^2 such that $S_1^2/S_2^2 > F_{m_1,m_2}$, where F_{m_1,m_2} is a number determined by the significance level, the degrees of freedom of S_1^2, and the degrees of freedom of S_2^2.

EXAMPLE 8.16.

Find the critical region for testing the hypothesis

$$H_0: \sigma_1^2 = \sigma_2^2$$

versus

$$H_1: \sigma_1^2 > \sigma_2^2$$

TESTS OF HYPOTHESES   523

at the 1% significance level if S_1^2 is based upon 8 observations, and thus has 7 degrees of freedom and S_2^2 has 5 degrees of freedom.

SOLUTION:

The critical region is determined by the value of $F_{7,5;.01}$. This value is found in Appendix G to be 10.456.

Thus, if the ratio $S_1^2/S_2^2 > 10.456$, the null hypothesis should be rejected. However, if $S_1^2/S_2^2 \leq 10.456$, then the null hypothesis should not be rejected.

8.8.6 Chi-Square Tests

It was shown in Section 8.7.2 that the sum of the squares of zero-mean i.i.d. normal random variables has a chi-square distribution. It can be shown that the sum of squares of certain dependent normal random variables is also chi-square. We now argue that the sum of squares of binomial random variables is also chi-square in the limit as $n \to \infty$.

That is, if X_1 is binomial, n, p_1, that is

$$p(X_1 = i) = \binom{n}{i} p_1^i p_2^{n-i}, \qquad i = 0, 1, \ldots, n$$

where $p_1 + p_2 = 1$ and we define

$$Y_n = \frac{X_1 - np_1}{\sqrt{np_1 p_2}}; \qquad Y = \lim_{n \to \infty} Y_n \qquad (8.63)$$

then Y will have a distribution which is standard normal as shown by the central limit theorem.

If we now define

$$Z_{1,n} = Y_n^2; \qquad Z = Y^2 \qquad (8.64)$$

then, in the limit as $n \to \infty$, $Z_{1,n} = Y^2$ has a chi-square distribution with one degree of freedom as was shown in Section 8.7.2. In order to prepare for the multinomial case, we rearrange the expression for $Z_{1,n}$ by noting that

$$\frac{1}{np_1 p_2} = \frac{1}{n}\left(\frac{p_1 + p_2}{p_1 p_2}\right) = \frac{1}{n}\left(\frac{1}{p_1} + \frac{1}{p_2}\right)$$

and we define $X_2 \triangleq n - X_1$; thus

$$X_1 - np_1 = n - X_2 - n(1 - p_2) = np_2 - X_2$$

Then Equation 8.64, using Equation 8.63, can be written

$$Z_{1,n} = \frac{(X_1 - np_1)^2}{np_1} + \frac{(X_2 - np_2)^2}{np_2} \tag{8.65}$$

Now, let $X_1, \ldots, X_m$ have a multinomial distribution, that is

$$p_{X_1,\ldots,X_m}(i_1, \ldots, i_m)$$

$$= \frac{n!}{i_1! \cdots i_m!} p_1^{i_1} \cdots p_m^{i_m} \tag{8.66}$$

where

$$\sum_{j=1}^{m} p_j = 1 \quad \text{and} \quad \sum_{j=1}^{m} i_j = n \tag{8.67}$$

It can be shown that for large n, $Z_{m-1,n}$ where

$$Z_{m-1,n} = \sum_{i=1}^{m} \frac{(X_i - np_i)^2}{np_i} \tag{8.68}$$

has approximately a χ^2_{m-1} distribution as $n \to \infty$. (The dependence of the X_i's require $m \geq 5$ for the approximation)

The random variable $Z_{m-1,n}$ can be used in testing the hypothesis that the i.i.d. observations $X_1, \ldots, X_n$ came from an assumed distribution. A typical null hypothesis is

$$H_0: \begin{cases} p_1 = p_{1,0} \\ p_2 = p_{2,0} \\ \vdots \\ p_m = p_{m,0} \end{cases} \tag{8.69}$$

The alternative hypothesis then is

$$H_1: \begin{cases} p_1 \neq p_{1,0} \\ \vdots \\ p_m \neq p_{m,0} \end{cases} \quad (8.70)$$

If the null hypothesis is true, then $Z_{m-1,n}$ will have an approximate chi-square distribution with $m - 1$ degrees of freedom. If $p_{1,0}, p_{2,0}, \ldots, p_{m,0}$ are the true values, then we expect that $Z_{m-1,n}$ will be small. Under the alternative hypotheses, $Z_{m-1,n}$ will be larger. Thus, using a table of the chi-square distribution, given the significance level α, we can find a number $\chi^2_{m-1;\alpha}$ such that

$$P(Z_{m-1,n} \geq \chi^2_{m-1;\alpha}) = \alpha$$

and values of $Z_{m-1,n}$ larger than $\chi^2_{m-1;\alpha}$ constitute the critical region.

EXAMPLE 8.17.

A die is to be tossed 60 times and the number of times each of the six faces occurs is noted. Test the null hypothesis that the die is fair at the 5% significance level given the following observed frequencies

Up face	1	2	3	4	5	6
No. of observances	7	9	8	11	12	13

SOLUTION:

Under the null hypothesis the expected number of observations of each face is 10 (np_i), and hence

$$Z_{5,60} = \sum_{i=1}^{6} \frac{(X_i - 10)^2}{10} = \frac{3^2}{10} + \frac{1^2}{10} + \frac{2^2}{10} + \frac{1^2}{10} + \frac{2^2}{10} + \frac{3^2}{10} = 2.8$$

Note that this variable has $6 - 1 = 5$ degrees of freedom because the number of observations in the last category is fixed given the sample size and the observed frequencies in the first five categories.

Since from Appendix E, $\chi^2_{5;.05} \approx 11.07 > 2.8$, we decide not to reject the null hypothesis that the die is fair.

526 STATISTICS

In testing whether certain observations are consistent with an assumed probability density function, the chi-square goodness of fit test is often used. The procedure is to divide the observations into m intervals. The number of observations in interval i corresponds with X_i in Equation 8.68 and the expected number in interval i corresponds with np_i.

The number d of degrees of freedom in this case is

$$d = m - 1 - k \qquad (8.71)$$

where $m - 1$ is the number of degrees of freedom associated with the m intervals and k is the number of parameters of the assumed pdf that are estimated from these observations.

EXAMPLE 8.18.

Test the hypothesis at the 1% significance level that X is a normal random variable when the 73 observations are (these are examination scores of students).

Score	Number of Observations
99	1
98	1
94	1
92	1
91	1
90	2
89	2
88	3
85	1
84	4
83	3
82	2
81	2
80	3
79	2
78	2
77	1
76	4
75	5
74	1
73	2

Score	Number of Observations
72	4
71	2
70	3
69	4
68	5
67	2
66	1
62	3
60	1
59	1
54	1
48	1
44	1

SOLUTION:

The sample mean is

$$\overline{X} \approx 75.74$$

and

$$S^2 \approx \frac{426873 - 73(75.74)^2}{72} \approx 112.57$$

$$S \approx 10.61$$

The observation intervals, expected values (each is 73 times the probability that a Gaussian random variable with mean 75.74 and variance 112.5 is within the observation interval) and observed frequencies are

Observation Interval	Expected	Observed
95.5–100	2.3	2
90.5–95.5	3.7	3
85.5–90.5	7.1	7
80.5–85.5	10.8	12
75.5–80.5	13.3	12
70.5–75.5	13.1	14
65.5–70.5	10.5	15

Observation Interval	Expected	Observed
60.5–65.5	6.7	3
55.5–60.5	3.7	2
0–55.5	1.8	3

Because two parameters, that is, μ and σ^2, of the normal distribution were estimated, there are $10 - 1 - 2 = 7$ degrees of freedom, and

$$\chi_7^2 \approx \frac{(2.3 - 2)^2}{2.3} + \frac{(.7)^2}{3.7} + \frac{(.1)^2}{7.1} + \frac{(1.2)^2}{10.8} + \frac{(1.3)^2}{13.3}$$
$$+ \frac{(.9)^2}{13.1} + \frac{(4.5)^2}{10.5} + \frac{(3.7)^2}{6.7} + \frac{(1.7)^2}{3.7} + \frac{(1.2)^2}{1.8}$$
$$\approx .039 + .132 + .001 + .133 + .127 + .062$$
$$+ 1.93 + 2.04 + .781 + .800$$
$$= 6.045$$

The value of $\chi_{7;.01}^2$ is 18.47; thus, 6.045 is not within the critical region (>18.47). Therefore, the null hypothesis of normality is not rejected.

8.8.7 Summary of Hypothesis Testing

The basic decision to be made in hypothesis testing is whether or not to reject a hypothesis based on i.i.d. observations $X_1, X_2, \ldots, X_n$. The first step in hypothesis testing is to find an appropriate statistic $Y = g(X_1, X_2, \ldots, X_n)$ on which to base the test. In many cases, the test statistic will be a normal, t, F, or χ^2 variable. The selection of a test statistic, while intuitively obvious in some cases (e.g., $\overline{X}$ for tests involving μ when σ^2 is known) is often guided by the maximum likelihood estimators of the parameters associated with the null and alternate hypothesis.

After a test statistic, say Y, is selected, the next step is to find $f_{Y|H_0}$ and $f_{Y|H_1}$. These densities are used to find the critical region for a given significance level and also to find the power (probability of a type II error) of the test. If any of the parameters of $f_{Y|H_0}$ and $f_{Y|H_1}$ are not known, they must be estimated from data and the estimated values are used to set up the test.

The introduction of composite alternative hypotheses (e.g., $\mu \neq 0$) enlarged hypothesis testing from the form introduced in Chapter 6. With a composite alternative hypothesis, we do not consider accepting the alternative hypothesis (what value are we accepting?); rather, we either reject or fail to reject the null hypothesis.

For testing against a one-sided (e.g., $\mu > 0$) alternate hypothesis, we use a one-sided test. When the alternate hypothesis is two-sided ($\mu \neq 0$) a two-sided test is used.

8.9 SIMPLE LINEAR REGRESSION

In this section we seek an equation that "best" characterizes the relation between X, a "controlled" variable, and Y, a "dependent" random variable. It is said that we *regress Y on X*, and the equation is called *regression*. In many applications we believe that a change in X causes Y to vary "dependently." However, we do not assume that X causes Y; we only assume that we can measure X without error and then observe Y. The equation is to be determined by the experimental data such as shown in Figure 8.12.

Examples are the following:

X	Y
Height of father	Height of son
Time in years	Electrical energy consumed
Voltage	Current

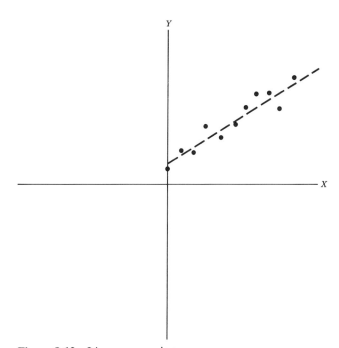

Figure 8.12 Linear regression.

X	Y
Impurity density	Strength
Height	Weight
Quantity purchased	Price
Sales	Profit

There are three parts of the problem to be solved: (1) What is the definition of the "best" characterization of the relation? (2) What is the "best" equation? and (3) How good is the "best" relation?

The accompanying model is $Y = \hat{Y} + e$, where e is the error. Our initial approach to this problem will restrict consideration to a "best" fit of the form $\hat{Y} = b_0 + b_1 X$, that is, a straight line or linear regression. Thus, errors between this "best" fit and the observed values will occur as shown in Figure 8.13.

The errors are defined by

$$e_1 = Y_1 - \hat{Y}_1 = Y_1 - (b_0 + b_1 X_1)$$

$$\vdots$$

$$e_n = Y_n - \hat{Y}_n = Y_n - (b_0 + b_1 X_n) \qquad (8.72)$$

Thus, we can from Equations 8.72 derive the set

$$Y_i = b_0 + b_1 X_i + e_i, \qquad i = 1, \ldots, n \qquad (8.73)$$

As was done in estimation in Chapter 7, we choose to minimize the sum of the squared errors. We have now assumed the answer to the first problem by defining

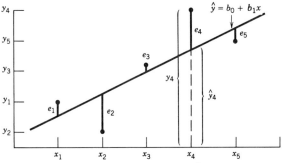

Figure 8.13 Error terms $e_i = Y_i - \hat{Y}_i$.

that the best fit is one that minimizes e^2 where

$$e^2 = \sum_{i=1}^{n} e_i^2 = \sum_{i=1}^{n} [Y_i - (b_0 + b_1 X_i)]^2 \tag{8.74}$$

and n is the number of observations ($n = 5$ in the example of Figure 8.13). In the following part of this section, all sums are from 1 to n.

We cannot change the values of X_i and Y_i; they are the data. We can minimize e^2 only by choosing b_0 and b_1, that is, the intercept and slope of the line. We now proceed to choose the best values of b_0 and b_1, which we will call $\hat{b}_0$ and $\hat{b}_1$, respectively, and we will show that

$$\hat{b}_1 = \frac{\sum (X_i Y_i - \overline{XY})}{\sum (X_i^2 - \overline{X}^2)} \tag{8.75}$$

$$\hat{b}_0 = \overline{Y} - \hat{b}_1 \overline{X} \tag{8.76}$$

where

$$\overline{X} = \frac{1}{n} \sum X_i \tag{8.77}$$

$$\overline{Y} = \frac{1}{n} \sum Y_i \tag{8.78}$$

First, we differentiate Equation 8.74 partially with respect to b_0 and b_1 and set the result equal to zero.

$$\frac{\partial e^2}{\partial b_0} = -2 \sum (Y_i - b_0 - b_1 X_i) \tag{8.79}$$

$$\frac{\partial e^2}{\partial b_1} = -2 \sum (X_i)(Y_i - b_0 - b_1 X_i) \tag{8.80}$$

Setting these two equations equal to zero and assuming the solution produces the values $\hat{b}_0$ and $\hat{b}_1$ which minimize e^2. The value of e^2 when $\hat{b}_0$ and $\hat{b}_1$ are used will be called ϵ^2. Thus

$$\sum (Y_i - \hat{b}_0 - \hat{b}_1 X_i) = 0$$

$$\sum (X_i)(Y_i - \hat{b}_0 - \hat{b}_1 X_i) = 0$$

We then rearrange these two equations into what is usually called the *normal form*:

$$\sum Y_i = n\hat{b}_0 + \hat{b}_1 \sum X_i \qquad (8.81)$$

$$\sum X_i Y_i = \hat{b}_0 \sum X_i + \hat{b}_1 \sum X_i^2 \qquad (8.82)$$

Dividing Equation 8.81 by n and rearranging results in

$$\hat{b}_0 = \overline{Y} - \hat{b}_1 \overline{X}$$

Using this result in Equation 8.82 produces

$$\sum (X_i Y_i) = (\overline{Y} - \hat{b}_1 \overline{X}) \sum X_i + \hat{b}_1 \sum X_i^2$$

or

$$\hat{b}_1 = \frac{\sum (X_i Y_i) - \overline{Y} \sum (X_i)}{\sum X_i^2 - \overline{X} \sum X_i} = \frac{\sum (X_i Y_i) - n\overline{X}\overline{Y}}{\sum X_i^2 - n\overline{X}^2} \qquad (8.83.\text{a})$$

To show that Equation 8.83.a is equivalent to Equation 8.75, consider the numerator of Equation 8.75.

$$\sum (X_i Y_i - \overline{X}\overline{Y}) = \sum (X_i Y_i) - \sum (\overline{X}\overline{Y}) = \sum (X_i Y_i) - n\overline{X}\overline{Y}$$

It can also be shown in the same fashion that the denominators are equal (see Problem 8.39).

We have shown that differentiating Equation 8.74 and setting the result equal to zero results in the solution given by Equations 8.75 and 8.76. We now prove that it is indeed a minimum. Define

$$Y_i^* = a_0 + a_1 X_i$$

We now demonstrate that

$$e^{2*} \stackrel{\Delta}{=} \sum (Y_i - Y_i^*)^2 \geq \sum [Y_i - (\hat{b}_0 + \hat{b}_1 X_i)]^2 = \epsilon^2$$

thus showing that $\hat{b}_0$ and $\hat{b}_1$ as defined by Equations 8.75 and 8.76 do indeed minimize the mean squared error:

$$\begin{aligned}e^{2*} &= \sum [Y_i - (a_0 + a_1 X_i)]^2 \\ &= \sum [(Y_i - \hat{b}_0 - \hat{b}_1 X_i) + (\hat{b}_0 - a_0) + (\hat{b}_1 - a_1)X_i]^2 \\ &= \sum (Y_i - \hat{b}_0 - \hat{b}_1 X_i)^2 + \sum [(\hat{b}_0 - a_0) + (\hat{b}_1 - a_1)X_i]^2 \\ &\quad + 2(\hat{b}_0 - a_0) \sum (Y_i - \hat{b}_0 - \hat{b}_1 X_i) \\ &\quad + 2(\hat{b}_1 - a_1) \sum X_i(Y_i - \hat{b}_0 - \hat{b}_1 X_i)\end{aligned}$$

The last two terms are equal to 0 by Equation 8.81 and Equation 8.82. Thus

$$\begin{aligned}e^{2*} &= \sum (Y_i - \hat{b}_0 - \hat{b}_1 X_i)^2 + \sum [(\hat{b}_0 - a_0) + (\hat{b}_1 - a_1)X_i]^2 \\ &= \epsilon^2 + \sum [(\hat{b}_0 - a_0) + (\hat{b}_1 - a_1)X_i]^2 \\ &= \epsilon^2 + C\end{aligned}$$

where $C \geq 0$ because it is the sum of nonnegative terms.

Thus, we have shown that $\hat{b}_0$ and $\hat{b}_1$ as given by Equations 8.75 and 8.76 do in fact minimize the sum of the squared errors.

An alternate form of Equation 8.75 can be obtained by noting that

$$\begin{aligned}\sum (X_i - \overline{X})(Y_i - \overline{Y}) &= \sum X_i Y_i - \overline{X} \sum Y_i - \overline{Y} \sum X_i + n\overline{X}\overline{Y} \\ &= \sum X_i Y_i - n\overline{X}\overline{Y}\end{aligned}$$

Thus

$$\hat{b}_1 = \frac{\sum (X_i - \overline{X})(Y_i - \overline{Y})}{\sum (X_i - \overline{X})^2} \qquad (8.83.\text{b})$$

where the equivalence of the denominators of Equation 8.75 and Equation 8.83.b is shown similarly.

To summarize, the minimum sum of squared errors occurs when the linear regression is

$$\hat{Y} = \hat{b}_0 + \hat{b}_1 X \qquad (8.84.\text{a})$$

where $\hat{b}_0$ is given by Equation 8.76 and $\hat{b}_1$ is given by Equation 8.75. The errors using this equation are

$$\epsilon_i = Y_i - \hat{Y}_i = Y_i - (\hat{b}_0 + \hat{b}_1 X_i) \qquad (8.84.b)$$

and are called *residual* errors. The minimum sum of residual squared errors is

$$\epsilon^2 = \sum_{i=1}^{n} \epsilon_i^2 \qquad (8.84.c)$$

EXAMPLE 8.19.

For the following observations shown in Figure 8.14,

X	Y
-2	-4
-1	-1
0	0
+1	+1
+2	+4

find the "best" straight-line fit and plot the errors.

SOLUTION:

$$\overline{X} = 0 \quad \overline{Y} = 0$$
$$\sum X_i Y_i = 8 + 1 + 0 + 1 + 8 = 18$$
$$\sum X_i^2 = 4 + 1 + 1 + 4 = 10$$
$$\hat{b}_1 = \frac{18}{10} = 1.8$$
$$\hat{b}_0 = 0$$

Thus using Equation 8.84.a

$$\hat{Y}_i = 1.8 \, X_i$$

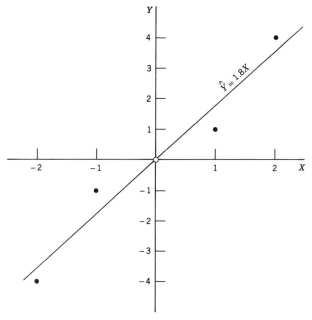

Figure 8.14 Scatter diagram of data for Example 8.19.

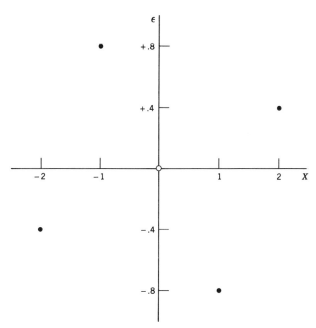

Figure 8.15 Plot of ϵ versus x, Example 8.19.

A table of the equation values and the residual errors follows:

X_i	Y_i	$\hat{Y}_i$	ϵ_i	ϵ_i^2
-2	-4	-3.6	$-.4$	.16
-1	-1	-1.8	$+.8$	.64
0	0	0	0	0
1	1	1.8	$-.8$	.64
2	4	3.6	$+.4$	.16

The residual errors can be plotted versus X as shown in Figure 8.15.

We have found a best fit and shown the errors that result from this fit. Are the errors too large or are they small enough? The next subsection considers this important question.

8.9.1 Analyzing the Estimated Regression

Note that the linear regression

$$\hat{Y}_i = \hat{b}_0 + \hat{b}_1 X_i$$

can be rewritten using Equation 8.76 as

$$\hat{Y}_i - \overline{Y} = \hat{b}_1(X_i - \overline{X}) \qquad (8.85)$$

Continuing to look at the residuals errors, we show the sum of the residual errors is zero. Using Equation 8.85, we have

$$\hat{Y}_i = \overline{Y} + \hat{b}_1(X_i - \overline{X})$$

Subtracting this equation from Y_i and summing, we obtain

$$\sum_{i=1}^{n} \epsilon_i = \sum (Y_i - \hat{Y}_i) = \sum (Y_i - \overline{Y}) - \hat{b}_1 \sum (X_i - \overline{X}) = 0$$

We now show an important result for analyzing residual errors:

$$\sum (Y_i - \overline{Y})^2 = \sum (Y_i - \hat{Y}_i)^2 + \sum (\hat{Y}_i - \overline{Y})^2 \qquad (8.86)$$

The left-hand term is the sum of the squared error about the mean; the first right-hand term is the sum of the squared error about the linear regression; and the last term is the sum of the squared difference between the regression line and the mean.

Equation 8.86 can be derived as follows;

$$Y_i - \hat{Y}_i = Y_i - \overline{Y} + \overline{Y} - \hat{Y}_i$$
$$= (Y_i - \overline{Y}) - (\hat{Y}_i - \overline{Y})$$

Squaring and summing both sides

$$\sum (Y_i - \hat{Y}_i)^2 = \sum [(Y_i - \overline{Y}) - (\hat{Y}_i - \overline{Y})]^2$$
$$= \sum (Y_i - \overline{Y})^2 + \sum (\hat{Y}_i - \overline{Y})^2$$
$$- 2 \sum (Y_i - \overline{Y})(\hat{Y}_i - \overline{Y}) \qquad (8.87)$$

Using Equation 8.85, the third (last) term in the foregoing equation may be rewritten as

$$-2 \sum (Y_i - \overline{Y})(\hat{Y}_i - \overline{Y}) = -2 \sum (Y_i - \overline{Y})\hat{b}_1(X_i - \overline{X})$$
$$= -2\hat{b}_1 \sum (Y_i - \overline{Y})(X_i - \overline{X})$$

Equation 8.83b produces

$$-2 \sum (Y_i - \overline{Y})(\hat{Y}_i - \overline{Y}) = -2\hat{b}_1^2 \sum (X_i - \overline{X})^2$$

Using Equation 8.85 again results in

$$-2 \sum (Y_i - \overline{Y})(\hat{Y}_i - \overline{Y}) = -2 \sum (\hat{Y}_i - \overline{Y})^2$$

Substituting this result in Equation 8.87, we have

$$\sum (Y_i - \hat{Y}_i)^2 = \sum (Y_i - \overline{Y})^2 - \sum (\hat{Y}_i - \overline{Y})^2$$

Adding $\sum (\hat{Y}_i - \overline{Y})^2$ to both sides produces Equation 8.86 and completes the proof.

From Equation 8.86 we have a very useful way of evaluating how good the fit (regression) is. If the fit is good, then most of the original variance about $\overline{Y}$ will be contained in the second term and the residual sum of squares will be relatively small. Thus we look at

$$R^2 = \frac{\text{sum of squares due to regression}}{\text{sum of squares about the mean}} = \frac{\sum (\hat{Y}_i - \overline{Y})^2}{\sum (Y_i - \overline{Y})^2}$$

$R^2 \times 100$ represents the percentage variation accounted for by $\hat{Y}$.

It is obvious that both numerator and denominator of R^2 are nonnegative, and if the denominator is zero, all Y_is are equal. Thus assuming all Y_is are not equal

$$R^2 \geq 0$$

Further, from Equation 8.86 and the fact that sums of squares are nonnegative

$$\sum (Y_i - \overline{Y})^2 \geq \sum (\hat{Y}_i - \overline{Y})^2$$

Thus $R^2 \leq 1$, and hence we have now shown that

$$0 \leq R^2 \leq 1$$

Furthermore, a large R^2 implies a good fit and a small R^2 implies a fit that is of little value. A very *rough* rule of thumb is that $R^2 > .8$ implies a good fit. For a better measure of fit, the concept of degrees of freedom is needed.

8.9.2 Goodness of Fit Test

Returning to Equation 8.86 we can rewrite it as

$$S = S_1 + S_2 \qquad (8.88)$$

where

$$S = \sum_{i=1}^{n} (Y_i - \overline{Y})^2 \qquad (8.89)$$

$$S_1 = \sum_{i=1}^{n} (Y_i - \hat{Y}_i)^2 \tag{8.90}$$

$$S_2 = \sum_{i=1}^{n} (\hat{Y}_i - \overline{Y})^2 \tag{8.91}$$

The degrees of freedom associated with a sum of squares indicates how many *independent* pieces of information are contained in the sum of squares. For example

$$\sum_{i=1}^{n} Y_i^2$$

has n degrees of freedom if it has n independent observations $Y_1, Y_2, \ldots, Y_n$. On the other hand

$$\sum_{i=1}^{n} (Y_i - \overline{Y})^2$$

has $(n - 1)$ degrees of freedom because given $Y_1, Y_2, \ldots, Y_{n-1}$, and $\overline{Y}$ we can determine Y_n from the fact that

$$\sum_{i=1}^{n} (Y_i - \overline{Y}) = 0$$

Thus, the sum S has $(n - 1)$ degrees of freedom. The sum S_1 has $(n - 2)$ degrees of freedom since two "parameters" $\hat{b}_0$ and $\hat{b}_1$ are included in $\hat{Y}_i$. Since $S = S_1 + S_2$, this leaves S_2 with one degree of freedom. It is the degree of freedom necessary to determine $\hat{b}_1$. We can use the sum of squares S_1 and S_2 to formulate a ratio that can be used to test the goodness of fit of a regression equation.

Let us first examine the terms in Equation 8.88 under the assumption that we have a very good fit (i.e., X and Y have a strong linear relationship). In this case $\hat{Y}_i \to Y_i$ and hence the term S_1 will be small and the terms S and S_2 will be of the order of $n\sigma_Y^2$. On the other hand, if the fit is poor (i.e., X and Y are not linearly related), $\hat{b}_1$ in Equation 8.75 $\to 0$, $\hat{Y}_i \to \overline{Y}$ and S_2 will be very small ($\to 0$) compared to S_1 and S.

Now, if we form the ratio, F, where

$$F = \frac{S_2/1}{S_1/(n - 2)} = \frac{\sum (\hat{Y}_i - \overline{Y})^2}{\sum (Y_i - \hat{Y}_i)^2/(n - 2)} \tag{8.92}$$

then, F will have a larger value when the fit is good and a smaller value when the fit is poor.

If $\hat{b}_1 = 0$, then the ratio given in Equation 8.92 has an F distribution (assuming that the underlying variables are normal) with degrees of freedom $n_1 = 1$ and $n_2 = n - 2$, we can use an F test for testing if the fit is good or not. Specifically, if H_0 is the null hypothesis that $\hat{b}_1 = 0$ (indicates a bad fit), then

Accept H_0: Bad fit   if $F < F_{1,n-2;\alpha}$
Accepts H_1: Good fit   if $F > F_{1,n-2;\alpha}$

where $F_{1,n-2;\alpha}$ is obtained from the table of F distribution and α is the significance level.

EXAMPLE 8.20.

Test whether the fit obtained in Example 8.19 is good at a significance level of 0.1.

SOLUTION:

From the data given in Example 8.19, we calculate $S_2 = 32.4$ with one degree of freedom and $S_1 = 1.6$ with three degrees of freedom. Hence

$$F = \frac{32.4/1}{1.6/3} = 60.75 \text{ with } n_1 = 1 \text{ and } n_2 = 3.$$

From Appendix G we obtain $F_{1,3;0.1} \approx 5.54$ and since $F = 60.75 > 5.54$ we judge the fit to be good.

Statisticians use an *analysis of variance* table (ANOVA) to analyze the sum of squared errors (S, S_1, and S_2) and judge the goodness of fit. The interested reader may find details of ANOVA in Reference [3]. Commercially available software packages, in addition to finding the best linear fit, also produce ANOVA tables.

8.10 MULTIPLE LINEAR REGRESSION

In this section we generalize the results of the previous section by seeking a linear equation that best characterizes the relation between $X_1, X_2, \ldots, X_n$, a set of n "controlled" variables, and Y, a "dependent" random variable. Experimental data are again the basis of the equation. Examples are as follows:

X_1	X_2	Y
Height of father	Height of mother	Height of child
Time in years	Heating degree days	Electric energy consumed
Voltage	Resistance	Current
Impurity density	Cross-sectional area	Strength
Height	Age	Weight

We first assume that there are two "controlled" variables and minimize the mean-square error just as was done in the case of simple linear regression. This results in a set of three linear equations in three unknowns that must be solved for $\hat{b}_0$, $\hat{b}_1$, and $\hat{b}_2$. Next we solve the simple linear regression problem again, this time by using matrix notation so that the technique can be used to solve the general case of n "controlled" variables.

8.10.1 Two Controlled Variables

We assume that

$$\hat{Y} = b_0 + b_1 X_1 + b_2 X_2 \tag{8.93}$$

$$Y_i = b_0 + b_1 X_{1,i} + b_2 X_{2,i} + e_i \tag{8.94}$$

$$e_i = Y_i - \hat{Y}_i = Y_i - b_0 - b_1 X_{1,i} - b_2 X_{2,i} \tag{8.95}$$

$$e^2 = \sum_{i=1}^{n} e_i^2 = \sum_{i=1}^{n} (Y_i - b_0 - b_1 X_{1,i} - b_2 X_{2,i})^2 \tag{8.96}$$

Just as in the last section, the sum of the squared errors is differentiated with respect to the unknowns and the results are set equal to zero. The b's are replaced by the $\hat{b}$'s, which represent the values of the b's that minimize e^2.

$$\frac{\partial e^2}{\partial b_0} = -2 \sum (Y_i - \hat{b}_0 - \hat{b}_1 X_{1,i} - \hat{b}_2 X_{2,i}) = 0$$

$$\frac{\partial e^2}{\partial b_1} = -2 \sum (X_{1,i})(Y_i - \hat{b}_0 - \hat{b}_1 X_{1,i} - \hat{b}_2 X_{2,i}) = 0$$

$$\frac{\partial e^2}{\partial b_2} = -2 \sum (X_{2,i})(Y_i - \hat{b}_0 - \hat{b}_1 X_{1,i} - \hat{b}_2 X_{2,i}) = 0 \tag{8.97}$$

Rewriting these three equations produces:

$$\hat{b}_0 + \hat{b}_1 \overline{X}_1 + \hat{b}_2 \overline{X}_2 = \overline{Y} \tag{8.98}$$

$$\hat{b}_0 \sum X_{1,i} + \hat{b}_1 \sum X_{1,i}^2 + \hat{b}_2 \sum X_{1,i} X_{2,i} = \sum X_{1,i} Y_i \tag{8.99}$$

$$\hat{b}_0 \sum X_{2,i} + \hat{b}_1 \sum X_{1,i} X_{2,i} + \hat{b}_2 \sum X_{2,i}^2 = \sum X_{2,i} Y_i \tag{8.100}$$

These three linear equations with three unknown coefficients define the best fitting plane. They are solved in Problem 8.43. We now introduce matrix notation in order to solve the general n controlled variable regression problem. In order to introduce the notation, the simple (one "controlled" variable) linear regression problem is solved again.

8.10.2 Simple Linear Regression In Matrix Form

Assume that there is a single "controlled" variable X and 10 (paired) observations, (X_i, Y_i)

$$\mathbf{Y} = \begin{bmatrix} Y_1 \\ Y_2 \\ \vdots \\ Y_{10} \end{bmatrix} \quad \mathbf{X} = \begin{bmatrix} 1 & X_1 \\ 1 & X_2 \\ \vdots & \vdots \\ 1 & X_{10} \end{bmatrix} \quad \mathbf{b} = \begin{bmatrix} b_0 \\ b_1 \end{bmatrix} \quad \mathbf{e} = \begin{bmatrix} e_1 \\ \vdots \\ e_{10} \end{bmatrix} \quad \hat{\mathbf{b}} = \begin{bmatrix} \hat{b}_0 \\ \hat{b}_1 \end{bmatrix}$$

Y is a 10×1 vector
X is a 10×2 matrix
b and $\hat{\mathbf{b}}$ are 2×1 vectors
e is a 10×1 vector

The equations that correspond with Equations 8.73 and 8.74 are

$$\mathbf{Y} = \mathbf{Xb} + \mathbf{e} \tag{8.101}$$

$$\mathbf{e}^T \mathbf{e} = \sum e_i^2 \tag{8.102}$$

The resulting normal equations that correspond with Equations 8.81 and 8.82 are

$$\mathbf{X}^T \mathbf{Y} = \mathbf{X}^T \mathbf{X} \hat{\mathbf{b}} \tag{8.103}$$

The solutions for $\hat{b}_0$ and $\hat{b}_1$ as given in Equations 8.75 and 8.76 are

$$\hat{\mathbf{b}} = (\mathbf{X}^T \mathbf{X})^{-1} \mathbf{X}^T \mathbf{Y} \tag{8.104}$$

In this case, the inverse $(\mathbf{X}^T\mathbf{X})^{-1}$ will exist if there is more than one value of X. In the case of more than one controlled variable, the inverse may not exist. We will discuss this difficulty in the next subsection.

The resulting simple linear regression is

$$\hat{\mathbf{Y}} = \mathbf{X}\hat{\mathbf{b}} \qquad (8.105.\text{a})$$

the resulting errors using $\hat{\mathbf{b}}$ are

$$\boldsymbol{\epsilon} = \begin{bmatrix} \epsilon_1 \\ \vdots \\ \epsilon_{10} \end{bmatrix} \qquad (8.105.\text{b})$$

where

$$\epsilon_i = Y_i - \hat{Y}_i \qquad (8.105.\text{c})$$

and

$$\epsilon^2 = \boldsymbol{\epsilon}^T\boldsymbol{\epsilon} \qquad (8.105.\text{d})$$

In checking the matrix forms versus the earlier scaler results, observe that

$$\mathbf{X}^T\mathbf{X} = \begin{bmatrix} 1 & \cdots & 1 \\ X_1 & \cdots & X_{10} \end{bmatrix} \begin{bmatrix} 1 & X_1 \\ \vdots & \vdots \\ 1 & X_{10} \end{bmatrix}$$

$$= \begin{bmatrix} n & \Sigma X_i \\ \Sigma X_i & \Sigma X_i^2 \end{bmatrix}$$

$$\mathbf{X}^T\mathbf{Y} = \begin{bmatrix} \Sigma Y_i \\ \Sigma X_i Y_i \end{bmatrix}$$

Note that the solution is given in Equations 8.104 and 8.105 and the reader can verify this result is the same as Equations 8.75, 8.76 and 8.84.

8.10.3 General Linear Regression

Assume that there are $(p - 1)$ controlled variables plus a constant, such that p parameters must be estimated. The matrix equation describing this multiple linear regression is

$$\mathbf{Y} = \mathbf{Xb} + \mathbf{e} \quad (8.106)$$

The best (minimum mean-square error) estimator will be

$$\hat{\mathbf{Y}} = \mathbf{X}\hat{\mathbf{b}} \quad (8.107)$$

where

$\mathbf{Y}$ and $\hat{\mathbf{Y}}$ are $n \times 1$ vectors
$\mathbf{X}$ is an $n \times p$ matrix of observations
$\mathbf{b}$ and $\hat{\mathbf{b}}$ are $p \times 1$ vectors of parameters
$\mathbf{e}$ is an $n \times 1$ vector of errors

$$\begin{aligned}\mathbf{e}^T\mathbf{e} &= (\mathbf{Y} - \mathbf{Xb})^T(\mathbf{Y} - \mathbf{Xb}) \\ &= \mathbf{Y}^T\mathbf{Y} - \mathbf{b}^T\mathbf{X}^T\mathbf{Y} - \mathbf{Y}^T\mathbf{Xb} + \mathbf{b}^T\mathbf{X}^T\mathbf{Xb} \\ &= \mathbf{Y}^T\mathbf{Y} - 2\mathbf{b}^T\mathbf{X}^T\mathbf{Y} + \mathbf{b}^T\mathbf{X}^T\mathbf{Xb} \end{aligned} \quad (8.108)$$

Note that Equation 8.108 is a scalar equation. The normal equations are found by partially differentiating Equation 8.108 and setting the results equal to zero, that is

$$\left.\frac{\partial \mathbf{e}^T\mathbf{e}}{\partial \mathbf{b}}\right|_{\mathbf{b}=\hat{\mathbf{b}}} = 0$$

This results in the normal equations (see Equations 8.98, 8.99, and 8.100):

$$\mathbf{X}^T\mathbf{X}\hat{\mathbf{b}} = \mathbf{X}^T\mathbf{Y} \quad (8.109)$$

Finally the solution is

$$\hat{\mathbf{b}} = (\mathbf{X}^T\mathbf{X})^{-1}\mathbf{X}^T\mathbf{Y} \quad (8.110)$$

In this case the inverse of the matrix $(\mathbf{X}^T\mathbf{X})$ may not exist if some of the "controlled" variables X_i are linearly related. In some practical situations the dependence between X_i and X_j cannot be avoided and a colinear effect may cause $(\mathbf{X}^T\mathbf{X})$ to be singular, which precludes a solution, or to be nearly singular, which can cause significant problems in estimating the components of $\hat{\mathbf{b}}$ and in interpreting the results. In this case it is usually best to delete one of the linearly dependent variables from the set of controlled variables. See Reference [3].

8.10.4 Goodness of Fit Test

A goodness of fit test for the multiple linear regression can be derived from Equation 8.86:

$$\sum_{i=1}^{n} (Y_i - \bar{Y})^2 = \sum_{i=1}^{n} (Y_i - \hat{Y}_i)^2 + \sum_{i=1}^{n} (\hat{Y}_i - \bar{Y})^2$$

| $(n - 1)$ degrees of freedom | $(n - p)$ degrees of freedom | $(p - 1)$ degrees of freedom |

Proceeding along the lines of discussion given in Section 8.9.2, we can form the test statistic

$$F = \frac{\sum (\hat{Y}_i - \bar{Y})^2/(p - 1)}{\sum (Y_i - \hat{Y}_i)^2/(n - p)}$$

and test the goodness of fit as

Accept H_0: Poor fit if $F < F_{p-1, n-p; \alpha}$
Accept H_1: Good fit if $F > F_{p-1, n-p; \alpha}$

where α is the significance level of the test.

8.10.5 More General Linear Models

Suppose one tried to fit the model $\hat{Y} = b_0 + b_1 X$ to the data

X	Y
-2	4
-1	1
0	0
1	1
2	4

One would find that $b_1 = 0$ and that the fit is useless (check this). It seems obvious that we should try to fit the model $\hat{Y} = b_0 + b_1 X + b_2 X^2$. But is this *linear* regression that we have studied? The answer is "yes" because the model is *linear in the parameters:* b_0, b_1, and b_2. We can define $X_1 = X$ and $X_2 = X^2$ and this is a multiple linear regression model. There is no requirement that the controlled variables X_i be statistically independent; however colinear controlled variables tend to cause problems, as previously mentioned.

Thus, standard computer programs which perform multiple linear regression include the possibility of fitting curves of the form $\hat{Y} = b_0 + b_1 X + b_2 X^2 + b_3 X^3$ or $\hat{Y} = b_0 + b_1 X_1 + b_2 X_2 + b_3 X_1^2 + b_4 X_1 X_2 + b_5 X_2^2$, and so on. These models are called linear regression because they are linear in the parameters. We present an example.

EXAMPLE 8.21.

For the data given here:

X	Y
−5	51
−3	18
−3	19
−2	11
−1	4
−1	5
0	1
1	8
2	15
2	13
3	28
4	41
4	40
5	63
5	57

(a) Plot the points (X_i, Y_i), $i = 1, 2, \ldots, 15$ and compute $\Sigma(X_i - \bar{X})^2$ and $\Sigma(Y_i - \bar{Y})^2$.

(b) Fit $\hat{Y} = \hat{b}_0 + \hat{b}_1 X$ and find $\Sigma(Y_i - \hat{Y}_i)^2$ and compare it with $\Sigma(Y_i - \bar{Y})^2$. Sketch the regression line.

(c) Fit $\hat{Y} = \hat{b}'_0 + \hat{b}'_1 X + \hat{b}'_2 X^2$ and find $\Sigma(Y_i - \hat{Y}_i)^2$ and compare it with $\Sigma(Y_i - \bar{Y})^2$. Sketch the regression line.

(d) Fit $\hat{Y} = \hat{b}''_0 + \hat{b}''_1 X + \hat{b}''_2 e^{2|X|}$. Find $\Sigma(Y_i - \hat{Y}_i)^2$ and compare with $\Sigma(Y_i - \bar{Y})^2$. Sketch the regression line.

(e) Comment on which is the best fit.

SOLUTION:

(a) The points are shown plotted in Figure 8.16 and
$$\Sigma(X_i - \bar{X})^2 \approx 140.6$$
$$\Sigma(Y_i - \bar{Y})^2 \approx 5862.8$$

(b)
$$\hat{Y} = 22.848 + 2.843 X$$
$$\Sigma(\hat{Y}_i - Y_i)^2 = 4725$$

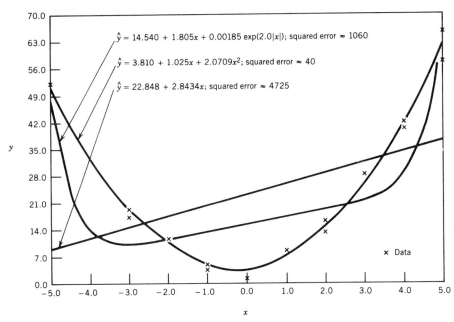

Figure 8.16 General linear regression, Example 8.21.

(c) $$\hat{Y} = 3.81 + 1.025X + 2.071X^2$$
$$\sum (\hat{Y}_i - Y_i)^2 = 40$$

(d) $$\hat{Y} = 14.54 + 1.805X + 0.00185 \exp(2.0\,|X|)$$
$$\sum (Y_i - \hat{Y}_i)^2 = 1060$$

(e) The data, along with the fits, are shown in Figure 8.16. It is clear from the plot and from the squared errors given above that (c) is the best fit.

8.11 SUMMARY

After measurements and statistics were discussed, this chapter presented theory and applications of three basic statistical problems. The first was how to estimate an unknown parameter from a sample of i.i.d. observations or measurements. We emphasized point (rather than interval) estimators, and we also emphasized classical (rather than Bayesian) estimators. The maximum likelihood criterion was used to derive the form of the point estimators, and the bias and variance or MSE were used to judge the "quality" of the estimators.

The second statistical problem discussed was how to test a hypothesis using observations or measurements. Specific tests discussed were tests of means, difference between means, variances, and chi-square tests of goodness of fit. All of these tests involved the following steps: (1) finding the statistic $Y = g(X_1, X_2, \ldots, X_n)$ on which to base the test, (2) finding $f_{Y|H_0}$ and $f_{Y|H_1}$, and (3) given the significance level, finding the critical region for the test and evaluating the power of the test. This differed from the hypothesis testing of Chapter 6 primarily because the alternative hypotheses are composite.

Finally, the problem of fitting curves to data was discussed using simple linear regression and multiple linear regression. This is another case of finding linear (affine) estimators. It differs from Chapter 7 because here we assume that the required means and covariances are unknown; thus they must be estimated from data. This estimation, that is, curve fitting, differs from the parameter estimation discussed at the beginning of the chapter because in curve fitting the value of the dependent random variable is to be estimated based upon observation(s) of one or more related (controlled) random variables. Parameters of a distribution are estimated from a sample (usually more than one observation) from that distribution.

8.12 REFERENCES

Many books on statistics are available on the topics covered in this chapter; we give a short list. Hogg and Craig [7] is an excellent introductory level test on mathematical statistics. The second half of this text covers the topics of point estimation, hypothesis testing and regression. References [2] and [4] provide similar coverage of these topics; which are presented with emphasis on engineering applications in [5], [6], and [8].

Bendat and Piersol [1] provide an excellent treatment of the application of statistical techniques to the measurement and analysis of random data with an emphasis on random signal analysis. Finally, an extensive treatment of regression may be found in [3].

[1] J. S. Bendat and A. G. Piersol, *Random Data: Analysis and Measurement Procedures,* John Wiley & Sons, New York, 1971.

[2] P. J. Bickel and K. A. Doksum, *Mathematical Statistics: Basic Ideas and Selected Topics,* Holden-Day, San Francisco, 1977.

[3] N. R. Draper and H. Smith, *Applied Regression Analysis,* 2nd ed., John Wiley & Sons, New York, 1981.

[4] D. A. S. Fraser, *Statistics: An Introduction,* John Wiley & Sons, New York, 1958.

[5] I. Guttman and S. S. Wilks, *Introductory Engineering Statistics,* John Wiley & Sons, New York, 1965.

[6] W. W. Hines and D. C. Montgomery, *Probability and Statistics in Engineering and Management Science,* 2nd ed., John Wiley & Sons, New York, 1980.

[7] R. B. Hogg and A. T. Craig, *Introduction to Mathematical Statistics,* 4th ed., Macmillan, New York, 1978.

[8] J. B. Kennedy and A. M. Neville, *Basic Statistical Methods For Engineers and Scientists,* 3rd ed., Harper and Row, New York, 1986.

[9] E. Parzen, "Estimation of a Probability Density Function and Its Mode," *Annals of Mathematical Statistics,* Vol. 33, 1962, pp. 1065–1076.

8.13 APPENDIX 8-A

Theorem: If $X_i, i = 1, \ldots, n$, are i.i.d. Gaussian random variables, then for $n \geq 2$; (1) $\overline{X}$ and S^2 are independent random variables; (2) $(n - 1)S^2/\sigma^2$ has the standard chi-square distribution with $m = (n - 1)$ degrees of freedom; and (3) $\sqrt{n}\,(\overline{X} - \mu)/S$ has the (Student's) t distribution with $(n - 1)$ degrees of freedom.

The proof consists of finding the conditional characteristic function of $(n - 1)S^2/\sigma^2$ given $\overline{X}$. We show that this conditional characteristic function is independent of $\overline{X}$ demonstrating the first result. Second we will note that this characteristic function agrees with Equation 8.39, demonstrating the second result. Finally, we use the fact that S^2 is independent of $\overline{X}$, and a simple ratio shows that $\sqrt{n}\,(\overline{X} - \mu)/S$ satisfies the conditions of Section 8.7.3 and is t with $(n - 1)$ degrees of freedom.

Let

$$Y_1 = \overline{X}, \quad Y_2 = X_2, \ldots, Y_n = X_n \qquad (A8.1)$$

then

$$X_1 = n\overline{X} - \sum_{j=2}^{n} X_j = nY_1 - \sum_{j=2}^{n} Y_j$$

The joint density of the n i.i.d. Gaussian random variables is

$$f_{X_1,\ldots,X_n}(x_1, \ldots, x_n) = \left(\frac{1}{\sqrt{2\pi}\,\sigma}\right)^n \exp\left\{-\sum_{i=1}^{n} \frac{(x_i - \mu)^2}{2\sigma^2}\right\} \qquad (A8.2)$$

Now

$$\sum_{i=1}^{n} (x_i - \mu)^2 = \sum_{i=1}^{n} (x_i - \overline{x} + \overline{x} - \mu)^2$$

$$= \left(\sum_{i=1}^{n} (x_i - \overline{x})^2\right) + n(\overline{x} - \mu)^2 + 2(\overline{x} - \mu)\sum_{i=1}^{n}(x_i - \overline{x})$$

550 STATISTICS

The last sum is zero, thus

$$\sum_{i=1}^{n} (x_i - \mu)^2 = \left(\sum_{i=1}^{n} (x_i - \bar{x})^2\right) + n(\bar{x} - \mu)^2$$

Using this in Equation A8.2

$$f_{X_1,\ldots,X_n}(x_1, \ldots, x_n) = \left(\frac{1}{\sqrt{2\pi}\,\sigma}\right)^n$$

$$\times \exp\left\{-\frac{\left[\sum_{i=1}^{n}(x_i - \bar{x})^2 + n(\bar{x} - \mu)^2\right]}{2\sigma^2}\right\} \quad \text{(A8.3)}$$

The transformation A8.1 has a Jacobian of n; thus, using Equations A8.1 and A8.3

$$f_{Y_1,\ldots,Y_n}(y_1, \ldots, y_n) = n \left(\frac{1}{\sqrt{2\pi}\,\sigma}\right)^n$$

$$\times \exp\left\{-\frac{1}{2\sigma^2}\left[(ny_1 - y_2 - \cdots - y_n - y_1)^2 \right.\right.$$

$$\left.\left. + \sum_{i=2}^{n}(y_i - y_1)^2 + n(y_1 - \mu)^2\right]\right\} \quad \text{(A8.4)}$$

$f_{Y_2,\ldots,Y_n|Y_1}$, is Equation A8.4 divided by $f_{Y_1}(y_1)$, that is

$$f_{Y_1}(y_1) = f_{\bar{X}}(y_1) = \frac{1}{\sqrt{2\pi}\,\sigma/\sqrt{n}} \exp\left\{-\frac{(y_1 - \mu)^2}{2\sigma^2/n}\right\}$$

Thus

$$f_{Y_2,Y_3,\ldots,Y_n|Y_1}(y_2, \ldots, y_n | y_1)$$

$$= f_{X_2,X_3,\ldots,X_n|\bar{X}}(y_2, \ldots, y_n | y_1) = \sqrt{n}\left(\frac{1}{\sqrt{2\pi}\,\sigma}\right)^{n-1}$$

$$\times \exp\left\{-\frac{1}{2\sigma^2}\left[(ny_1 - y_2 - \cdots - y_n - y_1)^2 + \sum_{i=2}^{n}(y_i - y_1)^2\right]\right\}$$

$$\quad \text{(A8.5)}$$

The conditional characteristic function of $(n - 1)S^2/\sigma^2$ given $Y_1 = y_1$ is

$$E\left\{\exp\left[\frac{j\omega(n-1)S^2}{\sigma^2}\right]\bigg|Y_1 = y_1\right\} = \int_{-\infty}^{\infty}\cdots\int_{-\infty}^{\infty}\sqrt{n}\left(\frac{1}{\sqrt{2\pi}\,\sigma}\right)^{n-1}$$
$$\times \exp\left(j\frac{2q\omega}{2\sigma^2}\right)\exp\left(-\frac{1}{2\sigma^2}q\right)dy_2,\ldots,dy_n$$

where

$$q = (ny_1 - y_2 - \cdots - y_n - y_1)^2 + \sum_{i=2}^{n}(y_i - y_1)^2$$
$$= \sum_{i=1}^{n}(x_i - \bar{x})^2 = (n-1)s^2$$

Thus

$$E\left\{\exp\left[\frac{j\omega(n-1)S^2}{\sigma^2}\right]\bigg|Y_1 = y_1\right\}$$
$$= \frac{1}{(1-j2\omega)^{(n-1)/2}}\int_{-\infty}^{\infty}\cdots\int_{-\infty}^{\infty}\sqrt{n}\left(\frac{1-j2\omega}{2\pi\sigma^2}\right)^{(n-1)/2}$$
$$\times \exp\left\{-\frac{1}{2\sigma^2}(1-j2\omega)q\right\}dy_2,\ldots,dy_n$$

The multiple integral can be shown to be equal to 1 because it is the $(n-1)$ dimensional volume under the conditional density of Equation A8.5 with σ^2 replaced by $[\sigma^2/(1-j2\omega)]$. Thus

$$E\left\{\exp\left[\frac{j\omega(n-1)S^2}{\sigma^2}\right]\bigg|Y_1 = y_1\right\} = \left[\frac{1}{(1-j2\omega)}\right]^{(n-1)/2} \quad (A8.6)$$

Note that Equation A8.6 is the same as Equation 8.39 with $n - 1 = m$. Thus $(n-1)S^2/\sigma^2$ has a conditional standard chi-square distribution with $(n-1)$ degrees of freedom.

Furthermore, because the conditional characteristic function of $(n-1)S^2/\sigma^2$ given $Y_1 = y_1$ or $\bar{X} = y_1$, does not depend upon y_1, we have shown that $\bar{X}$ and $(n-1)S^2/\sigma^2$ are independent. Thus, $(n-1)S^2/\sigma^2$ is also unconditionally chi-square. We have thus proved assertions (1) and (2).

Finally, because $\overline{X}$ and $(n - 1)S^2/\sigma^2$ are independent and $(n - 1)S^2/\sigma^2$ has a standard chi-square distribution with $(n - 1)$ degrees of freedom and because $\sqrt{n}\,(\overline{X} - \mu)/\sigma$ will be a standard normal random variable, then for n i.i.d. measurements from a Gaussian random variable

$$T_{n-1} = \frac{\sqrt{n}\,(\overline{X} - \mu)/\sigma}{\sqrt{\dfrac{(n - 1)S^2/\sigma^2}{(n - 1)}}} \tag{A8.7}$$

is the ratio defined in Equation 8.44. Thus

$$T_{n-1} = \frac{\sqrt{n}\,(\overline{X} - \mu)}{S} \tag{A8.8}$$

will have the density defined in Equation 8.45, thus proving assertion (3).

8.14 PROBLEMS

8.1 The diameters in inches of 50 rivet heads are given in the table. Construct (a) an empirical distribution function, (b) a histogram, and (c) Parzen's estimator using $g(y)$ and $h(n)$ as suggested in Equation 8.6.

.338	.342
.341	.350
.337	.346
.353	.354
.351	.348
.330	.349
.340	.335
.340	.336
.328	.360
.343	.335
.346	.344
.354	.334
.355	.333
.329	.325
.324	.328
.334	.349
.355	.333
.366	.326

.343	.326
.341	.355
.338	.354
.334	.337
.357	.331
.326	.337
.333	.333

8.2 Generate a sample of size 100 from a uniform (0, 1) random variable by using a random number table. Plot the histogram. Generate a second sample of size 100 and compare the resulting histogram with the first.

8.3 Generate a sample of size 1000 from a normal (2, 9) random variable and plot the empirical distribution function and histogram. (A computer program is essential.)

8.4 Using the data from Example 8-1, compute $\overline{X}$ and S^2 from

    a. The data in column 1

    b. The data in column 2

    c. The data in column 3

    d. The data in column 4

    e. The data in column 5

    f. The data in column 6

    g. The data in column 7

Compare the results. Is there more variation in the estimates of the mean or the estimates of the variance?

8.5 When estimating the mean in Problem 8.4, how many estimators were used? How many estimates resulted?

8.6 Let p represent the probability that an integrated circuit is good. Show that the maximum likelihood estimator of p is N_G/n where N_G is the number of good circuits in n independent trials (tests).

8.7 Find the maximum likelihood estimators of μ and σ^2 from n i.i.d. measurements from a normal distribution.

8.8 Find the maximum likelihood estimator of λ based on the sample $X_1, X_2, \ldots, X_n$ if

$$P(X = x_i; \lambda) = \exp(-\lambda)\frac{(\lambda)^x}{x!}, \quad x = 0, 1, \ldots; \quad \lambda > 0.$$

8.9 Assume that the probability that a thumbtack when tossed comes to rest with the point up is P, a random variable. Assuming that a thumbtack is selected from a population of thumbtacks that are equally likely to have P be any value between zero and one, find the a posteriori probability density of P for the selected thumbtack, which has been tossed 100 times with point up 90 times.

8.10 The number of defects in 100 yards of metal is Poisson distributed; that is

$$P(X = k|\Lambda = \lambda) = \exp(-\lambda)\frac{\lambda^k}{k!}$$

Given one observation was 10 defects, find

$$f_{\Lambda|X}(\lambda|10)$$

if f_Λ is equally likely between 0 and 15.

8.11 The noise in a communication channel is Gaussian with a mean of zero. The variance is assumed to be fixed for a given day but varies from day to day. Assume that it can be $\frac{1}{2}$, 1, or 2 and that each value is equally likely. One observation of the noise is available and its value is $\frac{1}{4}$. Find the a posteriori probabilities of each of the possible values of σ^2.

8.12 X is uniformly distributed between 0 and b but b is unknown. Assume that the a priori distribution on b is uniform between 2 and 4. Find the a posteriori distribution of b after one observation, $X = 2.5$.

8.13 Does $\epsilon = \epsilon_b + \epsilon_r$? If not, find an equation that relates these quantities.

8.14 If K is the number of times an event A occurs in n independent trials, show that K/n is an unbiased estimator of $P(A)$.

8.15 Use Tchebycheff's inequality to show for all $\epsilon > 0$

$$\lim_{n\to\infty} P[|\overline{X}_n - \mu| > \epsilon] = 0$$

8.16 a. Why do you average "laboratory readings" of one parameter?

b. Would one ever want to use a biased estimator rather than an unbiased one?

8.17 Find the bias of the maximum likelihood estimator of θ that was found in Example 8.6.

8.18 If an unknown parameter is to be estimated using (noisy) measurements, give examples where the mean of the measurements would not be a good estimator of the parameter.

8.19 The median m is defined by
$$F_X(m) = .5$$
Find $\hat{m}$ the estimator of the median derived from the empirical distribution function.

8.20 a. If X is uniform $(0, 10)$ and 20 cells are used in a histogram with 200 samples, find the bias, MSE, and normalized RMS error in the histogram.

b. Repeat part (a) if X is normal with a mean of 5 and a standard deviation of 1.5.

8.21 $\hat{\theta}_1$ and $\hat{\theta}_2$ are two independent unbiased estimators of θ with variances σ_1^2 and σ_2^2, respectively. Suppose we want to find a new unbiased estimator for θ of the form $\hat{\theta} = \alpha\hat{\theta}_1 + \beta\hat{\theta}_2$. (a) Find α and β that will minimize the variance of $\hat{\theta}$. (b) Find the variance of $\hat{\theta}$.

8.22 If $X_1, \ldots, X_n$ are i.i.d. observations from a normal random variable with mean μ and variance σ^2, show that
$$\overline{X} = \frac{1}{n}\sum_{i=1}^{n} X_i$$
is normal with
$$E[\overline{X}] = \mu$$
$$\sigma_{\overline{X}}^2 = \frac{\sigma^2}{n}$$

8.23 Let $Y = X^2$, where X is a normal random variable with mean 0 and variance 1. Show that Y has the density as given in Equation 8.35 by the usual transformation of variable technique. Also find the characteristic function of Y (Equation 8.39), starting with Equation 8.40.

8.24 Let
$$Z = \frac{1}{n}\sum_{i=1}^{n} X_i^2$$
where the X_i's are normal and independent with mean zero and variance σ^2. Show that
$$\text{Var}[Z] = \frac{2\sigma^4}{n}$$

8.25 Show that the product of f_{Y_1} as given in the equation following Equation A8.4 and $f_{Y_2 \cdots Y_n | Y_1}$ as given in Equation A8.5 produce Equation A8.4.

8.26 Find the variance of S^2.

8.27 Explain why $\Sigma(X_i - \overline{X})^2$ has only $(n - 1)$ degrees of freedom when there are n independent X_i's. Try $n = 1$, and a numerical example for $n = 3$.

8.28 With five i.i.d. samples from a normal R.V., find the probability that

   a. $S^2/\sigma^2 \geq .839$
   b. $S^2/\sigma^2 \geq 1.945$
   c. $.839 \leq S^2/\sigma^2 \leq 1.52$

8.29 With 31 i.i.d. samples from a normal R.V., find approximately

   a. $P[(\overline{X} - \mu)/(S/\sqrt{31})] > 1.7$
   b. $P[(\overline{X} - \mu)/(S/\sqrt{31})] > 2.045$

8.30 If σ^2 is known rather than being estimated by S^2, repeat Problem 8.29 and compare results.

8.31 Refer to Example 8.16. Find α such that $P[F_{7,5;\alpha} > 4.876]$.

8.32 If one suspected that the two samples in Example 8.16 might not have come from the same random variable, then one computed the ratio, $S_1^2/S_2^2 = 7$, what is the conclusion?

8.33 X is a normal random variable with $\sigma^2 = 8$. If we wish to test the hypothesis that $\mu = 0$ versus the alternative $\mu \neq 0$, at the 5% level of significance, and we want the critical region to be the region outside of ± 1, how many observations will be required?

8.34 If X is normal and the variance is unknown, find the critical region for testing the null hypothesis $\mu = 10$ versus the alternative hypothesis $\mu \neq 10$, if the significance level of the test is 1%. Find the answer if the number of observations is 1, 5, 10, 20, 50.

8.35 Refer to Problem 8.34. If $\overline{X} = 12$ and $S = 2$, when there are 21 observations, should the null hypothesis be accepted or rejected?

8.36 Test at the 10%, 5%, and 1% levels of significance the hypothesis that $\sigma_1^2 = \sigma_2^2$ if both S_1^2 and S_2^2 have 20 degrees of freedom, that is, find the three critical regions for F.

8.37 Test the hypothesis that the following data came from an exponential distribution. Use 1% significance level.

Interval	Number Observed
0–1	20
1–2	16
2–3	14
3–4	11
4–5	8
5–6	8
6–7	4
7–8	2
8–9	3

8.38 Test the hypothesis that the following data came from a uniform distribution (8, 10). Use 5% significance level.

Interval	Number Observed
8–8.2	7
8.2–8.4	8
8.4–8.6	9
8.6–8.8	10
8.8–9.0	9
9.0–9.2	11
9.2–9.4	9
9.4–9.6	8
9.6–9.8	7
9.8–10.0	8

8.39 Show that

$$\sum X_i^2 - n\overline{X}^2 = \sum (X_i - \overline{X})^2$$

8.40 Show that the sum of squares accounted for by regression is

$$\sum (\hat{Y}_i - \overline{Y})^2 = \hat{b}_1 \left\{ \sum X_i Y_i - \frac{\sum X_i \sum Y_i}{n} \right\}$$

8.41 Compare Equations 8.75 and 8.76 with Equations 7.4 and 7.5.

8.42 Show that

$$\text{Var}(\hat{b}_0) = \frac{\sum X_i^2 \sigma^2}{n \sum (X_i - \overline{X})^2}$$

where σ^2 is the variance of Y.

558 STATISTICS

8.43 Solve Equations 8.98, 8.99, and 8.100 for $\hat{b}_0$, $\hat{b}_1$, and $\hat{b}_2$.

8.44 Show that

    a. Equations 8.101 and 8.73 are identical.

    b. Equations 8.102 and 8.74 are identical.

    c. Equation 8.103 produces Equations 8.81 and 8.82.

    d. Equation 8.104 produces Equations 8.75 and 8.76.

8.45 The height and weight of 10 individuals are given below:

Height X (inches)	68	74	67	69	68	71	72	70	65	76
Weight Y (lbs)	148	170	150	140	148	160	170	165	135	180

    a. Find the best fit of the form $\hat{Y} = \hat{b}_0 + \hat{b}_1 X$ and compute the normalized RMS error of the fit. The normalized RMS error is defined as

$$\left[\frac{\sum (Y_i - \hat{Y}_i)^2}{\sum (Y_i - \bar{Y})^2} \right]^{1/2}$$

    b. Test if the fit is good at a significance level of 0.1.

8.46 The input–output data for a (nonlinear) amplifier is given:

Input X	0.1	0.2	0.4	0.5	0.6	0.8	0.9
Output Y	0.95	0.22	0.39	0.44	0.52	0.56	0.50

    a. Find the best fit of the form

$$\hat{Y} = \hat{b}_1 X - \hat{b}_3 X^3$$

    b. Compare the fit obtained in (a) with the best linear fit.

8.47 For the data given:

X_1	1	4	9	11	3	8	5	10	2	7	6
X_2	8	2	−8	−10	6	−6	0	−12	4	−2	−4
Y	6	8	1	0	5	3	2	−4	10	−3	5

    a. Find the best fit of the form

$$\hat{Y} = \hat{b}_0 + \hat{b}_1 X_1$$

and compute the normalized RMS error of the fit.

b. Find the best fit of the form

$$\hat{Y} = \hat{b}'_0 + \hat{b}'_1 X_1 + \hat{b}'_2 X_2$$

and compute the normalized RMS error of the fit and compare with (a).

c. At a significance level of 0.01, test if the fits obtained in (a) and (b) are good.

8.48 It is hypothesized that the peak load of an electrical utility is dependent on the real (adjusted for inflation) personal income (I) of its service territory, the population (P) in its service territory and the cooling degree-days (C) in the time period of interest. Given the following data, show that

$$\hat{L} \approx -23771 + .277I + 5.109P + 1.115C$$
$$\text{and} \quad R^2 \approx .9, \quad F_{3,16} \approx 51.85$$

Year and Quarter	Peak Load	Real Personal Income	Population	Cooling Degree Days
1982				
Q1	2885.00	36983.72	3200.00	11.00
Q2	3762.00	37006.05	3231.00	414.00
Q3	4218.00	36563.06	3251.00	1421.00
Q4	3004.00	36611.00	3271.00	105.00
1983				
Q1	2739.00	35574.63	3291.00	1.00
Q2	3618.00	35677.61	3311.00	378.00
Q3	4464.00	35267.69	3311.75	1440.00
Q4	3292.00	36453.88	3305.50	64.00
1984				
Q1	3164.00	36145.13	3310.25	0.00
Q2	4047.00	36151.49	3310.00	585.00
Q3	4651.00	36082.95	3307.75	1354.00
Q4	2813.00	36453.88	3305.50	71.00
1985				
Q1	3400.00	36353.26	3303.25	11.00
Q2	3813.00	36152.60	3301.00	548.00
Q3	4439.00	35944.74	3305.25	1334.00
Q4	3137.00	35965.61	3305.00	39.00
1986				
Q1	3023.00	35910.29	3303.36	26.00
Q2	4033.00	36017.64	3303.95	673.00
Q3	4698.00	35428.08	3305.29	1432.50
Q4	2960.00	35551.60	3306.01	105.00

CHAPTER NINE

Estimating the Parameters of Random Processes from Data

9.1 INTRODUCTION

In the first part of the book (Chapters 2 through 7), we developed probabilistic models for random signals and noise and used these models to derive signal extraction algorithms. We assumed that the models as well as the parameters of the models such as means, variances, and autocorrelation functions are known. Although this may be the case in some applications, in a majority of practical applications we have to estimate (or identify) the model structure as well as estimate parameter values from data. In Chapter 8 we developed procedures for estimating unknown values of parameters such as means, variances, and probability density functions. In this chapter we focus on the subject of estimating autocorrelation and power spectral density functions.

Autocorrelation and power spectral density functions describe the (second-order) time domain and frequency domain structure of stationary random processes. In any design using the MSE criteria, the design will be specified in terms of the correlation functions or equivalently the power spectral density functions of the random processes involved. When these functions are not known, we estimate them from data and use the estimated values to specify the design.

We emphasize both model-based (or parametric) and model-free (or nonparametric) estimators. Model-based estimators assume definite forms for autocorrelation and power spectral density functions. Parameters of the assumed model are estimated from data, and the model structure is also tested using the data. If the assumed model is correct, then accurate estimates can be obtained using relatively few data. Model-free estimators are based on relatively general assumptions and require more data. Thus, models can be used to reduce the amount of data required to obtain "satisfactory" estimates. However, a model that substitutes assumptions for data will produce "satisfactory" estimates only if the assumptions are correct.

In the first part of this chapter we emphasize model-free estimators for autocorrelation and power spectral density functions. The remainder of the chapter is devoted to model-based estimators using ARIMA (autoregressive integrated moving average) models. We use this Box–Jenkins type of parametric estimator because we feel that autoregressive and moving-average models are convenient and useful parametric models for random sequences and because Box–Jenkins algorithms provide well-developed and systematic procedures for identifying the model, estimating the parameters, and examining the adequacy of the fit. Software packages for processing data using the Box–Jenkins procedure are commercially available.

Throughout this chapter we will emphasize digital processing techniques for estimation.

9.2 TESTS FOR STATIONARITY AND ERGODICITY

In order to estimate any of the unknown parameters of a random process (e.g., mean, autocorrelation, and spectral density functions), the usual practice is to estimate these parameters from one sample function of the random process. *Thus, ergodicity is assumed.* Also, it was shown in Chapter 3 that stationarity is necessary for ergodicity. Thus, stationarity is also assumed. The standard practice is to test for stationarity but not to test for ergodicity. The reason for this apparent oversight is simply that many sample functions from the ensemble are needed to test for ergodicity, and many sample functions are usually either impossible or very expensive to obtain. This reason or excuse does not negate the fact that in both analysis and design the important sample function of the random process is the one(s) that the system will see in use. Thus, the sample function(s) from which the parameters of the random process are estimated must have the same parameters as the random process the system will see in its use. This is the essence of the ergodic assumption. Because in the rest of this chapter we assume ergodicity, this important assumption must be considered when estimating parameters of a random process and when using the estimated parameters in analysis and design. If we estimate parameters from one sample function, then that one sample certainly must be representative or a "good sample."

In the case of Gaussian random processes that are used to model a variety of phenomena, wide-sense stationarity implies strict-sense stationarity. Furthermore, we have shown in Section 3.8.2 that if the autocorrelation function of a zero-mean Gaussian random process satisfies

$$\int_{-\infty}^{\infty} |R_{XX}(\tau)| \, d\tau < \infty$$

then the process is ergodic. Thus, if we can assume that the process is Gaussian and that this condition is met, then testing for stationarity is equivalent to testing for ergodicity.

562 ESTIMATING THE PARAMETERS OF RANDOM PROCESSES

Various tests for stationarity are possible. For instance, one can test the hypothesis that the mean is stationary by a test of hypothesis introduced in Section 8.8.4. In addition, the ARIMA model estimation procedure suggested later in this chapter provides some automatic tests related to stationarity, and in fact suggests ways of creating stationary functions from nonstationary ones. Thus, stationarity can and should be tested when estimating the parameters of a random process. However, practical considerations militate against completely stationary processes. Rather, random sequences encountered in practice may be classified into three categories:

1. Those that are stationary over long periods of time: The underlying process seems stationary, for example, as with thermal-noise and white-noise generators, and the resulting data do not fail standard stationarity tests such as those mentioned later. No series will be stationary indefinitely; however, conditions may be stationary for the purpose of the situation being analyzed.
2. Those that may be considered stationary for short periods of time: For example, the height of ocean waves may be stationary for short periods when the wind is relatively constant and the tide does not change significantly. In this case, the data and their subsequent interpretation and use must be limited to these periods.
3. Sequences that are obviously nonstationary: Such series possibly may be transformed into (quasi-)stationary series as suggested in a later section.

9.2.1 Stationarity Tests

Given a sample from a random sequence such as shown in Figure 9.1, we wish to decide whether $X(n)$ is stationary. Of course, one would hope to decide this on the basis of knowledge of the underlying process; however, too often such is not the practical case. We now discuss some elementary tests for stationarity.

One requirement of stationarity is that $f_{X(n)}$ does not vary with n. In particular, the mean and the variance should not vary with n. A reasonable method of determining whether this is true is to divide the data into two or more sequential sections and calculate the sample mean and the sample variance from each section. The sample means may be compared informally by plotting or they may be formally tested for change using t tests as described in Section 8.8.4 of this book. Similarly the sample variances may be compared by plotting or by using F tests as described in Section 8.8.5.

9.2.2 Run Test for Stationarity

The tests suggested before used the t and the F distribution; thus they required the underlying distribution to be Gaussian. A test that does not require knowledge of the underlying distribution is the *run test*. It is used to test for randomness

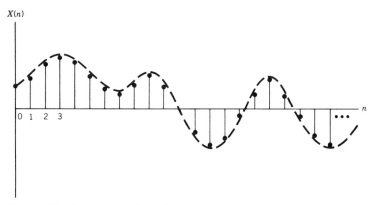

Figure 9.1 Sample function of a random sequence.

or for nonstationary trends in the data. First we will describe the test and then we will suggest how it can be applied to test for stationarity. In Section 9.4.7, we use this test for "whiteness" of a noise or error sequence.

Assume that we have $2N$ samples. N of these samples will be above (a) the median and N of these samples will be below (b) the median. (Ties and an odd number of samples can be easily coped with.)

The total number of possible distinct arrangements of the N a's and the N b's is $\binom{2N}{N}$. If the sample is random and i.i.d., then each of the possible arrangements is equally likely. A possible sequence with $(N = 8)$ is

$$a\ a\ b\ b\ b\ a\ b\ b\ b\ b\ a\ a\ a\ b\ a\ a$$

The total number of clusters of a's and b's will be called R for the number of *runs*. $R = 7$ in the foregoing sequence.

It can be shown [12] that the probability mass function for R is

$$P(R = n) = \begin{cases} 2\dfrac{\binom{N-1}{\frac{n}{2}-1}^2}{\binom{2N}{N}}, & n = 2, 4, \ldots, 2N \\[2ex] 2\dfrac{\binom{N-1}{\frac{n}{2}-1}\binom{N-1}{\frac{n}{2}-\frac{3}{2}}}{\binom{2N}{N}}, & n = 3, 5, \ldots, 2N-1 \end{cases} \quad (9.1)$$

564 ESTIMATING THE PARAMETERS OF RANDOM PROCESSES

This random variable has a mean and variance given by

$$E\{R\} = N + 1 \tag{9.2}$$

$$\sigma_R^2 = \frac{N(N-1)}{2N-1} \tag{9.3}$$

If the observed value of R is close to the mean, then randomness seems reasonable. On the other hand, if R is either too small or too large, then the null hypothesis of randomness should be rejected. Too large or too small is decided on the basis of the distribution of R. This distribution is tabulated in Appendix H.

EXAMPLE 9.1.

Test the given sequence for randomness at the 10% significance level.

SOLUTION:

In this sequence $N = 8$,

$$E[R] = 9$$

and

$$\sigma_R^2 = \frac{8(7)}{15} \approx 3.73$$

Using R as the statistic for the run test, the limits found at .95 and .05 from Appendix H with $N = 8$ are 5 and 12. Because the observed value of 7 falls between 5 and 12, the hypothesis of randomness is not rejected at the 10% significance level.

In order to test stationarity of the mean square values for either a random sequence or a random process, the time-average mean square values are computed for each of $2N$ nonoverlapping equal time intervals by (continuous data)

$$\widehat{X_i^2} = \frac{1}{T_i - T_{i-1}} \int_{T_{i-1}}^{T_i} X^2(t)\, dt$$

or by (sample data)

$$\widehat{X_i^2} = \frac{1}{N_i} \sum_{j=1}^{N_i} X^2(t_j); \quad t_j \in (T_{i-1}, T_i)$$

where T_{i-1} and T_i are respectively the lower and the upper time limits of the ith interval and N_i is the number of samples in the ith interval. The resulting sample mean square values can be the basis of a run test. That is, the sequence $\widehat{X_i^2}$, $i = 1, 2, \ldots, 2N$ is classified as above (a) or below (b) the median and the foregoing procedure is followed. It is important that the interval length ($T_i - T_{i-1}$) should be significantly longer than the correlation time of the random process. That is if

$$\tau > \frac{1}{10}(T_i - T_{i-1})$$

then $R_{XX}(\tau) \approx 0$ is a requirement to use the run test as a test for stationarity of mean square values as described.

9.3 MODEL-FREE ESTIMATION

In this section we address the problem of estimating the mean, variance, autocorrelation, and the power spectral density functions of an ergodic (and hence stationary) random process $X(t)$, given the random variables $X(0), X(1), \ldots, X(N-1)$, which are sampled values of $X(t)$. A normalized sampling rate of one sample per second is assumed. Although the estimators of this section require some assumptions (or model), they are relatively model-free as compared with the estimators introduced in the next section.

We will first specify an estimator for each of the unknown parameters or functions and then examine certain characteristics of the estimators including the bias and mean squared error (MSE). The estimators that we describe in this section will be similar in form to the "time averages" described in Chapter 3, except we will use discrete-time versions of the "time averages" to obtain these estimators. Problems at the end of this chapter cover some continuous versions.

9.3.1 Mean Value Estimation

With discrete samples, the mean is estimated by

$$\overline{X} = \frac{1}{N} \sum_{i=0}^{N-1} X(i) \tag{9.4}$$

It was shown in Chapter 8 that $\overline{X}$ is unbiased. In addition, the variance of $\overline{X}$ is

$$\text{Var}[\overline{X}] = \frac{1}{N^2} \sum_{i=0}^{N-1} \sum_{j=0}^{N-1} \sigma_{ij} \qquad (9.5.\text{a})$$

where

$$\sigma_{ij} = \text{covariance of } X(i), X(j)$$

In the stationary case

$$\text{Var}[\overline{X}] = \frac{NC_{XX}(0)}{N^2} + \frac{2(N-1)}{N^2} C_{XX}(1) + \frac{2(N-2)}{N^2} C_{XX}(2)$$
$$+ \cdots + \frac{2}{N^2} C_{XX}(N-1) \qquad (9.5.\text{b})$$

If $C_{XX}(i) = 0, i \neq 0$, that is the process is white, then

$$\text{Var}[\overline{X}] = \frac{C_{XX}(0)}{N}$$

9.3.2 Autocorrelation Function Estimation

With digital data we can estimate the autocorrelation function using

$$\hat{R}_{XX}(k) = \frac{1}{N-k} \sum_{i=0}^{N-k-1} X(i)X(i+k), \qquad k = 0, 1, 2, \ldots, N-1 \quad (9.6)$$

$$\hat{R}_{XX}(-k) = \hat{R}_{XX}(k)$$

Note that because of finite sample size (N), we have the following effects (see Figures 9.2 and 9.3).

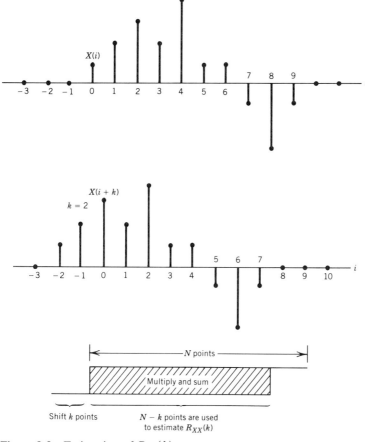

Figure 9.2 Estimation of $R_{xx}(k)$.

1. With N data points we can only estimate $R_{xx}(k)$ for values of k less than N, that is

$$\hat{R}_{xx}(k) = \begin{cases} \dfrac{1}{N-k} \sum_{i=0}^{N-k-1} X(i)X(i+k), & k < N \\ 0 & k \geq N \end{cases} \quad (9.7)$$

$$\hat{R}_{xx}(-k) = \hat{R}_{xx}(k)$$

This is equivalent to truncating the estimator for $|k| \geq N$.

2. As $k \to N$, we are using fewer and fewer points to obtain the estimate of $R_{xx}(k)$. This leads to larger variances in the estimated value of $R_{xx}(k)$ for $k \to N$.

568 ESTIMATING THE PARAMETERS OF RANDOM PROCESSES

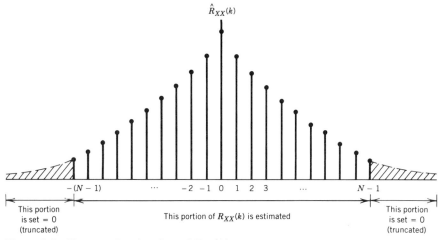

Figure 9.3 Truncated estimation of $R_{xx}(k)$.

It is easy to show that

$$E\{\hat{R}_{xx}(k)\} = R_{xx}(k), \quad k < N$$

that is, the estimator is unbiased if $k < N$. However, it is difficult to compute the variance of $\hat{R}_{xx}(k)$ since it will involve the fourth moments of the form

$$E\{X(n)X(n+m)X(k)X(k+m)\}$$

In the case of Gaussian processes, these moments can be evaluated (see Section 2.5.3) and the variance of $\hat{R}_{xx}(k)$ can be computed. (Problem 9.8.)

EXAMPLE 9.2.

Assume that we are estimating the autocorrelation function of a zero-mean white Gaussian random process with

$$R_{xx}(k) = \begin{cases} \sigma_X^2, & k = 0 \\ 0 & k \neq 0 \end{cases}$$

Find the variance of $\hat{R}_{xx}(k)$.

SOLUTION:

For $k = 0$

$$\hat{R}_{XX}(0) = \frac{1}{N} \sum_{i=0}^{N-1} [X(i)]^2$$

$$E\{\hat{R}_{XX}(0)\} = \frac{1}{N} \sum_{i=0}^{N-1} E\{[X(i)]^2\} = R_{XX}(0)$$

$$\text{Var}\{\hat{R}_{XX}(0)\} = E\{[\hat{R}_{XX}(0)]^2\} - [R_{XX}(0)]^2$$

$$E\{[\hat{R}_{XX}(0)]^2\} = \frac{1}{N^2} E\left\{\sum_{i=0}^{N-1}\sum_{j=0}^{N-1} [X(i)]^2[X(j)]^2\right\}$$

$$= \frac{1}{N^2}\left[\sum_{i=0}^{N-1} [E\{[X(i)]^4\} + (N-1)\sigma_X^2 E\{[X(i)]^2\}]\right]$$

$$= \frac{1}{N^2}[N(3\sigma_X^4) + N(N-1)\sigma_X^4]$$

$$= \left(\frac{N+2}{N}\right)\sigma_X^4$$

and hence

$$\text{Var}[\hat{R}_{XX}(0)] = \frac{N+2}{N}\sigma_X^4 - (\sigma_X^2)^2 = \frac{2\sigma_X^4}{N}$$

Similarly, it can be shown that

$$\text{Var}[\hat{R}_{XX}(k)] = \frac{1}{N-|k|}\sigma_X^4, \quad 1 \le |k| < N$$

Note that as $k \to N$, the variance increases rapidly.

9.3.3 Estimation of Power Spectral Density (psd) Functions

The psd function of a stationary random process is defined as

$$S_{XX}(f) = \int_{-\infty}^{\infty} R_{XX}(\tau)\exp(-j2\pi f\tau)\, d\tau$$

Based on the foregoing equation, we can define an estimator for the psd as

$$\hat{S}_{XX}(f) = \int_{-\infty}^{\infty} \hat{R}_{XX}(\tau)\exp(-j2\pi\tau f)\, d\tau \tag{9.8}$$

where $\hat{R}_{XX}(\tau)$ is an estimator of $R_{XX}(\tau)$. In the discrete case we can estimate $\hat{S}_{XX}(f)$ using the estimator

$$\hat{S}_{XX}(f) = \sum_{k=-(N-1)}^{N-1} \hat{R}_{XX}(k)\exp(-j2\pi k f), \qquad |f| < \frac{1}{2} \tag{9.9}$$

where

$$\hat{R}_{XX}(k) = \frac{1}{N-k} \sum_{i=0}^{N-k-1} X(i)X(i+k), \qquad k = 0, 1, 2, \ldots, N-1$$

is the estimator of the autocorrelation function defined in the preceding section.

The estimator given in Equation 9.9 has two major drawbacks. First, Equation 9.7 is analogous to time domain convolution, which is computationally intensive. Second, the variance of the estimator is too large; and there is no guarantee that Equation 9.9 will yield a nonnegative value for the estimate of the psd. These two problems can be overcome by using a DFT-based estimator of the psd called a *periodogram*.

Discrete Fourier Transform (DFT). We have relied on the sampling theorem (Section 3.10) to sample in the time domain at a sampling interval T_S if $T_S < 1/2B$, where B is the bandwidth of the random process, that is, $S_{XX}(f) = 0$, $|f| > B$. In subsequent work we have normalized the sampling interval T_S, to be 1, which requires that $B < \frac{1}{2}$. This has resulted in the often-used restriction, $|f| < \frac{1}{2}$ for Fourier transforms of sequences.

If $x(t) = 0$ for $t < t_1$ or $t > t_1 + T_M$, then by an identical type of argument, we can sample in the frequency domain at an interval $f_S < 1/T_M$. If we have normalized, $T_S = 1$, and if $x(n) = 0$ for $n < 0$ or $n > N - 1$, then if we choose $f_s = 1/N$, then we can completely represent the signal $x(t)$. If this is the case, then we have the usual Fourier transform of a sequence

$$X_F(f) = \sum_{n=0}^{N-1} x(n)\exp(-j2\pi n f), \qquad 0 \le f \le 1$$

where we have now taken the principle part of the cyclical $X_F(f)$ to be $0 \le f \le 1$ (rather than $|f| < \frac{1}{2}$ as before). Now set $f = k/N$, $k = 0, 1, 2, \ldots, N-1$

and we have

$$X_F\left(\frac{k}{N}\right) = \sum_{n=0}^{N-1} x(n)\exp\left(-j2\pi\frac{kn}{N}\right), \quad k = 0, \ldots, N-1 \quad (9.10.a)$$

and $x(n)$ can be recovered from $X_F(k/N)$ by the inverse transform

$$x(n) = \frac{1}{N}\sum_{k=0}^{N-1} X_F\left(\frac{k}{N}\right)\exp\left(+j2\pi\frac{kn}{N}\right), \quad n = 0, \ldots, N-1 \quad (9.10.b)$$

Equations 9.10.a and 9.10.b define the discrete Fourier transform (DFT). It can be shown that a time-limited signal cannot be perfectly band-limited; however, for practical purposes, we can record a signal from 0 to T_M, which is approximately limited to no frequency component at or above B. Discrete Fourier transforms (DFTs) based on such algorithms are readily available (see Appendix B). If N is chosen to be a power of 2, then fast Fourier transform (FFT) algorithms, which are computationally efficient, can be used to implement Equation 9.10.

Periodogram Estimator of the psd. In order to define this estimator, let us introduce another estimator for the autocorrelation function,

$$\hat{\hat{R}}_{XX}(k) = \frac{1}{N}\sum_{i=0}^{N-k-1} X(i)X(i+k), \quad k = 0, 1, \ldots, N-1 \quad (9.11.a)$$

Note that this estimator differs from $\hat{R}_{XX}(k)$ given in Equation 9.6 by the factor $(N-k)/N$, that is

$$\hat{\hat{R}}_{XX}(k) = \left(\frac{N-k}{N}\right)\hat{R}_{XX}(k), \quad k = 0, \ldots, N-1 \quad (9.11.b)$$

Now if we define

$$d(n) = \begin{cases} 1, & n = 0, \ldots, N-1 \\ 0 & \text{elsewhere} \end{cases} \quad (9.12)$$

then it can be shown (Problem 9.21) that

$$\hat{\hat{R}}_{XX}(k) = \frac{1}{N}\sum_{n=-\infty}^{\infty} [d(n)X(n)][d(n+k)X(n+k)], \quad k = 0, \ldots, N-1$$

572 ESTIMATING THE PARAMETERS OF RANDOM PROCESSES

The Fourier transform of $\hat{R}_{XX}(k)$, denoted by $\hat{S}_{XX}(f)$, is

$$\hat{S}_{XX}(f) = \sum_{k=-\infty}^{\infty} \hat{R}_{XX}(k)\exp(-j2\pi kf), \quad |f| < \frac{1}{2}$$

$$= \sum_{k=-\infty}^{\infty}\left[\frac{1}{N}\sum_{n=-\infty}^{\infty} d(n)X(n)d(n+k)X(n+k)\right]\exp(-j2\pi kf)$$

$$= \frac{1}{N}\left[\sum_{n=-\infty}^{\infty} d(n)X(n)\exp(j2\pi nf)\right]$$

$$\times \left[\sum_{k=-\infty}^{\infty} d(n+k)X(n+k)\exp(-j2\pi(n+k)f)\right]$$

Substituting $n + k = m$ in the second summation, we obtain

$$\hat{S}_{XX}(f) = \frac{1}{N} X_F^*(f)X_F(f) = \frac{1}{N}|X_F(f)|^2, \quad |f| < \frac{1}{2} \quad (9.13)$$

where $X_F(f)$ is the Fourier transform of the data sequence

$$X_F(f) = \sum_{n=-\infty}^{\infty} d(n)X(n)\exp(-j2\pi nf)$$

$$= \sum_{n=0}^{N-1} X(n)\exp(-j2\pi nf)$$

The estimator $\hat{S}_{XX}(f)$ defined in Equation 9.13 is called the *periodogram* of the data sequence $X(0), X(1), \ldots, X(N-1)$.

Bias of the Periodogram. The periodogram estimator is a biased estimator of the psd, and we can evaluate the bias by calculating $E\{\hat{S}_{XX}(f)\}$

$$E\{\hat{S}_{XX}(f)\} = E\left\{\sum_{k=-\infty}^{\infty} \hat{R}_{XX}(k)\exp(-j2\pi kf)\right\}$$

$$= \sum_{k=-\infty}^{\infty} E\{\hat{R}_{XX}(k)\}\exp(-j2\pi kf)$$

$$= \sum_{k=-\infty}^{\infty} \frac{1}{N}\left[\sum_{n=-\infty}^{\infty} d(n)d(n+k)\right.$$

$$\left.\times E\{X(n)X(n+k)\}\exp(-j2\pi kf)\right]$$

$$= \sum_{k=-\infty}^{\infty} R_{XX}(k)\exp(-j2\pi kf) \frac{\left[\sum_{n=-\infty}^{\infty} d(n)d(n+k)\right]}{N}$$

$$= \sum_{k=-\infty}^{\infty} q_N(k)R_{XX}(k)\exp(-j2\pi kf), \quad |f| < \frac{1}{2} \qquad (9.14)$$

where

$$q_N(k) = \frac{1}{N}\sum_{n=-\infty}^{\infty} d(n)d(n+k)$$

$$= \begin{cases} 1 - \frac{|k|}{N}, & |k| < N \\ 0 & \text{elsewhere} \end{cases} \qquad (9.15)$$

The sum on the right-hand side of Equation 9.14, is the Fourier transform of the product of $q_N(k)$ and $R_{XX}(k)$; thus

$$E\{\hat{S}_{XX}(f)\} = \int_{-1/2}^{1/2} S_{XX}(\alpha)Q_N(f-\alpha)\,d\alpha \qquad (9.16.\text{a})$$

where $Q_N(f)$ is the Fourier transform of $q_N(k)$, that is

$$Q_N(f) = \frac{1}{N}\left[\frac{\sin(\pi fN)}{\sin \pi f}\right]^2, \quad |f| < \frac{1}{2} \qquad (9.16.\text{b})$$

Equations 9.14 and 9.15 show that $\hat{S}_{XX}(f)$ is a biased estimator and the bias results from the truncation and the triangular window function [as a result of using $1/N$ instead of $1/(N - |k|)$ in the definition of $\hat{R}_{XX}(k)$]. Equation 9.16 shows that S_{XX} is convolved with a (sinc)² function in the frequency domain. Plots of $q_N(k)$, its transform $Q_N(f)$, and the windowing effect are shown in Figure 9.4.

The convolution given in Equation 9.16.a has an "averaging" effect and it produces a "smeared" estimate of $S_{XX}(f)$. The effect of this smearing is known as *spectral leakage*. If $S_{XX}(f)$ has two closely spaced spectral peaks, the periodogram estimate will smooth these peaks together. This reduces the spectral resolution as shown in Figure 9.4.b.

574   ESTIMATING THE PARAMETERS OF RANDOM PROCESSES

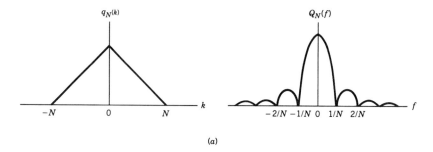

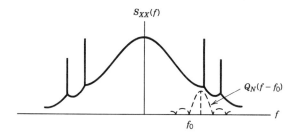

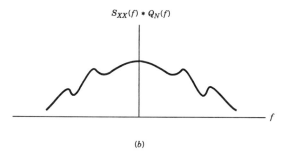

Figure 9.4 Periodogram estimator. (a) Window functions $q_N(k)$ and $Q_N(f)$. (b) Effect of windowing.

However, as N is increased, $Q_N(f) \to \delta(f)$ resulting in

$$E\{\hat{S}_{XX}(f)\} = \int_{-1/2}^{1/2} S_{XX}(\alpha)\delta(f - \alpha)\, d\alpha = S_{XX}(f)$$

Thus $\hat{S}_{XX}(f)$ is asymptotically unbiased.

Variance of the Periodogram. In order to compute the variance of the periodogram we need to make some assumptions about the psd that is being estimated. Assuming $X(n)$ to be a zero-mean white Gaussian sequence with variance σ^2, we can compute the variance of the periodogram at $f = p/N$ as follows. The periodogram at $f = p/N$ is given by

$$\hat{S}_{XX}\left(\frac{p}{N}\right) = \frac{1}{N}\left|\sum_{n=0}^{N-1} X(n)\exp\left(-\frac{j2\pi np}{N}\right)\right|^2, \quad p = 0, \ldots, [N/2]$$
$$= A_p^2 + B_p^2 \qquad (9.17)$$

where

$$A_p = \frac{1}{\sqrt{N}} \sum_{n=0}^{N-1} X(n)\cos\left(\frac{2\pi np}{N}\right) \qquad (9.18.\text{a})$$

and

$$B_p = \frac{1}{\sqrt{N}} \sum_{n=0}^{N-1} X(n)\sin\left(\frac{2\pi np}{N}\right) \qquad (9.18.\text{b})$$

Both A_p and B_p are linear combinations of Gaussian random variables and hence are Gaussian. Also because $X(n)$ is zero-mean

$$E\{A_p\} = E\{B_p\} = 0$$

and

$$\text{Var}\{A_p\} = E\{A_p^2\}$$
$$= \frac{\sigma^2}{N} \sum_{n=0}^{N-1} \cos^2\left(\frac{2\pi np}{N}\right)$$
$$= \begin{cases} \dfrac{\sigma^2}{2}, & p \neq 0, \left[\dfrac{N}{2}\right] \\ \sigma^2, & p = 0, \left[\dfrac{N}{2}\right] \end{cases} \qquad (9.19)$$

(See Problem 9.9.) Thus

$$A_p \sim \begin{cases} N\left(0, \dfrac{\sigma^2}{2}\right), & p \neq 0, \left[\dfrac{N}{2}\right] \\ N(0, \sigma^2), & p = 0, \left[\dfrac{N}{2}\right] \end{cases} \qquad (9.20)$$

Similarly, it can be shown that

$$B_p \sim \begin{cases} N\left(0, \dfrac{\sigma^2}{2}\right), & p \neq 0, \left[\dfrac{N}{2}\right] \\ 0 & p = 0, \left[\dfrac{N}{2}\right] \end{cases}$$

and

$$\text{Covar}\{A_p, B_p\} = \text{Covar}\{A_p, B_q\} = \text{Covar}\{A_p, A_q\}$$
$$= \text{Covar}\{B_p, B_q\} = 0 \text{ for all } p \neq q$$

Since A_p and B_p are independent Gaussian random variables, $\hat{S}_{XX}(p/N)$ (from Equation 9.17) is the sum of chi-square random variables. We note that

$$\frac{\hat{S}_{XX}\left(\dfrac{p}{N}\right)}{\dfrac{\sigma^2}{2}} \approx \chi_2^2, \quad p \neq 0, \left[\dfrac{N}{2}\right] \tag{9.21}$$

and

$$\frac{\hat{S}_{XX}\left(\dfrac{p}{N}\right)}{\sigma^2} \approx \chi_1^2, \quad p = 0, \left[\dfrac{N}{2}\right] \tag{9.22}$$

and hence (see Equation 8.42)

$$\text{Var}\left\{\hat{S}_{XX}\left(\dfrac{p}{N}\right)\right\} = \begin{cases} \sigma^4, & p \neq 0, \left[\dfrac{N}{2}\right] \\ 2\sigma^4, & p = 0, \left[\dfrac{N}{2}\right] \end{cases} \tag{9.23}$$

Equation 9.23 shows that, for most values of f, $\hat{S}_{XX}(f)$ has a variance of σ^4. Since we have assumed that $S_{XX}(f) = \sigma^2$, the normalized standard error, ϵ_r, of the periodogram estimator is

$$\epsilon_r = \frac{\sqrt{\text{Var } \hat{S}_{xx}(f)}}{S_{xx}(f)} = \frac{\sigma^2}{\sigma^2} = 100\%$$

EXAMPLE 9.3.

If $X(n)$ is a zero-mean white Gaussian sequence with unity variance, find the expected value and the variance of the periodogram estimator of the psd.

SOLUTION:

$$R_{xx}(k) = 1, \quad k = 0$$
$$= 0 \quad \text{elsewhere}$$

Using Equation 9.14

$$E\{\hat{S}_{xx}(f)\} = q_N(0) \cdot 1 = 1$$

Using Equation 9.23

$$\text{Var}\left\{\hat{S}_{xx}\left(\frac{p}{N}\right)\right\} = \begin{cases} 1, & p \neq 0, \left[\frac{N}{2}\right] \\ 2, & p = 0, \left[\frac{N}{2}\right] \end{cases}$$

Note that because of the white Gaussian assumption, both the mean and the variance are essentially (except at the end points) constant with p.

Thus, the periodogram estimator has a normalized error of 100%—a relatively poor estimator. In addition, the variance (and hence the normalized error) does not depend on the sample size N. Unlike most estimation problems, where the variance of the estimator is reduced as the sample size is increased, the

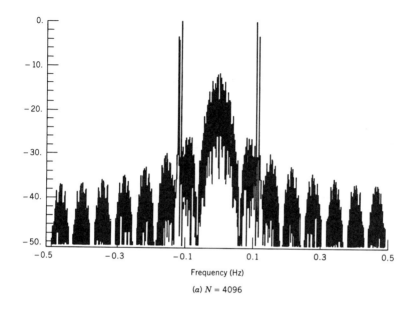
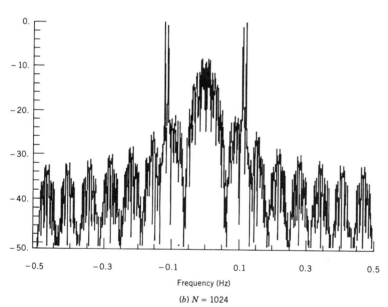

Figure 9.5 Periodogram of a random binary waveform plus two tones.

variance of the periodogram cannot be reduced by increasing the sample size. However, increasing the sample size N will produce better resolution in the frequency domain. (See Figure 9.5 for an example.)

The periodogram estimator can be improved by averaging or smoothing. Two (weighted) averaging techniques that are widely used involve averaging estimates obtained from nonoverlapping sections of the data or averaging the estimates in the frequency domain. Appropriate weighting (or window) functions are applied to control the bias and variance of the averaged estimators.

Trade-off Between Bias and Variance. Before we develop averaging techniques, let us examine the source of the bias and variance in the periodogram. Since the periodogram may be viewed as the transform of $\hat{R}_{XX}(k)$, let us examine the estimator $\hat{R}_{XX}(k)$. With reference to Figure 9.3, suppose that we estimate $R_{XX}(k)$ for $|k| = 0, 1, \ldots, M$ and use the estimated values of $R_{XX}(k)$ to form an estimate of $S_{XX}(f)$. Now, when $M \ll N$, and $N \gg 1$, we obtain "good" estimators of $R_{XX}(k)$ for $|k| \leq M$. However, the bias of $\hat{S}_{XX}(f)$ will be larger since the estimated autocorrelation function is truncated (set equal to zero) beyond $k > M$. As we increase $M \to N$, the bias of $\hat{S}_{XX}(f)$ will become smaller, but the variance of the estimator of $R_{XX}(k)$ will be larger as $k \to N$ since fewer and fewer points are used in the estimator. Thus, for a finite sample size, we cannot completely control both bias and variance; when we attempt to reduce one, the other one increases.

When the sample size is very large, we can reduce both the bias and variance to acceptable levels by using appropriate "windowing" (or averaging) techniques as explained in the following section.

9.3.4 Smoothing of Spectral Estimates

We can take the N measurements $X(0), X(1), \ldots, X(N-1)$, divide them into n sections, each of which contains N/n points, form n different estimators of the psd, and average the n estimators to form an averaged spectral estimator of the form

$$\bar{S}_{XX}(f) = \frac{1}{n} \sum_{k=1}^{n} \hat{S}_{XX}(f)_k \tag{9.24}$$

where $\hat{S}_{XX}(f)_k$ is the spectral estimate obtained from the kth segment of the data. If we assume that the estimators $\hat{S}_{XX}(f)_k$ are independent (which is not completely true if the data segments are adjacent), the variance of the averaged estimator will be reduced by the factor n. However, since fewer points are used to obtain the estimator $\hat{S}_{XX}(f)_k$, the function $Q_{N/n}(f)$ (Equation 9.16.b) will be wider than $Q_N(f)$ in the frequency domain, and from Equation 9.16.a it can be seen that the bias will be larger.

A similar form of averaging can also be done by averaging spectral estimates in the frequency domain. This averaging can be done simply as

$$\bar{\bar{S}}_{XX}\left(\frac{p}{N}\right) = \frac{1}{2m+1} \sum_{i=-m}^{m} \hat{S}_{XX}\left(\frac{p+i}{N}\right) \tag{9.25}$$

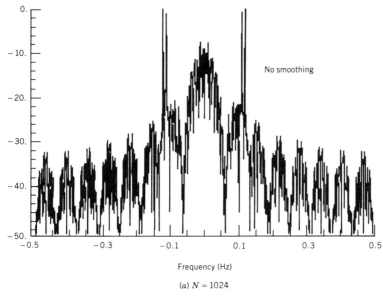

(a) $N = 1024$

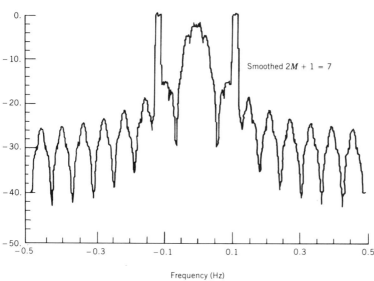

(b) $N=1024$

Figure 9.6 Effect of smoothing. (*a*) Unsmoothed; (*b*) smoothed, window size = 7.

Equation 9.25 represents a running average in the frequency domain using a sliding rectangular window of width $2m + 1$ points. Once again, if we assume that the estimators of adjacent spectral components are independent (this will not be true in general), then the averaging will reduce the variance. However, as in the case of averaging in the time domain, the bias will increase (as shown in Figure 9.6). Note that averaging will reduce spectral resolution and closely spaced spectral components will be merged together.

The averaging techniques described in the preceding paragraphs are simple running averages that use rectangular (or uniformly weighted) averaging windows. The variance is reduced while the bias increases. By using nonuniformly weighted window functions, we can control the trade-off between bias and variance and produce asymptotically unbiased and consistent estimators for the psd. In the following section we describe some of the windowing techniques that are widely used.

Windowing Procedure. Windowed or smoothed estimators of power spectral density functions are implemented using the following three steps:

Step 1. Compute $\hat{R}_{XX}(k)$ using Equation 9.6 or the following DFT/FFT operations.

1.a. Pad $X(n)$ with N zeroes and create a padded sequence $X_P(n)$ whose length is at least $2N$ points. The following FFT method will be computationally most efficient if $2N$ is chosen equal to a power of 2. This padding is necessary to avoid the circular (periodic) nature of the DFT, which can cause errors in convolution and correlation operations.

1.b. Compute $X_{P,F}(m/2N)$ according to

$$X_{P,F}\left(\frac{m}{2N}\right) = \sum_{n=0}^{2N-1} X_P(n) \exp\left(-\frac{j2\pi mn}{2N}\right)$$

$$m = 0, 1, 2, \ldots, 2N - 1$$

1.c. Obtain $\hat{\hat{R}}_{XX}(k)$ from

$$\hat{\hat{R}}_{XX}(k) = \frac{1}{N}\left[\frac{1}{2N}\sum_{m=0}^{2N-1}\left|X_{P,F}\left(\frac{m}{2N}\right)\right|^2 \exp\left(\frac{j2\pi km}{2N}\right)\right]$$

$$k = 0, 1, 2, \ldots, N - 1 \quad (9.26.a)$$

1.d. Compute $\hat{R}_{XX}(k)$ as

$$\hat{R}_{XX}(k) = \begin{cases} \dfrac{N}{N-k}\hat{\hat{R}}_{XX}(k), & k = 0, 1, \ldots, N-1 \\ 0 & k > N - 1 \end{cases}$$

$$\hat{R}_{XX}(-k) = \hat{R}_{XX}(k) \quad (9.26.b)$$

582 ESTIMATING THE PARAMETERS OF RANDOM PROCESSES

Step 2. Apply a weighted window and truncate $\hat{R}_{XX}(k)$ to $2M + 1$ points

$$\overline{R}_{XX}(k) = \hat{R}_{XX}(k)\lambda(k)$$
$$|k| = 0, 1, 2, \ldots, M, \qquad M \ll N \quad (9.26.c)$$

where $\lambda(k)$ is a window function to be discussed later.

Step 3. Pad $\overline{R}_{XX}(k)$ with zeroes for $|k| > M$ and take the DFT to obtain the smoothed estimate $\overline{S}_{XX}(f)$ as

$$\overline{S}_{XX}\left(\frac{p}{N}\right) = \sum_{k=-(N-1)}^{N-1} \overline{R}_{XX}(k) \exp\left(-\frac{j2\pi kp}{N}\right)$$

$$|p| = 0, 1, \ldots, \frac{N}{2} \quad (9.26.d)$$

Steps 1.b and 1.c, if completed via FFT algorithms, produce a computationally efficient method of calculating an estimate of $\hat{R}_{XX}(k)$. Truncating in Step 2, with $M \ll N$, is equivalent to deleting the "end points" of $\hat{R}_{XX}(k)$, which are estimates produced with fewer data points and hence are subjected to large variations. Deleting them will reduce the variance, but also will reduce the resolution, of the spectral estimator. Multiplying with $\lambda(k)$ and taking the Fourier transform (Equation 9.26.c) has the effect

$$E\{\overline{S}_{XX}(f)\} = \int_{-1/2}^{1/2} S_{XX}(\alpha) W_M(f - \alpha)\, d\alpha$$

where $W_M(f)$ is the Fourier transform of the window function $\lambda(k)$. In order to reduce the bias (and spectral leakage), $\lambda(k)$ should be chosen such that $W_M(f)$ has most of its energy in a narrow "main lobe" and has smaller "side lobes." This reduces the amount of "leakage." Several window functions have been proposed and we briefly summarize their properties. Derivations of these properties may be found in References [10] and [11]. It should be noted that most of these windows introduce a scale factor in the estimator of the power spectral density.

Rectangular Window.

$$\lambda(k) = \begin{cases} 1, & |k| \le M, \quad M < N \\ 0 & \text{otherwise} \end{cases} \quad (9.27.a)$$

$$W_M(f) = \frac{\sin\left[\left(M + \frac{1}{2}\right) 2\pi f\right]}{\sin(\pi f)} \quad (9.27.b)$$

Since $W_M(f)$ is negative for some f, this window might produce a negative estimate for the psd. This window is seldom used in practice.

Bartlett Window.

$$\lambda(k) = \begin{cases} 1 - \dfrac{|k|}{M}, & |k| \le M, \ M < N \\ 0 & \text{otherwise} \end{cases} \quad (9.28.\text{a})$$

$$W_M(f) = \dfrac{1}{M} \left[\dfrac{\sin(\pi f M)}{\sin(\pi f)} \right]^2 \quad (9.28.\text{b})$$

Since $W_M(f)$ is always positive, the estimated value is always positive. When $M = N - 1$, this window produces the unsmoothed periodogram estimator given in Equation 9.11.

Blackman–Tukey Window.

$$\lambda(k) = \begin{cases} \dfrac{1}{2}\left[1 + \cos\left(\dfrac{\pi k}{M}\right)\right], & |k| \le M, \ M < N \\ 0, & \text{otherwise} \end{cases} \quad (9.29.\text{a})$$

$$W_M(f) = \dfrac{1}{4}\left[D_M\left(2\pi f - \dfrac{\pi}{M}\right) + D_M\left(2\pi f + \dfrac{\pi}{M}\right)\right] + \dfrac{1}{2} D_M(2\pi f) \quad (9.29.\text{b})$$

where

$$D_M(2\pi f) = \dfrac{\sin\left[\left(M + \dfrac{1}{2}\right)2\pi f\right]}{\sin(\pi f)} \quad (9.29.\text{c})$$

Parzen Window.

$$\lambda(k) = \begin{cases} 1 - 6\left(\dfrac{k}{M}\right)^2 + 6\left(\dfrac{|k|}{M}\right)^3, & |k| \le \dfrac{M}{2} \\ 2\left(1 - \dfrac{|k|}{M}\right)^3, & \dfrac{M}{2} < k \le M < N \\ 0 & \text{otherwise} \end{cases} \quad (9.30.\text{a})$$

$$W_M(f) = \dfrac{3}{4M^3} \left[\dfrac{\sin\left(\dfrac{\pi M f}{2}\right)}{\dfrac{1}{2}\sin(\pi f)} \right]^4 \left[1 - \dfrac{2}{3}\sin^2(\pi f)\right] \quad (9.30.\text{b})$$

Note that $W_M(f)$ is nonnegative and hence Parzen's window produces a nonnegative estimate of the psd.

9.3.5 Bias and Variance of Smoothed Estimators

Expressions for the bias and variances of smoothed estimators are derived in Reference [11]. We present the results here.

For the Bartlett, Blackman-Tukey, and the Parzen windows, the asymptopic bias and variance are given by

$$\text{Bias}\{\overline{S}_{XX}(f)\} = C_1 \frac{S''_{XX}(f)}{M^2} \tag{9.31}$$

and

$$\text{Var}\{\overline{S}_{XX}(f)\} = C_2 \left(\frac{M}{N}\right) S^2_{XX}(f) \tag{9.32}$$

where C_1 and C_2 are constants that have different values for the different windows and $S''_{XX}(f)$ is the second derivative of $S_{XX}(f)$.

If we choose $M = 2\sqrt{N}$ as suggested in the literature on spectral estimation, then as the sample size $N \to \infty$, we can see from Equations 9.31 and 9.32 that both the bias and variance of the smoothed estimators approach zero. Thus, we have a family of asymptotically unbiased and consistent estimators for the psd.

9.4 MODEL-BASED ESTIMATION OF AUTOCORRELATION FUNCTIONS AND POWER SPECTRAL DENSITY FUNCTIONS

The model-free methods for estimating autocorrelation and spectral density functions previously discussed in this chapter are often useful. However, if a simple analytical expression for $R_{XX}(\tau)$ or $S_{XX}(f)$ is required such as in Wiener or Kalman filtering, then model-based estimators are used. In addition, fewer data are required to estimate a few parameters in a model than are needed for a model-free estimate; thus a model is used as a partial substitute for more data. In the remainder of this chapter, we discuss such estimators.

In this section a simple form of the autocorrelation function (or spectral density function) is assumed, and the parameters of this function are estimated. When one adopts such an approach, there is the concomitant obligation to investigate whether the assumed model is consistent with the data. Thus, the iterative steps in this procedure are as follows:

1. Assume a form of the autocorrelation function (or a model structure or type).
2. Estimate the parameters of this model.
3. Check to see whether the assumed model is consistent with the data. If not, return to Step 1 and start with a revised model.

In this part of the chapter, we propose to use *a priori* knowledge or assumptions to select the form of the model of the random sequence. If this knowledge (or assumptions) results in a good approximate model of the actual sequence, then the resulting estimators of both the autocorrelation function and power density spectrum are usually superior (i.e., smaller MSE with equivalent data) to the model-free estimators that were described previously.

Many random sequences encountered in engineering practice can be approximated by a rational transfer function model. In such a model, an input driving sequence $e(n)$ is related to the output sequence $X(n)$ by a linear difference equation:

$$X(n) = \sum_{i=1}^{p} \phi_{p,i} X(n-i) + \sum_{k=0}^{q} \theta_{q,k} e(n-k) \qquad (9.33)$$

where $\theta_{q,0} = 1$, $\phi_{p,p} \neq 0$, $\theta_{q,q} \neq 0$, and $e(n)$ is a zero-mean Gaussian white noise sequence.

This is the autoregressive moving average [ARMA (p, q)] model discussed in Section 5.2. These models can be reduced to state models as needed in Kalman filtering and to a rational transfer function

$$H(f) = \frac{\Theta(f)}{\Phi(f)} \qquad (9.34)$$

where

$$\Theta(f) = \sum_{k=0}^{q} \theta_{q,k} \exp(-j2\pi f k) \quad \text{(Moving average part)} \qquad (9.35)$$

$$\Phi(f) = 1 - \sum_{i=1}^{p} \phi_{p,i} \exp(-j2\pi f i) \quad \text{(Autoregressive part)} \qquad (9.36)$$

ESTIMATING THE PARAMETERS OF RANDOM PROCESSES

Then

$$S_{XX}(f) = |H(f)|^2 S_{ee}(f) = H(f)H(-f)S_{ee}(f) \tag{9.37}$$

where $S_{XX}(f)$ is the power spectral density function of the random sequence $X(n)$ and

$$S_{ee}(f) = \sigma_N^2, \qquad |f| < \frac{1}{2} \tag{9.38}$$

Thus, the remainder of this chapter is devoted to estimating the parameters of an autoregressive moving-average model. The estimates can be used in the model to obtain estimates of $R_{XX}(k)$ or $S_{XX}(f)$. The procedure for estimating the parameters of ARMA models has been well-developed by Box and Jenkins [3], and a large number of computer programs have been developed based on their work. We will introduce the Box–Jenkins iterative procedure, which consists of the four steps shown in Figure 9.7. These steps are discussed in the following sections.

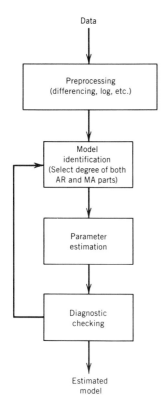

Figure 9.7 Steps in Box–Jenkins method of estimation of random sequence models.

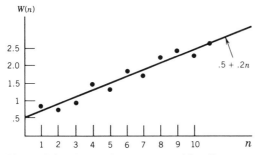

Figure 9.8 A sample sequence with a linear trend.

9.4.1 Preprocessing (Differencing)

The first step in model identification is testing for stationarity. Such tests were discussed in Section 9.2. We now consider as a part of this "Box–Jenkins" estimation process the possibility of transforming nonstationary sequences into quasistationary sequences.

Transformation of Nonstationary Series. In this section, we emphasize differencing to create stationarity. However, we mention that other types of transformations might be equally valuable. For example, assume the sequence $X(n)$ is such that the mean and variance both increase with time and that the standard deviation is proportional to the mean. (The standard deviation of electrical load of a utility is usually proportional to the mean.) Thus, if we transform $X(n)$ to $Y(n) = \ln X(n)$, then although the variation in mean still exists, the variance is now (approximately) constant (see Problem 9.22).

Differencing to Remove Linear Mean Trends. Assume $W(n)$ has a linear trend superimposed upon a white noise sequence. A plot of a sample sequence is shown in Figure 9.8. $W(n)$ is clearly a nonstationary sequence. However, its first difference $Y(n)$, where

$$Y(n) = W(n) - W(n-1), \quad n > 1 \quad (9.39)$$

will be stationary.

EXAMPLE 9.4.

Let

$$W(n) = .5 + .2n + A(n) \quad (9.40)$$

where $A(n)$ is a zero-mean stationary sequence. Find the difference $W(n) - W(n - 1)$, and show how to recover the original sequence from the resulting stationary model.

SOLUTION: Defining $Y(n) = W(n) - W(n - 1)$ we have

$$Y(n) = .5 + .2n + A(n) - .5 - .2(n - 1) - A(n - 1)$$

or

$$Y(n) = .2 + A(n) - A(n - 1) \qquad (9.41)$$

This sequence is stationary because the difference of two jointly stationary sequences is stationary. Furthermore the nonzero mean, .2 in the example, can be removed as follows

$$X(n) \triangleq Y(n) - .2 \qquad (9.42)$$

This is stationary with zero mean, and can be modeled as an ARMA sequence.
If a reasonable model is found for $X(n)$ [e.g., $X(n) = \frac{1}{2}X(n - 1) + \frac{1}{4}e(n - 1) + e(n)$], then we can find the model for the original sequence, $W(n)$, by using Equations 9.42 and 9.41. Indeed

$$Y(n) = .2 + X(n) \qquad (9.43)$$

and from the Equation 9.39

$$\begin{aligned} W(n) &= W(0) + Y(1) + Y(2) + \cdots + Y(n), \quad n > 1 \\ &= W(n - 1) + Y(n) \end{aligned} \qquad (9.44)$$

That is, the original series is a summation of the differenced series plus perhaps a constant as in Example 9.4. Because of the analogy with continuous processes, $W(n)$ is usually called the integrated version of $Y(n)$, and if $Y(n)$ (or $X(n)$ in the example) is represented by an ARMA model, then $W(n)$ is said to have an ARIMA model where the I stands for "integrated."

A general procedure for trend removal calls for first or second differencing of nonstationary sequences in order to attempt to produce stationary series. The resulting ARMA stationary models are then integrated (summed) the corresponding number of times to produce a model for the original sequence. The total model is then called ARIMA (p, d, q), where p is the order of the autoregressive (AR) part of the model, d is the order of differencing (I), and q is the order of the moving average (MA) part of the model.

Differencing to Remove Periodic Components. If $W(n)$ represents the average monthly temperature, then it seems reasonable that this series will have a periodic component of period 12. That is, January 1988 temperature should be close to January temperatures of other years. This series will definitely not be stationary. However, consider the sequence $X(n)$ where

$$X(n) = W(n) - W(n - 12), \quad n > 12 \qquad (9.45)$$

This series may well be stationary.

EXAMPLE 9.5.

Let

$$W(n) = 1 - \cos 2\pi \frac{n}{12} + A(n) \qquad (9.46)$$

where $A(n)$ is stationary. A sample function of $W(n)$ is shown in Figure 9.9. Find the periodic difference and recover the original series from the stationary series.

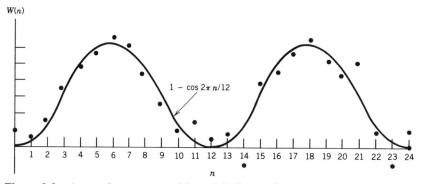

Figure 9.9 A sample sequence with a periodic trend.

590 ESTIMATING THE PARAMETERS OF RANDOM PROCESSES

SOLUTION:

If $X(n) = W(n) - W(n - 12)$, then $X(n) = A(n) - A(n - 12)$, which is stationary. Further $W(n)$ may be recovered from $X(n)$ by

$$W(n) = W(n - 12) + X(n) \tag{9.47}$$

Periodic components can best be detected by knowledge of the underlying process (e.g., weather or tides); however, they can also be detected by observing the data, either directly or through the spectral density function estimator discussed in section 9.3.3. Then they can be removed by the proper differencing and the resulting series can be checked for stationarity.

9.4.2 Order Identification

The purpose of "model identification" within the context of ARIMA (p, d, q) models is to obtain a reasonable guess of the specific values of p, the order of the autoregressive part of the model; d, the order of differencing; and q, the order of the moving average part of the model. Note that "reasonable guess" was used, because knowledge of the underlying process should also influence the model and because the total procedure of model estimation is iterative.

Identifying the Order of Differencing. We start by differencing the process $W(n)$ as many times as necessary (however, practically seldom more than two) in order to produce stationarity. We use plots of the process itself, the sample autocorrelation coefficient, and the sample partial autocorrelation coefficient to judge stationarity as will be explained.

Recall that an ARMA (p, q) model has an autocorrelation coefficient $r_{XX}(m)$ that satisfies a pth order linear difference equation for $m > q$. Thus, for distinct roots of the characteristic equation, obtained when solving the pth order difference equation satisfied by the autocorrelation function (see Section 5.2), we have

$$r_{XX}(m) = A_1\lambda_1^m + \cdots + A_p\lambda_p^m, \quad m > q \tag{9.48}$$

and all of the roots λ_i must lie within the unit circle. If $|\lambda_i| \approx 1$ then the autocorrelation coefficient will not decay rapidly with m. Thus, failure of the autocorrelation to decay rapidly suggests that a root with magnitude near 1 exists.

This indicates a nonstationarity of a form that simple differencing might transform into a stationary model because differencing removes a root of mag-

TABLE 9.1 AUTOCORRELATION FUNCTIONS BEHAVIOR AS A FUNCTION OF ARMA ORDER

ARMA (p, q) Order	Behavior of $r_{xx}(k)$	Behavior of $\phi_{k,k}$
(1, 0)	Exponential decay (sign may alternate)	$\phi_{1,1}$ is the only nonzero partial autocorrelation coefficient
(0, 1)	$r_{xx}(1)$ is the only nonzero autocorrelation coefficient	Exponential decay
(1, 1)	Exponential decay for $k \geq 2$	Exponential decay for $k \geq 2$
(2, 0)	Sum of two exponential decays or damped sinewave	Only $\phi_{1,1}$ and $\phi_{2,2}$ nonzero
(0, 2)	Only $r_{xx}(1)$ and $r_{xx}(2)$ are nonzero	Dominated by mixtures of exponentials or damped sinewave

nitude one. That is, if $r_{XX}(m)$ has a root of 1, then it can be shown that $W(n)$, where

$$W(n) = (1 - z^{-1})X(n) \tag{9.49}$$

(z^{-1} is the backshift operator) has a unit root of $r_{XX}(m)$ factored out, and thus may be stationary. Thus, it is usually assumed that the order of differencing necessary to produce stationarity has been reached when the differenced series has a sample autocorrelation coefficient that decays fairly rapidly.

Identifying the Order of Resultant Stationary ARMA Model. Having tentatively identified d, we now observe the estimated autocorrelation coefficients (a.c.c.) and partial autocorrelation coefficients (p.a.c.c.) of the differenced data. Recall from Chapter 5 that the partial autocorrelation coefficient $\phi_{m,m}$ of an autoregressive sequence of order p is zero for $m > p$. Also the autocorrelation coefficient, $r_{XX}(m)$ of a moving average sequence of order q is zero for $m > q$. Table 9.1 describes the behavior of ARMA (p, q) autocorrelation and partial autocorrelation coefficients as a function of p and q. This can be used with the estimated a.c.c. and p.a.c.c. in order to identify models in a trial and error procedure.

However, it should be noted that estimated autocorrelation coefficients and partial autocorrelation coefficients can have rather large variances, and the errors for different lags can also be highly correlated. Thus, exact adherence to the foregoing rules cannot be expected. In particular, estimated a.c.c.'s and p.a.c.c.'s cannot be expected to be exactly zero. More typical plots are those shown in

592 ESTIMATING THE PARAMETERS OF RANDOM PROCESSES

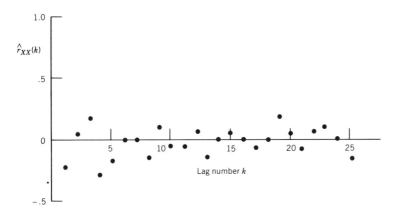

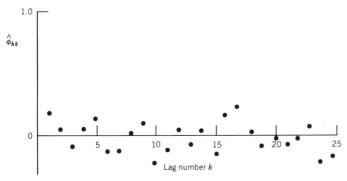

Figure 9.10 Sample autocorrelation and partial autocorrelation coefficients. (a) Sample $r_{xx}(k)$; (b) Sample $\hat{\phi}_{xx}(k)$.

Figure 9.10. Judgement and experience are required in order to decide when a.c.c.'s or p.a.c.c.'s are approximately equal to zero.

EXAMPLE 9.6.

Identify p and q tentatively from the sample autocorrelation coefficients and partial autocorrelation coefficients in Figure 9.11.

SOLUTION:

(a) ARMA (1, 0). The sample autocorrelation coefficient decays exponentially and the sample partial autocorrelation coefficient appears to be nearly zero for $i \geq 2$.
(b) ARMA (0, 1). Reverse of (a).
(c) ARMA (2, 0). $\hat{r}_{xx}$ is sinusoidal while $\hat{\phi}_{i,i} \approx 0$, $i \geq 3$.

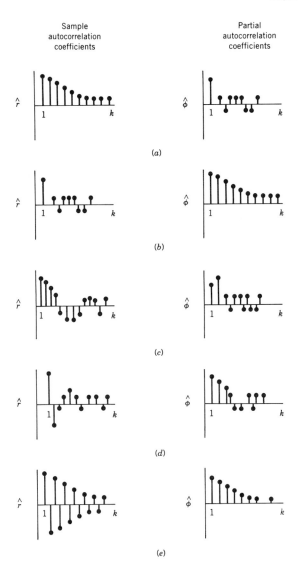

Figure 9.11 Sample a.c.c. and p.a.c.c. for Example 9.6. Cumulative power spectral density functions of residuals.

(d) ARMA (0, 2).
(e) Questionable. It could be ARMA (1, 0) with $\phi_{i,i}$ negative or ARMA (0, 1) or higher order and mixed. A trial-and-error procedure is necessary.

Thus, the order of both the autoregressive and the moving average parts of the model can be guessed as described. In practice, usually both p and q are small, say $p + q < 10$. Commercial computer algorithms that plot p.a.c.c.'s and a.c.c.'s for various trial models are readily available, and some packages automatically "identify" p, d, and q.

9.4.3 Estimating the Parameters of Autoregressive Process

We now assume that the first and second steps of Figure 9-7 have been accomplished, that is, that differencing of order d has been accomplished in order to create a stationary model and that we have identified the order of the resulting ARMA (p, q) model. In this and the following sections we consider the third step, that is, estimating the parameters of the identified model, starting with the AR model.

We will seek the maximum likelihood or approximate maximum likelihood estimators of the parameters of a function of the form

$$X(n) = \sum_{i=1}^{p} \phi_{p,i} X(n - i) + e(n), \qquad n = 1, 2, \ldots \qquad (9.50)$$

where $X(n)$ is the random sequence $\phi_{p,i}$, $i = 1, \ldots, p$ are the autoregressive parameters to be estimated, $e(n)$ is stationary Gaussian white noise with

$$E\{e(n)\} = 0$$

$$E\{e(n)e(j)\} = \begin{cases} \sigma^2, & n = j \\ 0 & n \neq j \end{cases}$$

and

$$f_{e(n)}(\lambda) = \frac{1}{\sqrt{2\pi}\sigma} \exp\left\{\frac{-\lambda^2}{2\sigma^2}\right\}$$

If we use the maximum likelihood approach for estimating $\phi_{p,i}$ and σ^2 using the data $x(1), x(2), \ldots, x(N)$, we first obtain the likelihood function

$$L = f[x(1), x(2), \ldots, x(N); \phi_{p,1}, \phi_{p,2}, \ldots, \phi_{p,p}; \sigma^2]$$

and then find the values of $\phi_{p,i}$ and σ^2 that maximize L. Although this approach is conceptionally simple, the resulting estimators are somewhat difficult to obtain.

Instead of using the likelihood function L, if we use a conditional likelihood function

$$L_C = f_{\mathbf{X}|\mathbf{X}_p}(x(p+1), x(p+2), \ldots, x(N);$$
$$\phi_{p,1}, \ldots, \phi_{p,n}, \sigma^2 | x(1), x(2), \ldots, x(p)) \quad (9.51)$$

then the resulting estimators are easy to compute. Note that

$$L = L_C f_{\mathbf{X}_p}(x(1), x(2), \ldots, x(p)) \quad (9.52)$$

and hence by using L_C to find the estimators, we are essentially ignoring the effects of the first p data points.

If p, the order of the autoregressive model, is small compared with N, the number of observations, then this "starting transient" can be neglected and Equation 9.51 can be used to approximate the unconditional density and thus the likelihood function.

From Equation 9.50 with $X(i) = x(i)$, $i = 1, 2, \ldots, p$, we obtain

$$e(p+1) = x(p+1) - \sum_{i=1}^{p} \phi_{p,i} x(p+1-i)$$

$$e(p+2) = x(p+2) - \sum_{i=1}^{p} \phi_{p,i} x(p+2-i)$$

$$\vdots$$

$$e(N) = x(N) - \sum_{i=1}^{p} \phi_{p,i} x(N-i), \quad N \gg p$$

The conditional likelihood function L_C is obtained from the linear transformation defined as

$$L_C \stackrel{\Delta}{=} \prod_{j=p+1}^{N} f_{e(j)}\left(x(j) - \sum_{i=1}^{p} \phi_{p,i} x(j-i)\right) |J| \quad (9.53)$$

where $|J|$ is the Jacobian of the transformation.

The reader can show that the Jacobian of the transformation is 1, and hence

$$L_C = \frac{1}{(\sqrt{2\pi}\sigma)^{N-p}} \exp\left\{-\frac{1}{2\sigma^2} \sum_{j=p+1}^{N} \left[x(j) - \sum_{i=1}^{p} \phi_{p,i} x(j-i)\right]^2\right\} \quad (9.54)$$

596 ESTIMATING THE PARAMETERS OF RANDOM PROCESSES

Equation 9.54 can be written as the logarithm l_c, of the conditional likelihood function

$$l_c(\phi_{p,1}, \phi_{p,2}, \ldots, \phi_{p,p}, \sigma^2) = \frac{-(N-p)}{2} \ln(2\pi) - \frac{(N-p)}{2} \ln \sigma^2$$
$$- \frac{1}{2\sigma^2} \sum_{j=p+1}^{N} \left[x(j) - \sum_{i=1}^{p} \phi_{p,i} x(j-i) \right]^2 \qquad (9.55)$$

In order to estimate $\phi_{p,1}, \ldots, \phi_{p,p}$, the important quantity in Equation 9.55 is the sum of squares function (it is the only part of l_c affected by the choice of $\phi_{p,i}$)

$$S(\phi_{p,1}, \ldots, \phi_{p,p}) = \sum_{j=p+1}^{N} \left[x(j) - \sum_{i=1}^{p} \phi_{p,i} x(j-i) \right]^2 \qquad (9.56)$$

This can be minimized (it is a negative term in Equation 9.54) by partially differentiating with respect to the $\phi_{p,i}$ just as was done in the earlier section on regression. A conditional maximum likelihood estimator of σ^2 is obtained by differentiating Equation 9.55 partially with respect to σ^2. We illustrate with some special cases.

First-order Autoregressive Model. We consider the model given in Equation 5.4 with power spectral density given in Equation 5.11, and autocorrelation function given in Equation 5.9. We seek maximum likelihood estimates of $\phi_{1,1}$ and σ^2, which is the variance of the noise, that is, σ_N^2. We note that these parameters determine the autocorrelation function, autocorrelation coefficient, and power spectral density.

As stated earlier, we will often use just the conditional likelihood function because of the difficulty in starting the difference equation. The logarithm l_c of this function is

$$l_c(\phi_{1,1}, \sigma^2) = -\frac{N-1}{2} \ln 2\pi - \frac{N-1}{2} \ln \sigma^2$$
$$- \frac{1}{2\sigma^2} \sum_{j=2}^{N} [x(j) - \phi_{1,1} x(j-1)]^2 \qquad (9.57)$$

We will maximize l_c by differentiating Equation 9.57 partially with respect to σ^2 and $\phi_{1,1}$, and setting the derivatives equal to zero, trusting that this procedure will result in estimators that fall within the allowable range [($-1 < \phi_{1,1} < +1$); $\sigma^2 > 0$].

Because $\phi_{1,1}$ appears in only the last or the sum of squares terms, the maximum likelihood estimator of $\phi_{1,1}$ is the one that minimizes

$$S(\phi_{1,1}) = \sum_{j=2}^{N} [x(j) - \phi_{1,1}x(j-1)]^2 \qquad (9.58)$$

This results in the estimator (the observations are now considered to be random variables)

$$\hat{\phi}_{1,1} = \frac{\sum_{j=2}^{N} X(j-1)X(j)}{\sum_{j=2}^{N} X^2(j-1)} = \hat{r}_{xx}(1) \qquad (9.59)$$

Note that this is simply the sample correlation coefficient at lag 1. In order to find an estimator for σ^2, we differentiate Equation 9.57 with respect to σ^2 and set the result equal to zero. This produces an estimator for σ^2 in which the residual sum of squares

$$\epsilon^2 \stackrel{\Delta}{=} \sum_{j=2}^{N} [X(j) - \hat{\phi}_{1,1}X(j-1)]^2 \qquad (9.60)$$

is divided by $N - 1$. However ϵ^2 has $(N - 2)$ degrees of freedom since there are $(N - 1)$ observations and one parameter has been estimated. Therefore an unbiased estimator of σ^2 in the first-order autoregressive model is

$$\widehat{\sigma^2} = \frac{1}{N-2} \sum_{n=2}^{N} \epsilon^2(n) = \frac{1}{N-2} \sum_{n=2}^{N} [X(n) - \hat{\phi}_{1,1}X(n-1)]^2 \qquad (9.61)$$

That is, $\epsilon(n)$ is the observed value of $e(n)$ with the estimated value $\hat{\phi}_{1,1}$ used for $\phi_{1,1}$.

Thus, Equations 9.59 and 9.61 can be used to estimate the parameters in a first-order autoregressive model. The estimated values of σ^2 and $\phi_{1,1}$ can be used to obtain an estimate of the psd of the process (according to Equation 5.11) as

$$\hat{S}_{xx}(f) = \frac{\widehat{\sigma^2}}{1 - 2\hat{\phi}_{1,1}\cos(2\pi f) + \hat{\phi}_{1,1}^2}, \quad |f| < \frac{1}{2} \qquad (9.62)$$

An example is given in Problem 9.28. Note that $\hat{S}_{XX}(f)$ will be biased even if $\hat{\phi}_{1,1}$ and $\widehat{\sigma^2}$ are unbiased. This will be a good estimator only if the variances of $\hat{\phi}_{1,1}$ and $\widehat{\sigma^2}$ are small. Similarly the autocorrelation function can be estimated by replacing the parameter in Equation 5.10 by its estimate.

Second-order Autoregressive Model. We now differentiate the sum of the squares in order to find estimators of $\phi_{2,1}$ and $\phi_{2,2}$ in the model

$$X(n) = \phi_{2,1}X(n-1) + \phi_{2,2}X(n-2) + e(n) \tag{9.63}$$

Differentiating and setting the resulting derivatives of the sum of squares function equal to zero results in

$$\sum X(j)X(j-1) = \hat{\phi}_{2,1} \sum X^2(j-1) + \hat{\phi}_{2,2} \sum X(j-1)X(j-2)$$
$$\sum X(j)X(j-2) = \hat{\phi}_{2,1} \sum X(j-1)X(j-2) + \hat{\phi}_{2,2} \sum X^2(j-2) \tag{9.64}$$

where all sums in Equation 9.64 are from $j = 3$ to $j = N$.
If N is large, then dividing both equations by $N - 2$ results in

$$\hat{R}_{XX}(1) \approx \hat{\phi}_{2,1}\hat{R}_{XX}(0) + \hat{\phi}_{2,2}\hat{R}_{XX}(1) \tag{9.65}$$
$$\hat{R}_{XX}(2) \approx \hat{\phi}_{2,1}\hat{R}_{XX}(1) + \hat{\phi}_{2,2}\hat{R}_{XX}(0) \tag{9.66}$$

The sample autocorrelation functions can be estimated from data by any of the standard methods; here we use

$$\hat{R}_{XX}(n) = \frac{1}{N-n} \sum_{i=1}^{N-n} X(i)X(i+n)$$

Equations 9.65 and 9.66 can be solved for $\hat{\phi}_{2,1}$ and $\hat{\phi}_{2,2}$ as follows:

$$\hat{\phi}_{2,1} = \frac{\begin{vmatrix} \hat{R}_{XX}(1) & \hat{R}_{XX}(1) \\ \hat{R}_{XX}(2) & \hat{R}_{XX}(0) \end{vmatrix}}{\begin{vmatrix} \hat{R}_{XX}(0) & \hat{R}_{XX}(1) \\ \hat{R}_{XX}(1) & \hat{R}_{XX}(0) \end{vmatrix}} = \frac{\hat{R}_{XX}(1)[\hat{R}_{XX}(0) - \hat{R}_{XX}(2)]}{[\hat{R}_{XX}(0)]^2 - [\hat{R}_{XX}(1)]^2} \tag{9.67}$$

Note that if Equation 9.67 is divided by $R_{XX}(0)$ in both numerator and denominator, and if the estimators are assumed to be true values, then this equation becomes a special case of the inverted Yule–Walker equation, (Equation 5.32). Also

$$\hat{\phi}_{2,2} = \frac{\hat{R}_{XX}(0)\hat{R}_{XX}(2) - [\hat{R}_{XX}(1)]^2}{[\hat{R}_{XX}(0)]^2 - [\hat{R}_{XX}(1)]^2} \quad (9.68)$$

The residual sum of squared error can be used to estimate σ^2, that is

$$\widehat{\sigma^2} = \frac{1}{N-4}\left\{\sum_{j=3}^{N}[X(j) - \hat{\phi}_{2,1}X(j-1) - \hat{\phi}_{2,2}X(j-2)]^2\right\} \quad (9.69)$$

where $N - 4$ is used because there are $N - 2$ observations and 2 degrees of freedom were used in estimating $\hat{\phi}_{2,1}$ and $\hat{\phi}_{2,2}$. Estimates of σ^2, $\phi_{2,1}$, and $\phi_{2,2}$ can be used to obtain the estimate of $S_{XX}(f)$. Note that the variance estimated is the variance of $e(n)$, that is, σ_N^2.

Procedure for General Autoregressive Models. For the autoregressive model

$$X(n) = \sum_{i=1}^{p} \phi_{p,i}X(n-i) + e(n)$$

the usual procedure is to approximate the maximum likelihood estimator by the conditional maximum likelihood estimator, which reduces to the usual regression or minimum MSE estimator (see Problem 9.32).

In general, minimizing the conditional sum of squares will result in a set of equations of the form

$$\begin{bmatrix} \hat{R}_{XX}(0) & \cdots & \hat{R}_{XX}(p-1) \\ \vdots & & \vdots \\ \hat{R}_{XX}(p-1) & \cdots & \hat{R}_{XX}(0) \end{bmatrix}\begin{bmatrix} \hat{\phi}_{p,1} \\ \vdots \\ \hat{\phi}_{p,p} \end{bmatrix} = \begin{bmatrix} \hat{R}_{XX}(1) \\ \vdots \\ \hat{R}_{XX}(p) \end{bmatrix} \quad (9.70)$$

where

$$\hat{R}_{XX}(n) = \frac{1}{N-p}\sum_{j=p+1}^{N} X(j)X(j-n), \quad n \le p$$

The set of equations (9.70) can be solved recursively for the $\hat{\phi}_{p,i}$ using the *Levinson recursive algorithm*. Levinson recursion starts with a first-order AR process and uses Gram–Schmidt orthogonalization to find the solutions through a pth order AR process. (See Reference [8].) One valuable feature of the Levinson recursive solution algorithm is that the residual sum of squared errors at each iteration can be monitored and used to decide when the order of the model is sufficiently large.

9.4.4 Estimating the Parameters of Moving Average Processes

Estimating the parameters of a moving average process is more complicated than estimating the parameters of an autoregressive process because the likelihood function is more complicated. As in the case of an autoregressive process, both the conditional and the unconditional likelihood functions can be considered. Note that in a first-order moving average process

$$e(n) = -\theta e(n-1) + X(n) \qquad (9.71.a)$$

or using the z^{-1} operator and assuming that this model is invertible (see Equation 5.38)

$$e(n) = \sum_{j=0}^{n} (-\theta)^j X(n-j) \qquad (9.71.b)$$

For example, using Equation 9.71.a

$$e(1) = -\theta e(0) + X(1)$$
$$e(2) = X(2) - \theta e(1) = X(2) - \theta X(1) + \theta^2 e(0)$$
$$e(3) = X(3) - \theta e(2) = X(3) - \theta X(2) + \theta^2 X(1) - \theta^3 e(0) \qquad (9.71.c)$$

Equation 9.71 shows that not only is the error a sum of the observations, but is a nonlinear function of θ. Thus, the sum of squares of the $e(i)$ is nonlinear in θ. Differentiating this function and setting it equal to zero will also lead to a set of nonlinear equations that will be difficult to solve. Higher-order moving average processes result in the same kind of difficulties. Thus, the usual regression techniques are not applicable for minimizing the sum of the squared errors. We will resort to an empirical technique for finding the best estimators for moving average processes.

First-order Moving Average Models. In this case of a moving average model, as in the autoregressive case, we maximize the likelihood function by minimizing the sum of the squared error terms, but the error terms are complicated as illustrated in Equation 9.71.

To minimize this sum of squared errors, we use a "trial and error" process. That is, we will assume a value, say θ_1, and a value of $e(0)$, then calculate the

TABLE 9.2 $X(n)$ FOR EXAMPLE 9.7

n	$X(n)$
1	.50
2	.99
3	−.48
4	−.20
5	−1.31
6	.81
7	1.82
8	2.46
9	1.07
10	−1.29

sum of the squared errors. Then, with the same starting value $e(0)$, we assume another value, say θ_2, and calculate the sum of the squared errors. This is done for a number of values of θ and the θ_i that minimizes the sum of the squared errors is the estimate $\hat{\theta}$. The observed value of error $e(n)$ will be denoted by $\epsilon(n)$ where $\epsilon(n) \stackrel{\Delta}{=} X(n) - \hat{\theta}\epsilon(n-1)$.

EXAMPLE 9.7.

Assuming that the correct model is

$$X(n) = \theta e(n-1) + e(n) \qquad (9.72)$$

Estimate θ from the data in Table 9.2.

TABLE 9.3 $\epsilon(n)$ IF $\theta = .5$ FOR EXAMPLE 9.7

n	$X(n)$	$\epsilon(n) = X(n) - .5\,\epsilon(n-1)$
0		0
1	.50	.50
2	.99	.74
3	−.48	−.85
4	−.20	+.22
5	−1.31	−1.42
6	.81	+1.52
7	1.82	1.06
8	2.46	1.93
9	1.07	.10
10	−1.29	−1.34

602 ESTIMATING THE PARAMETERS OF RANDOM PROCESSES

TABLE 9.4 SUM OF SQUARED ERRORS AND θ FOR EXAMPLE 9.7

θ	−.5	−.4	−.3	−.2	−.1	0	.1
$\sum_{i=1}^{10} \epsilon^2(n)\|\theta$	26.5	23.5	21.1	19.1	17.5	16.0	14.9

θ	.2	.3	.4	.5	.6	.7	.8	.9
$\sum_{i=1}^{10} \epsilon^2(n)\|\theta$	13.9	13.1	12.7	12.6	12.9	14.1	16.8	22.7

SOLUTION:

For illustration we start with an initial guess of θ = .5 and calculate the sum of the squared errors as shown in Table 9.3. The value $\epsilon(0)$, of $e(0)$ is assumed to be zero. This reduces the likelihood function to be the conditional [on $e(0) = 0$] likelihood function.

Table 9.4 gives the sum of squared errors for various values of θ. From this table it is clear that the sum of squared errors is minimum when $\theta \approx .5$. Thus, the best or (conditional) maximum likelihood estimate is $\hat{\theta} \approx .5$, if only one significant digit is used.

Note that the estimate in Example 9.7 is based on, or conditional on, $\epsilon(0) = 0$. A different assumption for $\epsilon(0)$ might result in a slightly different estimate. However, when $N \gg 1$, then the assumed value of $e(0)$ does not significantly affect the estimate of θ.

Problem 9.33 asks for an estimate based on the procedure described, and Problem 9.34 asks for an estimate when the initial value $\epsilon(0)$ is estimated by a backcasting procedure.

Second-order Moving Average Model. Estimation of the parameters of the second-order moving average model

$$X(n) = \theta_{2,1} e(n-1) + \theta_{2,2} e(n-2) + e(n)$$

is carried out in basically the same manner as for the first-order moving average process. Once again, the likelihood function is a nonlinear function of the parameters. Thus, the same trial calculation of the sums of squares is used. There are two modifications required.

First, two values of e, $\epsilon(0)$ and $\epsilon(-1)$, must be assumed, and they are usually assumed to be zero. Second, the plot of the sum of squares versus values of $\theta_{2,1}$ and $\theta_{2,2}$ is now two-dimensional.

EXAMPLE 9.8.

Using the data from Table 9.5, estimate $\theta_{2,1}$, $\theta_{2,2}$, σ_N^2, and σ_X^2 in a second-order moving average model.

SOLUTION:

The conditional sum of squared errors is shown in Table 9.6 for selected values of $\theta_{2,1}$ and $\theta_{2,2}$. From this table, the best estimators of $\theta_{2,1}$ and $\theta_{2,2}$ are

$$\hat{\theta}_{2,1} = .50 \quad \text{and} \quad \hat{\theta}_{2,2} = -.10$$

The variance of e, σ_N^2 can be estimated from

$$\widehat{\sigma^2}_N = \sum_{n=1}^{N} \frac{\epsilon^2(n)}{N-2} = \frac{97.481}{98} \approx .995$$

where the summation is the minimum sum of squares (97.481 in the example) and $N - 2$ is the number of observations minus two, the degrees of freedom used in estimating $\theta_{2,1}$ and $\theta_{2,2}$.

TABLE 9.5 $X(n)$, $n = 1, \ldots, 100$ (READ ACROSS) FOR EXAMPLE 9.8

1.102	0.699	2.336	1.026	−0.206
1.486	0.768	0.690	−0.001	−1.150
−0.273	−0.941	1.746	0.300	−1.494
0.192	0.023	1.633	1.188	−2.243
−0.961	−0.770	0.535	1.263	0.377
−0.480	0.091	1.007	1.169	0.346
−0.052	−0.186	0.033	0.106	1.402
0.879	−1.490	1.031	1.127	0.824
−0.478	−1.707	0.877	1.257	0.830
−1.534	−1.411	0.468	−0.066	−1.560
−1.803	−1.100	0.274	2.362	1.834
0.778	1.172	−0.503	−0.100	−1.527
−2.019	−1.498	−2.259	−1.272	−0.033
−0.096	2.723	−0.243	−2.256	−0.462
0.509	−1.240	−0.194	0.217	0.899
−0.683	−0.621	0.001	0.978	0.808
0.419	0.667	−0.694	0.355	1.121
−0.130	−0.895	−1.366	−0.247	0.741
0.047	−0.662	−1.225	−0.988	−0.313
0.312	−0.747	−1.816	−1.898	−1.592

TABLE 9.6 SUM OF SQUARED ERRORS FOR VARIOUS VALUES OF $\theta_{2,1}$ AND $\theta_{2,2}$

$\theta_{2,1}$ \ $\theta_{2,2}$	−0.14	−0.13	−0.12	−0.11	−0.10	−0.09	−0.08
0.45	97.749	97.730	97.732	97.756	97.801	97.863	97.944
0.46	97.681	97.648	97.640	97.654	97.690	97.746	97.820
0.47	97.640	97.593	97.572	97.576	97.602	97.649	97.716
0.48	97.629	97.566	97.531	97.522	97.537	97.575	97.633
0.49	97.650	97.568	97.517	97.493	97.496	97.523	97.572
0.50	97.706	97.602	97.532	97.492	97.481	97.495	97.534
0.51	97.798	97.671	97.579	97.521	97.493	97.493	97.519
0.52	97.932	97.777	97.661	97.581	97.535	97.518	97.530
0.53	98.109	97.923	97.779	97.675	97.607	97.572	97.568
0.54	98.336	98.113	97.938	97.806	97.714	97.658	97.635
0.55	98.615	98.351	98.140	97.977	97.857	97.777	97.732
0.56	98.954	98.643	98.391	98.192	98.040	97.932	97.864

Then using Equation 5.45.a

$$\widehat{\sigma_X^2} = (\hat{\theta}_{2,1}^2 + \hat{\theta}_{2,2}^2 + 1)\sigma_N^2 \approx (.5^2 + .1^2 + 1)(.995) \approx 1.25$$

An estimate of σ_X^2 can also be obtained from $(1/99) \sum_{i=1}^{100} [X(i) - \overline{X}]^2$ and it is approximately 1.25.

Procedure for General Moving Average Models. The procedure for estimating the parameters of the general moving average model

$$X(n) = \sum_{i=1}^{q} \theta_{q,i} e(n-i) + e(n)$$
$$= \sum_{i=0}^{q} \theta_{q,i} e(n-i), \qquad \theta_{q,0} = 1$$

is the same as that given previously. However, when q is 3 or larger, the optimization procedure is much more difficult and time consuming. We do not consider this case in this introductory text. We should note that the optimization is aided if reasonable first guesses of the parameters are available (see Section 9.4.6).

One can obtain a slightly better estimate of the sum of the squared errors by a reverse estimation of $\epsilon(0), \epsilon(-1), \ldots, \epsilon(-q + 1)$ (see Problem 9.34), but for reasonably long data sequences, this extra effort is usually unnecessary because the initial values have little significant effect on the final estimates.

Note that again in this case, the maximum likelihood estimator is closely approximated by the minimum MSE estimator based on the conditional sum of squares and the deviation is significant only for small N. However, one cannot minimize the sum of the squares by differentiating and setting the derivatives equal to zero as was done in the autoregressive case.

9.4.5 Estimating the Parameters of ARMA (p, q) Processes

If the order p of the numerator and the order q of the denominator of an ARMA (p, q) model are identified, then the conditional likelihood function is maximized by minimizing the sums of the squared errors of a sequence of observations, starting with, $\epsilon(1 - q), \epsilon(2 - q), \ldots, \epsilon(0)$ all assumed to be zero. The procedure including the trial-and-error minimization of $\Sigma_{n=p+1}^{N} \epsilon^2(n)$ is the same as with a moving average process. An example of an ARMA (1, 1) process illustrates the procedure.

TABLE 9.7 $X(n), n = 1, \ldots, 100$ (READ BY ROWS)

0.203	0.225	−1.018	−0.249	−0.545
1.015	1.064	0.323	−1.740	−0.600
−1.874	−1.255	0.211	−1.097	−1.936
1.284	−1.261	−0.993	1.393	0.127
−0.494	−2.913	0.278	−0.069	0.569
0.810	0.900	0.956	1.119	−0.231
−0.394	0.227	−0.765	0.884	0.461
1.084	0.998	1.370	−0.248	−1.234
−1.467	−1.628	−1.135	−0.922	0.450
−1.065	0.791	1.659	−0.055	−0.273
−2.336	−0.370	0.289	−0.504	0.239
−0.516	−1.346	−1.321	−0.467	0.738
−0.671	0.961	−0.771	0.561	0.163
−0.136	0.214	0.150	1.734	1.366
0.100	−0.250	0.945	−0.053	1.297
1.041	0.647	−1.531	1.348	−0.063
0.300	0.347	1.352	−0.913	0.477
−1.384	−0.471	1.464	−0.118	1.156
0.156	−0.016	0.696	−0.338	0.909
0.150	0.242	2.595	−0.477	−0.483

TABLE 9.8 SUM OF SQUARED ERRORS FOR VARIOUS VALUES OF $\theta_{1,1}$ AND $\phi_{1,1}$

$\theta_{1,1}$ \ $\phi_{1,1}$	−0.40	−0.36	−0.32	−0.31	−0.30	−0.29	−0.25
0.40	98.998	97.647	96.745	96.584	96.448	96.336	96.122
0.43	97.936	96.916	96.317	96.228	96.163	96.121	96.176
0.44	97.636	96.724	96.223	96.158	96.116	96.097	96.240
0.45	97.364	96.558	96.155	96.113	96.094	96.097	96.327
0.46	97.120	96.418	96.111	96.092	96.095	96.121	96.437
0.47	96.902	96.304	96.092	96.096	96.121	96.169	96.570
0.48	96.712	96.215	96.097	96.124	96.171	96.240	96.725
0.49	96.549	96.153	96.128	96.176	96.246	96.336	96.904
0.55	96.147	96.323	96.829	97.004	97.199	97.413	98.461
0.60	96.563	97.178	98.093	98.367	98.659	98.969	100.391

EXAMPLE 9.9.

Consider the data in Table 9.7. Find the estimates of $\phi_{1,1}$, $\theta_{1,1}$, and σ_N^2.

SOLUTION:

The sum of squares for various values of $\phi_{1,1}$ and $\theta_{1,1}$ are shown in Table 9.8. The resulting estimates are $\hat{\theta}_{1,1} = .46$ and $\hat{\phi}_{1,1} = -.31$. (Note $\hat{\theta}_{1,1} = .47$ and $\hat{\phi}_{1,1} = -.32$ are also possible choices.) The variance σ_N^2 of e is estimated from

$$\widehat{\sigma_N^2} = \frac{\sum_{i=2}^{N} \epsilon^2(n)}{N-3} = \frac{96.092}{97} \approx .99 \qquad (9.73)$$

9.4.6 ARIMA Preliminary Parameter Estimation

In sections 9.4.3, 9.4.4, and 9.4.5 we presented a general method of estimating ARMA parameters after differencing in order to produce stationarity and after deciding the order p of the autoregressive part of the model, and the order q of the moving average part. However, the trial and error procedure involved when $q \neq 0$ is easier when preliminary estimates of the parameters are available. Here we discuss how to obtain such preliminary estimates from the sample a.c.c. and the sample p.a.c.c.

Briefly consider the first-order moving average model. For zero-mean random variables the sample autocorrelation coefficient at lag 1 can be estimated by

$$\hat{r}_{XX}(1) = \frac{\sum_{n=2}^{N} X(n)X(n-1)}{\sum_{n=2}^{N} X^2(n)}$$

and from Equation 5.41.a it can be seen that

$$r_{XX}(1) = \frac{\theta_{1,1}}{1 + \theta_{1,1}^2}$$

Then a preliminary estimator of $\theta_{1,1}$ is

$$\hat{\theta}_{1,1} = \frac{1}{2\hat{r}_{XX}(1)} [1 \pm \sqrt{1 - 4\hat{r}_{XX}^2(1)}] \qquad (9.74)$$

It can be shown that only one of these values will fall within the invertibility limits, $-1 < \theta_{1,1} < +1$, if the correlation coefficients are true values. If the correlation coefficients are estimated, then usually Equation 9.74 will produce only one solution within the allowable constraints.

In a second-order moving average model (see Equation 5.45)

$$r_{XX}(1) = \frac{\theta_{2,1} + \theta_{2,1}\theta_{2,2}}{1 + \theta_{2,1}^2 + \theta_{2,2}^2} \qquad (9.75)$$

and

$$r_{XX}(2) = \frac{\theta_{2,2}}{1 + \theta_{2,1}^2 + \theta_{2,2}^2} \qquad (9.76)$$

These two equations can be solved for the initial estimates, $\hat{\theta}_{2,1}$ and $\hat{\theta}_{2,2}$ by using the estimated values for $r_{XX}(1)$ and $r_{XX}(2)$.

Initial estimates for a mixed autoregressive moving average process are found using the same general techniques. The autocorrelation coefficients for the first $p + q$ values of lag are estimated from the data, and equations that relate these values to the unknown parameters are used to arrive at preliminary estimates for the unknown parameters.

For instance for the ARMA (1, 1) process (see Equations 5.56 and 5.54.b)

$$r_{xx}(1) = \frac{(1 + \phi_{1,1}\theta_{1,1})(\phi_{1,1} + \theta_{1,1})}{(1 + \theta_{1,1}^2 + 2\phi_{1,1}\theta_{1,1})} \tag{9.77}$$

and

$$r_{xx}(2) = \phi_{1,1} r_{xx}(1) \tag{9.78}$$

These two equations can be solved in order to find initial estimates of $\phi_{1,1}$ and $\theta_{1,1}$ in terms of the estimates, $\hat{r}_{xx}(1)$ and $\hat{r}_{xx}(2)$.

In all cases the allowable regions for the unknown parameters must be observed. Then these initial estimates should be used in the more efficient (i.e., smaller variance) estimating procedures described in previous sections. (If N is very large, then the initial estimates may be sufficiently accurate.)

9.4.7 Diagnostic Checking

We have discussed identifying ARIMA models and estimating the parameters of the identified models. There is always the possibility that the identified model with the estimated parameters is not an acceptable model. Thus, the third and indispensable step in this iterative Box–Jenkins procedure is diagnostic checking of the model (see Figure 9.7).

One cannot decide via statistical tests that a model is "correct." The model may fail a statistical test because it is actually not a representative model (wrong p or wrong d or wrong q, or poorly estimated parameters) or because the process has changed between that realization that was tested and the realization that exists at the time of use. Similarly, serendipity could result in a model that is not representative of the realization at the time of test, but is sufficiently representative of the process being modeled to produce useful analysis and successful designs. However, the model must be checked. Furthermore, if the model "fails" the tests then a new and improved model should be proposed and the stages of identification, parameter estimation, and diagnostic testing should be repeated.

In this section, we introduce several diagnostic tests of a proposed ARIMA (p, d, q) model. All tests are based on the residual errors. First, if the residual errors are white, then they are random, and the run test described in Section 9.2 can be used directly on the residual errors without any additional processing. Tests based on the autocorrelation function of the residual errors are introduced in the next section. We also propose another test based on the power spectral distribution function of the errors. In all cases, the first essential check of a model is to observe errors. If one identifies, for example, an ARIMA (1, 0, 1) model and has estimated the parameters $\phi_{1,1}$ and $\theta_{1,1}$, then using these estimates and the same data, the observed errors $\epsilon(n)$, where

$$\epsilon(n) = X(n) - \hat{\phi}_{1,1}X(n-1) - \hat{\theta}_{1,1}\epsilon(n-1) \tag{9.79}$$

should be plotted and inspected. If possible and economical, the errors of a different realization or different time sample of the same process using the identified model and estimated parameters from the previous realization should be examined.

In addition, the sample autocorrelation coefficients and the sample partial autocorrelation coefficients of the errors should be plotted and inspected. They should look like sample functions from white noise.

A Test Using the Sample Autocorrelation Function of Residual Errors. In autoregressive moving-average models, $e(n)$ is white noise. Thus, if the model is correct and the parameters are well established, then the residual errors $\epsilon(n)$ should be uncorrelated except at lag 0. However, the question is, how much deviation from zero is too much to discredit the model?

There are a number of tests of the errors based on the sample autocorrelation coefficients $\hat{r}_{\epsilon\epsilon}$ of the errors. One is based on the sum of squares of the observed correlation coefficient of the errors.

The rather complicated and significant correlation between the sample correlation coefficients $\hat{r}_{\epsilon\epsilon}(m)$ of residual errors has led to tests based on the sum of the squares of the first k (say 20) values of $r_{\epsilon\epsilon}(m)$.

It is possible to show* that if the fitted model is appropriate, then

$$C(k) = (N - d) \sum_{m=1}^{k} \hat{r}_{\epsilon\epsilon}^2(m) \tag{9.80}$$

is approximately distributed as

$$\chi^2_{k-p-q}$$

where $\hat{r}_{\epsilon\epsilon}$ is the sample correlation coefficient of the observed errors $\epsilon(m)$, for example, $\epsilon(m) = X(m) - \hat{X}(m)$, $N - d$ is the number of samples (after differencing) used to fit the model, and p and q are the number of autoregressive and moving average terms respectively in the model.

Thus, a table of chi-square with $(k - p - q)$ degrees of freedom can be used to compare the observed value of C with the value $\chi^2_{k-p-q;\alpha}$ found in a table

---

*See Section 8.2.2 of Reference [5].

ESTIMATING THE PARAMETERS OF RANDOM PROCESSES

of a chi-square distribution. An observed C larger than that from the table indicates that the model is inadequate at the α significance level.

EXAMPLE 9.10.

Using the data of Table 9.9, which are the observed errors $\epsilon(n)$, after an assumed ARIMA (1, 0, 2) model has been fit, test for randomness using a chi-square test with the first 20 sample autocorrelation coefficients, at the 5% significance level.

SOLUTION:

Using Equation 9.80

$$C(20) = 100 \sum_{m=1}^{20} \hat{r}_{\epsilon\epsilon}^2(m) \approx 341$$

From Appendix E, $\chi_{17,.05}^2 \approx 27.6$. Thus, the hypothesis of random errors is rejected.

TABLE 9.9 $\epsilon(n)$ (READ ACROSS)

0.092	0.390	-0.728	-0.886	-1.097
0.128	1.244	1.142	-0.979	-1.449
-2.751	-3.454	-2.363	-2.936	-4.526
-2.396	-3.063	-3.932	-1.782	-1.353
-2.021	-4.727	-3.674	-2.730	-1.923
-0.920	-0.119	0.598	1.373	0.697
-0.019	0.276	-0.450	0.487	1.014
1.765	2.408	3.209	2.306	0.532
-0.862	-2.029	-2.501	-2.702	-1.636
-2.339	-1.328	0.655	0.240	-0.428
-2.708	-2.595	-1.486	-1.760	-1.352
-1.634	-2.835	-3.658	-3.360	-1.939
-2.342	-1.254	-1.816	-1.241	-0.819
-1.016	-0.725	-0.493	1.232	2.320
1.695	0.945	1.766	1.516	2.452
3.166	3.121	0.988	2.178	2.140
1.928	2.082	3.122	1.723	1.780
0.380	-0.146	1.687	1.409	2.115
1.999	1.497	2.003	1.408	2.041
1.996	1.817	4.177	3.036	1.690

MODEL-BASED ESTIMATION

Durbin–Watson Test. If we wish to examine the residual errors $\epsilon(n)$ after fitting an ARIMA model, the Durbin–Watson test is often used. The observed errors $\epsilon(n)$ are assumed to follow a first-order autoregressive model, that is,

$$\epsilon(n) = \rho\epsilon(n-1) + a(n) \tag{9.81}$$

where $a(n)$ is zero-mean Gaussian white noise. Under this assumption if $\rho = 0$, then the $\epsilon(n)$ would also be white noise. If $\rho = 0$, then

$$D = \frac{\sum_{n=2}^{N} [\epsilon(n) - \epsilon(n-1)]^2}{\sum_{j=1}^{N} \epsilon^2(j)} \tag{9.82}$$

would have an expected value of approximately 2 (for large N). (See Problem 9.41.)

However, if $\rho > 0$ then D would tend to be smaller than 2. On the other hand, if $\rho < 0$, then D would tend to be larger than 2.

In order to test the null hypothesis that $\rho = 0$ versus the one-sided alternative $\rho > 0$, the procedure is as follows and uses a table of the Durbin–Watson statistic (see Appendix I).

1. If $D < d_L$ (where D is from Equation 9.82, and d_L is found in Appendix I corresponding with the significance level of the test, the sample size N and the number of coefficients that have been estimated, $k = p + q$), reject H_0, and conclude that errors are positively correlated.
2. If $D > d_U$ (from Appendix I), accept H_0.
3. If $d_L \leq D \leq d_U$, the test is inconclusive.

To test the null hypothesis $\rho = 0$ versus the alternative $\rho < 0$ the procedure is

1. Use $4 - D$ rather than D.
2. Then follow the previous procedure.

EXAMPLE 9.11.

Use the data from Example 9.10 and test the hypothesis $\rho = 0$ at the 5% significance level versus the alternative, $\rho > 0$.

612 ESTIMATING THE PARAMETERS OF RANDOM PROCESSES

SOLUTION:

From Equation 9.82 we obtain

$$D = \frac{\sum_{n=2}^{100}[\epsilon(n) - \epsilon(n-1)]^2}{\sum_{j=1}^{100}\epsilon^2(j)} \approx .246$$

This is considerably below 1.61 which is d_L in Appendix I for 100 degrees of freedom with 3 coefficients (called regressors in Appendix I) having been estimated. Therefore we can conclude that $\rho > 0$, and the errors are correlated. Thus, a different model is needed.

A Test Using the Cumulative Estimated Power Spectral Density Function of the Residual Errors. When the aim of the model is to estimate a power spectral density function, we must check whether or not the model has adequately captured the frequency domain characteristics. If the residual errors are white, as they will be with a "good" model, then

$$S_{\epsilon\epsilon}(f) = \sigma_N^2, \quad |f| < \frac{1}{2} \qquad (9.83)$$

It has been shown by Bartlett* that the cumulative spectral density function $C(f)$ of the residuals provides an efficient test for whiteness, where

$$C(f) = \int_{-f}^{f} S_{\epsilon\epsilon}(\lambda)\, d\lambda, \quad 0 < f < \frac{1}{2} \qquad (9.84)$$

Note that if $S_{\epsilon\epsilon}(f)$ is as given in Equation 9.83 then $C(f)$ would be a straight line between $(0, 0)$ and $(\frac{1}{2}, \sigma_N^2)$. Because σ_N^2 is not known, then its estimator

$$\widehat{\sigma^2} = \frac{1}{N - d - p - q} \sum_{n=1}^{N} \epsilon^2(n)$$

---

*See Section 8.2.4 of Reference [3].

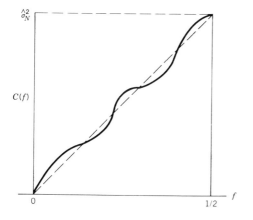

(a) Approximately white residuals

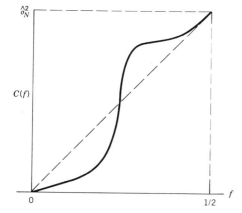

(b) Nonwhite residuals.

Figure 9.12 Cumulative spectral density functions of residuals.

is used where N, d, p, q, are as defined in the foregoing chi-square test. The plot shown in Figure 9.12a represents a reasonable fit. On the other hand, lack of whiteness can be seen in Figure 9.12b. This figure indicates that there is significant deviation from white noise with an excess of power at about $f = \frac{1}{4}$.

9.5 SUMMARY

The primary purpose of this chapter was to introduce the subject of estimating the unknown parameters of a random sequence from data. The basic ideas of statistics that were introduced in Chapter 8 were applied to estimating the mean value, the autocorrelation function, and the spectral density function. Throughout this chapter, the random sequence is assumed to be sta-

614 ESTIMATING THE PARAMETERS OF RANDOM PROCESSES

tionary, and if only one sample function is available, which is often the case, then ergodicity is also assumed. Tests for stationary were discussed in Section 9.2, and simple transformations to induce stationarity were discussed in Section 9.4.1.

Two basic approaches to estimating unknown parameters of random processes were introduced in this chapter. The first, which is called model-free estimation, basically calculates estimates of the mean, autocorrelation function, and power spectral density function from a time-limited sample function(s), that is, realization(s), of the random process. The advantage of this method is that the data primarily determine the important descriptors of the random process. The disadvantage is that more data are required in order to reduce the variance of the estimator to an acceptable level. In addition, it was shown that simple estimators of the power spectral density function will have a variance that does not decrease as the amount of data increases. Thus, some averaging or smoothing is necessary, and window functions for smoothing were introduced.

The second basic approach to estimating unknown parameters of random processes is called a model-based approach. This approach was presented by introducing the Box–Jenkins method of estimating the parameters of ARIMA model. After differencing (the I in ARIMA) the data in order to produce (quasi-) stationarity, the orders of the resulting autoregressive (AR) moving-average (MA) model were "identified" using the sample autocorrelation and partial autocorrelation functions. Then the $(p + q)$ parameters of the ARMA model were estimated. These estimates of the parameters can be substituted in mathematical expressions to obtain estimates of psd or autocorrelation functions. Finally, diagnostic checking of the residual errors was introduced.

This model-based approach has the advantage of needing fewer data for estimators with acceptably small variance, that is, the model is a substitute for data. If the assumed model is adequate, this is a definite savings; if not, the use of the estimated parameters in the assumed model is likely to result in poor designs and analysis based on the model. Another advantage of the ARMA parametric models is that the form of the resulting model is very convenient for both analysis and design of communication and control systems.

9.6 REFERENCES

The literature on spectral estimation is vast and is expanding at a rapid rate. References [8] and [9] present comprehensive coverage of recent developments in parameter estimation, most of which emphasize autoregressive models. In certain important applications, these recently developed techniques result in more efficient estimators than the general methods introduced here. References [7], [10] and [11] cover both parametric and nonparametric methods of spectral estimation. References [8], [9] and [10] are recent textbooks that have excellent treatments of spectral estimation methods. Reference [7] compares a variety of techniques that are not covered in this chapter and contains several interesting examples.

The book by Box and Jenkins [3] contains an excellent coverage of fitting ARIMA models to data. Much of the material in the second half of this chapter is derived from [3]. Reference [6] covers similar material.

Reference [2] is the classic reference on nonparametric estimation of power spectra, and [1] has been a widely used text on this material.

Current articles dealing with spectral estimation may be found in the recent issues of the *IEEE Transactions on Acoustics, Speech and Signal Processing*.

[1] J. S. Bendat and A. G. Piersol, *Random Data: Analysis and Measurement Procedures,* Second Edition, Wiley Interscience, New York, 1986.

[2] R. B. Blackman and J. W. Tukey, *The Measurement of Power Spectra,* Dover, New York, 1958.

[3] G. E. P. Box and G. M. Jenkins, *Time Series Analysis: Forecasting and Control,* Holden-Day, San Francisco, 1976.

[4] N. R. Draper and H. Smith, *Applied Regression Analysis,* 2nd ed., John Wiley & Sons, New York, 1981.

[5] J. Durbin, "Testing for Serial Correlation in Least-Squares Regression when Some of the Regressors are Lagged Dependent Variables," *Econometrica,* Vol. 38, 1970, p. 410.

[6] G. M. Jenkins and D. G. Watts, *Spectral Analysis and Its Applications,* Holden-Day, San Francisco, 1968.

[7] S. M. Kay and S. L. Marple, Jr., "Spectrum Analysis—A Modern Perspective," *Proceedings of the IEEE,* Vol. 69, No. 11, Nov. 1981.

[8] S. M. Kay, *Modern Spectral Estimation—Theory and Applications,* Prentice-Hall, Englewood Cliffs, N.J., 1986.

[9] S. L. Marple, Jr., *Digital Spectral Analysis,* Prentice-Hall, Englewood Cliffs, N.J., 1986.

[10] N. Mohanty, *Random Signals, Estimation and Identification,* Van Nostrand and Reinhold, New York, 1986.

[11] M. B. Priestly, *Spectral Analysis and Time Series,* Academic Press, New York, 1981.

[12] S. S. Wilks, *Mathematical Statistics,* John Wiley & Sons, New York, 1962.

9.7 PROBLEMS

9.1 a. Test the stationarity of the mean of the sample

.1; $-.2$; $-.3$; .2; .4; .1; $-.1$; .5; $-.3$;
.2; $-.2$; $-.3$; $-.4$; $-.1$; 1; $-.2$; .1; $-.2$.

by comparing the sample mean of the first half with the sample mean of the second half and testing, at the 1% significance level, the null hypothesis that they are the same.

b. Test for stationarity of the variance by using the F test on the sample variances of the first and second part.

9.2 $X(n)$ is a stationary zero-mean Gaussian random sequence with an autocorrelation function

$$R_{XX}(k) = \exp(-0.2k^2)$$

Show that $X(n)$ is ergodic.

9.3 Generate (using a computer program) a sequence of 100 independent Gaussian random variates, $W(n)$, $n = 1, 2, \ldots, 100$, with $E\{W(n)\} = 0$ and $\text{Var}\{W(n)\} = 1$. Using $W(n)$, generate 100 samples of

$$X(n) = 10 + (0.01)(n)W(n), \quad n = 1, 2, \ldots, 100$$

Divide the data $X(n)$ into 10 groups of 10 contiguous samples each and test for stationarity of mean-square values using the run test at a significance level of 0.1.

9.4 If X_1 and X_2 are jointly normal with zero means, then show that

a.

$$E\{X_1^2 X_2^2\} = E\{X_1^2\}E\{X_2^2\} + 2E^2\{X_1 X_2\}$$

Hint: $E\{X_1^2 X_2^2\} = E\{E[X_1^2 X_2^2 | X_2]\} = E\{X_2^2 E\{X_1^2 | X_2]\}$

$$= E\{X_2^2[\sigma_1^2(1 - r^2) + r^2 \frac{\sigma_1^2}{\sigma_2^2} X_2^2]\}$$

b.

$$E\{X_2^4\} = 3\sigma_2^4$$

Hint: See Example 2.13.

9.5 $X(n)$ is a zero-mean Gaussian random sequence with

$$R_{XX}(k) = \exp(-0.2k^2)$$

Assume that $R_{XX}(k)$, at $k = 10$, is estimated using 100 samples of $X(n)$ according to

$$\hat{R}_{XX}(10) = \frac{1}{90} \sum_{i=1}^{90} X(i)X(i + 10)$$

Find the variance of $\hat{R}_{XX}(10)$.

9.6 Show that Equation 9.5.b follows from Equation 9.5.a in the stationary case.

9.7 If $R_{XX}(n) = a^n$, $n \geq 0$, $0 < a < 1$, find $\text{Var}[\overline{X}]$.

9.8 Show that the variance of $\hat{R}_{XX}(k)$ as given in Equation 9.7 for a Gaussian random sequence is for $0 \leq k \leq N$

$$\frac{1}{(N-k)^2} \sum_{m=-(N-k-1)}^{N-k-1} (N - k - |m|)[R_{XX}^2(m) + R_{XX}^2(m-k)R_{XX}(m+k)] - \mu^4$$

9.9 Show that

$$\sum_{n=0}^{N-1} \cos^2\left(\frac{2\pi np}{N}\right) = \frac{N}{2}, \quad p \neq 0, \left[\frac{N}{2}\right]$$

$$= N, \quad p = 0, \left[\frac{N}{2}\right]$$

9.10 Find the mean, variance, and covariance of A_p and B_p defined in Equation 9.18, that is find $E\{A_p\}$, $E\{B_p\}$, $\text{Var}\{A_p\}$, $\text{Var}\{B_p\}$, $\text{Covar}\{A_p, A_q\}$ and $\text{Covar}\{A_p B_q\}$.

9.11 With $N = 500$ and $M = 25$,

  a. Plot the Barlett, Blackman–Tukey, and the Parzen windows in the time domain (i.e., plot $\lambda(k)$, $|k| = 1, 2, \ldots, M$).

  b. Plot $W_N(f)$ for these windows, for $|f| = p/N$, $p = 0, 1, 2, \ldots, [N/2]$.

9.12 Using a computer program, generate 1000 samples of the realization of the process

$$X(n) = \frac{[W(n) + W(n-1)]}{2}, \quad n = 1, 2, 3, \ldots, 1000$$

where $W(n)$ is an independent Gaussian sequence with zero-mean and unit variance.

  a. Find $R_{XX}(k)$ and $S_{XX}(f)$.

  b. Using the data, obtain the periodogram based estimate of $S_{XX}(f)$ and compare the estimated value of $S_{XX}(f)$ with the true value of $S_{XX}(f)$ obtained in part (a), that is, sketch $S_{XX}(f)$ and $\hat{S}_{XX}(f)$.

9.13 With reference to Problem 9.12, divide the data into 10 contiguous segments of 100 points each. Obtain $\hat{S}_{XX}(f)$ for each segment and average the estimates. Plot the average of the estimates $\overline{\overline{S}}_{XX}(f)$ and $S_{XX}(f)$.

9.14 Smooth the spectral estimate in the frequency domain obtained in Problem 9.12 using a rectangular window of size $2m + 1 = 9$. Compare the smoothed estimate $\overline{S}_{XX}(f)$ with $S_{XX}(f)$.

9.15 By changing the seed to the random-number generator, obtain 10 sets of 100 data points using the model given in Problem 9.12. Obtain the spectral estimate $\hat{S}_{XX}(f)$ for each of the 10 data sets. Using these estimates, compute the sample variance of $\hat{S}_{XX}(f)$ as

$$\text{Var}\{\hat{S}_{XX}(f)\} = \frac{1}{9} \sum_{i=1}^{10} [\hat{S}_{XX}(f)_i - \overline{S}_{XX}(f)]^2$$

where $\overline{S}_{XX}(f)$ is the average of the 10 estimates. Plot $\text{Var}\{\hat{S}_{XX}(f)\}$ versus f for $|f| = p/N, p = 0, 1, 2, \ldots, 50$.

9.16 With reference to Problem 9.12, apply the Barlett, Blackman–Tukey, and Parzen windows with $M = 25$ and plot the smoothed estimates of $S_{XX}(f)$.

9.17 With continuous data and

$$\hat{\mu} = \frac{1}{2T} \int_{-T}^{T} X(t) \, dt$$

Show that

a. $\hat{\mu}$ is unbiased

b. $\sigma_{\hat{\mu}}^2 = \frac{1}{T} \int_{0}^{2T} \left(1 - \frac{\tau}{2T}\right) C_{XX}(\tau) \, d\tau$

9.18 Refer to Problem 9.17. Let X be band-limited white noise with

$$C_{XX}(\tau) = \sigma^2 \frac{\sin 2\pi B\tau}{2\pi B\tau}$$

Show that

$$\sigma_{\hat{\mu}}^2 = \frac{\sigma^2}{4BT}$$

Note that there are $4BT$ independent samples in the sampling time $2T$.

9.19 If we estimate $E\{X^2(t)\}$ of an ergodic process by

$$\widehat{X^2} = \frac{1}{T} \int_{0}^{T} X^2(t) \, dt$$

Show that

a. $\widehat{X^2}$ is unbiased.

b. $\text{Var}\{\widehat{X^2}\} = \frac{2}{T} \int_{0}^{T} \left(1 - \frac{\tau}{T}\right) [R_{X'X'}(\tau) - \mu_X^4 - 2\mu_X^2 \sigma_X^2 - \sigma_X^4] \, d\tau$

and if $\mu_X = 0$ and X is Gaussian, then

$$\text{Var}\{\widehat{X^2}\} = \frac{4}{T} \int_{0}^{T} \left(1 - \frac{\tau}{T}\right) R_{XX}^2(\tau) \, d\tau$$

9.20 Refer to Problem 9.19. If X is zero-mean Gaussian band-limited (B) white noise and $B \gg 1/T$, show that

$$\text{Var}\{\widehat{X^2}\} \approx \frac{\sigma^4}{BT}$$

if C_{XX} is as in Problem 9.18.

9.21 Using Equations 9.11, 9.12 and 9.13, show that

a. $\hat{R}_{XX}(k) = \dfrac{1}{N} \displaystyle\sum_{i=-\infty}^{\infty} d(i)X(i)d(i+k)X(i+k),$
$k = 0, \ldots, N-1$

b. $\hat{S}_{XX}(f) = \displaystyle\sum_{m=-\infty}^{\infty} \hat{R}_{XX} \exp(-j2\pi mf)$

9.22 $X(n)$ is a random sequence with

$$\sigma_{X(n)} = c\mu_{X(n)}$$
$$Y(n) = \ln X(n)$$

TABLE 9.10 DATA FOR PROBLEM 9.27 (READ ACROSS)

1.065	0.711	−1.079	−0.731	−1.165
0.115	0.263	−0.478	−2.006	−0.068
−0.581	0.022	1.611	−0.309	−1.883
1.177	−1.773	−2.176	0.891	−0.250
−0.802	−2.409	1.446	1.577	1.400
1.311	0.709	0.205	−0.031	−1.539
−1.512	−0.207	−0.698	1.392	1.442
1.853	1.685	1.542	−0.582	−2.043
−2.207	−2.253	−1.541	−0.951	0.796
−0.366	1.643	2.850	0.516	−0.143
−2.323	−0.658	−0.016	−1.422	−0.634
−0.918	−1.364	−0.607	0.799	2.085
0.269	1.353	−0.687	−0.107	−0.567
−1.280	−0.676	−0.341	1.675	1.711
0.473	0.419	1.782	0.427	1.088
0.459	−0.595	−2.821	0.423	−0.321
0.002	0.847	2.114	−0.087	1.095
−0.721	−0.488	1.318	−0.975	−0.153
−0.771	−0.802	0.656	0.105	1.640
1.159	0.887	3.042	−0.724	−1.542

TABLE 9.11 DATA FOR PROBLEM 9.28 (READ ACROSS)

0.199	0.304	−0.937	−0.669	−0.612
0.828	1.517	0.732	−1.664	−1.348
−2.077	−1.950	−0.166	−0.887	−2.298
0.602	−0.624	−1.468	1.071	0.752
−0.450	−3.120	−0.880	0.166	0.613
1.097	1.223	1.277	1.433	0.131
−0.586	0.019	−0.689	0.569	0.842
1.246	1.402	1.698	0.212	−1.444
−2.017	−2.197	−1.703	−1.247	0.212
−0.766	0.419	2.046	0.608	−0.359
−2.491	1.325	0.230	0.314	0.085
−0.378	−1.540	−1.827	−0.992	0.653
−0.290	0.718	−0.351	0.229	0.406
−0.090	0.148	0.229	1.782	2.045
0.564	−0.317	0.767	0.285	1.221
1.530	0.987	−1.359	0.653	0.484
0.229	−0.454	1.460	−0.404	0.040
−1.203	−1.051	1.326	−0.513	1.086
0.603	−0.019	0.646	−0.092	0.730
0.500	0.255	2.661	0.522	−0.799

Find approximately the standard deviation of $Y(n)$ using a Taylor series approximation.

9.23 If an estimated ARMA model is

$$X(n) = .9X(n-1) - .2X(n-2) + .5e(n-1) + e(n)$$

find the corresponding autocorrelation function, autocorrelation coefficient, and power density spectrum.

9.24 Examine the data set given in Table 9.9 for stationarity by

    a. Plotting the data.

    b. Plotting $\overline{X}$ versus n.

    c. Plotting S^2 versus n.

    d. Plotting $\hat{r}_{XX}$ versus n.

    e. Plotting $X(n) \times X(n-1)$ versus n

9.25 If $X_1(n)$ and $X_2(n)$ are jointly wide-sense stationary sequences
  a. Show that $Y(n) = X_1(n) - X_2(n)$ is wide-sense stationary.
  b. Find the mean and autocorrelation function of $Y(n)$

9.26. Find the first difference of the sequence
  a. 1, 1.5, 2.0, 3.5, 3.0, 4.0.
  b. Create the original series from the first difference.

9.27 The data $W(n)$ in Table 9.10 (read across) have a periodic component. Plot the data and find a reasonable guess for m in the model
$$X(n) = (1 - z^{-m})W(n) = W(n) - W(n - m)$$
Plot $X(n)$. Recover $W(n)$ from $X(n)$.

9.28 Estimate the parameters $\phi_{1,1}$ and σ^2 of a first-order autoregressive model from the data set in Table 9.11 (read across).

TABLE 9.12 DATA FOR PROBLEM 9.31 (READ ACROSS)

0.208	−0.979	−0.729	−0.558	0.916
1.633	0.729	−1.814	−1.529	−2.013
−1.807	0.068	−0.592	−2.175	0.677
−0.357	−1.414	1.104	0.931	−0.456
−3.225	−0.949	0.436	0.846	1.182
1.203	1.173	1.286	−0.031	−0.793
−0.093	−0.665	0.577	0.924	1.246
1.335	1.568	0.047	−1.679	−2.168
−2.145	−1.504	−0.967	0.478	−0.532
0.473	2.135	0.646	−0.541	−2.654
−1.402	0.430	−0.069	0.159	−0.333
−1.541	−1.825	−0.804	0.877	−0.085
0.727	−0.323	0.163	0.410	−0.096
0.102	0.219	1.771	2.054	0.432
−0.578	0.587	0.268	1.159	1.497
0.886	−1.540	0.446	0.548	0.226
0.402	1.421	−0.435	−0.126	−1.241
−1.082	1.414	0.691	1.044	0.535
−0.146	0.529	−0.123	0.660	0.492
0.195	2.587	0.518	−1.049	−0.561

622 ESTIMATING THE PARAMETERS OF RANDOM PROCESSES

TABLE 9.13 DATA FOR PROBLEM 9.33 (READ ACROSS)

0.302	0.284	2.439	0.922	0.039
1.466	0.708	0.828	−0.034	−1.051
−0.268	−1.035	1.813	0.076	−1.215
0.135	−0.080	1.696	1.067	−2.072
−0.827	−0.996	0.604	1.102	0.455
−0.427	0.134	0.938	1.167	0.382
0.029	−0.186	0.045	0.080	1.414
0.810	−1.358	1.121	0.895	0.996
−0.516	−1.565	0.840	1.075	0.942
−1.524	−1.245	0.303	−0.131	−1.494
−1.771	1.176	0.189	2.273	1.780
0.941	1.230	−0.496	0.044	−1.644
−1.879	−1.631	−2.305	−1.299	−0.187
−0.147	2.735	−0.400	−1.892	−0.571
0.397	−1.267	−0.079	0.043	0.967
−0.758	−0.453	−0.128	0.997	0.736
0.514	0.672	−0.679	0.449	0.988
−0.075	−0.817	−1.367	−0.260	0.623
0.043	−0.600	−1.218	−0.990	−0.385
0.264	−0.777	−1.737	−1.924	−1.658

9.29 Differentiate Equation 9.57 partially with respect to σ^2 and set the resulting derivative equal to zero in order to find the maximum likelihood estimator of σ^2.

9.30 Why are "Equations" 9.65 and 9.66 only approximations? Show the exact form that follows from Equation 9.64.

9.31 Estimate the parameters $\phi_{2,1}$, $\phi_{2,2}$, and σ^2 of a second-order autoregressive model from the data set in Table 9.12 (read across).

9.32 Give the matrix form of the estimators for an order p autoregressive model. *Hint:* See Section 8.10.3.

9.33 Given the data in Table 9.13, estimate θ in the model $X(n) = \theta e(n-1) + e(n)$.

9.34 In the previous problem, estimate the unconditional sum of squares using the following procedure:

  a. Assume that $a(101) = 0$.

b. Backcast the model (assuming a value for θ) using $a(n) = X(n) - \theta a(n + 1)$, $n = 100, 99, \ldots, 1$. where $a(n)$ is a white-noise sequence. It is different from $e(n)$ but it has the same probabilistic characteristics.

c. Assume that $a(0)$ is equal to zero, that is, its unconditional expectation because it is independent of $X(n)$, $n \geq 1$.

d. At the end of the backcast, $X(0)$ can be found.

e. Using this $X(0)$, as opposed to assuming it is equal to zero, find $X(1)$ from $X(n) = \theta \epsilon(n - 1) + \epsilon(n)$. That is, $\epsilon(0) = X(0) - \theta \epsilon(-1)$ and because $\epsilon(-1) = 0$, $\epsilon(0) = X(0)$. Then $\epsilon(1) = X(1) - \theta \epsilon(0)$ and subsequently all $\epsilon(n)$ and the sum of squares can be calculated.

9.35 Compare the estimated unconditional sum of squares in Problem 9.33 with the conditional sum of squares from Problem 9.34.

9.36 Estimate $\phi_{1,1}$, $\theta_{1,1}$, and σ_N^2 in the model

$$X(n) = \phi_{1,1} X(n - 1) + \theta_{1,1} e(n - 1) + e(n)$$

using the data in Table 9.14.

TABLE 9.14 DATA FOR PROBLEM 9.36 (READ ACROSS)

0.199	0.340	−0.843	−0.771	−0.890
0.560	1.471	1.090	−1.253	−1.411
−2.527	−2.600	−1.049	−1.483	−2.745
−0.167	−1.095	−1.683	0.541	0.593
−0.255	−3.108	−1.485	−0.594	0.305
1.023	1.443	1.687	1.962	0.757
−0.130	0.119	−0.664	0.483	0.820
1.487	1.786	2.260	0.904	−0.915
−2.038	−2.688	−2.491	−2.119	−0.571
−1.224	0.090	1.835	0.963	0.086
−2.344	−1.719	−0.446	−0.636	−0.119
−0.524	−1.655	−2.225	−1.589	0.019
−0.540	0.606	−0.356	0.268	0.376
0.038	0.204	0.273	1.870	2.427
1.307	0.290	1.030	0.543	1.505
1.884	1.585	−0.779	0.768	0.509
0.486	0.600	1.657	−0.005	0.306
−1.160	−1.183	0.919	0.513	1.323
0.895	0.365	0.845	0.132	0.901
0.681	0.538	2.858	1.136	−0.128

624 ESTIMATING THE PARAMETERS OF RANDOM PROCESSES

9.37 Let $e(n)$ be a zero-mean unit variance white Gaussian sequence, and
$$X(n) = .3X(n-1) - .2e(n-1) + e(n)$$

a. Find $r_{XX}(k)$.

b. Generate a sequence $X(1), \ldots, X(100)$ using a random number generator. Find the sample autocorrelation coefficient $\hat{r}_{XX}(k)$.

c. Change the seed in the random number generator and repeat part (b). Compare the results.

d. Use either sequence and estimate the ARMA model.

e. Calculate and plot $\epsilon(n)$ using the estimated model.

f. Plot $\hat{r}_{\epsilon\epsilon}(k)$ from the results of part (e).

9.38 Solve Equations 9.75 and 9.76 for $\theta_{2,1}$ and $\theta_{2,2}$ if $r_{XX}(1) = .4$ and $r_{XX}(2) = .1$.

9.39 Solve Equations 9.77 and 9.78 for $\phi_{1,1}$ and $\theta_{1,1}$ if $r_{XX}(1) = .4$ and $r_{XX}(2) = .1$.

TABLE 9.15 ERROR SEQUENCE FOR PROBLEM 9.42 (READ ACROSS)

0.543	-0.122	1.178	-0.405	-1.257
0.611	-0.553	0.543	-0.847	0.131
-0.806	-1.710	-0.677	-0.159	-0.207
-1.496	-1.214	-0.476	0.388	1.537
0.539	0.289	-1.602	0.202	2.797
1.244	0.549	-0.229	0.909	-1.479
-0.411	0.470	-1.077	0.698	0.328
-0.235	0.579	-0.966	0.658	0.367
0.576	-0.444	-0.152	-0.260	-0.814
1.604	-1.479	1.154	1.166	-0.368
-0.352	0.751	-0.046	-1.757	1.999
-0.033	-0.197	-0.515	2.229	0.194
-0.514	1.530	0.115	0.181	-0.583
0.536	0.863	-0.313	-1.216	-0.352
-0.323	-1.215	-0.542	-2.164	-0.544
-0.320	-0.364	-0.199	-1.376	-0.853
-1.527	2.056	-0.651	0.742	-2.107
-1.716	0.516	-0.916	-1.266	-1.600
0.863	-0.878	0.090	0.992	1.020
0.438	-0.741	0.789	0.640	-0.352

9.40 At the 5% level of significance, what is the acceptable range of $C(20)$ as given by Equation 9.80 in order to fail to reject the hypothesis that the residuals of an ARIMA $(0, 1, 2)$ model are white when $\sigma^2 = 1$?

9.41 Show that if $\epsilon(n)$ is a zero-mean white noise sequence with variance σ^2, then the expected value of D as given by Equation 9.82 is approximately equal 2.

9.42 Test the error sequence shown in Table 9.15 for whiteness using the run test.

9.43 Test the error sequence shown in Table 9.15 for whiteness, that is, $\rho = 0$, using the Durbin–Watson test.

9.44 Test the error sequence shown in Table 9.15 for whiteness using a cumulative spectral density plot.

9.45 Test the error sequence shown in Table 9.15 for whiteness using a chi-square test on the sum of the sample autocorrelation coefficients.

APPENDIX A

Fourier Transforms

$$X(f) = \int_{-\infty}^{\infty} x(t)\exp(-j2\pi ft)\, dt$$

$$x(t) = \int_{-\infty}^{\infty} X(f)\exp(j2\pi ft)\, df$$

$$\int_{-\infty}^{\infty} |x(t)|^2\, dt = \int_{-\infty}^{\infty} |X(f)|^2\, df$$

TABLE A.1 TRANSFORM THEOREMS

Name of Theorem	Signal	Fourier Transform
(1) Superposition	$a_1 x_1(t) + a_2 x_2(t)$	$a_1 X_1(f) + a_2 X_2(f)$
(2) Time delay	$x(t - t_0)$	$X(f)\exp(-j2\pi f t_0)$
(3) Scale change	$x(at)$	$|a|^{-1} X(f/a)$
(4) Frequency translation	$x(t)\exp(j2\pi f_0 t)$	$X(f - f_0)$
(5) Modulation	$x(t)\cos 2\pi f_0 t$	$\tfrac{1}{2}X(f - f_0) + \tfrac{1}{2}X(f + f_0)$
(6) Differentiation	$\dfrac{d^n x(t)}{dt^n}$	$(j2\pi f)^n X(f)$
(7) Integration	$\int_{-\infty}^{t} x(t')\, dt'$	$(j2\pi f)^{-1} X(f) + \tfrac{1}{2}X(0)\delta(f)$
(8) Convolution	$\int_{-\infty}^{\infty} x_1(t - t') x_2(t')\, dt'$ $= \int_{-\infty}^{\infty} x_1(t') x_2(t - t')\, dt'$	$X_1(f) X_2(f)$
(9) Multiplication	$x_1(t) x_2(t)$	$\int_{-\infty}^{\infty} X_1(f - f') X_2(f')\, df'$ $= \int_{-\infty}^{\infty} X_1(f') X_2(f - f')\, df'$

Source: K. Sam Shanmugan, *Digital and Analog Communication Systems*, John Wiley & Sons, New York, 1979, p. 581.

FOURIER TRANSFORMS

TABLE A.2   TRANSFORM PAIRS

Signal $x(t)$	Transform $X(f)$
(1) Rectangular pulse of amplitude A from $-\tau/2$ to $\tau/2$	$A\tau \dfrac{\sin \pi f \tau}{\pi f \tau} \triangleq A\tau \operatorname{sinc} f\tau$
(2) Triangular pulse of amplitude B from $-\tau$ to τ	$B\tau \dfrac{\sin^2 \pi f \tau}{(\pi f \tau)^2} \triangleq B\tau \operatorname{sinc}^2 f\tau$
(3) $e^{-\alpha t} u(t)$	$\dfrac{1}{\alpha + j2\pi f}$
(4) $\exp(-|t|/\tau)$	$\dfrac{2\tau}{1 + (2\pi f \tau)^2}$
(5) $\exp[-\pi(t/\tau)^2]$	$\tau \exp[-\pi(f\tau)^2]$
(6) $\dfrac{\sin 2\pi W t}{2\pi W t} \triangleq \operatorname{sinc} 2Wt$	Rectangular pulse of amplitude $1/(2W)$ from $-W$ to W
(7) $\exp[j(2\pi f_c t + \phi)]$	$\exp(j\phi)\delta(f - f_c)$
(8) $\cos(2\pi f_c t + \phi)$	$\tfrac{1}{2}\delta(f - f_c)\exp(j\phi) + \tfrac{1}{2}\delta(f + f_c)\exp(-j\phi)$
(9) $\delta(t - t_0)$	$\exp(-j2\pi f t_0)$
(10) $\sum_{m=-\infty}^{\infty} \delta(t - mT_s)$	$\dfrac{1}{T_s} \sum_{n=-\infty}^{\infty} \delta\left(f - \dfrac{n}{T_s}\right)$
(11) $\operatorname{sgn} t = \begin{cases} +1, & t > 0 \\ -1, & t < 0 \end{cases}$	$-\dfrac{j}{\pi f}$
(12) $u(t) = \begin{cases} 1, & t > 0 \\ 0, & t < 0 \end{cases}$	$\dfrac{1}{2}\delta(f) + \dfrac{1}{j2\pi f}$

Source: K. Sam Shanmugan, *Digital and Analog Communication Systems*, John Wiley & Sons, New York, 1979, p. 582.

APPENDIX B

Discrete Fourier Transforms

The DFT and the inverse DFT are computed by

$$X_F\left(\frac{k}{N}\right) = \sum_{n=0}^{N-1} x(n)\exp\left(\frac{-j2\pi kn}{N}\right), \quad k = 0, 1, 2, \ldots, N-1$$

$$x(n) = \frac{1}{N}\sum_{k=0}^{N-1} X_F\left(\frac{k}{N}\right)\exp\left(\frac{j2\pi kn}{N}\right), \quad n = 0, 1, 2, \ldots, N-1$$

If $x(n)$ is even,

$$X_F\left(\frac{k}{N}\right) = X_F^*\left(\frac{N-k}{N}\right), \quad k = 1, 2, \ldots, \left[\frac{N}{2}\right]$$
$$= X_F^*(-k/N)$$

TABLE OF DFT PAIRS

Time Sequence $x(n)$	DFT $X_F\left(\dfrac{k}{N}\right)$
$\exp(2\pi j f_0 n)$ $f_0 = \dfrac{m}{N}$, $m < \dfrac{N}{2}$	$X_F\left(\dfrac{k}{N}\right) = \begin{cases} 1 & k = m \\ 0 & \text{otherwise} \end{cases}$
$\cos 2\pi f_0 n$ $f_0 = \dfrac{m}{N}$, $m < \dfrac{N}{2}$	$X_F\left(\dfrac{k}{N}\right) = \begin{cases} \dfrac{1}{2} & k = m \\ 0 & \text{otherwise} \end{cases}$
$\sin 2\pi f_0 n$ $f_0 = \dfrac{m}{N}$, $m < \dfrac{N}{2}$	$X_F\left(\dfrac{k}{N}\right) = \begin{cases} \dfrac{1}{2j} & k = m \\ 0 & \text{otherwise} \end{cases}$
$x(n) = \begin{cases} 1 & \text{for } n = 0 \\ 0 & \text{otherwise} \end{cases}$	$X_F\left(\dfrac{k}{N}\right) = 1$ for all k
$x(n) = \begin{cases} 1 & 0 \leq |n| < N_1 < \dfrac{N}{2} \\ 0 & \text{elsewhere} \end{cases}$	$X_F\left(\dfrac{k}{N}\right) = \dfrac{\sin\left[\left(N_1 + \dfrac{1}{2}\right)\dfrac{2\pi k}{N}\right]}{\sin\left(\dfrac{\pi k}{N_1}\right)}$
$x(n) = \begin{cases} 1 & \text{for } n = n_0 \\ 0 & \text{otherwise} \end{cases}$	$X_F\left(\dfrac{k}{N}\right) = \exp\left(\dfrac{-j2\pi k n_0}{N}\right)$

APPENDIX C

Z Transforms

f_k	$F(z)$
1. δ_k	1, all z
2. 1, $k \geq 0$	$(1 - z^{-1})$, $1 < \|z\|$
3. k, $k \geq 0$	$z^{-1}(1 - z^{-1})^{-2}$, $1 < \|z\|$
4. k^n, $k \geq 0$	$\left(-z\dfrac{d}{dz}\right)^n (1 - z^{-1})^{-1}$, $1 < \|z\|$
5. $\binom{k}{n}$, $n \leq k$	$z^{-n}(1 - z^{-1})^{n+1}$, $0 < \|z\|$
6. $\binom{n}{k}$, $0 \leq k \leq n$	$(1 + z^{-1})^n$, $0 < \|z\|$
7. α^k, $k \geq 0$	$(1 - \alpha z^{-1})^{-1}$, $\|\alpha\| < \|z\|$
8. $k^n \alpha^k$, $k \geq 0$	$\left(-z\dfrac{d}{dz}\right)^n (1 - \alpha z^{-1})^{-1}$, $\|\alpha\| < \|z\|$
9. α^k, $k < 0$	$-(1 - \alpha z^{-1})^{-1}$, $\|z\| < \|\alpha\|$
10. $k^n \alpha^k$, $k < 0$	$-\left(-z\dfrac{d}{dz}\right)^n (1 - \alpha z^{-1})^{-1}$, $\|z\| < \|\alpha\|$
11. $\alpha^{\|k\|}$, all k	$(1 - \alpha^2)[(1 - \alpha z)(1 - \alpha z^{-1})]^{-1}$, $\|\alpha\| < \|z\| < \left\|\dfrac{1}{\alpha}\right\|$
12. $\dfrac{1}{k}$, $k > 0$	$-\ln(1 - z^{-1})$, $1 < \|z\|$
13. $\cos \alpha k$, $k \geq 0$	$(1 - z^{-1}\cos \alpha)(1 - 2z^{-1}\cos \alpha + z^{-2})^{-1}$, $1 < \|z\|$
14. $\sin \alpha k$, $k \geq 0$	$z^{-1}\sin \alpha (1 - 2z^{-1}\cos \alpha + z^{-2})^{-1}$, $1 < \|z\|$
15. $a \cos \alpha k + b \sin \alpha k$, $k \geq 0$	$[a + z^{-1}(b \sin \alpha - a \cos \alpha)](1 - 2z^{-1}\cos \alpha + z^{-2})^{-1}$, $1 < \|z\|$
16. $a \cos \alpha k + \left(\dfrac{d + c \cos \alpha}{\sin \alpha}\right)\sin \alpha k$, $k \geq 0$	$(c + dz^{-1})(1 - 2z^{-1}\cos \alpha + z^{-2})^{-1}$, $1 < \|z\|$
17. $\cosh \alpha k$, $k \geq 0$	$(1 - z^{-1}\cosh \alpha)(1 - 2z^{-1}\cosh \alpha + z^{-2})^{-1}$, $\max\left\{\|\alpha\|, \left\|\dfrac{1}{\alpha}\right\|\right\} < \|z\|$
18. $\sinh \alpha k$, $k \geq 0$	$(z^{-1}\sinh \alpha)(1 - 2z^{-1}\cosh \alpha + z^{-2})^{-1}$, $\max\left\{\|\alpha\|, \left\|\dfrac{1}{\alpha}\right\|\right\} < \|z\|$

Source: R. A. Gabel and R. A. Roberts, *Signals and Linear Systems*, Third Edition John Wiley & Sons, New York, 1980, p. 186.

APPENDIX D

Gaussian Probabilities

(1) $P(X > \mu_X + y\sigma_X) = Q(y) = \int_y^\infty \frac{1}{\sqrt{2\pi}} \exp\left(\frac{-z^2}{2}\right) dz$

(2) $Q(0) = \frac{1}{2}; \quad Q(-y) = 1 - Q(y), \quad \text{when } y \geq 0$

(3) $Q(y) \approx \frac{1}{y\sqrt{2\pi}} \exp\left(\frac{-y^2}{2}\right)$ when $y > 4$

(4) $\text{erfc}(y) \triangleq \frac{2}{\sqrt{\pi}} \int_y^\infty \exp(-z^2) \, dz = 2Q(\sqrt{2}y), \, y > 0.$

TABLE D.1 GUASSIAN PROBABILITIES

y	Q(y)	y	Q(y)	y	Q(y)	Q(y)	y
.05	.4801	1.05	.1469	2.10	.0179		
.10	.4602	1.10	.1357	2.20	.0139		
.15	.4405	1.15	.1251	2.30	.0107	10^{-3}	3.10
.20	.4207	1.20	.1151	2.40	.0082		
.25	.4013	1.25	.1056	2.50	.0062		
.30	.3821	1.30	.0968	2.60	.0047	$\frac{10^{-3}}{2}$	3.28
.35	.3632	1.35	.0885	2.70	.0035		
.40	.3446	1.40	.0808	2.80	.0026		
.45	.3264	1.45	.0735	2.90	.0019		
.50	.3085	1.50	.0668	3.00	.0013	10^{-4}	3.70
.55	.2912	1.55	.0606	3.10	.0010		
.60	.2743	1.60	.0548	3.20	.00069		
.65	.2578	1.65	.0495	3.30	.00048	$\frac{10^{-4}}{2}$	3.90
.70	.2420	1.70	.0446	3.40	.00034		
.75	.2266	1.75	.0401	3.50	.00023		
.80	.2119	1.80	.0359	3.60	.00016	10^{-5}	4.27
.85	.1977	1.85	.0322	3.70	.00010		
.90	.1841	1.90	.0287	3.80	.00007		
.95	.1711	1.95	.0256	3.90	.00005	10^{-6}	4.78
1.00	.1587	2.00	.0228	4.00	.00003		

Source: K. Sam Shanmugan, *Digital and Analog Communication Systems*, John Wiley & Sons, New York, 1979, pp. 583–84.

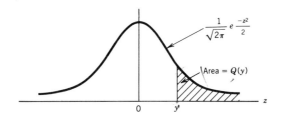

APPENDIX E

Table of Chi-square Distributions

Percentage Points of the χ_m^2 Distribution

That is, values of $\chi_{m;\alpha}^2$ where m represents degrees of freedom and

$$\int_0^{\chi_{m;\alpha}^2} \frac{1}{2\Gamma(m/2)} \left(\frac{y}{2}\right)^{(m/2)-1} \exp\{-y/2\} \, dy = 1 - \alpha.$$

For $m < 100$, linear interpolation is adequate. For $m > 100$, $\sqrt{2\chi_m^2}$ is approximately normally distributed with mean $\sqrt{2m-1}$ and unit variance, so that percentage points may be obtained from Appendix D.

m \ α	.995	.990	.975	.950	.050	.025	.010	.005
1	392704 × 10⁻¹⁰	157088 × 10⁻⁹	982069 × 10⁻⁹	393214 × 10⁻⁸	3.84146	5.02389	6.63490	7.87944
2	.0100251	.0201007	.0506356	.102587	5.99147	7.37776	9.21034	10.5966
3	.0717212	.114832	.215795	.351846	7.81473	9.34840	11.3449	12.8381
4	.206990	.297110	.484419	.710721	9.48773	11.1433	13.2767	14.8602
5	.411740	.554300	.831211	1.145476	11.0705	12.8325	15.0863	16.7496
6	.675727	.872085	1.237347	1.63539	12.5916	14.4494	16.8119	18.5476
7	.989265	1.239043	1.68987	2.16735	14.0671	16.0128	18.4753	20.2777
8	1.344419	1.646482	2.17973	2.73264	15.5073	17.5346	20.0902	21.9550
9	1.734926	2.087912	2.70039	3.32511	16.9190	19.0228	21.6660	23.5893
10	2.15585	2.55821	3.24697	3.94030	18.3070	20.4831	23.2093	25.1882
11	2.60321	3.05347	3.81575	4.57481	19.6751	21.9200	24.7250	26.7569
12	3.07382	3.57056	4.40379	5.22603	21.0261	23.3367	26.2170	28.2995
13	3.56503	4.10691	5.00874	5.89186	22.3621	24.7356	27.6883	29.8194
14	4.07468	4.66043	5.62872	6.57063	23.6848	26.1190	29.1413	31.3193
15	4.60094	5.22935	6.26214	7.26094	24.9958	27.4884	30.5779	32.8013
16	5.14224	5.81221	6.90766	7.96164	26.2962	28.8454	31.9999	34.2672
17	5.69724	6.40776	7.56418	8.67176	27.5871	30.1910	33.4087	35.7185
18	6.26481	7.01491	8.23075	9.39046	28.8693	31.5264	34.8053	37.1564

Percentage Points of the χ_m^2 Distribution (Continued)

That is, values of $\chi_{m;\alpha}^2$ where m represents degrees of freedom and

$$\int_0^{\chi_{m;\alpha}^2} \frac{1}{2\Gamma(m/2)} \left(\frac{y}{2}\right)^{(m/2)-1} \exp\{-y/2\}\, dy = 1 - \alpha.$$

For $m < 100$, linear interpolation is adequate. For $m > 100$, $\sqrt{2\chi_n^2}$ is approximately normally distributed with mean $\sqrt{2m-1}$ and unit variance, so that percentage points may be obtained from Appendix D.

m \ α	.995	.990	.975	.950	.050	.025	.010	.005
19	6.84398	7.63273	8.90655	10.1170	30.1435	32.8523	36.1908	38.5822
20	7.43386	8.26040	9.59083	10.8508	31.4104	34.1696	37.5662	39.9968
21	8.03366	8.89720	10.28293	11.5913	32.6705	35.4789	38.9321	41.4010
22	8.64272	9.54249	10.9823	12.3380	33.9244	36.7807	40.2894	42.7956
23	9.26042	10.19567	11.6885	13.0905	35.1725	38.0757	41.6384	44.1813
24	9.88623	10.8564	12.4011	13.8484	36.4151	39.3641	42.9798	45.5585

25	10.5197	11.5240	13.1197	14.6114	37.6525	40.6465	44.3141	46.9278
26	11.1603	12.1981	13.8439	15.3791	38.8852	41.9232	45.6417	48.2899
27	11.8076	12.8786	14.5733	16.1513	40.1133	43.1944	46.9630	49.6449
28	12.4613	13.5648	15.3079	16.9279	41.3372	44.4607	48.2782	50.9933
29	13.1211	14.2565	16.0471	17.7083	42.5569	45.7222	49.5879	52.3356
30	13.7867	14.9535	16.7908	18.4926	43.7729	46.9792	50.8922	53.6720
40	20.7065	22.1643	24.4331	26.5093	55.7585	59.3417	63.6907	66.7659
50	27.9907	29.7067	32.3574	34.7642	67.5048	71.4202	76.1539	79.4900
60	35.5346	37.4848	40.4817	43.1879	79.0819	83.2976	88.3794	91.9517
70	43.2752	45.4418	48.7576	51.7393	90.5312	95.0231	100.425	104.215
80	51.1720	53.5400	57.1532	60.3915	101.879	106.629	112.329	116.321
90	59.1963	61.7541	65.6466	69.1260	113.145	118.136	124.116	128.299
100	67.3276	70.0648	74.2219	77.9295	124.342	129.561	135.807	140.169

Source: I. Guttman and S. S. Wilks, *Introduction to Engineering Statistics*, John Wiley & Sons, New York, 1965, pp. 316–17.

APPENDIX F

Table of t_m Distributions

TABLE F.1

Values of $t_{m;\alpha}$ where m equals degrees of freedom and

$$\int_{-\infty}^{t_{m;\alpha}} \frac{\Gamma[(m+1)/2]}{\sqrt{\pi m}\,\Gamma(m/2)} \left(1 + \frac{t^2}{m}\right)^{-(m+1)/2} dt = 1 - \alpha$$

Percentage Points of the t_m Distribution

m \ α	.25	.1	.05	.025	.01	.005
1	1.000	3.078	6.314	12.706	31.821	63.657
2	.816	1.886	2.920	4.303	6.965	9.925
3	.765	1.638	2.353	3.182	4.541	5.841
4	.741	1.533	2.132	2.776	3.747	4.604
5	.727	1.476	2.015	2.571	3.365	4.032
6	.718	1.440	1.943	2.447	3.143	3.707
7	.711	1.415	1.895	2.365	2.998	3.499
8	.706	1.397	1.860	2.306	2.896	3.355
9	.703	1.383	1.833	2.262	2.821	3.250
10	.700	1.372	1.812	2.228	2.764	3.169
11	.697	1.363	1.796	2.201	2.718	3.106
12	.695	1.356	1.782	2.179	2.681	3.055
13	.694	1.350	1.771	2.160	2.650	3.012
14	.692	1.345	1.761	2.145	2.624	2.977

TABLE F.1 (*Continued*)

Values of $t_{m;\alpha}$ where m equals degrees of freedom and

$$\int_{-\infty}^{t_{m;\alpha}} \frac{\Gamma[(m+1)/2]}{\sqrt{\pi m}\,\Gamma(m/2)} \left(1 + \frac{t^2}{m}\right)^{-(m+1)/2} dt = 1 - \alpha$$

Percentage Points of the t_m Distribution

m \ α	.25	.1	.05	.025	.01	.005
15	.691	1.341	1.753	2.131	2.602	2.947
16	.690	1.337	1.746	2.120	2.583	2.921
17	.689	1.333	1.740	2.110	2.567	2.898
18	.688	1.330	1.734	2.101	2.552	2.878
19	.688	1.328	1.729	2.093	2.539	2.861
20	.687	1.325	1.725	2.086	2.528	2.845
21	.686	1.323	1.721	2.080	2.518	2.831
22	.686	1.321	1.717	2.074	2.508	2.819
23	.685	1.319	1.714	2.069	2.500	2.807
24	.685	1.318	1.711	2.064	2.492	2.797
25	.684	1.316	1.708	2.060	2.485	2.787
26	.684	1.315	1.706	2.056	2.479	2.779
27	.684	1.314	1.703	2.052	2.473	2.771
28	.683	1.313	1.701	2.048	2.467	2.763
29	.683	1.311	1.699	2.045	2.462	2.756
30	.683	1.310	1.697	2.042	2.457	2.750
40	.681	1.303	1.684	2.021	2.423	2.704
60	.679	1.296	1.671	2.000	2.390	2.660
120	.677	1.289	1.658	1.980	2.358	2.617
∞	.674	1.282	1.645	1.960	2.326	2.576

Source: I. Guttman and S. S. Wilks, *Introduction to Engineering Statistics,* John Wiley & Sons, New York, 1965, p. 319.

APPENDIX G

Table of F Distributions

Values of $F_{m_1,m_2;\alpha}$, where (m_1, m_2) is the pair of degrees of freedom in F_{m_1,m_2} and

$$\frac{\Gamma((m_1 + m_2)/2)}{\Gamma(m_1/2)\Gamma(m_2/2)} \left(\frac{m_1}{m_2}\right)^{m_1/2} \int_0^{F_{m_1,m_2;\alpha}} \lambda^{(m_1/2)-1} \left(1 + \frac{m_1}{m_2}\lambda\right)^{-(m_1+m_2)/2} d\lambda = 1 - \alpha$$

Percentage Points of the F_{m_1,m_2} Distribution

$\alpha = .10$

m_2 \ m_1	1	2	3	4	5	6	7	8	9
1	39.864	49.500	53.593	55.833	57.241	58.204	58.906	59.439	59.858
2	8.5263	9.0000	9.1618	9.2434	9.2926	9.3255	9.3491	9.3668	9.3805
3	5.5383	5.4624	5.3908	5.3427	5.3092	5.2847	5.2662	5.2517	5.2400
4	4.5448	4.3246	4.1908	4.1073	4.0506	4.0098	3.9790	3.9549	3.9357
5	4.0604	3.7797	3.6195	3.5202	3.4530	3.4045	3.3679	3.3393	3.3163
6	3.7760	3.4633	3.2888	3.1808	3.1075	3.0546	3.0145	2.9830	2.9577
7	3.5894	3.2574	3.0741	2.9605	2.8833	2.8274	2.7849	2.7516	2.7247
8	3.4579	3.1131	2.9238	2.8064	2.7265	2.6683	2.6241	2.5893	2.5612
9	3.3603	3.0065	2.8129	2.6927	2.6106	2.5509	2.5053	2.4694	2.4403
10	3.2850	2.9245	2.7277	2.6053	2.5216	2.4606	2.4140	2.3772	2.3473
11	3.2252	2.8595	2.6602	2.5362	2.4512	2.3891	2.3416	2.3040	2.2735
12	3.1765	2.8068	2.6055	2.4801	2.3940	2.3310	2.2828	2.2446	2.2135
13	3.1362	2.7632	2.5603	2.4337	2.3467	2.2830	2.2341	2.1953	2.1638
14	3.1022	2.7265	2.5222	2.3947	2.3069	2.2426	2.1931	2.1539	2.1220
15	3.0732	2.6952	2.4898	2.3614	2.2730	2.2081	2.1582	2.1185	2.0862
16	3.0481	2.6682	2.4618	2.3327	2.2438	2.1783	2.1280	2.0880	2.0553
17	3.0262	2.6446	2.4374	2.3077	2.2183	2.1524	2.1017	2.0613	2.0284
18	3.0070	2.6239	2.4160	2.2858	2.1958	2.1296	2.0785	2.0379	2.0047
19	2.9899	2.6056	2.3970	2.2663	2.1760	2.1094	2.0580	2.0171	1.9836
20	2.9747	2.5893	2.3801	2.2489	2.1582	2.0913	2.0397	1.9985	1.9649
21	2.9609	2.5746	2.3649	2.2333	2.1423	2.0751	2.0232	1.9819	1.9480
22	2.9486	2.5613	2.3512	2.2193	2.1279	2.0605	2.0084	1.9668	1.9327
23	2.9374	2.5493	2.3387	2.2065	2.1149	2.0472	1.9949	1.9531	1.9189
24	2.9271	2.5383	2.3274	2.1949	2.1030	2.0351	1.9826	1.9407	1.9063
25	2.9177	2.5283	2.3170	2.1843	2.0922	2.0241	1.9714	1.9292	1.8947
26	2.9091	2.5191	2.3075	2.1745	2.0822	2.0139	1.9610	1.9188	1.8841
27	2.9012	2.5106	2.2987	2.1655	2.0730	2.0045	1.9515	1.9091	1.8743
28	2.8939	2.5028	2.2906	2.1571	2.0645	1.9959	1.9427	1.9001	1.8652
29	2.8871	2.4955	2.2831	2.1494	2.0566	1.9878	1.9345	1.8918	1.8560
30	2.8807	2.4887	2.2761	2.1422	2.0492	1.9803	1.9269	1.8841	1.8498
40	2.8354	2.4404	2.2261	2.0909	1.9968	1.9269	1.8725	1.8289	1.7929
60	2.7914	2.3932	2.1774	2.0410	1.9457	1.8747	1.8194	1.7748	1.7380
120	2.7478	2.3473	2.1300	1.9923	1.8959	1.8238	1.7675	1.7220	1.6843
∞	2.7055	2.3026	2.0838	1.9449	1.8473	1.7741	1.7167	1.6702	1.6315

APPENDIX G (Continued)

$$\alpha = .10$$

m_2 \ m_1	10	12	15	20	24	30	40	60	120	∞
1	60.195	60.705	61.220	61.740	62.002	62.265	62.529	62.794	63.061	63.328
2	9.3916	9.4081	9.4247	9.4413	9.4496	9.4579	9.4663	9.4746	9.4829	9.4913
3	5.2304	5.2156	5.2003	5.1845	5.1764	5.1681	5.1597	5.1512	5.1425	5.1337
4	3.9199	3.8955	3.8689	3.8443	3.8310	3.8174	3.8036	3.7896	3.7753	3.7607
5	3.2974	3.2682	3.2380	3.2067	3.1905	3.1741	3.1573	3.1402	3.1228	3.1050
6	2.9369	2.9047	2.8712	2.8363	2.8183	2.8000	2.7812	2.7620	2.7423	2.7222
7	2.7025	2.6681	2.6322	2.5947	2.5753	2.5555	2.5351	2.5142	2.4928	2.4708
8	2.5380	2.5020	2.4642	2.4246	2.4041	2.3830	2.3614	2.3391	2.3162	2.2926
9	2.4163	2.3789	2.3396	2.2983	2.2768	2.2547	2.2320	2.2085	2.1843	2.1592
10	2.3226	2.2841	2.2435	2.2007	2.1784	2.1554	2.1317	2.1072	2.0818	2.0554
11	2.2482	2.2087	2.1671	2.1230	2.1000	2.0762	2.0516	2.0261	1.9997	1.9721
12	2.1878	2.1474	2.1049	2.0597	2.0360	2.0115	1.9861	1.9597	1.9323	1.9036
13	2.1376	2.0966	2.0532	2.0070	1.9827	1.9576	1.9315	1.9043	1.8759	1.8462
14	2.0954	2.0537	2.0095	1.9625	1.9377	1.9119	1.8852	1.8572	1.8280	1.7973
15	2.0593	2.0171	1.9722	1.9243	1.8990	1.8728	1.8454	1.8168	1.7867	1.7551
16	2.0281	1.9854	1.9399	1.8913	1.8656	1.8388	1.8108	1.7816	1.7507	1.7182
17	2.0009	1.9577	1.9117	1.8624	1.8362	1.8090	1.7805	1.7506	1.7191	1.6856
18	1.9770	1.9333	1.8868	1.8368	1.8103	1.7827	1.7537	1.7232	1.6910	1.6567
19	1.9557	1.9117	1.8647	1.8142	1.7873	1.7592	1.7298	1.6988	1.6659	1.6308
20	1.9367	1.8924	1.8449	1.7938	1.7667	1.7382	1.7083	1.6768	1.6433	1.6074
21	1.9197	1.8750	1.8272	1.7756	1.7481	1.7193	1.6890	1.6569	1.6228	1.5862
22	1.9043	1.8593	1.8111	1.7590	1.7312	1.7021	1.6714	1.6389	1.6042	1.5668
23	1.8903	1.8450	1.7964	1.7439	1.7159	1.6864	1.6554	1.6224	1.5871	1.5490
24	1.8775	1.8319	1.7831	1.7302	1.7019	1.6721	1.6407	1.6073	1.5715	1.5327
25	1.8658	1.8200	1.7708	1.7175	1.6890	1.6589	1.6272	1.5934	1.5570	1.5176
26	1.8550	1.8090	1.7596	1.7059	1.6771	1.6468	1.6147	1.5805	1.5437	1.5036
27	1.8451	1.7989	1.7492	1.6951	1.6662	1.6356	1.6032	1.5686	1.5313	1.4906
28	1.8359	1.7895	1.7395	1.6852	1.6560	1.6252	1.5925	1.5575	1.5198	1.4784
29	1.8274	1.7808	1.7306	1.6759	1.6465	1.6155	1.5825	1.5472	1.5090	1.4670
30	1.8195	1.7727	1.7223	1.6673	1.6377	1.6065	1.5732	1.5376	1.4989	1.4564
40	1.7627	1.7146	1.6624	1.6052	1.5741	1.5411	1.5056	1.4672	1.4248	1.3769
60	1.7070	1.6574	1.6034	1.5435	1.5107	1.4755	1.4373	1.3952	1.3476	1.2915
120	1.6524	1.6012	1.5450	1.4821	1.4472	1.4094	1.3676	1.3203	1.2646	1.1926
∞	1.5987	1.5458	1.4871	1.4206	1.3832	1.3419	1.2951	1.2400	1.1686	1.0000

APPENDIX G (Continued)

$$\alpha = .05$$

m_2 \ m_1	1	2	3	4	5	6	7	8	9
1	161.45	199.50	215.71	224.58	230.16	233.99	236.77	238.88	240.54
2	18.513	19.000	19.164	19.247	19.296	19.330	19.353	19.371	19.385
3	10.128	9.5521	9.2766	9.1172	9.0135	8.9406	8.8868	8.8452	8.8123
4	7.7086	6.9443	6.5914	6.3883	6.2560	6.1631	6.0942	6.0410	5.9988
5	6.6079	5.7861	5.4095	5.1922	5.0503	4.9503	4.8759	4.8183	4.7725
6	5.9874	5.1433	4.7571	4.5337	4.3874	4.2839	4.2066	4.1468	4.0990
7	5.5914	4.7374	4.3468	4.1203	3.9715	3.8660	3.7870	3.7257	3.6767
8	5.3177	4.4590	4.0662	3.8378	3.6875	3.5806	3.5005	3.4381	3.3881
9	5.1174	4.2565	3.8626	3.6331	3.4817	3.3738	3.2927	3.2296	3.1789
10	4.9646	4.1028	3.7083	3.4780	3.3258	3.2172	3.1355	3.0717	3.0204
11	4.8443	3.9823	3.5874	3.3567	3.2039	3.0946	3.0123	2.9480	2.8962
12	4.7472	3.8853	3.4903	3.2592	3.1059	2.9961	2.9134	2.8486	2.7964
13	4.6672	3.8056	3.4105	3.1791	3.0254	2.9153	2.8321	2.7669	2.7144
14	4.6001	3.7389	3.3439	3.1122	2.9582	2.8477	2.7642	2.6987	2.6458
15	4.5431	3.6823	3.2874	3.0556	2.9013	2.7905	2.7066	2.6408	2.5876
16	4.4940	3.6337	3.2389	3.0069	2.8524	2.7413	2.6572	2.5911	2.5377
17	4.4513	3.5915	3.1968	2.9647	2.8100	2.6987	2.6143	2.5480	2.4943
18	4.4139	3.5546	3.1599	2.9277	2.7729	2.6613	2.5767	2.5102	2.4563
19	4.3808	3.5219	3.1274	2.8951	2.7401	2.6283	2.5435	2.4768	2.4227
20	4.3513	3.4928	3.0984	2.8661	2.7109	2.5990	2.5140	2.4471	2.3928
21	4.3248	3.4668	3.0725	2.8401	2.6848	2.5727	2.4876	2.4205	2.3661
22	4.3009	3.4434	3.0491	2.8167	2.6613	2.5491	2.4638	2.3965	2.3419
23	4.2793	3.4221	3.0280	2.7955	2.6400	2.5277	2.4422	2.3748	2.3201
24	4.2597	3.4028	3.0088	2.7763	2.6207	2.5082	2.4226	2.3551	2.3002
25	4.2417	3.3852	2.9912	2.7587	2.6030	2.4904	2.4047	2.3371	2.2821
26	4.2252	3.3690	2.9751	2.7426	2.5868	2.4741	2.3883	2.3205	2.2655
27	4.2100	3.3541	2.9604	2.7278	2.5719	2.4591	2.3732	2.3053	2.2501
28	4.1960	3.3404	2.9467	2.7141	2.5581	2.4453	2.3593	2.2913	2.2360
29	4.1830	3.3277	2.9340	2.7014	2.5454	2.4324	2.3463	2.2782	2.2229
30	4.1709	3.3158	2.9223	2.6896	2.5336	2.4205	2.3343	2.2662	2.2107
40	4.0848	3.2317	2.8387	2.6060	2.4495	2.3359	2.2490	2.1802	2.1240
60	4.0012	3.1504	2.7581	2.5252	2.3683	2.2540	2.1665	2.0970	2.0401
120	3.9201	3.0718	2.6802	2.4472	2.2900	2.1750	2.0867	2.0164	1.9588
∞	3.8415	2.9957	2.6049	2.3719	2.2141	2.0986	2.0096	1.9384	1.8799

TABLE OF F DISTRIBUTIONS

APPENDIX G (Continued)

$\alpha = .05$

m_2 \ m_1	10	12	15	20	24	30	40	60	120	∞
1	241.88	243.91	245.95	248.01	249.05	250.09	251.14	252.20	253.25	254.32
2	19.396	19.413	19.429	19.446	19.454	19.462	19.471	19.479	19.487	19.496
3	8.7855	8.7446	8.7029	8.6602	8.6385	8.6166	8.5944	8.5720	8.5494	8.5265
4	5.9644	5.9117	5.8578	5.8025	5.7744	5.7459	5.7170	5.6878	5.6581	5.6281
5	4.7351	4.6777	4.6188	4.5581	4.5272	4.4957	4.4638	4.4314	4.3984	4.3650
6	4.0600	3.9999	3.9381	3.8742	3.8415	3.8082	3.7743	3.7398	3.7047	3.6688
7	3.6365	3.5747	3.5108	3.4445	3.4105	3.3758	3.3404	3.3043	3.2674	3.2298
8	3.3472	3.2840	3.2184	3.1503	3.1152	3.0794	3.0428	3.0053	2.9669	2.9276
9	3.1373	3.0729	3.0061	2.9365	2.9005	2.8637	2.8259	2.7872	2.7475	2.7067
10	2.9782	2.9130	2.8450	2.7740	2.7372	2.6996	2.6609	2.6211	2.5801	2.5379
11	2.8536	2.7876	2.7186	2.6464	2.6090	2.5705	2.5309	2.4901	2.4480	2.4045
12	2.7534	2.6866	2.6169	2.5436	2.5055	2.4663	2.4259	2.3842	2.3410	2.2962
13	2.6710	2.6037	2.5331	2.4589	2.4202	2.3803	2.3392	2.2966	2.2524	2.2064
14	2.6021	2.5342	2.4630	2.3879	2.3487	2.3082	2.2664	2.2230	2.1778	2.1307
15	2.5437	2.4753	2.4035	2.3275	2.2878	2.2468	2.2043	2.1601	2.1141	2.0658
16	2.4935	2.4247	2.3522	2.2756	2.2354	2.1938	2.1507	2.1058	2.0589	2.0096
17	2.4499	2.3807	2.3077	2.2304	2.1898	2.1477	2.1040	2.0584	2.0107	1.9604
18	2.4117	2.3421	2.2686	2.1906	2.1497	2.1071	2.0629	2.0166	1.9681	1.9168
19	2.3779	2.3080	2.2341	2.1555	2.1141	2.0712	2.0264	1.9796	1.9302	1.8780
20	2.3479	2.2776	2.2033	2.1242	2.0825	2.0391	1.9938	1.9464	1.8963	1.8432
21	2.3210	2.2504	2.1757	2.0960	2.0540	2.0102	1.9645	1.9165	1.8657	1.8117
22	2.2967	2.2258	2.1508	2.0707	2.0283	1.9842	1.9380	1.8895	1.8380	1.7831
23	2.2747	2.2036	2.1282	2.0476	2.0050	1.9605	1.9139	1.8649	1.8128	1.7570
24	2.2547	2.1834	2.1077	2.0267	1.9838	1.9390	1.8920	1.8424	1.7897	1.7331
25	2.2365	2.1649	2.0889	2.0075	1.9643	1.9192	1.8718	1.8217	1.7684	1.7110
26	2.2197	2.1479	2.0716	1.9898	1.9464	1.9010	1.8533	1.8027	1.7488	1.6906
27	2.2043	2.1323	2.0558	1.9736	1.9299	1.8842	1.8361	1.7851	1.7307	1.6717
28	2.1900	2.1179	2.0411	1.9586	1.9147	1.8687	1.8203	1.7689	1.7138	1.6541
29	2.1768	2.1045	2.0275	1.9446	1.9005	1.8543	1.8055	1.7537	1.6981	1.6377
30	2.1646	2.0921	2.0148	1.9317	1.8874	1.8409	1.7918	1.7396	1.6835	1.6223
40	2.0772	2.0035	1.9245	1.8389	1.7929	1.7444	1.6928	1.6373	1.5766	1.5089
60	1.9926	1.9174	1.8364	1.7480	1.7001	1.6491	1.5943	1.5343	1.4673	1.3893
120	1.9105	1.8337	1.7505	1.6587	1.6084	1.5543	1.4952	1.4290	1.3519	1.2539
∞	1.8307	1.7522	1.6664	1.5705	1.5173	1.4591	1.3940	1.3180	1.2214	1.0000

APPENDIX G (Continued)

$$\alpha = .025$$

m_2 \ m_1	1	2	3	4	5	6	7	8	9
1	647.79	799.50	864.16	899.58	921.85	937.11	948.22	956.66	963.28
2	38.506	39.000	39.165	39.248	39.298	39.331	39.355	39.373	39.387
3	17.443	16.044	15.439	15.101	14.885	14.735	14.624	14.540	14.473
4	12.218	10.649	9.9792	9.6045	9.3645	9.1973	9.0741	8.9796	8.9047
5	10.007	8.4336	7.7636	7.3879	7.1464	6.9777	6.8531	6.7572	6.6810
6	8.8131	7.2598	6.5988	6.2272	5.9876	5.8197	5.6955	5.5996	5.5234
7	8.0727	6.5415	5.8898	5.5226	5.2852	5.1186	4.9949	4.8994	4.8232
8	7.5709	6.0595	5.4160	5.0526	4.8173	4.6517	4.5286	4.4332	4.3572
9	7.2093	5.7147	5.0781	4.7181	4.4844	4.3197	4.1971	4.1020	4.0260
10	6.9367	5.4564	4.8256	4.4683	4.2361	4.0721	3.9498	3.8549	3.7790
11	6.7241	5.2559	4.6300	4.2751	4.0440	3.8807	3.7586	3.6638	3.5879
12	6.5538	5.0959	4.4742	4.1212	3.8911	3.7283	3.6065	3.5118	3.4358
13	6.4143	4.9653	4.3472	3.9959	3.7667	3.6043	3.4827	3.3880	3.3120
14	6.2979	4.8567	4.2417	3.8919	3.6634	3.5014	3.3799	3.2853	3.2093
15	6.1995	4.7650	4.1528	3.8043	3.5764	3.4147	3.2934	3.1987	3.1227
16	6.1151	4.6867	4.0768	3.7294	3.5021	3.3406	3.2194	3.1248	3.0488
17	6.0420	4.6189	4.0112	3.6648	3.4379	3.2767	3.1556	3.0610	2.9849
18	5.9781	4.5597	3.9539	3.6083	3.3820	3.2209	3.0999	3.0053	2.9291
19	5.9216	4.5075	3.9034	3.5587	3.3327	3.1718	3.0509	2.9563	2.8800
20	5.8715	4.4613	3.8587	3.5147	3.2891	3.1283	3.0074	2.9128	2.8365
21	5.8266	4.4199	3.8188	3.4754	3.2501	3.0895	2.9686	2.8740	2.7977
22	5.7863	4.3828	3.7829	3.4401	3.2151	3.0546	2.9338	2.8392	2.7628
23	5.7498	4.3492	3.7505	3.4083	3.1835	3.0232	2.9024	2.8077	2.7313
24	5.7167	4.3187	3.7211	3.3794	3.1548	2.9946	2.8738	2.7791	2.7027
25	5.6864	4.2909	3.6943	3.3530	3.1287	2.9685	2.8478	2.7531	2.6766
26	5.6586	4.2655	3.6697	3.3289	3.1048	2.9447	2.8240	2.7293	2.6528
27	5.6331	4.2421	3.6472	3.3067	3.0828	2.9228	2.8021	2.7074	2.6309
28	5.6096	4.2205	3.6264	3.2863	3.0625	2.9027	2.7820	2.6872	2.6106
29	5.5878	4.2006	3.6072	3.2674	3.0438	2.8840	2.7633	2.6686	2.5919
30	5.5675	4.1821	3.5894	3.2499	3.0265	2.8667	2.7460	2.6513	2.5746
40	5.4239	4.0510	3.4633	3.1261	2.9037	2.7444	2.6238	2.5289	2.4519
60	5.2857	3.9253	3.3425	3.0077	2.7863	2.6274	2.5068	2.4117	2.3344
120	5.1524	3.8046	3.2270	2.8943	2.6740	2.5154	2.3948	2.2994	2.2217
∞	5.0239	3.6889	3.1161	2.7858	2.5665	2.4082	2.2875	2.1918	2.1136

APPENDIX G (Continued)

$$\alpha = .025$$

m_2 \ m_1	10	12	15	20	24	30	40	60	120	∞
1	968.63	976.71	984.87	993.10	997.25	1001.4	1005.6	1009.8	1014.0	1018.3
2	39.398	39.415	39.431	39.448	39.456	39.465	39.473	39.481	39.490	39.498
3	14.419	14.337	14.253	14.167	14.124	14.081	14.037	13.992	13.947	13.902
4	8.8439	8.7512	8.6565	8.5599	8.5109	8.4613	8.4111	8.3604	8.3092	8.2573
5	6.6192	6.5246	6.4277	6.3285	6.2780	6.2269	6.1751	6.1225	6.0693	6.0153
6	5.4613	5.3662	5.2687	5.1684	5.1172	5.0652	5.0125	5.9589	4.9045	4.8491
7	4.7611	4.6658	4.5678	4.4667	4.4150	4.3624	4.3089	4.2544	4.1989	4.1423
8	4.2951	4.1997	4.1012	3.9995	3.9472	3.8940	3.8398	3.7844	3.7279	3.6702
9	3.9639	3.8682	3.7694	3.6669	3.6142	3.5604	3.5055	3.4493	3.3918	3.3329
10	3.7168	3.6209	3.5217	3.4186	3.3654	3.3110	3.2554	3.1984	3.1399	3.0798
11	3.5257	3.4296	3.3299	3.2261	3.1725	3.1176	3.0613	3.0035	2.9441	2.8828
12	3.3736	3.2773	3.1772	3.0728	3.0187	2.9633	2.9063	2.8478	2.7874	2.7249
13	3.2497	3.1532	3.0527	2.9477	2.8932	2.8373	2.7797	2.7204	2.6590	2.5955
14	3.1469	3.0501	2.9493	2.8437	2.7888	2.7324	2.6742	2.6142	2.5519	2.4872
15	3.0602	2.9633	2.8621	2.7559	2.7006	2.6437	2.5850	2.5242	2.4611	2.3953
16	2.9862	2.8890	2.7875	2.6808	2.6252	2.5678	2.5085	2.4471	2.3831	2.3163
17	2.9222	2.8249	2.7230	2.6158	2.5598	2.5021	2.4422	2.3801	2.3153	2.2474
18	2.8664	2.7689	2.6667	2.5590	2.5027	2.4445	2.3842	2.3214	2.2558	2.1869
19	2.8173	2.7196	2.6171	2.5089	2.4523	2.3937	2.3329	2.2695	2.2032	2.1333
20	2.7737	2.6758	2.5731	2.4645	2.4076	2.3486	2.2873	2.2234	2.1562	2.0853
21	2.7348	2.6368	2.5338	2.4247	2.3675	2.3082	2.2465	2.1819	2.1141	2.0422
22	2.6998	2.6017	2.4984	2.3890	2.3315	2.2718	2.2097	2.1446	2.0760	2.0032
23	2.6682	2.5699	2.4665	2.3567	2.2989	2.2389	2.1763	2.1107	2.0415	1.9677
24	2.6396	2.5412	2.4374	2.3273	2.2693	2.2090	2.1460	2.0799	2.0099	1.9353
25	2.6135	2.5149	2.4110	2.3005	2.2422	2.1816	2.1183	2.0517	1.9811	1.9055
26	2.5895	2.4909	2.3867	2.2759	2.2174	2.1565	2.0928	2.0257	1.9545	1.8781
27	2.5676	2.4688	2.3644	2.2533	2.1946	2.1334	2.0693	2.0018	1.9299	1.8527
28	2.5473	2.4484	2.3438	2.2324	2.1735	2.1121	2.0477	1.9796	1.9072	1.8291
29	2.5286	2.4295	2.3248	2.2131	2.1540	2.0923	2.0276	1.9591	1.8861	1.8072
30	2.5112	2.4120	2.3072	2.1952	2.1359	2.0739	2.0089	1.9400	1.8664	1.7867
40	2.3882	2.2882	2.1819	2.0677	2.0069	1.9429	1.8752	1.8028	1.7242	1.6371
60	2.2702	2.1692	2.0613	1.9445	1.8817	1.8152	1.7440	1.6668	1.5810	1.4822
120	2.1570	2.0548	1.9450	1.8249	1.7597	1.6899	1.6141	1.5299	1.4327	1.3104
∞	2.0483	1.9447	1.8326	1.7085	1.6402	1.5660	1.4835	1.3883	1.2684	1.0000

APPENDIX G (Continued)

$$\alpha = .01$$

m_2 \ m_1	1	2	3	4	5	6	7	8	9
1	4052.2	4999.5	5403.3	5624.6	5763.7	5859.0	5928.3	5981.6	6022.5
2	98.503	99.000	99.166	99.249	99.299	99.332	99.356	99.374	99.388
3	34.116	30.817	29.457	28.710	28.237	27.911	27.672	27.489	27.345
4	21.198	18.000	16.694	15.977	15.522	15.207	14.976	14.799	14.659
5	16.258	13.274	12.060	11.392	10.967	10.672	10.456	10.289	10.158
6	13.745	10.925	9.7795	9.1483	8.7459	8.4661	8.2600	8.1016	7.9761
7	12.246	9.5466	8.4513	7.8467	7.4604	7.1914	6.9928	6.8401	6.7188
8	11.259	8.6491	7.5910	7.0060	6.6318	6.3707	6.1776	6.0289	5.9106
9	10.561	8.0215	6.9919	6.4221	6.0569	5.8018	5.6129	5.4671	5.3511
10	10.044	7.5594	6.5523	5.9943	5.6363	5.3858	5.2001	5.0567	4.9424
11	9.6460	7.2057	6.2167	5.6683	5.3160	5.0692	4.8861	4.7445	4.6315
12	9.3302	6.9266	5.9526	5.4119	5.0643	4.8206	4.6395	4.4994	4.3875
13	9.0738	6.7010	5.7394	5.2053	4.8616	4.6204	4.4410	4.3021	4.1911
14	8.8616	6.5149	5.5639	5.0354	4.6950	4.4558	4.2779	4.1399	4.0297
15	8.6831	6.3589	5.4170	4.8932	4.5556	4.3183	4.1415	4.0045	3.8948
16	8.5310	6.2262	5.2922	4.7726	4.4374	4.2016	4.0259	3.8896	3.7804
17	8.3997	6.1121	5.1850	4.6690	4.3359	4.1015	3.9267	3.7910	3.6822
18	8.2854	6.0129	5.0919	4.5790	4.2479	4.0146	3.8406	3.7054	3.5971
19	8.1850	5.9259	5.0103	4.5003	4.1708	3.9386	3.7653	3.6305	3.5225
20	8.0960	5.8489	4.9382	4.4307	4.1027	3.8714	3.6987	3.5644	3.4567
21	8.0166	5.7804	4.8740	4.3688	4.0421	3.8117	3.6396	3.5056	3.3981
22	7.9454	5.7190	4.8166	4.3134	3.9880	3.7583	3.5867	3.4530	3.3458
23	7.8811	5.6637	4.7649	4.2635	3.9392	3.7102	3.5390	3.4057	3.2986
24	7.8229	5.6136	4.7181	4.2184	3.8951	3.6667	3.4959	3.3629	3.2560
25	7.7698	5.5680	4.6755	4.1774	3.8550	3.6272	3.4568	3.3239	3.2172
26	7.7213	5.5263	4.6366	4.1400	3.8183	3.5911	3.4210	3.2884	3.1818
27	7.6767	5.4881	4.6009	4.1056	3.7848	3.5580	3.3882	3.2558	3.1494
28	7.6356	5.4529	4.5681	4.0740	3.7539	3.5276	3.3581	3.2259	3.1195
29	7.5976	5.4205	4.5378	4.0449	3.7254	3.4995	3.3302	3.1982	3.0920
30	7.5625	5.3904	4.5097	4.0179	3.6990	3.4735	3.3045	3.1726	3.0665
40	7.3141	5.1785	4.3126	3.8283	3.5138	3.2910	3.1238	2.9930	2.8876
60	7.0771	4.9774	4.1259	3.6491	3.3389	3.1187	2.9530	2.8233	2.7185
120	6.8510	4.7865	3.9493	3.4796	3.1735	2.9559	2.7918	2.6629	2.5586
∞	6.6349	4.6052	3.7816	3.3192	3.0173	2.8020	2.6393	2.5113	2.4073

APPENDIX G (Continued)

$\alpha = .01$

m_2 \ m_1	10	12	15	20	24	30	40	60	120	∞
1	6055.8	6106.3	6157.3	6208.7	6234.6	6260.7	6286.8	6313.0	6339.4	6366.0
2	99.399	99.416	99.432	99.449	99.458	99.466	99.474	99.483	99.491	99.501
3	27.229	27.052	26.872	26.690	26.598	26.505	26.411	26.316	26.221	26.125
4	14.546	14.374	14.198	14.020	13.929	13.838	13.745	13.652	13.558	13.463
5	10.051	9.8883	9.7222	9.5527	9.4665	9.3793	9.2912	9.2020	9.1118	9.0204
6	7.8741	7.7183	7.5590	7.3958	7.3127	7.2285	7.1432	7.0568	6.9690	6.8801
7	6.6201	6.4691	6.3143	6.1554	6.0743	5.9921	5.9084	5.8236	5.7372	5.6495
8	5.8143	5.6668	5.5151	5.3591	5.2793	5.1981	5.1156	5.0316	4.9460	4.8588
9	5.2565	5.1114	4.9621	4.8080	4.7290	4.6486	4.5667	4.4831	4.3978	4.3105
10	4.8492	4.7059	4.5582	4.4054	4.3269	4.2469	4.1653	4.0819	3.9965	3.9090
11	4.5393	4.3974	4.2509	4.0990	4.0209	3.9411	3.8596	3.7761	3.6904	3.6025
12	4.2961	4.1553	4.0096	3.8584	3.7805	3.7008	3.6192	3.5355	3.4494	3.3608
13	4.1003	3.9603	3.8154	3.6646	3.5868	3.5070	3.4253	3.3413	3.2548	3.1654
14	3.9394	3.8001	3.6557	3.5052	3.4274	3.3476	3.2656	3.1813	3.0942	3.0040
15	3.8049	3.6662	3.5222	3.3719	3.2940	3.2141	3.1319	3.0471	2.9595	2.8684
16	3.6909	3.5527	3.4089	3.2588	3.1808	3.1007	3.0182	2.9330	2.8447	2.7528
17	3.5931	3.4552	3.3117	3.1615	3.0835	3.0032	2.9205	2.8348	2.7459	2.6530
18	3.5082	3.3706	3.2273	3.0771	2.9990	2.9185	2.8354	2.7493	2.6597	2.5660
19	3.4338	3.2965	3.1533	3.0031	2.9249	2.8442	2.7608	2.6742	2.5839	2.4893
20	3.3682	3.2311	3.0880	2.9377	2.8594	2.7785	2.6947	2.6077	2.5168	2.4212
21	3.3098	3.1729	3.0299	2.8796	2.8011	2.7200	2.6359	2.5484	2.4568	2.3603
22	3.2576	3.1209	2.9780	2.8274	2.7488	2.6675	2.5831	2.4951	2.4029	2.3055
23	3.2106	3.0740	2.9311	2.7805	2.7017	2.6202	2.5355	2.4471	2.3542	2.2559
24	3.1681	3.0316	2.887	2.7380	2.6591	2.5773	2.4923	2.4035	2.3099	2.2107
25	3.1294	2.9931	2.8502	2.6993	2.6203	2.5383	2.4530	2.3637	2.2695	2.1694
26	3.0941	2.9579	2.8150	2.6640	2.5848	2.5026	2.4170	2.3273	2.2325	2.1315
27	3.0618	2.9256	2.7827	2.6316	2.5522	2.4699	2.3840	2.2938	2.1984	2.0965
28	3.0320	2.8959	2.7530	2.6017	2.5223	2.4397	2.3535	2.2629	2.1670	2.0642
29	3.0045	2.8685	2.7256	2.5742	2.4946	2.4118	2.3253	2.2344	2.1378	2.0342
30	2.9791	2.8431	2.7002	2.5487	2.4689	2.3860	2.2992	2.2079	2.1107	2.0062
40	2.8005	2.6648	2.5216	2.3689	2.2880	2.2034	2.1142	2.0194	1.9172	1.8047
60	2.6318	2.4961	2.3523	2.1978	2.1154	2.0285	1.9360	1.8363	1.7263	1.6006
120	2.4721	2.3363	2.1915	2.0346	1.9500	1.8600	1.7628	1.6557	1.5330	1.3805
∞	2.3209	2.1848	2.0385	1.8783	1.7908	1.6964	1.5923	1.4730	1.3246	1.0000

TABLE OF F DISTRIBUTIONS

APPENDIX G (Continued)

$\alpha = .005$

m_2 \ m_1	1	2	3	4	5	6	7	8	9
1	16211	20000	21615	22500	23056	23437	23715	23925	24091
2	198.50	199.00	199.17	199.25	199.30	199.33	199.36	199.37	199.39
3	55.552	49.799	47.467	46.195	45.392	44.838	44.434	44.126	43.882
4	31.333	26.284	24.259	23.155	22.456	21.975	21.622	21.352	21.139
5	22.785	18.314	16.530	15.556	14.940	14.513	14.200	13.961	13.772
6	18.635	14.544	12.917	12.028	11.464	11.073	10.786	10.566	10.391
7	16.236	12.404	10.882	10.050	9.5221	9.1554	8.8854	8.6781	8.5138
8	14.688	11.042	9.5965	8.8051	8.3018	7.9520	7.6942	7.4960	7.3386
9	13.614	10.107	8.7171	7.9559	7.4711	7.1338	6.8849	6.6933	6.5411
10	12.826	9.4270	8.0807	7.3428	6.8723	6.5446	6.3025	6.1159	5.9676
11	12.226	8.9122	7.6004	6.8809	6.4217	6.1015	5.8648	5.6821	5.5368
12	11.754	8.5096	7.2258	6.5211	6.0711	5.7570	5.5245	5.3451	5.2021
13	11.374	8.1865	6.9257	6.2335	5.7910	5.4819	5.2529	5.0761	4.9351
14	11.060	7.9217	6.6803	5.9984	5.5623	5.2574	5.0313	4.8566	4.7173
15	10.798	7.7008	6.4760	5.8029	5.3721	5.0708	4.8473	4.6743	4.5364
16	10.575	7.5138	6.3034	5.6378	5.2117	4.9134	4.6920	4.5207	4.3838
17	10.384	7.3536	6.1556	5.4967	5.0746	4.7789	4.5594	4.3893	4.2535
18	10.218	7.2148	6.0277	5.3746	4.9560	4.6628	4.4448	4.2759	4.1410
19	10.073	7.0935	5.9161	5.2681	4.8526	4.5614	4.3448	4.1770	4.0428
20	9.9439	6.9865	5.8177	5.1743	4.7616	4.4721	4.2569	4.0900	3.9564
21	9.8295	6.8914	5.7304	5.0911	4.6808	4.3931	4.1789	4.0128	3.8799
22	9.7271	6.8064	5.6524	5.0168	4.6088	4.3225	4.1094	3.9440	3.8116
23	9.6348	6.7300	5.5823	4.9500	4.5441	4.2591	4.0469	3.8822	3.7502
24	9.5513	6.6610	5.5190	4.8898	4.4857	4.2019	3.9905	3.8264	3.6949
25	9.4753	6.5982	5.4615	4.8351	4.4327	4.1500	3.9394	3.7758	3.6447
26	9.4059	6.5409	5.4091	4.7852	4.3844	4.1027	3.8928	3.7297	3.5989
27	9.3423	6.4885	5.3611	4.7396	4.3402	4.0594	3.8501	3.6875	3.5571
28	9.2838	6.4403	5.3170	4.6977	4.2996	4.0197	3.8110	3.6487	3.5186
29	9.2297	6.3958	5.2764	4.6591	4.2622	3.9830	3.7749	3.6130	3.4832
30	9.1797	6.3547	5.2388	4.6233	4.2276	3.9492	3.7416	3.5801	3.4505
40	8.8278	6.0664	4.9759	4.3738	3.9860	3.7129	3.5088	3.3498	3.2220
60	8.4946	5.7950	4.7290	4.1399	3.7600	3.4918	3.2911	3.1344	3.0083
120	8.1790	5.5393	4.4973	3.9207	3.5482	3.2849	3.0874	2.9330	2.8083
∞	7.8794	5.2983	4.2794	3.7151	3.3499	3.0913	2.8968	2.7444	2.6210

APPENDIX G (Continued)

$$\alpha = .005$$

m_2 \ m_1	10	12	15	20	24	30	40	60	120	∞
1	24224	24426	24630	24836	24940	25044	25148	25253	25359	25465
2	199.40	199.42	199.43	199.45	199.46	199.47	199.47	199.48	199.49	199.51
3	43.686	43.387	43.085	42.778	42.622	42.466	42.308	42.149	41.989	41.829
4	20.967	20.705	20.438	20.167	20.030	19.892	19.752	19.611	19.468	19.325
5	13.618	13.384	13.146	12.903	12.780	12.656	12.530	12.402	12.274	12.144
6	10.250	10.034	9.8140	9.5888	9.4741	9.3583	9.2408	9.1219	9.0015	8.8793
7	8.3803	8.1764	7.9678	7.7540	7.6450	7.5345	7.4225	7.3088	7.1933	7.0760
8	7.2107	7.0149	6.8143	6.6082	6.5029	6.3961	6.2875	6.1772	6.0649	5.9505
9	6.4171	6.2274	6.0325	5.8318	5.7292	5.6248	5.5186	5.4104	5.3001	5.1875
10	5.8467	5.6613	5.4707	5.2740	5.1732	5.0705	4.9659	4.8592	4.7501	4.6385
11	5.4182	5.2363	5.0489	4.8552	4.7557	4.6543	4.5508	4.4450	4.3367	4.2256
12	5.0855	4.9063	4.7214	4.5299	4.4315	4.3309	4.2282	4.1229	4.0149	3.9039
13	4.8199	4.6429	4.4600	4.2703	4.1726	4.0727	3.9704	3.8655	3.7577	3.6465
14	4.6034	4.4281	4.2468	4.0585	3.9614	3.8619	3.7600	3.6553	3.5473	3.4359
15	4.4236	4.2498	4.0698	3.8826	3.7859	3.6867	3.5850	3.4803	3.3722	3.2602
16	4.2719	4.0994	3.9205	3.7342	3.6378	3.5388	3.4372	3.3324	3.2240	3.1115
17	4.1423	3.9709	3.7929	3.6073	3.5112	3.4124	3.3107	3.2058	3.0971	2.9839
18	4.0305	3.8599	3.6827	3.4977	3.4017	3.3030	3.2014	3.0962	2.9871	2.8732
19	3.9329	3.7631	3.5866	3.4020	3.3062	3.2075	3.1058	3.0004	2.8908	2.7762
20	3.8470	3.6779	3.5020	3.3178	3.2220	3.1234	3.0215	2.9159	2.8058	2.6904
21	3.7709	3.6024	3.4270	3.2431	3.1474	3.0488	2.9467	2.8408	2.7302	2.6140
22	3.7030	3.5350	3.3600	3.1764	3.0807	2.9821	2.8799	2.7736	2.6625	2.5455
23	3.6420	3.4745	3.2999	3.1165	3.0208	2.9221	2.8198	2.7132	2.6016	2.4837
24	3.5870	3.4199	3.2456	3.0624	2.9667	2.8679	2.7654	2.6585	2.5463	2.4276
25	3.5370	3.3704	3.1963	3.0133	2.9176	2.8187	2.7160	2.6088	2.4960	2.3765
26	3.4916	3.3252	3.1515	2.9695	2.8728	2.7738	2.6709	2.5633	2.4501	2.3297
27	3.4499	3.2839	3.1104	2.9275	2.8318	2.7327	2.6296	2.5217	2.4078	2.2867
28	3.4117	3.2460	3.0727	2.8899	2.7941	2.6949	2.5916	2.4834	2.3689	2.2469
29	3.3765	3.2111	3.0379	2.8551	2.7594	2.6601	2.5565	2.4479	2.3330	2.2102
30	3.3440	3.1787	3.0057	2.8230	2.7272	2.6278	2.5241	2.4151	2.2997	2.1760
40	3.1167	2.9531	2.7811	2.5984	2.5020	2.4015	2.2958	2.1838	2.0635	1.9318
60	2.9042	2.7419	2.5705	2.3872	2.2898	2.1874	2.0789	1.9622	1.8341	1.6885
120	2.7052	2.5439	2.3727	2.1881	2.0890	1.9839	1.8709	1.7469	1.6055	1.4311
∞	2.5188	2.3583	2.1868	1.9998	1.8983	1.7891	1.6691	1.5325	1.3637	1.0000

Source: I. Guttman and S. S. Wilks, *Introduction to Engineering Statistics,* John Wiley & Sons, New York, 1965, pp. 320–29.

APPENDIX H

Percentage Points of Run Distribution

VALUES OF $r_{N;\alpha}$ SUCH THAT Prob $[R > r_{N;\alpha}] = \alpha$

N	α					
	0.99	0.975	0.95	0.05	0.025	0.01
5	2	2	3	8	9	9
6	2	3	3	10	10	11
7	3	3	4	11	12	12
8	4	4	5	12	13	13
9	4	5	6	13	14	15
10	5	6	6	15	15	16
11	6	7	7	16	16	17
12	7	7	8	17	18	18
13	7	8	9	18	19	20
14	8	9	10	19	20	21
15	9	10	11	20	21	22
16	10	11	11	22	22	23
18	11	12	13	24	25	26
20	13	14	15	26	27	28
25	17	18	19	32	33	34
30	21	22	24	37	39	40
35	25	27	28	43	44	46
40	30	31	33	48	50	51
45	34	36	37	54	55	57
50	38	40	42	59	61	63
55	43	45	46	65	66	68
60	47	49	51	70	72	74
65	52	54	56	75	77	79
70	56	58	60	81	83	85
75	61	63	65	86	88	90
80	65	68	70	91	93	96
85	70	72	74	97	99	101
90	74	77	79	102	104	107
95	79	82	84	107	109	112
100	84	86	88	113	115	117

Source: J. S. Bendat and A. G. Piersol, *Random Data: Analysis and Measurement and Procedures,* John Wiley & Sons, New York, 1971, p. 396.

APPENDIX I

CRITICAL VALUES OF THE DURBIN–WATSON STATISTIC

TABLE I.1 CRITICAL VALUES OF THE DURBIN–WATSON STATISTIC

Sample Size	Probability in Lower Tail (Significance Level = α)	k = Number of Regressors (Excluding the Intercept)									
		1		2		3		4		5	
		d_L	d_U	d_L	d_U	d_L	d_U	d_L	d_U	d_L	d_U
15	.01	.81	1.07	.70	1.25	.59	1.46	.49	1.70	.39	1.96
	.025	.95	1.23	.83	1.40	.71	1.61	.59	1.84	.48	2.09
	.05	1.08	1.36	.95	1.54	.82	1.75	.69	1.97	.56	2.21
20	.01	.95	1.15	.86	1.27	.77	1.41	.63	1.57	.60	1.74
	.025	1.08	1.28	.99	1.41	.89	1.55	.79	1.70	.70	1.87
	.05	1.20	1.41	1.10	1.54	1.00	1.68	.90	1.83	.79	1.99
25	.01	1.05	1.21	.98	1.30	.90	1.41	.83	1.52	.75	1.65
	.025	1.13	1.34	1.10	1.43	1.02	1.54	.94	1.65	.86	1.77
	.05	1.20	1.45	1.21	1.55	1.12	1.66	1.04	1.77	.95	1.89
30	.01	1.13	1.26	1.07	1.34	1.01	1.42	.94	1.51	.88	1.61
	.025	1.25	1.38	1.18	1.46	1.12	1.54	1.05	1.63	.98	1.73
	.05	1.35	1.49	1.28	1.57	1.21	1.65	1.14	1.74	1.07	1.83
40	.01	1.25	1.34	1.20	1.40	1.15	1.46	1.10	1.52	1.05	1.58
	.025	1.35	1.45	1.30	1.51	1.25	1.57	1.20	1.63	1.15	1.69
	.05	1.44	1.54	1.39	1.60	1.34	1.66	1.29	1.72	1.23	1.79
50	.01	1.32	1.40	1.28	1.45	1.24	1.49	1.20	1.54	1.16	1.59
	.025	1.42	1.50	1.38	1.54	1.34	1.59	1.30	1.64	1.26	1.69
	.05	1.50	1.59	1.46	1.63	1.42	1.67	1.38	1.72	1.34	1.77
60	.01	1.38	1.45	1.35	1.48	1.32	1.52	1.28	1.56	1.25	1.60
	.025	1.47	1.54	1.44	1.57	1.40	1.61	1.37	1.65	1.33	1.69
	.05	1.55	1.62	1.51	1.65	1.48	1.69	1.44	1.73	1.41	1.77
80	.01	1.47	1.52	1.44	1.54	1.42	1.57	1.39	1.60	1.36	1.62
	.025	1.54	1.59	1.52	1.62	1.49	1.65	1.47	1.67	1.44	1.70
	.05	1.61	1.66	1.59	1.69	1.56	1.72	1.53	1.74	1.51	1.77
100	.01	1.52	1.56	1.50	1.58	1.48	1.60	1.45	1.63	1.44	1.65
	.025	1.59	1.63	1.57	1.65	1.55	1.67	1.53	1.70	1.51	1.72
	.05	1.65	1.69	1.63	1.72	1.61	1.74	1.59	1.76	1.57	1.78

Source: D. C. Montgomery and E. A. Peck, *Introduction to Linear Regression Analysis,* John Wiley & Sons, New York, 1982, p. 478.

INDEX

INDEX

Aliasing error (effect), 193, 195
Almost periodic process, 142
Almost sure convergence, 89
Alternate hypothesis, 514
Analog communication system, 325
  effects of noise, 323
  preemphasis filtering, 329
  receiver design, 325
Analog to digital converters, 196, 199
Analytical description of a random process, 121
Analytic signal, 319
A posteriori probability, 344, 345, 492
Approximation:
  of distribution of Y, 81
  of Gaussian probabilities, 87
A priori probability, 346, 351, 475, 492, 517
Arrival times, 295, 296
Assumptions for a point process, 297
Autocorrelation function, 122
  bounds on, 143, 340
  complex process, 122
  continuity, 144
  derivative of a process, 163, 165
  ergodicity, 179
  estimation of, 566
  input–output relations for, 221, 223, 229
  linear system response, 221, 228, 229
  Poisson process, 132, 302
  properties, 143, 206, 207
  random binary waveform, 135
  random walk, 129
  relationship to power spectrum, 145
  shot noise, 310
  time averaged, 169
  Wiener–Levy process, 130
Autocovariance function, 122
Autoregressive integrated moving average (ARIMA) model, 588
Autoregressive moving average (ARMA) process, 271
  ARMA (1, 1) model, 273
  ARMA (p, q) model, 273
  autocorrelation, 272
  estimators of, 585, 605
  model, 271
  power spectral densities, 271
  properties of, 271–275
  transfer function, 271
Autoregressive (AR) process, 245, 250, 421
  autocorrelation coefficient, 254, 259, 261
  autocorrelation function, 253, 257, 258, 261
  estimators of parameters, 594
  first-order, 252
  mean, 253, 257, 261

model, 250
  partial autocorrelation coefficient (p.a.c.c.), 263, 264
  power spectral density, 255, 260, 262
  preliminary parameter estimates, 606
  second-order, 256
  transfer function, 251
  variance, 253, 257, 259
  Yule–Walker equations, 261
Average value, *see* Expected values
Axioms of probability, 12, 13

Backshift operator, 255, 266
Baire functions, 56
Band-limited processes, *see* Bandpass processes
Band-limited white noise, 314
  properties, 315
Bandpass processes, 146, 209
  amplitude and phase, 322
  definition, 146
  Hilbert transforms, 319
  quadrature form, 317, 322
Bandwidth, 146, 209, 214, 240
  definition, 147
  effective, 147
  of filters, 209
  noise bandwidth, 248
  of a power spectrum, 147
  rms bandwidth, 209, 309
  of signals, 146
Bartlett window, 583
Bayes decision rule, 348, 349
Bayesian estimators, 491
Bayes' rule, 19, 25, 31, 468
  densities, 37
Bernoulli random variable, 206
Bias of an estimator, 493
  for correlation, 568
  definition, 494
  for histograms, 497
  of periodogram, 572
  of smoothed estimators, 584
  for spectral density, 572
Binary communication system, 31, 357
Binary detection, 343, 352, 514
  colored noise, 361
  continuous observations, 355
  correlation receiver, 356
  error probabilities, 346
  MAP decision rule, 345
  matched filter, 355
  multiple observations, 352
  single observation, 343
  unknown parameters, 364

654 INDEX

Binomial random variable, 29, 100
  mean, 29, 100
  probability, 29
  variance, 29, 100
Birth and death process, 292
Bivariate Gaussian, 46
  conditional mean, 103, 395
  conditional variance, 395
Blackman–Tukey window, 583
Boltzmann's constant, 314
Bounds on Probabilities, 78–81
  Chernoff bound, 78
  Union bound, 79
Box–Jenkins procedure, 586
Brownian motion, 127
Butterworth filters, see Filters

Campbell's theorem, 310
Cartesian product, 16
Cauchy criterion, 94, 108, 163
Cauchy density function, 102
Causal system, 216, 217
Central limit theorem, 89, 90, 91
Chain rule of probability, 17
Chapman–Kolmogorov equation, 205, 284, 289
Characteristic function, 39, 40, 102
  definition, 39
  inversion, 40
  moment generating property, 40
  normal random variable, 41, 42, 104
  for several random variables, 40
  for sum of independent random variables, 41, 42, 67, 68, 102
Chebychev's inequality, see Tchebycheff inequality
Chernoff bound, 78
Chi-square distribution, 505, 549, 576, 609
  definition, 506
  degrees of freedom, 506
  goodness-of-fit test using, 523, 545, 609
  mean and variance of, 507
  plot of, 508
  statistical table for, 632
Chi-square tests, 523
  degrees of freedom, 526
Classical definition of probabilities, 13
Classification of random processes, 117
Classification of systems, 216
Class of sets, 12
Coherence function, 149
Colored noise, 361
Combined experiments, 16
Combined sample space, 16
Communication systems:
  analog, 325
  digital, 330

Complement of an event (set), 11
Complex envelope, 118
Complex random variable, 46, 47
Complex-valued random process, 118, 210
Conditional expected value, 28, 38, 48, 101, 102, 395
  definition, 28
  as a random variable, 397
Conditional probability, 16
Conditional probability density function, 36
  definition, 37
  multivariate Gaussian, 51, 103
Conditional probability mass function, 25
Conditional variance, 395, 468
Confidence interval (limits), 503
Consistent estimator, 502
Continuity, 160
  of autocorrelation, 144
  in mean square, 161, 162
Continuous random process, 177
Continuous random sequence, 177
Continuous random variable, 33
Continuous-time process, 117
Convergence, 88–95
  almost everywhere, 88–95
  almost sure, 89
  in distribution, 89, 95
  everywhere, 88
  in mean square, 94, 95
  in probability, 93, 95
  sums, 91
Convolution, 326
  of density functions, 67
  discrete form, 219
  integral form, 227
Correlation, 27, 38
  of independent random variables, 27
  of orthogonal random variables, 27
  of random variables, 27
Correlation coefficient, 27, 122, 124, 398
Correlation function, 122
  autocorrelation, 122
  autocovariance, 122
  for complex valued process, 122
  cross-correlation, 122
  cross-covariance, 122
  of derivative of a process, 165
  properties, 143
  for stationary process, 143
  time averaged, 169
Correlation time, 148
Cost of errors, 348
Countable set, 9
Counting process, 297
Counting properties of a point process, 297

Covariance, 27, 38, 49
  complex random variables, 47
  definition, 27, 38
  of independent random variables, 27
  of orthogonal random variables, 27, 50
  partial, 399
Covariance function, 122, 124
Covariance matrix, 50, 313
  definition, 50
  after linear transformation, 70
  positive semidefiniteness, 105
  symmetry, 50
Critical region, 514, 515
Cross-correlation function, 124
  basic properties of, 144
  definition, 124
  of derivative of a process, 165
  Fourier transform of, 148, 149
  relation to cross-spectra, 148
  of response of linear systems, 221, 228
Cross-power spectral density, 148, 149
Cumulant generating function, 42, 68, 92
Cumulants, 42, 68, 92

Decision rule, 344
  Bayes, 348
  M-ary, 366
  MAP, 345
  min–max, 351
  Neyman–Pearson, 351
  square law, 366
  threshold, 346
Decision theory, 343
Decomposition, quadrature, 317
Degrees of freedom, 495, 506
  for chi-square distribution, 506, 508
  for chi-square goodness-of-fit test, 525, 526
  for F distribution, 511
  for periodogram estimate, 576
  for t distribution, 508, 510
Delta function, 39
DeMorgan's laws, 12
Density, see Power spectral density function, Probability density function (pdf)
Derivative of random process, 162
  autocorrelation, 165
  condition for existence, 163
  definition, 162
  MS derivative, 163
Detection of signals, 341
  binary, 343, 352, 514
  M-ary, 366
  square law, 366
Diagnostic checking 608
  Durbin–Watson test, 611
  using cumulative psd, 612
  using sample autocorrelation function, 609
Differencing, 587
  for periodic component removal, 589
  for trend removal, 587
Differentiation, 160, 162
Differentiator, 232
Digital communication system, 336
  effects of noise, 336
Digital filter, see Filters
Digital Wiener filter, 411
Dirac delta function, 39
Discrete Fourier Transform (DFT), 570
  table, 629
Discrete linear models, 250, 275
Discrete random process, 117
Discrete random sequence, 117
Discrete random variable, 24
Discrete-time process, 117
Disjoint sets, 11
Distribution free, 77
Distribution function, 22, 222
  convergence in, 89
  joint, 23
  properties, 22
  of random processes, 119
  of random vectors, 47
Distribution of $\overline{X}$ and S^2, 504, 510
Distributive law, 11
Durbin–Watson test, 611
  table for, 649

Effective bandwidth, 147
Effective noise temperature, 314
Efficiency, 503
Efficient estimator, 502
Eigenvalues, 106, 213, 362
Eigenvectors, 106, 213, 362
Empirical distribution function, 479, 480, 483
Empty set, 9
Ensemble:
  member, 114
  for random variable, 21
  for stochastic process, 111, 113, 114
Ensemble average, 122
Ergodicity, 166, 176, 177, 561
  autocorrelation, 179
  definition, 177
  in distribution, 182
  jointly, 182
  of mean, 178
  normal processes, 183, 185
  of power spectrum, 180
  tests for, 182
  wide sense, 182
Ergodic random process, 177, 178

656 INDEX

Error:
  covariance matrix, 435
  mean squared MSE, 496
  minimum MSE, 382, 389
  recursive expression for, 426
  residual MSE, 389
  RMS, 496
Estimate, 485
Estimating:
  multivariate, 384
  with one observation, 379
  random variable with a constant, 379
  vector space representation, 383
Estimators:
  for autocorrelation function, 566
  Bayesian, 491
  for covariance, 487
  distribution of, 504
  for distribution function, 479
  efficient, 502
  linear, 379
  linear minimum MSE, 380
  maximum likelihood, 488
  for mean, 486, 565
  minimum MSE, 379
  model based, 584
  model free, 565
  multivariate using innovations, 400
  nonlinear, 393
  nonparametric, 478
  notation for, 487
  for parameters of ARMA process, 606
  for parameters of AR process, 594
  for parameters of MA process, 600
  parametric, 478
  for pdf, 479, 481
  point, 485
  for power spectral density, 569
  for probability density, 481, 487
  for probability distribution, 479
  unbiased minimum variance, 497
  for variance, 486
Event, 12
  on combined sample space, 16
  complement, 11, 15
  independent, *see* Statistical independence
  intersection, 10
  joint, 16
  mutually exclusive, 11, 15
  probability of, 12, 13
  statistically independent, 20
  union, 10
Expected values:
  conditional, 27, 48
  definition, 26, 38
  of a function of random variables, 27, 38
  of a linear function, 101
  of a random process, 121
Exponential density function, 303

Factorial moments, 28
Failure rate, 291
False alarm probability, 346
Fast Fourier transform, 571
F distribution, 511
  applied to goodness-of-fit in regression, 539
  applied to sample variance, 513, 522
  definition, 511
  statistical table for, 638
FFT, 571
Filtering, 224, 230, 328, 377
  minimum MSE, 377
Filtering problem, 224
Filters, 239
  autoregressive, 251
  autoregressive moving average, 272
  bandpass, 231, 240
  bandwidth definition, 146, 239
  Butterworth, 239, 242, 248
  causal, 411
  digital, 251, 266, 272, 407, 412
  finite (response), 411
  highpass, 240
  idealized, 240
  Kalman, 419, 421, 432
  lowpass, 235, 240, 241
  matched, 355
  moving average, 266
  noise bandwidth, 241
  noncausal, 407
  prediction, 443
  quadrature, 319
  realizable, 411, 417
  recursive, 251
  relation between Kalman and Wiener, 465
  smoothing, 407, 443
  Wiener, 407, 442
Finite sums of random variables, 91
First-order autoregressive model, 252
Fourier series expansion, 185, 187, 213
Fourier transform, 39, 155, 219
  of autocorrelation function, 145
  of cross-correlation function, 148, 149
  discrete, 570
  inverse, 219, 232
  properties of, 626
  table of pairs, 627
Fourth-order moment, 55
  of a Gaussian random variable, 41
Frequency–domain analysis, 142, 145
Frequency smoothing, 579
Functions of random variables, 61

Gamma function, 507
Gaussian density function, 44
    conditional, 51
    marginal, 51
    m-dimensional (m-variate), 50
    standard, 45
    two-dimensional (bivariate), 41, 46, 51
    univariate, 43
Gaussian distribution:
    table of, 631
Gaussian random process, 127, 212, 312
    definition, 127, 313
    model for noise, 314
Gaussian random variable(s):
    characteristic function of, 41
    conditional density function of, 51
    linear transformation of, 52, 53, 69, 70
    marginal density function of, 52, 53
    moments, 41
    properties of, 50–53
Gaussian random vector, 43, 51, 102
Gauss–Markov process, 127
    sequence, 421
Geometric random variable, 100
Goodness-of-fit test, 523, 525, 545
Gram–Charlier series, 83

Hermite polynomials, *see* Tchebycheff–
    Hermite polynomials
Highpass filter, 240
Hilbert transforms, 319
Histograms, 479, 481, 482
    bias, 499
    definition, 481
    variance, 502
Homogeneous Markov chain, 282
Homogeneous process, 298
Hypothesis, 344, 475
    alternate, 344, 514
    composite, 371, 517
    null, 344, 514
    simple, 517
Hypothesis testing, 513–528
    critical region, 515
    for equality of means, 518, 520
    for equality of variances, 522
    for goodness of fit, 523
    power of a test, 517
    procedure for, 514
    type I, II error for, 514

Identically distributed random variables, 477
Impulse response of linear systems, 219, 227
Independence, *see* Statistical independence
Independent and identically distributed (i.i.d.), 477

Independent increments, 126, 301
Independent noise samples, 353
Independent random processes, 124
Independent random variables, 25
Inequalities:
    cosine, 102
    Schwartz, 102
    Tchebycheff, 27, 77
    triangle, 102
Innovations, 397, 471
    definition, 397
    estimators using innovations, 400
    filter, 402
    matrix definition, 401
    partial covariance, 399
Input–output relations:
    for multiple-input system, 230
    for single-input system, 223
Integral of a random process, 16
    autocorrelation, 165
    condition for existence, 166
    definition, 165
Integrals:
    table, 235
    time-average, 160, 165, 168
Intensity of transition, 289
Interarrival times, 296
Intersection of sets, 10
Interval estimates, 503

Jacobian, 59, 63, 71
Joint characteristic function, 54
    of Gaussian random variables, 54
Joint distribution function, 23
Joint event, 16
Jointly ergodic random processes, 182
Jointly Gaussian random variables, 395
Jointly stationary, 136
Joint probability, 16
Joint probability density function, 36, 37
Joint probability mass function, 25, 32, 33
Joint sample space, 15, 16

Kalman filter, 419, 471
    alternate forms, 471, 473
    comparision with Wiener filter, 465
    MS error, 427
    scalar, 421, 429, 471
    steady state, 431
    vector, 432, 439, 473
Kalman gain, 428, 437
    steady state, 43
Karhunen–Loeve expansion, 188, 363
Kolmogorov equations, 289

658 INDEX

Laplace transform (two-sided), 451
Lattice-type random variable, 28
Law of large numbers, 93, 94
Leakage, spectral, 573, 582
Levinson recursive algorithm, 599
Likelihood function, 488, 595
  conditional, 595
Likelihood ratio, 346, 350
Limitations of linear estimation, 391
Limiting state probabilities, 285
Limit in mean square, 161
Linear filter, 377
Linear MS estimation, 377–474
Linear regression, 529
Linear system, 215, 216
  causal, 215, 217
  impulse response, 219
  lumped parameter, 215, 216
Linear time invariant system:
  causal, 216
  continuous time, 227
  discrete time, 218
  output autocorrelation function, 221, 223, 229, 230
  output mean-squared value, 234
  output mean value, 221, 222, 228, 229
  output power spectrum, 224, 228, 230
  stable, 219, 220, 227
  stationarity of the output, 222, 229
  transfer function, 219, 230
Linear transformation, 402
  of Gaussian variables, 70
  general, 66
  random processes, see Linear system
  of random variable, 66
Linear trend, correction for, 587
Lowpass filter, see Filters
Lowpass processes, 146, 173
  sampling theorem, 190, 191

Marginal probability, 16, 25
Marginal probability density function, 36, 37
Markov chain, 276, 295
  continuous time, 276, 289
  homogeneous, 282
  limiting state probabilities, 285
  long-run behavior, 284
  state diagram, 277
  two-state, 286, 291
Markov processes, 126, 249, 276
  birth death process, 292
  Chapman–Kolmogorov equations, 289
  homogeneous, 282
  transition intensities, 289
Markov property, 126, 205

Markov sequences, 139, 249, 276, 295
  asymptotic behavior, 284
  chains, 276
  Chapman–Kolmogorov equations, 284
  homogeneous, 282
  state diagram, 277
  state probabilities, 279
  transition matrix, 281
  transition probabilities, 279
Martingale, 126, 129, 205, 206
M-ary detection, 366
Matched filter, 355
  for colored noise, 362
  for white noise, 355
Matrix, covariance, 50, 313
Maximum *a posteriori* (MAP) rule, 345, 350, 368
Maximum likelihood estimator, 488
  of probability, 553
  of σ^2 in normal, 553
  of λ in Poisson, 553
  of μ in normal, 553
Maxwell density, 57
Maxwell random variable, 507
Mean square (MS):
  continuity, 161, 162
  continuous process, 162, 211
  convergence, 94
  derivative, 162, 163
  differentiability, 162, 211
  integral, 165, 211
  limit, 94
  value, 234, 564
Mean squared error (MSE), 327, 379, 469, 496, 560
  definition, 327, 496
  for histograms, 500
  for Kalman filters, 429, 437
  recursive expression for, 427
  for Wiener filters, 456
Mean squared estimation, 377–474
  filtering, 407–466
  recursive filtering, 420
Mean value, 26
  of complex random variables, 47
  conditional, 27
  estimate of, 486
  of function of several random variables, 81
  normalized RMS error for, 497
  of random variable, 26
  of system response, 221, 228
  of time averages, 170
Mean vector, 49
Measurements, 476
  unbiased, 476
Median of a random variable, 467, 555

Memoryless nonlinear system, 217
Memoryless transformation, 217
Minimum of i.i.d. random variables, 70, 72
Minimum mean squared estimation, 377
Minimum MSE estimators:
 linear, 379
 multiple observations, 384
 nonlinear, 393
 orthogonality principle, 384
 using one observation, 379
Minimum MSE weight, 428
Minmax decision rule, 351
Mixed distribution, 61
Mixed-type probability law, 35
Mixed-type random variable, 35
Model identification:
 AR(1), 596
 AR(2), 598
 ARMA process, 605
 AR process, 594
 diagnostic checking, 608
 differencing, 587
 Durbin–Watson test, 611
 MA(1), 600
 MA(2), 602
 moving average process, 600
 order, 590
 order of ARMA models, 592
Modulation, 160
Moment generating function, 39, 40, 89, 90
Moments:
 central, 26
 factorial, 28
 of a multivariate Gaussian pdf, 53, 54, 55
Moments of a random variable:
 from characteristic function, 40, 54, 55
 from moment generating function, 40
Monte Carlo technique, 73–76
Moving Average (MA) process:
 autocorrelation coefficient, 268, 269
 autocorrelation function, 267, 269, 270
 estimation, 600
 first-order, 266
 mean, 267, 269, 270
 model, 265, 266
 partial correlation coefficient, 268
 power spectral density, 268, 269, 270, 271
 second-order, 269
 transfer function, 265
 variance, 267, 269, 270
MSE, *see* Mean squared error
MS limit, 94
Multidimensional distribution, Gaussian, 50
Multinomial probability mass function, 30
Multiple input–output systems, 238

Multivariate Gaussian, 50–55
 characteristic function, 54
 conditional densities, 51
 covariance matrix, 50
 density, 50
 fourth joint moment, 55
 independence, 51
 linear transformation, 51
 marginal densities, 51
 moments, 53
Mutually exclusive events, 11, 15
Mutually exhaustive sets, 11

Narrowband Gaussian process, 314, 322
Narrowband processes, 317
 envelope of, 322
 phase of, 322
 quadrature decomposition of, 317, 318
Neyman–Pearson rule, 351, 373
Noise, 314
 in analog communication systems, 322
 band-limited, Gaussian, 314, 315
 colored, 361
 in digital communication systems, 330
 white, 314
 white Gaussian, 314
Noise bandwidth, 241, 248
Noise temperature, 314
Nonlinear estimators, 393
Nonlinear systems, memoryless, 217
Nonlinear transformations, 73
Nonparametric estimation procedures, 478, 479, 560, 565
Nonstationary random process, 117
 examples of, 138
Normal density function, *see* Gaussian density function
Normal distribution function, *see* Gaussian distribution
Normal form in regression, 532
Normalized errors, 497
 bias, 497
 MSE, 497
 RMS error, 497
 standard error, 497
Normal processes, *see* Gaussian random process
Normal random variable, *see* Gaussian random variable(s)
Null hypothesis, 513
Null set, 9
Nyquist's sampling theorem, 190, 193

Observation model, 420
Observation noise, 420

Optimum unrealizable filter:
    continuous, 418
    discrete, 417
Order identification, 590
    of ARMA model, 591
    of differencing, 590
Order statistics, 70–73
Orthogonal, 27, 124
    random process, 124
    random variables, 27
Orthogonality conditions, 385, 389, 408, 423, 443
Orthogonality principle, 384
Output noise power, filters, 241
Output response, linear systems, 221, 228

Partial autocorrelation coefficient (p.a.c.c.), 262
    definition, 263
Partial covariance, 399
Parseval's theorem, 186
Parzen's estimator for pdf's, 479, 484, 485
    bias, 501
    variance, 501
Peak value distribution, 72
Periodic process, 186
Periodogram, 571
    bias, 572
    estimator for psd, 571
    smoothing, 579
    variance, 574
    windowing, 581
Point estimates, 485
Point estimators, 485
Point processes, 249, 295, 312
    assumptions, 297
    counting process, 297
    interarrival time, 296
    Poisson, 298
    shot noise, 307
    time of occurrence, 295
    waiting time, 296
Poisson increments, 300
Poisson point process, 298
    homogeneous, 298
    superposition, 302
Poisson process, 131, 206, 249
    applications to queues, 303
    assumptions, 131, 132, 298
    autocorrelation, 132
    mean, 132
    properties, 300
Poisson random variable, 29, 30, 32
    mean, 30, 100
    probability, 30
    variance, 30, 100

Positive semidefiniteness, 105
Power:
    from autocorrelation function, 146
    average, of a waveform, 146, 231
    from power spectral density, 146, 231
Power and bandwidth calculations, 146
Power spectral density function, 142, 207, 208, 209
    basic properties of, 143, 144, 146
    definition, 143, 145
    ergodicity, 180
    estimation of, 569
    examples of, 153–160
    input–output relations for, 223, 224, 230
    random sequence, 150
    two-sided, 146
Power of a test, 518
Power transfer function, 230
Prediction, 407, 443, 457
Prediction filter, 407
Preprocessing, 587
Probabilistic model, 13
Probabilistic structure, 116
Probability, 12
    a posteriori, 344, 345
    a priori, 346
    axioms of, 13
    classical definition, 13
    conditional, 15, 16, 17, 18, 19
    of detection, 351
    of error, 356, 348
    of false alarm, 346
    joint, 15, 16, 17, 18
    marginal, 15, 16, 17, 18
    measure, 12
    of miss, 346
    relative frequency, 13
    state, 279
    transition, 279
Probability axioms, 13
Probability density function (pdf), 33, 34
    approximation, 83
    basic properties of, 34
    Cauchy, 102
    chi-square, 505
    conditional, 37, 48
    of discrete random variables, 39
    estimation of, 481, 484
    examples of, 35, 43–46
    exponential, 303
    F, 511
    Gaussian, 43, 44, 50
    joint, 36, 47
    marginal, 36, 48
    normal, 43, 44
    of N random variables, 47

INDEX   661

properties, 34
  of random vectors, 49
  Rayleigh, 507
  of sums of random variables, 67
  t, 508
  transformation of, 59
  of two random variables, 36
  uniform, 43
Probability of detection, 351
Probability distribution, 13
Probability distribution function, 22
Probability of event, 12
Probability generating function, 28
Probability mass function, 24, 25, 29, 33
  binomial, 29
  multinomial, 29
  Poisson, 29
  uniform, 29
Probability measure, 12
Probability of miss, 346
Probability space, 12
Product space, two dimensional, 16
Pulse response, 219

Quadrature representation, 317, 321
Quadrature spectral density, 321, 322
Quantization, 189, 196, 197, 201
  for Guassian variable, 201
  MS error, 199
  nonuniform, 200
  uniform, 197
Quantization error, 197, 200
Quantizer, 196
Queues, 303
  state diagram of, 305

R^2, 538
Random binary waveform, 132, 133, 156, 175, 246
  autocorrelation, 135
  power spectral density, 156
Random experiment, 9
Randomness tests, 564, 608
Random process(es):
  autocorrelation function, 122, 142
  autocovariance function, 122
  bandpass, 147
  classification, 117
  coherence function, 149
  complex, 118, 210
  continuous-time, 117, 160
  correlation coefficient, 124
  cross-correlation function, 124
  cross-covariance function, 124
  cross-power spectrum, 148
  definition, 113, 119

  derivative of, 162
  discrete, 117
  distribution function of, 119
  equality of, 124
  ergodic, 178
  formal definition, 119
  Gaussian, 127, 212, 312, 313
  integral of, 165
  jointly ergodic, 182
  jointly stationary, 136
  lowpass, 147
  mean, 122
  methods of description, 119
  N-order stationary, 136
  notation, 114
  periodic component in, 144
  Poisson, 131
  random binary, 132
  sample function, 114
  stationary, 135
  strict-sense stationary, 135
  wide-sense stationary, 136
  Wiener, 127
  Wiener–Levy, 130
Random sequence, 117
  continuous, 117
  discrete, 117
  input–output, 221
  power spectral density, 149
Random telegraph process, 338
Random variable(s), 8, 21
  binomial, 29
  Cauchy, 102
  characteristic function of, 39
  chi-square, 505
  complex, 46
  conditional density, 37
  continuous, 33
  correlation, 27, 38
  covariance, 27, 38
  definition, 21
  discrete, 24
  expected value, 27, 38
  F, 511
  Gaussian, 43, 50
  independence, 24, 26, 37
  lattice type, 28
  marginal density, 36, 37
  marginal distribution, 25
  mean, 26, 38
  mixed, 35
  moments, 26
  N-dimensional, 47
  normal, 43, 44, 45
  orthogonal, 27
  Poisson, 29, 30

662 INDEX

Random variable(s)—(*Continued*)
  Rayleigh, 322, 507
  S^2, 516
  standard deviation, 27
  t, 508
  transformation of, 55, 56, 57, 66
  uncorrelated, 27, 38
  uniform, 29, 43
  variance of, 27, 38
  vector, 47
Random vectors, 47
Random walk, 127, 128, 205, 206
Rate function, 298
Rayleigh random variable, 507
Realizable linear system, 411, 417
Realizable Wiener filter, 417, 448, 453
  digital, 411
Realization of a process, 114
Receiver design, 325
Receiver operating characteristic (ROC), 352
Recursive estimators, 420
Recursive filter, 251, 419
Regression, 529
  analysis of linear, 536
  analysis of variance, 540
  general, 543
  goodness of fit, 538, 545
  matrix form, 542
  multivariable, 540
  nonlinear, 546
  normal equations, 532
  scalar, 529
  simple linear, 529
Relative frequency, 13
Residual errors, 534
RMS bandwidth, 209
Root mean squared (RMS) error, 497
Run test, 562
  description of 562, 563
  probability mass function, 563
  for stationarity, 562, 563, 564
  statistical table for, 648

Sample function, 114
Sample (set), 477
Sample space, 12, 16
Sampling of random signals, 189, 190
  aliasing effect, 193
  lowpass, 190
Sampling theorem, 191
Schwartz inequality, 102
Sequences of random variables, 88
Series approximations of pdf's, 83
  Gram–Charlier series, 83
Series expansion, 185, 187
Set operations, 10, 11

Set(s), 9
  complement, 11
  countable, 10
  disjoint, 11
  elements of, 9
  empty, 9
  equality of, 10
  finite, 10
  infinite, 10
  intersection, 10, 11
  mutually exclusive, 11
  null, 9
  uncountable, 10
  union, 10
Shot noise, 250, 307
  autocorrelation function, 310
  Campbell's theorem, 310
  mean, 309
  spectral density, 311
Signal detection, 341, 364
Signal estimation, 342
Signal model, 420
Signal to noise ratio, 330
Signal vector, 432
Significance level, 514
Smoothing filter, 407
Smoothing of spectral estimates, 579, 580
  bias and variance, 584
  over ensemble, 579
  over frequency, 579
Space (entire), 9
Spectral decomposition, 185
Spectral density, power, *see* Power spectral
  density function
Spectral factorization, 246
Spectral leakage, 573
Spectrum factorization, 450
Square law detection, 366
Stable linear system, 219, 227
Standard deviation, 27
Standard error, normalized, 497
Standard Gaussian (normal), 45
State diagram, 277, 280
State matrix, 432
State model, 420, 468
State vector, 432
Stationarity, 135, 229, 561
  asymptotically, 141
  cyclo, 142
  in an interval, 142
  strict sense, 135
  tests for, 142, 562
  wide-sense, 136
Stationary increments, 142
Stationary random process, 117, 135
  asymptotically stationary, 141

INDEX   663

autocorrelation of, 142
cyclostationary, 142
jointly stationary, 136
linear transformation of, 229
mean square value of, 231
mean value of, 136
Nth order stationarity, 136
output of LTIVC systems, 222, 229
stationary increments, 142
strongly (strict-sense), 135
tests for, 142
weakly (wide-sense), 136
Statistic, 477
definition, 478
Statistical average, *see* Expected values
Statistical independence, 20, 25, 37
Statistical tables, 631–649
chi-square, 633
Durbin–Watson, 649
F, 638
Gaussian, 631
run distribution, 648
t, 636
Steady–state behavior:
Kalman filter, 432
Markov sequence, 284
Steady-state distribution, 285
Stochastic process, *see* Random process
Strictly stationary process, 135
Strongly stationary process, 135
Student's t distribution, 508, 509, 549
applied to sample means, 519, 520
definition, 508, 509
percentage point of, 636
plot of, 510
statistical table for, 636
Subexperiments, 16, 17
Subset, 10
Sum of squared errors, 530, 597, 602
Sums of random variables, 67
density function of, 67
Superposition of, Poisson process, 302
Synthetic sampling, *see* Monte Carlo technique

t, *see* Student's t distribution
Tables of:
chi-square distribution, 633
Discrete Fourier transforms, 628–629
Durbin–Watson distribution, 649
F distribution, 638
Fourier transform pairs, 627
Gaussian distribution, 631
integrals, 235
run distribution, 648
t distribution, 636
Z transforms, 630

Taylor series, 81, 90, 498
Tchebycheff–Hermite polynomials, 82–86, 108
Tchebycheff inequality, 27, 38, 77–80, 94
Tests:
for ergodicity, 561
goodness-of-fit, 525
runs, 562
for stationarity, 561
Tests of hypothesis, 513
chi-square, 523
equality of two means, 520
goodness-of-fit, 538
mean value with known variance, 518
mean value with unknown variance, 519
variances, 522
Thermal noise, 314
Time averages, 166, 168
autocorrelation, 169, 213
convergence, 166
mean, 169, 213
mean and variance of, 170
power spectral density, 169
Time series, 110
Transfer function, 219, 459, 462
Transformation:
of Gaussian random variables, 51, 52, 69
linear, 66, 402
memoryless, 217
of nonstationary series, 587
of random variables, 55–65
linear, 66–70
nonlinear, 73, 81
Transition matrix, 282
Transition probabilities, 279
n-step, 279
one-step, 279
Trend removal, 587
Triangle inequality, 102
Two-sided psd, 146
Two-sided tests, 529
Two-state Markov chain, 286, 291
Type I, II error, 346, 514

Unbiased estimator, 493
Uncorrelated random processes, 124
Uncorrelated random variables, 27
Uncountable set, 10
Uniform density function, 43
mean, 43
variance, 43
Uniform quantizing, 198
Union bound, 79
Union of events, 10, 15
Union of sets, 10
Unrealizable filter, 417, 444

Variance:
  of autocorrelation function estimator, 569
  of complex random variables, 47
  definition, 27
  estimate of, 486
  of histograms, 498
  of mean estimator, 566
  of periodogram, 574
  of smoothed estimators, 584
  of time averages, 170
Vector measurement, 432
Vector-valued random variable, 47

Waiting time sequence, 296
Weakly ergodic process, 182
Weakly stationary process, *see* Wide sense stationarity
White Gaussian noise, 314
White noise, 314
  autocorrelation function of, 315
  band-limited, 315
  input to Wiener filters, 449
  power spectral density of, 314
  properties, 315
Wide sense ergodicity, 182
Wide sense stationarity, 136

Wiener filter, 407, 442
  continuous time, 442
  digital, 407
  discrete time, 407
  mean squared error, 409, 445
  minimum mean squared error of, 445
  nonrealizable, 408, 417, 444, 445
  prediction, 443
  realizable, 417, 448, 455
  real-time, 448
  smoothing, 443, 453
Wiener–Hopf equation, 449
  solution, 464
Wiener–Khinchine relationship, 145
Wiener–Levy process, 130
Wiener process, 127, 129, 130, 205
  properties, 130
Windowing, 581
  Bartlett, 583
  Blackman–Tukey, 583
  Parzen, 583
  Rectangular, 582

Yule–Walker equations, 262, 263, 387

Z transform, 220
  table, 630

(ua) 4331 69 2769 - work
(ua) 4331 343303

$$\int uv = u\int v\,dx - \int \left(\frac{du}{dx} \int v\,dx\right) dx$$